*Volcanic Hazards in Ce*

Edited by

William I. Rose
Geological Engineering & Sciences
Michigan Technological University
Houghton, Michigan 49931
USA

Gregg J.S. Bluth
Geological Engineering & Sciences
Michigan Technological University
Houghton, Michigan 49931
USA

Michael J. Carr
Rutgers University
Department of Geological Sciences, Wright Laboratory
610 Taylor Road
Piscataway, New Jersey 08854-8066
USA

John W. Ewert
U.S. Geological Survey
Cascades Volcano Observatory
Vancouver, Washington 98683
USA

Lina C. Patino
Geological Science Department
Michigan State University
East Lansing, Michigan 48824-1115
USA

James W. Vallance
U.S. Geological Survey
Cascades Volcano Observatory
Vancouver, Washington 98683
USA

THE
GEOLOGICAL
SOCIETY
OF AMERICA

# Special Paper 412

3300 Penrose Place, P.O. Box 9140 ▪ Boulder, Colorado 80301-9140 USA

2006

Copyright © 2006, The Geological Society of America, Inc. (GSA). All rights reserved. GSA grants permission to individual scientists to make unlimited photocopies of one or more items from this volume for noncommercial purposes advancing science or education, including classroom use. For permission to make photocopies of any item in this volume for other noncommercial, nonprofit purposes, contact the Geological Society of America. Written permission is required from GSA for all other forms of capture or reproduction of any item in the volume including, but not limited to, all types of electronic or digital scanning or other digital or manual transformation of articles or any portion thereof, such as abstracts, into computer-readable and/or transmittable form for personal or corporate use, either noncommercial or commercial, for-profit or otherwise. Send permission requests to GSA Copyright Permissions, 3300 Penrose Place, P.O. Box 9140, Boulder, Colorado 80301-9140, USA.

Copyright is not claimed on any material prepared wholly by government employees within the scope of their employment.

Published by The Geological Society of America, Inc.
3300 Penrose Place, P.O. Box 9140, Boulder, Colorado 80301-9140, USA
www.geosociety.org

Printed in U.S.A.

GSA Books Science Editor: Abhijit Basu

**Library of Congress Cataloging-in-Publication Data**

Volcanic hazards in Central America / [edited by] William I. Rose ... [et al.].
    p. cm.--(Special paper ; 412).
  Includes bibliographical references and index.
  ISBN-13 978-0-8137-2412-6 (pbk.)
  ISBN-10 0-8137-2412-0 (pbk.)
    1. Volcanic hazard analysis--Central America. 2. Volcanism--Central America. 3. Geology, Structural--Central America. I. Rose, William I. (William Ingersoll), 1944-. II. Special papers (Geological Society of America) ; 412.

QE524.2.C35 V65 2006
551.210972—dc22

2006045652

**Cover:** Lahar hazard map for Fuego Volcano, Guatemala. Lahars are one of the most devastating hazards associated with volcanoes. To forecast where these hazards extend, digital topography of the volcano is used together with a mathematical model. The different colors show the limits of hazard zones based on the total volume of the lahar. This shaded relief map is from Plate 1 of J.W. Vallance, S.P. Schilling, O. Matías, W.I. Rose, and M.M. Howell, 2001, Volcano Hazards at Fuego and Acatenango, Guatemala: U.S. Geological Survey Open File Report 01-431. Such maps are one tool that volcanologists used to communicate with the public about the nature an probability of volcanic hazards.

10 9 8 7 6 5 4 3 2 1

# Contents

*Preface* ................................................................................................................................. v

1. *Large-volume volcanic edifice failures in Central America and associated hazards* ............ 1
   L. Siebert, G.E. Alvarado, J.W. Vallance, and B. van Wyk de Vries

2. *Origin of the modern Chiapanecan volcanic arc in southern México inferred from
   thermal models* ................................................................................................................. 27
   V.C. Manea and M. Manea

3. *Geological evolution of the Tacaná Volcanic Complex, México-Guatemala* ..................... 39
   A. García Palomo, J.L. Macías, J.L. Arce, J.C. Mora, S. Hughes, R. Saucedo, J.M. Espíndola,
   R. Escobar, and P. Layer

4. *The chemistry of spring waters and fumarolic gases encircling Santa María volcano, Guatemala:
   Insights into regional hydrothermal activity and implications for volcano monitoring* ........ 59
   J.A. Walker, S. Templeton, and B.I. Cameron

5. *Downstream aggradation owing to lava dome extrusion and rainfall runoff at
   Volcán Santiaguito, Guatemala* ......................................................................................... 85
   A.J.L. Harris, J.W. Vallance, P. Kimberly, W.I. Rose, O. Matías, E. Bunzendahl, L.P. Flynn,
   and H. Garbeil

6. *The Escuintla and La Democracia debris avalanche deposits, Guatemala:
   Constraining their sources* ................................................................................................ 105
   C.A. Chesner and S.P. Halsor

7. *Diverse volcanism in southeastern Guatemala: The role of crustal contamination* ............ 121
   B.I. Cameron and J.A. Walker

8. *Volcanic hazards in Nicaragua: Past, present, and future* ................................................. 141
   A. Freundt, S. Kutterolf, H.-U. Schmincke, T. Hansteen, H. Wehrmann, W. Pérez, W. Strauch,
   and M. Navarro

9. *The A.D. 1835 eruption of Volcán Cosigüina, Nicaragua: A guide for assessing local
   volcanic hazards* ............................................................................................................... 167
   W. Scott, C. Gardner, G. Devoli, and A. Alvarez

10. *The youngest highly explosive basaltic eruptions from Masaya Caldera (Nicaragua):
    Stratigraphy and hazard assessment* ................................................................................. 189
    W. Pérez and A. Freundt

11. ***Fontana Tephra: A basaltic Plinian eruption in Nicaragua*** ............................. 209
    H. Wehrmann, C. Bonadonna, A. Freundt, B.F. Houghton, and S. Kutterolf

12. ***Tephra deposits for the past 2600 years from Irazú Volcano, Costa Rica*** ................ 225
    S.K. Clark, M.K. Reagan, and D.A. Trimble

13. ***The eruptive history of Turrialba volcano, Costa Rica, and potential hazards from future eruptions*** ............................................................................ 235
    M. Reagan, E. Duarte, G.J. Soto, and E. Fernández

14. ***Recent volcanic history of Irazú volcano, Costa Rica: Alternation and mixing of two magma batches, and pervasive mixing*** .................................................... 259
    G.E. Alvarado, M.J. Carr, B.D. Turrin, C.C. Swisher III, H.-U. Schmincke, and K.W. Hudnut

# *Preface*

Most Central Americans live in the midst of dramatic and obvious geological features. In spite of this, geology is not much studied in schools and universities in Central America. Likewise, in developing countries like most of those in Central America, natural hazards are far from the most immediate and demanding issues. The capitals and major cities of Central America, however, are all within volcanic hazard zones, and there is near certainty of volcanic calamity if one thinks—as geologists do—in a time scale of centuries. After Indonesia, Central America is the world's most volcanically active region and is therefore a region where improved understanding and planning for inevitable hazards could be most useful.

Visiting geologists come to Central America to pursue their particular research goals, attracted by the volcanic activity, which is accessible, diverse, and much more frequent and persistent than in most other volcanic regions. Geologists visiting from afar see the compelling natural hazards, whereas residents from Central America often have only limited awareness of the potential problem. There has been much effort at collaboration between foreign and Central American scientists, but this is typically hampered by their ironically different goals. Local scientists are focused on hazard mitigation, while foreign scientists often pursue basic science directions, or depend on resource-dependent methodologies that seem "esoteric" to Central Americans.

In this Special Paper, we highlight examples of work that bridges the goals of both foreign and local scientists. By collaborating, foreign and local scientists exchange information and insight, and all gain. Foreign scientists gain access to Central American volcanoes, and local scientists gain access to new advances and specialized labs. We have begun an era where the agencies responsible for geological hazards in Central America are rapidly developing and where the benefits of collaboration seem to help in developing better infrastructure for hazards studies. We hope this modest volume will encourage more scientific collaboration with a hazards focus at this auspicious time.

William I. Rose

# Large-volume volcanic edifice failures in Central America and associated hazards

**Lee Siebert**
Smithsonian Institution, Global Volcanism Program, National Museum of Natural History MRC-119, Washington, D.C. 20013-7012, USA

**Guillermo E. Alvarado**
Sismológia y Vulcanología, Instituto Costarricense de Electricidad, Apdo. 10032-1000, San José, Costa Rica

**James W. Vallance**
U.S. Geological Survey, Cascades Volcano Observatory, Bldg. 10, Suite 100, 1300 SE Cardinal Court, Vancouver, Washington 98683, USA

**Benjamin van Wyk de Vries**
Laboratoire Magmas et Volcans, Observatoire de Physique du Globe, Université Blaise Pascal, Clermont-Ferrand, France

## ABSTRACT

Edifice-collapse phenomena have, to date, received relatively little attention in Central America, although ~40 major collapse events (≥0.1 km$^3$) from about two dozen volcanoes are known or inferred in this volcanic arc. Volcanoes subjected to gravitational failure are concentrated at the arc's western and eastern ends. Failures correlate positively with volcano elevation, substrate elevation, edifice height, volcano volume, and crustal thickness and inversely with slab descent angle. Collapse orientations are strongly influenced by the direction of slope of the underlying basement, and hence are predominately perpendicular to the arc (preferentially to the south) at its extremities and display more variable failure directions in the center of the arc.

The frequency of collapse events in Central America is poorly constrained because of the lack of precise dating of deposits, but a collapse interval of ~1000–2000 yr has been estimated during the Holocene. These high-impact events fortunately occur at low frequency, but the proximity of many Central American volcanoes to highly populated regions, including some of the region's largest cities, requires evaluation of their hazards. The primary risks are from extremely mobile debris avalanches and associated lahars, which in Central America have impacted now-populated areas up to ~50 km from a source volcano. Lower probability risks associated with volcanic edifice collapse derive from laterally directed explosions and tsunamis. The principal hazards of the latter here result from potential impact of debris avalanches into natural or man-made lakes. Much work remains on identifying and describing debris-avalanche deposits in Central America. The identification of potential collapse sites and assessing and monitoring the stability of intact volcanoes is a major challenge for the next decade.

**Keywords:** volcano collapse, debris avalanche; hazards, Central America.

# INTRODUCTION AND NOMENCLATURE

Large-scale volcanic edifice collapse is a common process in volcanoes in a wide variety of tectonic settings and has been identified at more than 400 Quaternary volcanoes worldwide. Although steep-sided continental margin stratovolcanoes and lava-dome complexes are particularly susceptible to edifice failure, relatively little attention has been devoted to this topic in Central America. Previous reviews of debris avalanches have focused primarily on subduction-zone volcanic arcs such as Papua New Guinea (Johnson, 1987), Indonesia (MacLeod, 1989), Japan (Ui et al., 1986), Kamchatka (Melekestsev and Braitseva, 1984), the West Indies (Roobol et al., 1983), México (Capra et al., 2002), Guatemala (Vallance et al., 1995), Costa Rica (Alvarado et al., 2004), and the Central Andes (Francis and Wells, 1988), but also include intraplate areas such as the Hawaiian (Moore et al., 1989) and Canary Islands (Carracedo, 1994).

Because the study of edifice-failure events in Central America is in its relative infancy, some of the descriptions of deposits that follow represent preliminary work subject to revision with more detailed investigations. We review here known and inferred large collapse events (>0.1 km$^3$) in Central America to provide context as a guide for further research and to bring attention to the volcanic hazard implications of these low-frequency but high-impact events. More than 40 known or inferred edifice failures are described here, more than two-thirds of which are in Guatemala and Costa Rica. This primarily reflects geologic factors, but is also influenced by the relative numbers of avalanche field studies.

A debris avalanche is a flowing mixture of debris, rock, and water that moves rapidly downslope under the influence of gravity. Debris avalanches differ from debris flows (often referred to in volcanic terrain as lahars) in that they are not water-saturated and in that the load is supported by particle-particle interactions. A wide variety of nomenclature has been applied to deposits of volcanic mass movements; however, the terminology of Glicken (1991) has commonly been used, in which block-facies and mixed-facies material is distinguished. Block-facies material consists of intact to partially disaggregated, locally homogeneous segments of the edifice that were transported relatively intact and may contain both clast-supported and matrix-supported areas. This facies may contain individual blocks or megablocks, discrete segments of the pre-failure edifice that may contain both blocks and matrix. Large segments of the edifice known as toreva blocks (Reiche, 1937) may slide relatively short distances without disaggregating. Mixed-facies material consists of heterolithologic mixtures of crushed and disaggregated block-facies and accessory materials in which matrix dominates, although it may contain large clasts to more than meter-scale. A third component, the lahar facies (or debris-flow facies), represents a water-saturated facies of the avalanche transitional to and derived from the mixed facies. This was referred to by Palmer et al. (1991) as the "marginal facies" of New Zealand debris avalanches. Contacts between volcanic mass movement facies are gradational, and their proportions are highly variable from deposit to deposit.

# GUATEMALA

## Tacaná Volcano

Guatemalan volcanoes are some of the highest and largest in Central America and, along with Costa Rican volcanoes, have produced the majority of the region's known edifice failures (Fig. 1). Tacaná is a 4060-m-high stratovolcano that straddles the México–Guatemala border and is the highest peak in Central America. The summit of San Antonio volcano, the youngest of the Tacaná volcanic complex, is cut by a collapse scarp on the NW side (Fig. 2) that was the source of a debris avalanche that traveled at least 8 km northwest to the Coatán River (Macías et al., 2004). The Agua Caliente debris-avalanche deposit is confined within steep-walled valleys and is up to 200 m thick, with a volume of ~1 km$^3$. Hummocky topography with individual blocks as large as 70 m is present locally. The collapse scarp is partially filled by an andesitic lava dome, and the debris-avalanche deposit is overlain directly by a 20-m-thick sequence of at least four block-and-ash flow deposits emplaced as part of the same eruptive sequence. The collapse event ended with generation of 75-m-thick lahar deposits along the San Rafael River. The Agua Caliente debris-avalanche deposit lies between a 26,340 +910/−820 yr B.P. debris-flow deposit and an overlying pyroclastic-surge deposit radiocarbon dated at 10,610 +330/−315 yr B.P. (Macías et al., 2004).

An area of volcaniclastic hummocky topography that represents an earlier collapse event (Siebert et al., 2004; Macías, 2004, personal commun.) is located along the Río Las Majadas on the Guatemalan side of the border. Mora et al. (2004) noted several additional collapse scarps at the Tacaná volcanic complex, including a 1 × 2-km-wide horseshoe-shaped scarp on the SW side of the San Antonio edifice and other scarps north to east of San Antonio (Fig. 2). A massive, pinkish debris-avalanche deposit at Muxbal, SE of San Antonio volcano, contains yellow-to-orange hydrothermally altered zones and meter-sized blocks fractured in a jigsaw fashion, characteristic of avalanche deposits (Mora et al., 2004). This avalanche deposit originated from the collapse of Chichuj, an older volcano at the eastern side of the Tacaná complex. The debris-avalanche deposit is overlain by fluvial deposits and a block-and-ash flow deposit radiocarbon dated ca. 28,540 ± 260 yr B.P. More recent work (see García-Palomo et al., this volume) documents these and other edifice failure deposits from Tacaná.

## Cerro Quemado Volcano

The andesitic-to-dacitic Cerro Quemado lava-dome complex is one of several post-caldera domes of the Almolonga volcanic field in SW Guatemala (Conway et al., 1992). Cerro Quemado, located immediately NE of Santa María volcano, was constructed during the Holocene from eight distinct vents that produced lava domes and steep-sided lava flows. About 1150 $^{14}$C years ago, the summit of Cerro Quemado collapsed,

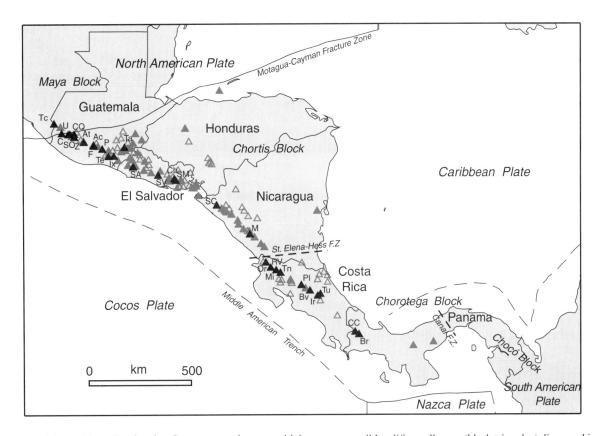

Figure 1. Map of Central America showing Quaternary volcanoes with known or possible edifice collapses (black triangles) discussed in the text. From left to right: Tc—Tacaná; U—Unnamed (near El Tumbador); C—Chicabal; SO—Siete Orejas; CQ—Cerro Quemado; Z—Zunil; At—Atitlán; Ac—Acatenango; F—Fuego; P—Pacaya; Te—Tecuamburro; Ix—Ixhuatán; Ta—Tahual; SA—Santa Ana; SV—San Vicente; Ch—Chinameca; SM—San Miguel; SC—San Cristóbal (Casita); M—Mombacho; Or—Orosí; RV—Rincón de la Vieja; Mi—Miravalles; Tn—Tenorio; Pl—Platanar; Bv—Barva; Ir—Irazú; Tu—Turrialba; CC—Cerro Colorado; Br—Barú. Pleistocene volcanoes are shown by open triangles and volcanoes with known or possible Holocene eruptions by solid gray triangles. Solid line marks transform plate boundary; dashed line, convergent plate boundary. F.Z.—Fracture zones.

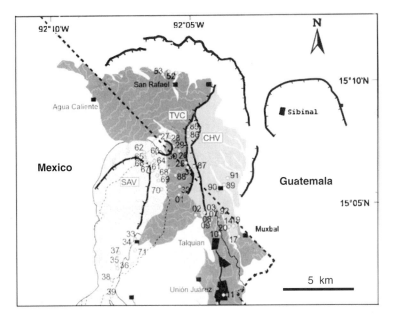

Figure 2. Map of collapse scarps at Tacaná volcano, after Mora et al. (2004). SAV—San Antonio volcanic complex; TVC—Tacaná volcanic complex; CHV—Chichuj volcanic complex. Numbers are samples numbers from Mora et al. (2004). Dashed line marks México-Guatemala border.

producing a debris avalanche that swept 6 km to the northwest (Fig. 3) (Conway et al., 1992). The avalanche deposit blankets ~13 km² of the Llano del Pinal valley and has an irregular surface with ~150 hummocks up to ~15 m high and 500 m wide. The 0.13 km³ deposit consists primarily of Cerro Quemado dome material but also includes soil and diatomite clasts ripped up from a dry lake bed on the valley floor. The collapse left a 1 × 1.5-km-wide, NW-facing scarp that truncated the summit of Cerro Quemado and was subsequently partially filled by a post-collapse lava dome (Fig. 3).

Overlying the debris-avalanche deposit is a fines-depleted pyroclastic-flow deposit that originated from a lateral blast (Fig. 3) that accompanied the debris avalanche (Conway et al., 1992; Vallance et al., 1995). The unstratified and unsorted deposit has textural features comparable to those of the 1980 Mount St. Helens lateral-blast deposit. It has a friable texture with fines-depleted openwork debris with angular-to-subangular lithic blocks to 1 m in diameter. It furthermore contains vitreous charcoal along with vesicular and prismatically jointed blocks and a few breadcrust bombs, all of which reflect a hot, juvenile component. The thin deposit (10–25 cm thick) covers a 40 km² area west of Cerro Quemado and is exposed on neighboring Siete Orejas volcano to an elevation of 3100 m. The deposit covers a 110° arc symmetrical about an axis originating from the center of the scarp at Cerro Quemado (Fig. 3). The associated eruption further resembled the 1980 Mount St. Helens eruption in that it produced juvenile pyroclastic flows interpreted as secondary flowage of the lateral blast from surrounding hillsides, and a lava dome formed in the new collapse scarp.

## Atitlán Volcano

Atitlán (Fig. 4) is the youngest of three post-caldera stratovolcanoes constructed on or near the southern rim of scenic Atitlán caldera (Newhall, 1987). Hummocks of a debris-avalanche deposit of possible Holocene age occur along the Pacific coastal plain up to 20 km south and SW of the volcano (Haapala et al., 2005). Renewed eruptions from the symmetrical volcano have restored the edifice, and no failure scarp is visible.

## Acatenango Volcano

A debris-avalanche lobe underlies ~120 km² of the Pacific coastal plain (Fig. 5) SW of the Acatenango-Fuego volcanic chain (Vallance et al., 1995). The exposed distal portion has a volume of 2.4 km³ (Vallance et al., 1995); inclusion of the estimated buried proximal area produces a volume of ~5 km³ (Siebert et al., 2004). Proximally, the deposit is buried by alluvium, but distal areas show numerous extensively weathered hummocks up to 40 m high with spheroidal weathering at the margin of large clasts. The hummocks also contain some fresher, non-juvenile, angular basaltic and andesitic clasts. The geometry of the deposit requires an origin from the Fuego-Meseta-Acatenango complex (Fig. 5). Vallance et al. (1995) inferred an origin from a buried or

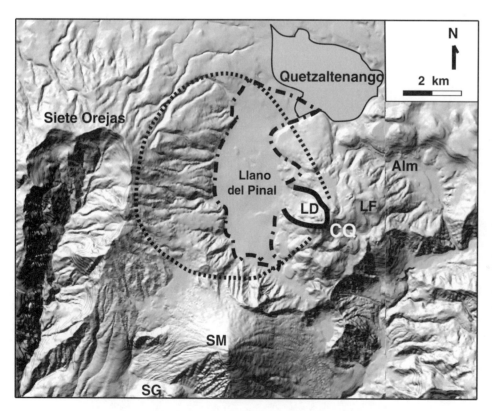

Figure 3. Digital elevation model map of Cerro Quemado volcano with horseshoe-shaped collapse scarp (solid line) and ca. 1150 yr B.P. debris-avalanche (dashed-dotted line) and lateral-blast deposits (dotted line). Secondary and primary pyroclastic-flow deposits overlying the debris-avalanche deposit are not distinguished. CQ—Cerro Quemado volcano; LD—Post-collapse lava dome; LF—1818 A.D. Cerro Quemado lava flow; SM—Santa María volcano; SG—Santiaguito lava-dome complex; Alm—town of Almolongo. After Conway et al. (1992) and Vallance et al. (1995).

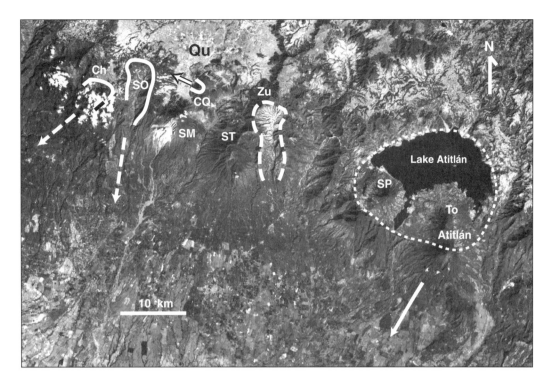

Figure 4. U.S. National Aeronautics and Space Administration Landsat image of southwestern Guatemala. Solid lines mark debris avalanche source scarps; dashed line, Zunil depression. Arrows mark debris avalanche travel direction; dashed arrows, direction of possible avalanches. Dashed circle shows boundary of Atitlán III caldera (Newhall, 1987). Ch—Chicabal; SO—Siete Orejas; Qu—Quetzaltenango City; CQ—Cerro Quemado; SM—Santa María; ST—Santo Tomás; Zu—Zunil; SP—San Pedro; To—Tolimán.

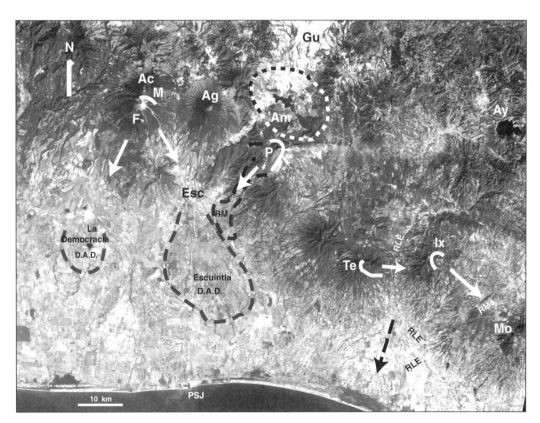

Figure 5. U.S. National Aeronautics and Space Administration Landsat image of south-central Guatemala. Solid white lines—debris avalanche source scarps; dashed lines—debris-avalanche deposits; solid arrows—avalanche travel direction; dashed arrow—possible avalanche travel direction; dotted white line—Amatitlán caldera (Wunderman and Rose, 1984). Volcanoes from west to east: Ac—Acatenango; M—Meseta; F—Fuego; Ag—Agua; P—Pacaya; Am—Amatitlán caldera; Te—Tecuamburro; Ix—Ixhuatán; Mo—Moyuta; Ay—Ayarza caldera. Other labeled features: D.A.D.—debris avalanche deposit; Gu—Guatemala City; Esc—Escuintla City; RM—distal extent of Río Metapa debris avalanche from Pacaya volcano; PSJ—Puerto San José; RLE—Río Los Esclavos; RMa—Río Margaritas.

eroded edifice of the Fuego-Acatenango complex based on clast geochemistry. Basset (1996) correlated a proximal avalanche deposit on the SW flank of the ancient Acatenango edifice with La Democracia deposit and attributed the deposit to collapse of the oldest edifice of Acatenango volcano between ~70,000 and 43,000 yr B.P., a conclusion supported by Chesner and Halsor (this volume).

## Fuego-Acatenango Complex

The largest debris-avalanche deposit in Guatemala forms a broad fan south of the city of Escuintla on the Pacific coastal plain (Fig. 5). The deposit forms a poorly vegetated fan that was identified by Rose et al. (1975) from Skylab photography as a possible volcanic lahar or avalanche fan, and Vallance et al. (1995) recognized its debris-avalanche origin. The exposed portion of the Escuintla debris-avalanche deposit covers an area of 300 km$^2$ and has an estimated volume of 9 km$^3$ (Vallance et al., 1995). Inclusion of the likely area of the buried proximal portion of the deposit gives an estimated volume of ~15 km$^3$ (Siebert et al., 2004).

Abundant hummocks in the deposit range between 5 and 50 m in height and contain highly fractured and deformed blocks of basalt, andesite, and dacite. Two distinct hummock clusters suggest the possibility of emplacement in two (or more) slide blocks. Fresh roadcut exposures revealed features such as step-like normal faulting away from the center of hummocks with offsets of several meters (Fig. 6), matrix color mottling, smearing of breccia fragments along internal fault planes, and preserved transported dike segments.

The Escuintla avalanche was considered by Vallance et al. (1995) to have originated from Fuego because of the orientation of the axis of the deposit and the presence of a potential source scarp on Meseta volcano (the predecessor to Fuego) and lithologic similarities to Fuego rocks. Additional analyses by Chesner and Halsor (this volume) showing amphibole-bearing dacitic rocks in the deposit not found at Meseta suggest an origin more consistent with Acatenango. This conflicts with the physical evidence of a possible avalanche source scarp on Meseta. The origin of this scarp is more consistent with catastrophic edifice failure than long-term erosional processes, and it would consequently be the youngest avalanche source area in the Acatenango-Fuego massif. The youngest exposed deposit from the massif (the Escuintla deposit) would then most logically derive from the Meseta scarp. The presence of dacitic deposit samples, however, requires either a dacitic component to Meseta not currently exposed or the incorporation of dacitic basement or volcaniclastic rocks in the avalanche, and uncertainties remain regarding the origin of the Escuintla avalanche.

The 84,000 yr B.P. Los Chocoyos ash from Atitlán caldera does not overlie the avalanche deposit, which was emplaced sometime after a ca. 30,000 yr B.P. lava flow from Meseta and before the estimated 8500 yr required to construct the post-Meseta Fuego edifice based on projections of historical eruption rates (Vallance et al., 2001a; Chesner and Halsor, this volume).

## Pacaya Volcano

The basaltic Pacaya stratovolcano, constructed near the southern rim of Amatitlán caldera south of Guatemala City, is one of the most active in Guatemala (Fig. 7). MacKenney cone, frequently active since 1961, was constructed within and partially overtops the avalanche scarp. The Río Metapa debris-avalanche

Figure 6. Hummock of Escuintla debris avalanche from the Acatenango-Fuego complex ~40 km from Fuego volcano. Dashed lines mark major outcrop scale step-like faults; LF—offset lava flow segments. The many parallel diagonal patterns on outcrop are grooves from roadcut excavation.

deposit extends 25 km south to reach the Pacific coastal plain (Vallance et al., 1995; Kitamura and Matías, 1995). The deposit underlies ~55 km² of the valley to an estimated thickness of 120 m. The predominant deposit lithology is basaltic clasts and fractured lava-flow segments from Pacaya volcano, but includes large, irregular clasts of fibrous, white, biotite-bearing pumice from Amatitlán caldera (Wunderman and Rose, 1984) that predate Pacaya volcano.

The deposit was estimated to have a volume of ~0.65 km³ by Vallance et al. (1995). Later discovery of the deposit in the headwaters of the Río Marinalá valley to the west (Kitamura and Matías, 1995) indicates that this is a minimum volume. With the inclusion of these additional western deposits, the volume could have exceeded 1 km³; however, this possible larger volume is nevertheless insufficient to account for the volume of the collapse scarp that extends south of MacKenney cone. Because the scarp intersected older Tertiary volcanic rocks on its northeastern side, it is asymmetrical (Fig. 7). The scarp headwall appears to be clearly related to collapse forming the Río Metapa debris avalanche, and other than a 200-m-long segment overtopped by MacKenney cone, is contiguous with the linear southeastern portion of the scarp, whose morphology resembles other edifice-collapse scarps. The deficient volume of the Río Metapa avalanche deposit, however, raises the possibility that the southeast portion of the scarp could have originated from an earlier collapse whose deposits are buried by proximal and other coastal plain deposits, and uncertainties persist about the origin of this scarp.

The Río Metapa avalanche was inferred by Vallance et al. (1995) to be between ~2000 and 400 yr old based on soil profiles and an absence of historical records of the collapse. Kitamura and Matías (1995) stratigraphically bracketed the avalanche deposit and its associated tephra layer between tephra layers radiocarbon dated at 1555 ± 80 and 595 ± 70 yr B.P.

A pyroclastic-surge deposit containing juvenile fragments blankets the proximal portion of the Río Metapa avalanche deposit over a 90° arc extending >12 km SSW of Pacaya volcano (Vallance et al., 1995; Kitamura and Matías, 1995). The surge deposits are thicker in depressions than on ridge tops and contain abundant low-angle dune and antidune cross bedding. Thicknesses reached 1.5 m in proximal areas, and the deposit is >20 cm thick at its most distal documented locality, 12 km SSW of the collapse scarp (Kitamura and Matías, 1995). The axes of the pyroclastic-surge deposit and the avalanche scarp are coincident and consistent with a lateral focusing of the surge. Stratigraphic evidence supports linkage of the surge and avalanche deposits, with the surge resulting from sudden decompression of a hot hydrothermal system and consequent magmatic eruption (Vallance et al., 1995).

**Tecuamburro Volcano**

The andesitic-to-dacitic Tecuamburro volcano SE of Pacaya was constructed within an east-facing scarp that truncates the older andesitic Miraflores volcano (Duffield et al., 1989, 1991).

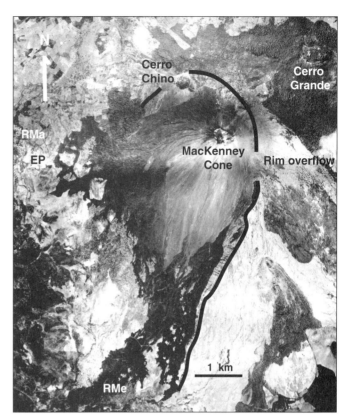

Figure 7. Instituto Geográfico Nacional 1:30,000 scale aerial photograph of Pacaya volcano. Solid black lines—collapse scarp at Pacaya; RMa—Río Marinalá drainage; RMe—Río Metapa drainage; EP—El Patrocinio village. Cerro Chino is a post-collapse pyroclastic cone overtopping the collapse scarp; MacKenney cone was constructed within the scarp. Dark areas inside scarp are historical lava flows from MacKenney cone.

The avalanche caldera at Miraflores is at least 4 km wide and 6 km long, and the Miraflores debris-avalanche deposit covers roughly 10 km² opposite the scarp on the flanks of Ixhuatán volcano across the Río Los Esclavos (Fig. 5). The deposit displays anomalously hummocky terrain with few surface outcrops of cohesive rocks. Dissection by the Río Los Esclavos perpendicular to the original avalanche travel direction exposes disrupted andesitic lava flow sequences of the avalanche deposit. The age of the edifice collapse is bracketed by overlying pyroclastic-flow deposits radiocarbon dated at 38,300 ± 1000 yr B.P. and the ~100,000 yr B.P. K-Ar age of Miraflores stratovolcano (Duffield et al., 1991).

The exposed deposit is of insufficient volume to account for the roughly 4 km³ collapse scarp, and it is likely that that the Miraflores avalanche was deflected downvalley and traveled many kilometers onto the Pacific coastal plain (Vallance et al., 1995). Although a corresponding deposit has not been confirmed, hummocks of uncertain origin are found on the coastal plain (Vallance et al., 1995). After reaching the coastal plain, the Río Los Esclavos is anomalously deflected to the southeast, parallel to the volcanic front, indicating a subtle topographic high to the west. Landsat images display an area of land-use patterns, comparable to that seen for the Escuintla debris-avalanche deposit to the west, that extends nearly to the coast and is consistent with a possible underlying avalanche deposit from Tecuamburro (Fig. 5).

## Other Possible Collapse Events

Several other scarps on Guatemalan volcanoes are suggestive of edifice failure, but extensive weathering and lack of exposures has hindered identification of associated collapse deposits (Vallance et al., 1995). Large horseshoe-shaped depressions at Zunil (Fig. 4) and Siete Orejas (Figs. 3 and 4) in western Guatemala and Tahual in SE Guatemala (Fig. 1) narrow at their breached ends, a morphology typical of depressions formed by erosional processes. However, a diamicton of possible (but uncertain) edifice-collapse origin was observed below Siete Orejas. Immediately west of Siete Orejas is Chicabal, a crater-lake–capped stratovolcano constructed within a 4.5-km-wide horseshoe-shaped scarp consistent with edifice-failure origin (Fig. 4). A broad fan below the scarp forces the Río Naranja to detour 30 km west before reaching the coast near the Mexican border. Extensive weathering, however, precludes determination of the origin of the deposit. Another horseshoe-shaped scarp cuts an unnamed volcano 12 km east of El Tumbador in western Guatemala (Fig. 1).

The SE side of the summit of Ixhuatán volcano, east of Tecuamburro volcano, is cut by a large horseshoe-shaped caldera that contains the Cerro Los Achiotes lava-dome complex (Fig. 5). Deposits extending SE were interpreted by Reynolds (1987) as originating from a lateral blast similar to that at Mount St. Helens in 1980. Hummocky terrain extending SE beyond the Río Margaritas toward the base of Moyuta volcano (Fig. 5), visible on a 1:50,000 scale topographic map, is consistent with a debris avalanche origin (Siebert et al., 2004), although these deposits have not been mapped in detail.

## EL SALVADOR

### Santa Ana Volcano

Collapse of Santa Ana volcano during the late Pleistocene produced a voluminous debris-avalanche deposit (Pullinger, 1998; Siebert et al., 2004) exposed over a broad area SSE of Santa Ana volcano to the coast (Fig. 8). The roughly 20-km-wide Acajutla Peninsula extends up to 7 km off the former coastline, and the submarine terminus of the deposit lies 46–50 km from the volcano. Pyroclastic-flow and airfall deposits from nearby Coatepeque caldera dated ca. 57,000 B.P. (Rose et al., 1999) do not overlie the Acajutla debris-avalanche deposit, providing an upper limit to its age. Soil profiles on the deposit suggest it is older than Holocene. The estimated volume of the deposit, including the substantial submarine portion, inferred from bathymetry and the inflection point of otherwise shore-parallel bathymetric contours, is $16 \pm 5$ km$^3$ (Siebert et al., 2004). Most of the avalanche source area is buried, but a 5-km-long arcuate segment of the scarp is exposed on the NW side of the modern Santa Ana edifice (Fig. 9).

The medial portion of the deposit below mantling lava flows from Santa Ana volcano contains abundant hummocks, which reach to ~60 m high and are typically ~100–300 m in longest dimension; hummocks are also found to the current shoreline and offshore. Some apparent hummocks are instead exposed islands of Tertiary Bálsamo Formation rocks in line with NE-SW–trending ridges of similar composition to the east. Four distinct hummock clusters in the medial portion of the deposit lie near these buried linear ridge extensions (Fig. 10). These ridges impeded avalanche movement and stranded block-facies hummocks on the proximal sides of the ridges.

Block-facies material is abundant in avalanche hummocks, where quarries expose spectacular sections of color-mottled units of the former edifice, which have been transported relatively intact from their source on the volcano. Large clasts up to 6 m in size are exposed, and jigsaw blocks are present. One hummock ~30 km from the volcano contains large segments of essentially

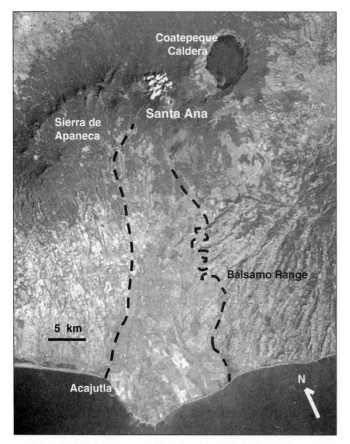

Figure 8. U.S. National Aeronautics and Space Administration Space Shuttle image STS61C-31-45 (12 Jan. 1986), aerial oblique image of Santa Ana and the Acajutla debris-avalanche deposit (dashed line); port city of Acajutla is labeled. Width of Acajutla Peninsula is 20 km.

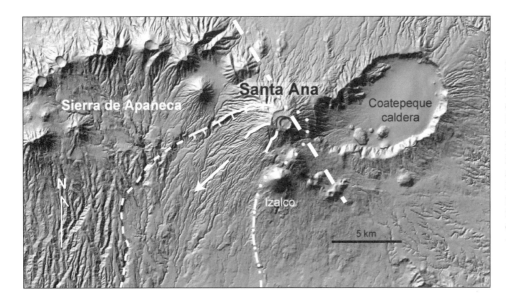

Figure 9. Shaded-relief digital elevation model view of the Santa Ana massif. Hachured line marks exposed portion of avalanche source area; dotted line, its inferred buried extent. Dashed-dot line marks distal margin of possible scarp from this or earlier collapse event. Arrow indicates failure direction. Dashed line at left is margin of Acajutla debris-avalanche deposit; diagonal dashed line at top marks axis of 20-km-long NW-trending chain of satellitic cones and vents cutting across the volcano and extending ~7 km north of the top of the image.

intact bedded ash-and-lapilli layers transported without significant disruption, as well as massive and brecciated lava flows and a 20 × 30 m segment of a basaltic block-and-ash flow deposit.

Mixed-facies or lahar-facies material in intra-hummock areas, which was the transport medium for block-facies material, comprises the volumetrically dominant component of the deposit. These materials consist of homogenized heterolithologic matrix containing clasts from Santa Ana volcano mixed at the margins of the deposit with significant amounts of accessory material from the Bálsamo Formation on the east and Tertiary ignimbrites on the west. Ripped-up soil clasts, clastic dikes, and large segments of block-facies material transported within the mixed facies were observed in stream exposures.

An indistinct second horseshoe-shaped depression, 1.6 × 2 km wide, cuts the upper SW flank of modern Santa Ana volcano (Fig. 9) and could represent the scarp of a smaller edifice failure that post-dated construction of the current summit cone of Santa Ana (Siebert et al., 2004). A light-brown dacitic airfall pumice deposit directly overlies a debris-avalanche deposit that could correspond to this later collapse at one proximal location, although their synchronicity is not certain.

## San Vicente Volcano

Another large debris-avalanche deposit in El Salvador originated from twin-peaked San Vicente volcano, east of Ilopango caldera (Major et al., 2004; Siebert et al., 2004). The Tecoluca debris avalanche from San Vicente volcano traveled at least 25 km southeast to the Río Lempa, a major river that transects the volcanic chain (Fig. 11). A large area of hummocky terrain is found southeast of the city of Tecoluca with abundant hills up to several tens of meters high.

The deposit is predominantly matrix supported, with large clasts up to several meters in size that are often highly fractured

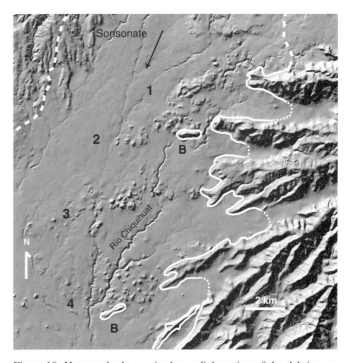

Figure 10. Hummock clusters in the medial portion of the debris-avalanche deposit between solid dashed lines at left and right are numbered; letters "B" indicate islands of Bálsamo Formation rocks along trend with Bálsamo ridges. Black arrow at top marks generalized flow direction of avalanche from Santa Ana volcano (~15 km NW). Digital elevation model is based on 10-m contours and does not show small hummocks.

and occasionally weathered to the extent that large clasts can be cut with a trowel. The matrix is a poorly sorted mixture of typically subangular to angular, medium to very fine sandy particles of variable induration and coloration, with lithic clasts of variable lithologies, weathering, and densities; the matrix also

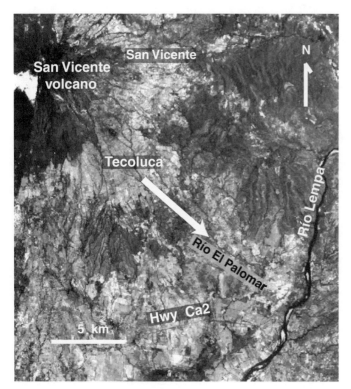

Figure 11. U.S. National Aeronautics and Space Administration Landsat image showing San Vicente volcano and location of Tecoluca debris avalanche and debris flow.

contains small pumice fragments. Stream beds along the Río El Palomar 16 km SE of the summit display angular clasts to more than a meter in diameter, matrix color variations, jigsaw blocks, and clast concentrations; a prismatically jointed block was also observed. The age of the deposit is not known precisely, but the presence of overlying soil layers up to several meters thick and the lack of a visible source area on the dominantly Pleistocene San Vicente volcano indicate a Pleistocene age.

## San Miguel–Chinameca Complex

Deposits of apparent debris-avalanche origin are found in the Lolotique area in eastern El Salvador, north of the city of Chinameca and north and west of the San Miguel–Chinameca volcanic complex (Fig. 1) (C. Pullinger, 2002, personal commun.). The Lolotique deposits appear to thicken toward the southeast and may have originated from a proto-Chinameca (also known as El Pacayal) or proto–San Miguel volcano (C. Pullinger, 2005, personal commun.). Escobar (2003) described a younger debris-avalanche deposit located southwest of San Miguel volcano, but sparse outcrops did not permit direct linkage to San Miguel.

## HONDURAS

No volcanoes with major edifice-collapse events are known to us in Honduras. Much of Honduras lies behind the volcanic front, and only small-volume basaltic shield volcanoes and monogenetic volcanoes are present in the central highlands (Williams and McBirney, 1969). Larger stratovolcanoes lie in Honduras along the Gulf of Fonseca, and these volcanoes are more susceptible to gravitational collapse, although none have been identified.

## NICARAGUA

### Mombacho Volcano

*Las Isletas Avalanche*

Mombacho volcano, rising above Lake Nicaragua southeast of Masaya and Apoyo calderas, is cut by two large horseshoe-shaped depressions and has been subject to at least four major edifice failures (Fig. 12). The most well-known debris avalanche from Mombacho produced the dramatic Las Isletas (Fig. 13), an arcuate peninsula and island chain (also referred to as the Isletas de Granada or the Aseses Peninsula) that extends into Lake Nicaragua (Ui, 1972; van Wyk de Vries and Borgia, 1996; van Wyk de Vries and Francis, 1997). The 6-km-long peninsula lies across the shallow 3-km-wide Aseses Bay and is flanked by more than 500 small island hummocks (Viramonte et al., 1997).

The proximal source area on the NE flank of Mombacho forms a wedge-shaped area ~1.2–2 km wide and 2.5 km long. The scarp deflects significantly outward at the base of the volcano and extends an additional 3 km or so to the lakeshore. The floor of the proximal portion of the scarp crudely parallels the slope of the edifice, suggesting a dip-slope component to the failure (van Wyk de Vries and Francis, 1997). The lower scarp cuts into the basement Las Sierras ignimbrites, which form a component of the distal deposit at Las Isletas. No evidence for associated eruptive products was found in or overlying the debris-avalanche deposit, and the avalanche may have been caused by failure in the substrata. Its age is not precisely known, but pottery fragments were found throughout a hummock, and a progressively sheared sarcophagus with human bones was found within the upper part of the avalanche deposit (van Wyk de Vries et al., 2005), which if allochthonous, implies that the avalanche is Holocene, perhaps just preconquest.

*El Cráter Avalanche*

A large, steep-sided depression cuts the southern side of Mombacho volcano and was the source of a major debris avalanche that swept onto low-angle slopes beyond the volcano (Ui, 1972; van Wyk de Vries and Borgia, 1996; van Wyk de Vries and Francis, 1997). The scarp is ~1.8 × 2.3 km wide and narrows somewhat at its breached end; its headwall rises ~700 m to Mombacho's eastern summit and partially truncates the headwall of Las Isletas scarp, indicating that the El Cráter collapse is younger. The scar contains highly altered material and hosts active superheated fumaroles. A hummocky deposit covering an area of ~60 km$^2$ extends ~13 km to the south, where the Río

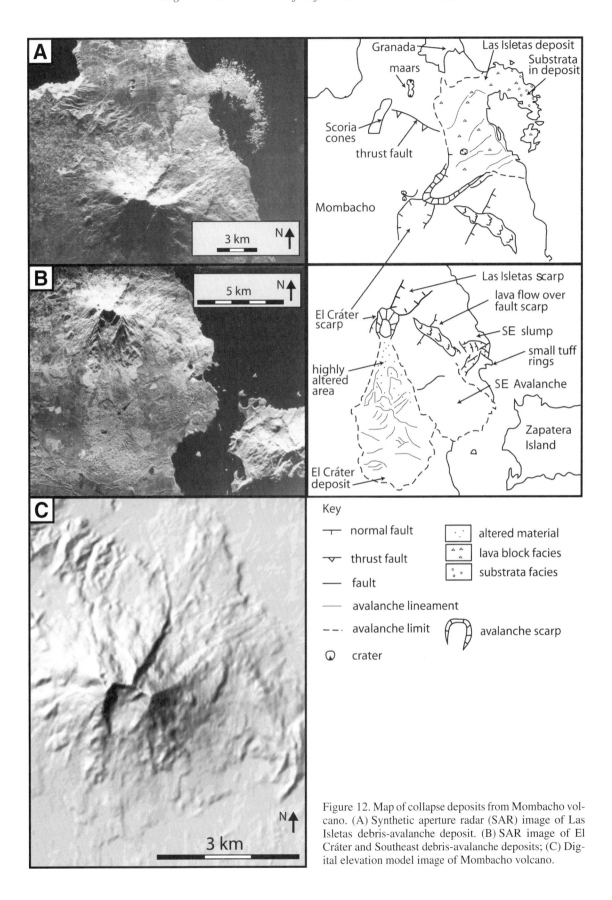

Figure 12. Map of collapse deposits from Mombacho volcano. (A) Synthetic aperture radar (SAR) image of Las Isletas debris-avalanche deposit. (B) SAR image of El Cráter and Southeast debris-avalanche deposits; (C) Digital elevation model image of Mombacho volcano.

Figure 13. The dramatic Las Isletas peninsula and island chain consist of debris-avalanche deposits ~10 km from their source at Mombacho volcano, whose lower slopes lie at the bottom of the image. Photo by Jaime Incer.

El Pital is deflected around its distal margin. The proximal part of the avalanche deposit contains greater proportions of hydrothermally altered rocks than the distal part. As with the Las Isletas deposit, no evidence for associated eruptive activity was found, and the collapse probably occurred due to weakening by progressive alteration (van Wyk de Vries and Francis, 1997).

Contemporary accounts indicate that following a major earthquake, Mombacho "burst" in ca. 1570 A.D. and produced an avalanche to the south that destroyed the Indian village of Mombacho, killing 400 persons (Incer, 1990; Feldman, 1993; Vallance et al., 2001b). One account stated that the collapse occurred during a stormy night with heavy rainfall the night *after* the night of a major earthquake (Vallance et al., 2001b). Another states that an avalanche of comparable size to the north would have reached the city of Granada, located a similar distance from the summit as the distal El Cráter deposit. The ca. 1570 event may have been a lahar that originated from a preexisting El Cráter depression (van Wyk de Vries and Francis, 1997; Vallance et al., 2001b). Contemporary accounts are also consistent with its origin in historical time (van Wyk de Vries et al., 2005). Overlying soils lack a prominent development of clay in the B horizon, a process that occurs rapidly in tropical environments, but their ages (and that of El Cráter avalanche) are not known with certainty.

*Southeast Avalanche*

A third debris avalanche at Mombacho forms a deposit (Fig. 12) extending into Lake Nicaragua across a narrow channel from Zapatera volcano. The Southeast debris avalanche (van Wyk de Vries and Borgia, 1996; Vallance et al., 2001b) extends at least 11 km from the summit of Mombacho to the coast, forming an irregular shoreline with small offshore islands. The deposit covers an area comparable to Las Isletas avalanche and has a maximum hummock density along the axis of the deposit several km inboard of the shoreline. This avalanche is older than the Las Isletas and El Cráter avalanches, and no source area is visible on the volcano.

A fourth edifice failure was noted by van Wyk de Vries and Borgia (1996) east of Mombacho (Fig. 12). In contrast to the other failures at Mombacho, the "East Megablock Slump" did not thoroughly disaggregate, but formed a series of linear toreva-like ridges (Reiche, 1937) a kilometer or more in length transverse to the direction of movement.

### Other Nicaraguan Events

Other edifice failures, smaller in volume than those considered here, have occurred in Nicaragua. The catastrophic collapse that rapidly transformed to a lahar in 1988 at Casita volcano at the eastern side of the San Cristóbal complex (Fig. 1) (van Wyk de Vries et al., 2000; Kerle and van Wyk de Vries, 2001; Scott et al., 2005) had devastating consequences. Large boulders and subdued hummocky topography are located beyond the NW base of El Chonco (J. Incer, 1998, personal commun.; Siebert et al., 2004), a small satellitic volcano at the western base of the San Cristóbal volcanic complex, west of Casita. No debris-avalanche deposits from El Chonco were mentioned by Hazlett (1987) or Van Wyk de Vries and Borgia (1996), and the El Chonco deposit may represent a small volume block-and-ash flow deposit from collapse of a growing lava dome. Collapse of hot lava domes such as occurred at the Santiaguito dome at Santa María volcano in Guatemala in 1929 (Mercado et al., 1988) and at Soufrière Hills volcano on Montserrat in July 1997 (Voight et al., 2002) can produce widespread deposits transitional to those of debris avalanches.

## COSTA RICA

### Orosí Volcano

Along with Guatemala, Costa Rica leads Central America in the number of identified edifice collapses. These occur along the central Cordillera, extending from Orosí to Turrialba volcanoes (Fig. 14) (Alvarado et al., 2004). The Orosí volcanic complex, consisting of four eroded forested edifices, lies at the NW end of the Guanacaste Range, near the border with Nicaragua. Cerro Cacao is a 1659-m-high stratovolcano at the southern end of the complex.

An avalanche deposit (Castillo, 1978; Tournon, 1984) on the SW side of Cerro Cacao has thicknesses that vary from >5 m in the Río Quebrada Grande and Salitral to 40 m in certain parts of the Río Ahogados. The Quebrada Grande deposit has a maximum area of 52 km$^2$, and hummocky topography is common. Rounded to subangular, decimeter to 3.5 m size clasts of andesitic, dacitic, and basaltic composition (Zamora et al., 2004) are present in a matrix of subrounded to angular sand-sized, poorly sorted pumice up to 2.5 cm in size and multicolored scoria fragments up to 3 cm in size. The deposit is frequently clast-supported, and jigsaw textures are sometimes present. The deposit originated from a 2.5 × 3.5-km horseshoe-shaped scarp that cuts the SW side of Cerro Cacao (Fig. 15). A morphologically similar 3-km-wide scarp cuts the eastern flank and was the source of debris flows originating from deep gullies cutting each side of the scarp (Kerle and van Wyk de Vries, 2001). Detailed analysis of the scarp morphology suggested it may be a gravitational slump similar to that proposed at Casita (Cecchi et al., 2005).

## Rincón de la Vieja Volcano

On the south flank of Rincón de la Vieja volcano, a debris-avalanche deposit covers an area up to 18 km$^2$ (Alvarado et al., 2004; Zamora et al., 2004). Hummocky topography with relief of up to 20 m is found near the Río Colorado canyon and in the Azufrales area within the Rincón de la Vieja National Park. The deposit is in part covered by Holocene pumice-fall deposits. This unit has been relatively well studied because surficial outcrops are supplemented by wells drilled to measure geothermal gradients. The deposit consists mainly of angular to subangular, centimeter-to-decameter clasts of varying compositions. Occasional reddish-gray pumice and fragments of lithic tuffs and ignimbrite clasts occur in a variably colored fine-sandy matrix. In areas near Azufrales, the deposit textures differ and consist of laminated, massive and/or vesicular aphanitic and porphyritic blocks with variable degrees of hydrothermal alteration. The apparent paucity of large clasts may reflect a more water-rich collapse or more easily disaggregated source materials.

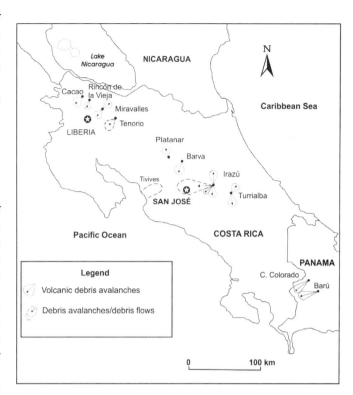

Figure 14. Map of avalanche deposits in Costa Rica and Panama (after Alvarado et al., 2004). Possible overlying deposits are not shown at Miravalles, Irazú, Turrialba, and Barú volcanoes.

## Miravalles Volcano

### La Fortuna Avalanche

Three edifice failures, La Fortuna, Las Mesas, and Upala, have been documented at Miravalles volcano. La Fortuna, the largest, is the most thoroughly studied avalanche deposit in Costa Rica as a result of at least 86 wells drilled in conjunction with an assessment of the country's largest geothermal development. Lahar deposits at Miravalles, covering much of Guayabo caldera, were formed by destruction of the SW part of the Paleo-Miravalles edifice (Fernández, 1984) and attributed by Alvarado (1987) and Melson (1988) to a debris-avalanche origin. The La Fortuna deposit covers an area of ~90–130 km$^2$ west and southwest of the volcano and is partly overlain by pyroclastic-flow deposits, lava flows, and orange-colored pumice-fall deposits. The deposit displays a typical hummocky morphology with numerous lagoons and small marshes tens of meters in diameter distributed between hills mostly meters to tens of meters in height. Hummock orientations are longitudinal, transverse, or arcuate. Toward the west, deposits are thinner and do not display clear avalanche morphology, and may represent predominantly mixed facies or lahar facies material.

The avalanche deposit consists of a chaotic mixture of angular clasts up to a meter in size, composed mostly of andesitic, basaltic-andesitic, and occasional dacitic lava up to 5 m

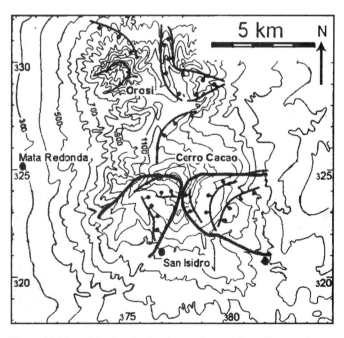

Figure 15. Map of the Orosí volcanic complex and Cerro Cacao volcano, modified from Kerle and van Wyk de Vries (2001). Solid lines mark avalanche source scarps; solid lines with dots represent normal faults.

in diameter, with some hydrothermally altered silicified blocks, soils, and pumice within a sandy matrix. Large transported lava-flow segments had been interpreted as in situ lava flows or lava domes during initial geothermal investigations. The deposit also contains entrained tuff deposits and fluvio-lacustrine sediments and displays multi-hued areas reflecting hydrothermal alteration. Pumice fragments generally <2 cm in size are extensively altered and have an uncertain origin, either contemporaneous with or predating the collapse event. The question of whether the collapse was accompanied by an eruption remains open.

Data from 73 well logs in conjunction with field observations produce an estimated volume of 6–8 km$^3$ for the avalanche deposit (Alvarado et al., 2004). Three radiocarbon dates for the La Fortuna avalanche deposit average ~7400 yr B.P. and give a calibrated age of 8275 yr B.P. Comparison of topographic contours with aerial photographs indicates that the La Fortuna avalanche(s) took place to the southwest of the Bajo de Chiqueros depression and thus pertains more to the Paleo-Miravalles edifice than the modern edifice. This would imply that the modern Miravalles edifice is very young, built in <8000 yr, following a quiescence of centuries to a few thousand years, as inferred from archaeological dates (Hurtado de Mendoza and Alvarado, 1988).

*Las Mesas Avalanche*

The Las Mesas debris-avalanche deposit consists of a chaotic mixture of lava blocks, Santa Rosa pyroclastic-flow materials, airfall ash deposits, sediments, and dominantly orange-colored, rounded argillaceous blocks (to 1.5 m diameter) and pumice-flow material. Andesitic lava fragments are angular to rounded and up to several meters in size. Ash lenses 5–20 cm thick and up to >2 m long are present, along with alluvial sediment with clasts <30 cm in diameter.

Deposits of the Las Mesas debris avalanche overlie those of the La Fortuna avalanche and the Santa Rosa pyroclastic flow and are interlayered with lava flows overlying the La Fortuna deposit. The Las Mesas avalanche deposit is consequently estimated to have been emplaced roughly between the 8000-yr-old La Fortuna avalanche and the lava flows roughly dated from archaeological evidence at ~2000 yr ago (Hurtado de Mendoza and Alvarado, 1988). Despite the apparent youthful age of the Las Mesas deposit, only a subtle source scarp, most of which has been filled by post-collapse eruptions, remains on the SW side of the modern edifice. Geothermal well logs indicate a Las Mesas avalanche deposit thickness of 15–100 m, and a volume of 0.1–0.25 km$^3$ has been estimated (Alvarado et al., 2004).

*Upala Avalanche*

Hummocky topography below the NE flank of Miravalles volcano, near the settlement of Upala, has been related to collapse of the Paleo-Miravalles volcano, which left a horseshoe-shaped crater visible from the Caribbean sector. The deposit, which has not been studied in detail, covers an area of at least 20 km$^2$ (Alvarado et al., 2004).

**Tenorio Volcano**

The Tierras Morenas deposit, one of the largest edifice-failure deposits currently known in Costa Rica, covers an area of ~80–120 km$^2$ southwest of the main cone of Tenorio and has an estimated volume of 2.0 ± 0.4 km$^3$ (Alvarado et al., 2004). The deposit has a hummocky topography only in the proximal facies, such as at Hacienda Tenorio. The lack of hummocks elsewhere may represent transformation into a debris flow.

**Platanar Volcano**

Alvarado and Carr (1993) noted a relatively old (>100,000? yr) gravitational collapse that originated from the volcanic massif of Platanar-Chocosuela and produced the Chocosuela debris avalanche. The deposit extends up to ~20 km NNW from the volcano to near the village of Vaca Blanca, west of the Aguas Zarcas cinder cones, but few details are available. The large horseshoe-shaped Chocosuela caldera, possibly associated with the collapse, is open to the northwest. Only the southern and eastern portions of its rim are visible; elsewhere, the Quaternary Platanar and Porvenir volcanoes overtop it.

**Barva Volcano Area**

*Western Central Valley Avalanche*

Barva volcano is the largest-volume volcano of Costa Rica, but is covered to a large extent by virgin forest and consequently is one of the least studied. It is possible that the pseudo lava fields that cover the Western Central Valley are products of debris avalanches. Remnants of possible hummocks and blocks, locally with hydrothermal alteration, are observed between the Juan Santamaría airport and Manolos, along with transported volcanic breccias, block-facies lava flow segments, and blocks with imbricated structures (Méndez and Hidalgo, 2004; Alvarado et al., 2004). The stratigraphic position of the deposit indicates a mid-Pleistocene age between ca. 150 and 230 ka (Méndez and Hidalgo, 2004).

*Tivives Avalanche*

Although the exact origin of the <0.65 Ma Tivives volcaniclastic deposit (Gans et al., 2003) is not known, it clearly originates from paleovolcanism of the Quaternary Cordillera Central (Alvarado et al., 2004). This deposit, interpreted originally as a Tertiary lahar (Madrigal, 1970), apparently corresponds to a late Pleistocene debris avalanche that transformed into a debris flow and was channelized by the paleo-course of the Río Grande de Tárcoles. It originated in the drainages of the Río Grande and Jesús María rivers and is exposed in the Pacific coastal cliffs of Guacalillo-Tivives (Fig. 14).

The deposit consists of poorly sorted clasts of lava, ignimbrites, stratified pyroclastic rocks (pumiceous tuffs and lithic and accretionary lapilli tuffs), hydrothermally altered lava, and deformed plastic clays. The deposit contains megablocks, some composed of pyroclastic rocks (stratified or with internal faults), and variably deformed vertically tilted lavas. The blocks are contained within a light-brown crystal-rich matrix with minor lithic fragments. The deposit has a thickness of several tens of meters and covers several tens of square kilometers.

## Irazú

### Prusia-Tierra Blanca Avalanche

Between the settlements of Sabanilla and San Juan de Chicoá, a semicircular escarpment ("Cráter del Derrumbo") extends to the SW. A deposit with a very irregular topography extends to near Tierra Blanca and is probably a debris-avalanche deposit (Alvarado, 1993; Pavanelli et al., 2003, 2004). The deposit (also known as the Reventado debris-avalanche deposit) has a maximum thickness of 100 m and displays hummocks as high as 20–30 m in the proximal sector. It has a chaotic texture that consists of fresh volcanic rocks (some possibly juvenile) and hydrothermally altered rocks in sharp contact within a sandy orange-to-brown matrix. The deposit has an estimated volume of 1–2 km$^3$ (Pavanelli et al., 2003). On the basis of local stratigraphy, its age is upper Pleistocene or even Holocene (Alvarado et al., this volume).

### Río Birrís Avalanche

An avalanche deposit along the Río Birrís on the SE flank of the Irazú has been covered by the ca. 17 ka Cervantes lava flow (Alvarado et al., 2004). The deposit consists of lava blocks several meters in diameter, with jigsaw textures, supported in a brown matrix. The absence of a hummocky morphology suggests the avalanche may have had a water-saturated component (Pavanelli et al., 2003).

### Río Costa Rica Avalanche

A debris-avalanche deposit is partially exposed on the Caribbean flank of Irazú volcano, near Guápiles, between the Ríos Corinto and Costa Rica (Alvarado, 1987, 1993). It originated from the upper northern flank of Irazú and is partially exposed over an area of at least 5.5 km$^2$ between the settlements of Flores (La Unión) and Rancho Redondo de Guápiles at the northern base of the volcano.

### Cabeza de Vaca Avalanche

A deposit named "Central Valley Lavina" appears to correspond to a debris-avalanche deposit from part of the Paleo-Irazú edifice, the Cabeza de Vaca stratovolcano on the western side of modern Irazú volcano (Fig. 16). This ~2 km$^3$ avalanche covered an area of 130 km$^2$ and was possibly diluted by incorporation of water from glacial melting and transformed into a lahar (Hidalgo et al., 2004; Alvarado et al., 2004). Preserved hummocky topography in the block-facies component was still observed until a few years ago near Zapote and at the intersection of the San José–Escazú and Hatillos freeways. The deposit underlies much of the capital city of San José (Fig. 16). Abundant wells allow determination of the area and extent of the deposit, which reaches ~25 km from its apparent source on Cabeza de Vaca, which is cut by a horseshoe-shaped escarpment 4 km long and 2.5 km wide, oriented to the WSW (Hidalgo et al., 2004).

## Turrialba Volcano

### Angostura Avalanche

The Angostura deposit (Alvarado et al., 2004) is exposed widely in the Turrialba Valley as far as the confluence of the Tuís

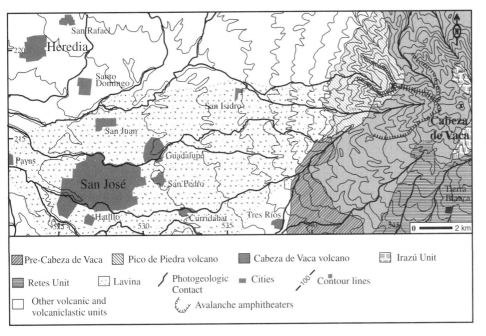

Figure 16. Map of debris-avalanche and debris-flow deposit from Cabeza de Vaca volcano on west flank of Irazú volcano (Hidalgo et al., 2004).

and Reventazón rivers. It is best seen in three outcrops several hundred meters long along the road to La Toma, La Ventana, and La Casa de Máquinas of the Angostura hydroelectric project. The deposit consists of angular blocks of aphanitic to porphyritic andesites and rare fossiliferous limestones along with minor rounded andesitic and sandstone blocks, within a matrix (65%–75%) of reddish brown to gray ash, mud, and plastic deformed clay fragments. Megablocks of intensely fractured fossiliferous limestones are present. Block-facies material includes matrix color variation, and jigsaws blocks are present. The lava clasts are mostly andesitic, but also include dacitic and basaltic clasts up to 1.5 m in diameter. Larger blocks, several meters in diameter, are composed mainly of relatively fresh lava but also include fractured blocks of weathered or hydrothermally altered lavas.

The age of the avalanche deposit is constrained by three independent radiometric dates by different investigators. These three dates have similar calibrated radiocarbon ages, averaging 17,035 yr B.P. The deposit also contains rare, mostly carbonized wood a few centimeters to several meters in length. It is not clear whether these indicate syneruptive activity or are incorporated older wood fragments. Although comparable in age to the Cervantes lava flow, the avalanche deposit contains no observed juvenile fragments.

The volume of the edifice failure can be estimated either from outcrop thicknesses, adjusted for erosion, or by calculating the in-place failure volume. Preliminary estimates suggest a deposit extent of 15 km$^2$ to perhaps as much as 25 km$^2$. Observed deposit thickness varies between 50 and 85 m. Assuming an average thickness of 50 m produces a volume between 0.75 and 1.25 km$^3$, although additional mapping is needed. The horseshoe-shaped Coliblanco depression along the SW flank of the Irazú-Turrialba massif, in the divide between the volcanoes that is drained by the Río Turrialba, has a volume comparable to that estimated for the Angostura debris-avalanche deposit, but its relationship to the deposit is uncertain.

### Bajos Avalanche

Turrialba volcano displays a clear asymmetry to the NE, with a graben that extends toward this flank from the summit escarpments. A hummocky volcanic avalanche deposit extends at least 3 km from the ~1 km$^3$ summit depression (Soto, 1988). A distal deposit of volume comparable to that of the summit depression may be covered by younger lava flows or hidden in lowland rain forest, and has not been identified. Reagan et al. (this volume) considered the summit amphitheater to be the result of several erosive and gravitational events, and its origin remains uncertain.

### Santa Rosa Avalanche

Another relatively young avalanche deposit from Turrialba volcano, possibly <17,000 radiocarbon years old, is observed near the settlement of Santa Rosa, 2 km northeast of the city of Turrialba. Here, a series of conical hills composed of blocks of lava of diverse composition is easily discernible in radar images (Alvarado et al., 2004).

## PANAMA

### Barú Volcano

Barú is a large, 3477-m-high andesitic stratovolcano in the western Cordillera of Panama, near the border with Costa Rica. The summit of the volcano, the highest in Panama, is cut by a 6 × 10 km horseshoe-shaped caldera that is breached widely to the west and was the source of perhaps the largest edifice collapse in Central America (Fig. 17). The massive Hato del Volcán debris-avalanche deposit blankets a large area below the volcano and extends beyond the Pan-American Highway, >20 km from the collapse scarp (Instituto de Recursos Hidraulicos y Electrificacion [IRHE], 1987; Universidad Tecnológica de Panamá [UTP], 1992; Siebert et al., 2004). A large area of hummocky topography with hills several tens of meters in height, closed depressions, and a group of small lakes (Las Lagunas) is found below the avalanche caldera beyond the settlements of Hato del Volcán (also known as Volcán) and Nueva California. Hummocky topography extends to the SW, deflecting the course of the Río Chiriquí Viejo to the west. Flat-lying proximal areas are underlain by pyroclastic-flow deposits perhaps associated with the post-collapse summit lava dome complex.

Restoration of inferred pre-failure contours within the ~45 km$^2$ avalanche scarp at Barú suggests that the pre-failure summit could have reached ~4000 m elevation and that as much as 30 km$^3$ may have collapsed (Siebert et al., 2004). About 5 km$^3$ of this volume remained within the avalanche caldera in the form of large toreva blocks that slid relatively intact without disaggregating. They form linear back-tilted ridges up to a kilometer high, occupying much of the basal portion of the collapse scarp. These blocks are distinct from the post-collapse lava dome complex that later grew above the central vent to form the present-day summit

Figure 17. Aerial view from the west of the Barú avalanche caldera (dashed line). Toreva blocks lie at the base of the avalanche source scarp, with the summit lava-dome complex beyond. HV—Hato de Volcán town. Width of caldera at breach is ~7 km. Photo by Kathleen Johnson.

of the volcano, ~150 m above the height of the caldera rim. The basal sediment layer cored from a lake characterized by Behling (2000) as a volcanic crater (but more likely a pond formed within a closed depression on the surface of the avalanche deposit) was dated at 2860 ± 50 yr B.P. and may provide a limiting age for the collapse. The maximum age limit for the collapse is not known precisely, but was estimated by IRHE (1987) to be <12.4 ka.

The geometry of the avalanche caldera suggests that more than one generation of collapse may have occurred. The NE part of the caldera rim continues to the east beyond the youngest scarp and may correlate with a 4-km-long scarp 2 km south of the SE side of the younger scarp (Fig. 18) and could be a remnant of an earlier collapse.

### Cerro Colorado Volcano

Cerro Colorado, located immediately NW of Barú, has a 7-km-wide horseshoe-shaped caldera (now partially filled by post-collapse lava domes of Cerro Totuma and Cerro Pelón) that is breached to the southwest (Fig. 18). Its area is comparable to that of the Barú caldera, suggesting that its original volume likely exceeded 10 km$^3$. The origin of this depression has been variously attributed to extensional tectonics, directed blasts, landslides, and erosion (IRHE, 1987). In the Cotitos–Los Pozos area within the southwest side of the depression, large toreva-like segments of the former edifice that slid intact are found (IRHE, 1987; UTP, 1992). Debris-avalanche and/or debris-flow deposits are found along the Río Chiriquí Vieja toward the coastal plain, west of deposits from Barú volcano (IRHE, 1987), and hummocky topography covers broad areas to the southwest of the scarp, extending across the border into Costa Rica. As with those from Barú, these deposits have not been studied in detail. Deposits of lateral explosions and pyroclastic-flow deposits were noted by UTP (1992), although their relationship to a possible edifice failure was not stated.

### Unknown Source

Deposits of possible debris-avalanche origin occur in the San Vito area of southeastern Costa Rica. A large area of volcaniclastic rocks with deformed blocks to 100 m or more in size was mapped as debris flows and volcanic breccias by Tournon and Alvarado (1997). These may represent deposits of large debris avalanches from volcanoes across the Panamanian border. K-Ar and $^{40}$Ar/$^{39}$Ar dates of volcanic rocks in the San Vito area range from 1.07 ± 0.04 to 1.71 ± 0.09 Ma (Drummond et al., 1995; MacMillan et al., 2004).

## DISCUSSION

The Central American volcanic arc can be considered in terms of two large tectonic blocks at the western end of the Caribbean plate—the Chortis block on the north, which underlies Guatemala to Nicaragua, and the Chorotega block, extending from

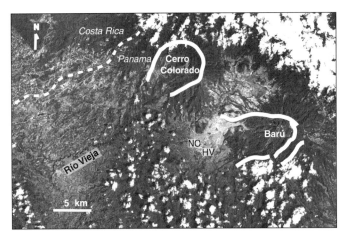

Figure 18. U.S. National Aeronautics and Space Administration Landsat image of Barú and Cerro Colorado volcanoes. NC—Nuevo California; HV—Hato de Volcán (also known as Volcán). Dashed white line marks approximate location of Costa Rica–Panama border from Landsat data file.

Costa Rica to Panama (Fig. 1). Subduction angles of the Cocos plate vary significantly and are shallower at either end of the arc (Burbach et al., 1984; Carr, 1984), ranging from low-angle subduction of 30°–10° in México (Pardo and Suárez, 1995), steepening to 65° beneath Nicaragua, and then flattening to 40° beneath Costa Rica (Rüpke et al., 2002). Although large-scale volcanic edifice collapse has affected about two dozen volcanoes in Central America (Table 1), almost 90% of these volcanoes lie at the margins of the arc, at the western end of the Chortis block, and along the Chorotega block to the east.

Gravitational collapse correlates closely with elevated heights of both the volcanic substrate and volcanic edifices. All volcanoes that rise >2000 m above their bases lie in the three countries at the ends of the arc, as do 75% of volcanoes rising >1500 m. The only volcano >2 km in height not known to have undergone major edifice failure is Agua volcano in Guatemala. Crustal thickness and elevation of the volcanic substrate decrease toward the center of the arc, where regional sediment-filled depressions and low-angle shield volcanoes are common and most volcanoes rise <1000 m.

### Mobility

Travel distances of volcanic debris avalanches in Central America have reached up to ~50 km. Avalanches in Nicaragua in the central part of the arc have tended to be less mobile, reflecting the lower relief and smaller edifice volumes of mid-arc volcanoes. H/L ratios (the vertical drop from the summit to the distal end of the deposit divided by the travel distance) have commonly been used to characterize the mobility of volcanic debris avalanches. The Acajutla avalanche in El Salvador, with an H/L ratio of 0.05, is the lowest currently known in Central America, comparable to other high mobility avalanches globally. The use of H/L ratios

TABLE 1. CENTRAL AMERICAN DEBRIS AVALANCHES (≥0.1 km³)

| Volcano* | Avalanche | Direction | Length† (km) | H/L§ | Area (km²) | Volume# (km³) | Age | References |
|---|---|---|---|---|---|---|---|---|
| **Guatemala** | | | | | | | | |
| Tacaná | Agua Caliente | NW | >8 | 0.35 | 6 | 1 | 26.3–10.6 ka | Macías et al., 2004 |
| Tacaná | Río Las Majadas | NNE | – | – | – | – | Pleistocene | Siebert et al., 2004; Macías, 2004 (personal commun.) |
| Tacaná | Muxbal area | South | – | – | – | – | >28.5 ka | Mora et al., 2004 |
| Unnamed? | – | South | – | – | – | – | – | Vallance et al., 1995 |
| Chicabal? | – | SW | – | – | – | [>1] | Pleistocene | Vallance et al., 1995 |
| Siete Orejas? | – | South | – | – | – | [>1] | Pleistocene | Vallance et al., 1995 |
| Almolonga | Cerro Quemado | NW | 6 | 0.13 | 13 | 0.1 | ~1.2 ka | Conway et al., 1992; Vallance et al., 1995 |
| Atitlán | – | S-SW | >20? | – | – | – | Holocene? | Haapala et al. (2005) |
| Acatenango | La Democracia | SSW | 42 | 0.09 | 210 | 5 | 70–43 ka | Vallance et al., 1995; Basset, 1996; Siebert et al., 2004; Chesner and Halsor, this volume |
| Fuego/Acatenango | Escuintla | SSE | 50 | 0.08 | 440 | 15 | 30–8.5 ka? | Vallance et al., 1995; Siebert et al., 2004; Chesner and Halsor, this volume |
| Pacaya | Río Metapa | SSW | 25 | 0.10 | 55 | >1 | 1.6–0.6 ka | Vallance et al., 1995; Siebert et al., 2004 |
| Tecuamburro | Miraflores | SE | 15 | 0.12 | – | (~4) | 100–38.3 ka | Duffield et al., 1989, 1991; Vallance et al., 1995 |
| Ixtahuan? | Los Achiotes | SE | 15? | 0.07 | – | – | Pleistocene | Reynolds, 1987; Siebert et al., 2004 |
| **El Salvador** | | | | | | | | |
| Santa Ana | Acajutla | SW | 48 | 0.04 | 540 | 16 ± 5 | <56.9 ka | Pullinger, 1998; Siebert et al., 2004 |
| Santa Ana? | Modern Santa Ana | SW | – | – | – | – | – | Siebert et al., 2004 |
| San Vicente | Tecoluca | SE | 24 | 0.09 | – | – | Pleistocene | Major et al., 2004; Siebert et al., 2004; this paper |
| San Miguel/Chinameca? | Lolotique | NW? | – | – | – | [>1] | – | Pullinger, 2002 (personal commun.); Pullinger, 2005 (personal commun.) |
| San Miguel | – | SW | – | – | – | – | – | Escobar, 2003 |
| **Nicaragua** | | | | | | | | |
| Mombacho | Las Isletas | NE | >12 | 0.11 | 45? | (1) | Holocene | Ui, 1972; Van Wyk de Vries and Borgia, 1996; Van Wyk de Vries and Francis, 1997; Vallance et al., 2001b; Van Wyk de Vries et al., 2005 |
| Mombacho | El Cráter | South | 13 | 0.10 | 60 | (1) | Holocene (1570 AD?) | Ui, 1972; Incer, 1990; Van Wyk de Vries and Borgia, 1996; Van Wyk de Vries and Francis, 1997; Vallance et al., 2001b; Van Wyk de Vries et al., 2005 |
| Mombacho | Southeast | SE | >11 | 0.12 | – | – | – | Van Wyk de Vries and Borgia, 1996; Vallance et al., 2001b |
| Mombacho | E. Megablock Slump | East | >11 | 0.12 | – | – | – | Van Wyk de Vries and Borgia, 1996 |

*Continued*

TABLE 1. CENTRAL AMERICAN DEBRIS AVALANCHES (≥0.1 km³) (continued)

| Volcano* | Avalanche | Direction | Length† (km) | H/L§ | Area (km²) | Volume# (km³) | Age | References |
|---|---|---|---|---|---|---|---|---|
| Costa Rica | | | | | | | | |
| Cerro Cacao | Quebrada Grande | SE | 16 | 0.07 | 36–52 | 1.7 ± 0.3 | Pleistocene | Castillo, 1978; Tournon, 1984; Alvarado et al., 2004 |
| Rincón de la Vieja | Azufrales | South | 9 | 0.12 | 11–18 | 0.3 ± 0.1 | Pleistocene | Alvarado et al., 2004 |
| Miravalles | La Fortuna | SW | 19 | 0.11 | 90–130 | 7 ± 1 | 8.3 ka | Fernández, 1984; Alvarado, 1987; Alvarado et al., 2004 |
| Miravalles | Upala | NE | 13 | 0.08 | >20 | – | Pleistocene | Alvarado et al., 2004 |
| Miravalles | Las Mesas | SW? | 6 | – | >3 | 0.1–0.25 | <8.3 ka | Alvarado et al., 2004 |
| Tenorio | Tierras Morenas | South | 17 | 0.07 | 80–120 | 2.0 ± 0.4 | Pleistocene | Alvarado et al., 2004 |
| Platanar | Chocosuela | NW? | 20 | 0.10 | – | – | >100 ka? | Alvarado and Carr, 1993 |
| Barva? | Coyol | SW | – | – | – | – | ~150–230 ka | Méndez and Hidalgo, 2004; Alvarado et al., 2004 |
| Unknown source | Tivives | – | – | – | – | – | <650 ka | Gans et al., 2003; Alvarado et al., 2004 |
| Irazú | Prusia-Tierra Blanca | WSW | 13? | 0.14 | 35 | 1.5 ± 0.5 | Holocene? | Alvarado, 1993; Pavanelli et al., 2003 |
| Irazú | Río Costa Rica | North | 27 | 0.11 | – | – | Pleistocene | Alvarado et al., 2004 |
| Irazú | Río Birrís | SE | – | – | – | – | >54 ka | Pavanelli et al., 2003 |
| Irazú | Cabeza de Vaca | SSW | 25? | – | 130 | 2 | ~150–250 ka | Alvarado et al., 2004; Hidalgo et al., 2004 |
| Turrialba | Angostura | ESE | 15 | 0.07 | 15–28 | 1.0 ± 0.3 | 17 ka | Soto, 1988; Alvarado et al., 2004 |
| Turrialba | Santa Rosa | SE | 5 | – | – | – | <17 ka | Alvarado et al., 2004 |
| Turrialba? | Bajos | NE | – | – | – | – | – | Soto, 1988 |
| Panamá | | | | | | | | |
| Unknown source | San Vito | SW? | – | – | – | – | Pleistocene | Tournon and Alvarado, 1997; this paper |
| Cerro Colorado | Río Chiriquí Viejo | SSW | – | – | – | (>10) | Pleistocene | Instituto Recursos Hidraulicos y Electrificacion (IRHE), 1987; Siebert et al., 2004 |
| Barú | Hato del Volcán | WSW | – | – | – | (25) | Holocene? | Instituto Recursos Hidraulicos y Electrificacion (IRHE), 1987; Siebert et al., 2004 |
| Barú | – | SW | – | – | – | – | Pleistocene | Siebert et al., 2004 |

*Volcano names followed by a "?" represent edifice collapse events inferred from source morphology on maps or aerial photographs, but without confirmed field identification of deposits.

†Length is distance from summit or headwall scarp to terminus of debris-avalanche deposit.

§H/L is ratio between vertical drop and travel length; in some cases may include lahar facies of edifice collapse event. The area/volume (A/V) ratios of Vallance and Scott (1997) were not displayed here because of the relative paucity of data, but all deposits for which data are available fall in their debris avalanche field.

#Volume data in parentheses are source area volumes; figures in square brackets are order-of-magnitude volume assessments.

is not always straightforward, however, because they decrease with increasing volume (Siebert et al., 1987), and mass movements with a significant lahar-facies component display greatly enhanced mobility (Vallance and Scott, 1997). Vallance and Scott proposed using the ratio between deposit area and volume to characterize volcanic mass movements, with a ratio of $V \propto A^{3/2}$ for debris avalanches and a lower value for lahars. Area and volume data are sparse for Central American deposits in Table 1, but all deposits with data would plot in the debris-avalanche field.

The travel distances of debris avalanches are influenced by factors such as topography (confined or unconfined), failure volume (which itself can be proportional to edifice volume), the degree of pre-failure hydrothermal alteration, and water content. Water saturated mass movements tend to display enhanced mobility with respect to less saturated flows (Scott et al., 2001; Capra, et al., 2002). The distinction between granular flow of debris avalanches sensu stricto and water-saturated flows (lahar facies) transformed from debris avalanches has been emphasized at edifice failures at Mount Rainier (Vallance and Scott, 1997) and in México (Carrasco-Núñez et al. 1993; Scott et al., 2001; Capra et al., 2002). Modeling by Iverson et al. (1997) noted that liquefaction of the flowing mass was partial to complete due to variable high pore-fluid pressures and granular temperature increases resulting from the translation of landslide energy to internal vibrational energy. The rheology of grain-fluid mixtures in debris avalanches or debris flows is variable and evolves during flow initiation, transport, and deposition (Iverson and Vallance, 2001).

For very large avalanches, the volume of water entrained during emplacement is trivial compared to the avalanche volume, inhibiting complete transformation. The occurrence of offshore hummocky block-facies material at avalanches from both relatively small-volume coastal stratovolcanoes and large-volume oceanic shield volcanoes furthermore implies incomplete incorporation of water even in a totally submerged environment. This suggests that the very large water volumes required for saturation (>15% water by volume; Glicken, 1991) derive primarily from meteoric or hydrothermal water within the edifice or in situ melting of glacier icecaps. The incorporation of large amounts of accessory soils and other materials during emplacement can promote mobility by increasing the clay content and thus enhancing pore-fluid pressures and cushioning grain interactions (Capra et al., 2002). Significant bulking of large mass movements through incorporation of accessory materials has been observed, although inclusion of incompletely saturated accessory materials would locally reduce the degree of water saturation. Glacier-clad volcanoes in México were noted to be more prone to avalanche-lahar transformation (Carrasco-Núñez, et al. 1993; Vallance, 2000; Capra et al., 2002), and this could have been a factor in Central America at high-elevation volcanoes such as Barva, Irazú, Turrialba, and Barú (see also Hidalgo et al., 2004).

Textural evidence at large avalanches such as the Acajutla debris-avalanche deposit in El Salvador suggest that large avalanches are not completely water saturated (Siebert et al., 2004). Existing nomenclature does not adequately characterize mass movements with varying degrees of water saturation, and use of terms like debris avalanche and debris flow (or lahar) may imply an unwarranted degree of certainty about process. Many large avalanches may consist of local domains with varying degrees of water saturation, and more work is required to identify textural evidence for distinguishing unsaturated (or partially saturated) mixed-facies material from saturated lahar-facies material.

**Factors Contributing to Collapse**

A wide variety of factors contribute to edifice failure (Campbell et al., 1995; McGuire, 1996; Siebert et al., 1987). One significant factor is hydrothermal alteration, which can significantly weaken an edifice and increase pore-fluid pressures, contributing to failure (López and Williams, 1993). Although more difficult at heavily vegetated volcanoes, detailed geotechnical analyses of hydrothermal alteration and edifice stability, such as conducted by Zimbelman et al. (2004) at Citlaltépetl (Pico de Orizaba) in México, can lead to refined estimates of areas prone to future failure and likely failure volumes.

Deformation structures resulting from gravitational settling of volcanoes have been noted at Costa Rican and Nicaraguan volcanoes by Borgia (1994) and van Wyk de Vries and Borgia (1996). Large, steep-sided cones produce sufficient load at their bases to exceed the tensile strength of basement rocks, and deformation styles have been related to basement lithology types (van Wyk de Vries and Borgia, 1996). Large amounts of edifice deformation have occurred at Concepción and Maderas volcanoes, which overlie less competent lacustrine sedimentary layers. Mombacho volcano overlies more competent volcanic and marine strata and is considered to have experienced slow spreading (van Wyk de Vries and Francis, 1997). Measurements at Arenal volcano in Costa Rica have documented a radius of active deformation extending as far as 5 km from the summit crater (Alvarado et al., 2003). Dominantly effusive activity at Arenal has created a structurally overloaded, brittle 15 km$^3$ lava cone overlying weak, weathered, volcanic rocks, contributing to basement deformation.

Compressive stresses released by thrust faulting in basal sediments may have precipitated several edifice collapses, although the relationship between edifice deformation and failure probability can be equivocal (van Wyk de Vries and Borgia, 1996; van Wyk de Vries and Francis, 1997). Steep-sided volcanoes with substantial substrate deformation like Concepción remain standing, while nearby Mombacho volcano has failed repeatedly. Slow volcano spreading may in some cases decrease the likelihood of collapse by relieving stress within the edifice and creating inward dipping faults, although sector spreading, as occurred at Mombacho, can lead to stress build up (van Wyk de Vries and Francis, 1997).

Volcanoes with summit lava-dome complexes, such as Cerro Quemado in Guatemala, are prone to collapse, and short-term deformation produced by endogenous and exogenous dome growth may contribute to instability, along with earthquakes, extension, and enhanced pore-fluid pressures accompanying dike intrusion (Elsworth and Voight, 1995; Siebert et al., 1987).

**Failure Direction**

Preferential volcano failure directions have been observed around the margins of the Caribbean plate in the West Indies (Boudon et al., 2002), Guatemala (Vallance et al., 1995), and the eastern Mexican Volcanic Belt (Carrasco-Núñez et al., 2006). In contrast to West Indies arc-slope failures, which occur largely in a direction opposite that of the ocean-floor trench, most of the Central American slope failures occur toward the trench side of the arc (Siebert et al., 2004). These preferred failure directions have been attributed to the construction of volcanoes on regional basement slopes. In the West Indies, this reflects the higher slope of islands toward the deep backarc Grenada Basin (Boudon et al., 2002) and the construction of the present Antilles volcanoes on basinward dipping volcaniclastic deltas (Oehler et al., 2005). In Guatemala, failure direction has been attributed to the construction of volcanoes over inclined basements at the trenchward margin of the arc (Vallance et al., 1995). A similar topographic effect is seen in México in the Cofre de Perote–Citlaltépetl range, which is constructed above the margin of the Altiplano. The basement drops dramatically toward the coastal plain, and all slope failures have occurred to the east (Carrasco-Núñez et al., 2006). In contrast, an isolated volcano without regional basement gradients, such as the island volcano of Augustine in Alaska, has been subject to radial failures on all sides of the volcano (Begét and Kienle, 1992; Siebert et al., 1995).

Recent modeling by Wooller et al. (2004) supports the observed significance of topographic effects by showing the influence of very minor basement slopes of as low as one half a degree on edifice deformation and failure direction. This sensitivity to basement geometry has implications for hazard assessments for future collapses in Central America. Volcanoes in Guatemala and Costa Rica, at opposite ends of the Central American arc, would tend to produce failures perpendicular to the trend of the arc; those in the central plains of El Salvador and Nicaragua, where the basement topographic effect is less pronounced, could have more variable failure directions (Fig. 19). In Guatemala, the regional slope will tend to send debris avalanches to the south; however, in Costa Rica, where the volcanoes straddle a topographic divide, the relationship is less clear.

**Frequency of Occurrence**

Large-scale volcanic edifice failure is a high-impact but relatively low-frequency event. Although globally major edifice failures during the past 500 yr have occurred at a rate of about four per century (Siebert et al., 1987), the recurrence interval on a regional or individual volcano basis is substantially longer. The paucity of precise dates for most Central American collapses precludes rigorous evaluation of their frequency. Of the roughly 40 possible failures in Central America (Table 1), five to as many as nine occurred during the Holocene. The data from Table 1 suggest a regional recurrence interval during the past 10,000 yr of ~1000–2000 yr. The age of Pleistocene collapses in Central America is not well constrained, but Alvarado et al. (2004) roughly estimated a recurrence interval for Costa Rican volcanoes during the past 50,000 yr of 500–2000 yr, with possible collapse intervals at individual volcanoes such as Irazú and Turrialba of 3000–6000 yr. Augustine volcano in Alaska, a low-volume volcano with a high magma effusion rate, collapsed a dozen times in the past 2000 yr (Begét and Kienle, 1992). However, this atypical frequency is not seen in Central America, where no volcano is currently known to have collapsed more than four times.

The number of major collapse events in Central America is comparable when normalized to arc length with those identified in México (Capra et al., 2002), but meaningful comparisons are problematic because of large uncertainties regarding the number

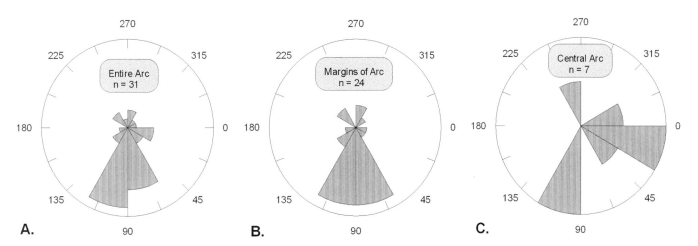

Figure 19. Collapse orientation at Central American volcanoes as measured from trend of arc. (A) Entire arc; (B) volcanoes at margins of arc; (C) volcanoes in El Salvador and Nicaragua in center of arc.

of deposits that have occurred at Colima volcano in México in particular, as well as the preliminary nature of mass movement studies in Central America. While large collapses are not so frequent, smaller landslides such as at Casita in 1998 (van Wyk de Vries et al., 2000; Kerle and van Wyk de Vries, 2001; Scott et al., 2005) are more common and suggest a possible negative power law relationship of frequency to size. Smaller events may cumulatively be as damaging as the infrequent large events.

## HAZARDS ASSOCIATED WITH EDIFICE FAILURE

### Debris Avalanches

Central American debris avalanches have had volumes ranging to more than 10 km$^3$ and have traveled to ~50 km from the volcano. The recurrence interval of major collapses (>0.1 km$^3$) is fortunately low, and only about a half dozen to less than a dozen of these events have to date been documented in Central America during the Holocene. However, ~75% of the avalanches in Table 1 with known travel distances have exceeded 10 km, distances potentially impacting populated regions during future collapses. Although many Central American volcanoes are located in sparsely populated regions where the impacts of large avalanches would be limited, others, such as San Salvador in El Salvador and the Poás to Turrialba volcano group in Costa Rica, rise in proximity to some of the region's largest cities. More than 286,000 people currently live on top of the Acajutla debris-avalanche deposit (Siebert et al., 2004), and one of Guatemala's largest cities, Escuintla, was constructed astride the avalanche deposit from the Acatenango-Fuego volcanic complex (Vallance et al., 1995).

Massive edifice failure on the scale of previous events in Central America will be truly catastrophic if it occurs in a densely populated area. However, these low frequency events typically occur at intervals far in excess of time frames considered by hazard planners. Hence, land-use zoning that precludes habitation of areas potentially affected is not viable. Smaller-scale collapses such as at Casita may be an exception. The 1998 lahar has cleared the potential collapse zone of habitation, thus reducing the political risk of hazard zonation. The high velocity (to 150 m/s) and widespread extent of volcanic debris avalanches from large-scale edifice collapse preclude hazard mitigation after the onset of the event. Consequently, a focus on the assessment of both long-term edifice stability and identification of short-term precursors to collapse needs to be accompanied by education of public officials and residents regarding potential hazards.

### Lateral Blasts and Associated Eruptive Activity

A collapse-related lateral blast has unequivocally been identified to date in Central America only at Cerro Quemado in Guatemala (Conway et al., 1992). This was an order of magnitude smaller in volume than the 1980 Mount St. Helens lateral blast and affected a proportionally smaller area. The occurrence of lateral blasts in association with edifice failure is contingent on the timing of collapse with respect to magmatic explosions and is a function of the proximity of the failure plane to the magma body within the edifice. When the failure plane truncates or nearly truncates an upper edifice magma body and its associated hydrothermal envelope, decompression-initiated explosions are synchronous with landsliding, and the newly formed steep horseshoe-shaped headwall scarp and the laterally moving avalanche in part deflect the explosions laterally (Siebert, 2002). When slope failure is completed prior to the onset of explosions (such as at Shiveluch volcano in Kamchatka in 1964), open-vent vertical columns result. Magmatic eruptions at Tacaná, Pacaya, and Santa Ana (see previous sections) post-dated collapse and apparently did not involve a lateral blast. Inspection of the literature regarding edifice failures indicates that lateral blasts accompanying edifice collapses are the exception rather than the rule. Most magmatic eruptions associated with edifice failure span a broad spectrum of styles ranging from modest post-collapse construction of Strombolian cinder cones to major Plinian eruptions producing pumiceous pyroclastic flows.

Commonly, eruptions associated with edifice failures lack any magmatic component, and increased pore water pressures resulting from upper level hydrothermal systems can be sufficient to initiate failure and trigger phreatic eruptions (Reid et al., 2001; Reid, 2004). Examination of major edifice failures globally over the past 500 yr, when contemporary accounts are more available, shows that collapses were divided roughly equally among those accompanied by magmatic eruptions, phreatic eruptions, or no eruptions (Siebert et al., 1987). Few eruptive products have to date been identified as being associated with Central American collapses; the lack of preservation or recognition of smaller-scale eruptive deposits can inhibit comparison with older events.

### Lahars (Debris Flows)

Lahars occur in association with debris avalanches under several circumstances, and lahar facies can be a significant component of volcanic mass movements. Glacial melting at high volcanoes (>3100 m) in the Cordillera Central of Costa Rica during the late Pleistocene may have contributed to transformation of avalanches into distal lahars (Alvarado et al., 2004), which can have broader areas of inundation than debris avalanches themselves (Vallance and Scott, 1997; Scott et al., 2001; Capra et al., 2002). Lahars can also form by localized consolidation and liquefaction shortly after emplacement of an incompletely saturated debris avalanche, such as at Mount St. Helens in 1980 (Fairchild, 1987). Debris avalanches can indirectly produce lahars when lakes formed behind avalanche-dammed drainages are overtopped and catastrophically drained, as at Bandai volcano in 1888 (Sekiya and Kikuchi, 1889). Hazards from lake-derived lahars can be mitigated by construction of stabilized spillways or diversion tunnels as were successfully implemented at Mount St. Helens (Glicken et al., 1989).

## Tsunamis

Only two known debris avalanches from Central American volcanoes were mobile enough to reach the Pacific coast and potentially produce tsunamis. A significant component of the Acajutla debris avalanche from Santa Ana volcano in El Salvador reached the coast, and the Tivives deposit in Costa Rica is exposed in coastal cliffs and could have been tsunamigenic. The magnitude of slope-failure related tsunamis is enhanced when the basal failure plane extends below sea level. Few other Central American volcanoes, other than those in the Gulf of Fonseca, are sufficiently proximal to the coastline to produce tsunamis. The tsunami potential at these relatively small-volume volcanoes, however, would be more limited.

One of the principal hazards from collapse-related tsunamis in Central America, however, derives from debris avalanches entering lakes. In Guatemala, Tolimán and San Pedro volcanoes lie near the southern rim of Atitlán caldera, and potential debris avalanches from these volcanoes could impact Lake Atitlán and produce tsunamis that would affect lakeshore towns and houses. A collapse to the NE at Santa Ana volcano in El Salvador could produce a tsunami in the Coatepeque caldera lake (Fig. 9), which would have a devastating effect on lakeshore dwellings (Siebert et al., 2004). Debris avalanches from Mombacho volcano, which swept into Lake Nicaragua, probably caused tsunamis. Future collapses of Mombacho and other volcanoes bordering Lake Nicaragua and Managua, such as Momotombo and Concepción, pose a similar risk. In Costa Rica, potential debris avalanches at Arenal volcano with estimated volumes of 0.03–0.75 km$^3$ could be sufficiently mobile to reach the artificial reservoir behind the Segregado dam, 6.5 km west of the summit, and cause potentially fatal tsunamis (Alvarado, et al., 2003).

## CONCLUSIONS

More than 40 major edifice-failure events (≥0.1 km$^3$) from about two dozen volcanoes are known or inferred in Central America. Volcanoes subjected to gravitational failure are strongly concentrated at the western and eastern ends of the arc. Failures correlate positively with volcano elevation, substrate elevation, edifice height, volcano volume, and crustal thickness, and inversely with slab descent angle. Collapse orientations are also strongly influenced by the direction of slope of the underlying basement. Failures occur predominately perpendicular to the arc at its margins and display more variable failure directions in the center of the arc.

Debris-avalanche deposits display evidence for varying degrees of water saturation, which can influence mobility. The frequency of collapse events in Central America is poorly constrained because of the lack of precise dating of older deposits, but an average collapse interval of ~1000–2000 yr is estimated during the Holocene. In addition to direct hazards from mobile debris avalanches and associated lahars, associated hazards derive from laterally directed explosions, contingent on the proximity of failure planes to magma bodies, and tsunamis from impact of debris avalanches into the sea, or (more likely with Central American volcanoes) natural or man-made lakes.

Description of the Central American debris avalanches represents an important step in assessing their hazards. Much work still has to be done on the deposits, and probably more will be discovered. Another equally important process is the location of potential collapse sites on intact volcanoes, their stability assessment, and monitoring. While some sites have been identified and some possible additions proposed here, none of these are yet monitored to any satisfactory degree, and none have been fully assessed for stability. This, and the subsequent challenge for risk assessment, is a major challenge for the next decade.

## ACKNOWLEDGMENTS

We thank Bill Rose for his long time interest in promoting research on volcanism in Central America. His National Science Foundation grant supported collaborative field studies in Central America by Siebert and Vallance on two occasions in 1988 and work on the Santa Ana debris avalanche by Siebert in 1999 and 2002. We thank José Luis Macías for discussion regarding debris avalanches at Tacaná. Jon Major, Gregg Bluth, Kevin Scott, and Tom Pierson provided reviews that led to the improvement of this manuscript.

## REFERENCES CITED

Alvarado, G.E., 1987, Algunos depósitos volcánicos de avalanchas calientes direccionales en Costa Rica: III Jornadas Geológicas de Costa Rica, 24–25 Septiembre 1987, San José, Costa Rica, p. 18 [abstract].

Alvarado, G.E., 1993, Volcanology and petrology of Irazú volcano, Costa Rica [PhD thesis]: University of Kiel, 261 p.

Alvarado, G.E., and Carr, M., 1993, The Platanar-Aguas Zarcas volcanic centers, Costa Rica: spatial-temporal association of Quaternary calc-alkaline and alkaline volcanism: Bulletin of Volcanology, v. 55, p. 443–453, doi: 10.1007/BF00302004.

Alvarado, G.E., Carboni, S., Cordero, M., Avilés, E., Valverde, M., and Leandro, C., 2003, Estabilidad del cono y comportamiento de la fundación del edificio volcánico del Arenal (Costa Rica): Boletín del Observatorio Sismológico y Vulcanológico de Arenal y Miravalles, v. 14, no. 26, p. 21–73.

Alvarado, G.E., Vega, E., Chaves, J., and Vásquez, M., 2004, Los grandes deslizamientos (volcánicos y no-volcánicos) de tipo *debris avalanche* en Costa Rica: Revista Geológica de América Central, no. 30, p. 83–99.

Alvarado, G.E., Carr, M.J., Turrin, Brent D., Swisher, C.C., Schmincke, H.-U., and Hudnut, K.W., 2006, this volume, Recent volcanic history of Irazú volcano, Costa Rica: Alternation and mixing of two magma batches, and pervasive mixing, *in* Rose, W.I., Bluth, G.J.S., Carr, M.J., Ewert, J.W., Patino, L., and Vallance, J.W., Volcanic hazards in Central America: Geological Society of America Special Paper 412, doi: 10.1130/2006.2412(14).

Basset, T., 1996, Histoire éruptive et évaluation des aléas du volcano Acatenango (Guatemala) [Ph.D. thesis]: Université de Genève, Terre & Environment, Section des Sciences de la Terre, v. 3, p. 1–240 and appendices.

Begét, J.E., and Kienle, J., 1992, Cyclic formation of debris avalanches at Mount St. Augustine volcano, Alaska: Nature, v. 356, p. 701–704, doi: 10.1038/356701a0.

Behling, H., 2000, A 2860-year high-resolution pollen and charcoal record from the Cordillera de Talamanca in Panama; a history of human and volcanic forest disturbance: The Holocene, v. 10, p. 387–393, doi: 10.1191/095968300668797683.

Borgia, A., 1994, The dynamic basis of volcanic spreading: Journal of Geophysical Research, v. 99, p. 17,791–17,804, doi: 10.1029/94JB00578.

Boudon, G., Le Friant, A., Deplus, C., Komorowski, J.-C., and Semet, M.P., 2002, Volcano flanks of the Lesser Antilles Arc collapse, sometimes repeatedly: How and why: Montagne Pelee 1902–2002, Explosive volcanism in subduction zones, St. Pierre, Martinique, May 12–16, 2002, abstract volume, p. 66.

Burbach, G.V., Frohlich, C., Pennington, W.D., and Matumoto, T., 1984, Seismicity and tectonics of the subducted Cocos Plate: Journal of Geophysical Research, v. 89, p. 7719–7735.

Campbell, C.S., Cleary, P.W., and Hopkins, M., 1995, Large-scale landslide simulations: global deformation, velocities, and basal friction: Journal of Geophysical Research, v. 100, p. 8267–8283, doi: 10.1029/94JB00937.

Capra, L., Macías, J.L., Scott, K.M., Abrams, M., and Garduño-Monroy, V.H., 2002, Debris avalanches and debris flows transformed from collapses in the Trans-Mexican Volcanic Belt, Mexico—Behavior and implications for hazard assessment: Journal of Volcanology and Geothermal Research, v. 113, p. 81–110, doi: 10.1016/S0377-0273(01)00252-9.

Carr, M.J., 1984, Symmetrical and segmented variation of physical and geochemical characteristics of the Central America volcanic front: Journal of Volcanology and Geothermal Research, v. 20, p. 231–252, doi: 10.1016/0377-0273(84)90041-6.

Carracedo, J.C., 1994, The Canary Islands: An example of structural control on the growth of large oceanic-island volcanoes: Journal of Volcanology and Geothermal Research, v. 60, p. 225–241, doi: 10.1016/0377-0273(94)90053-1.

Carrasco-Núñez, G., Vallance, J.W., and Rose, W.I., 1993, A voluminous avalanche-induced lahar from Citlaltépetl volcano, Mexico: Implications for hazard assessment: Journal of Volcanology and Geothermal Research, v. 59, p. 35–46, doi: 10.1016/0377-0273(93)90076-4.

Carrasco-Núñez, G., Díaz-Castellón, R., Siebert, L., Hubbard, R., Sheridan, M.A., and Rodríguez, S., 2006, Multiple edifice-collapse events in the Eastern Mexican Volcanic Belt: The role of sloping substrate and implications for hazard assessment: Journal of Volcanology and Geothermal Research (in press).

Castillo, R., 1978, Reconocimiento geológico preliminar de una parte de las faldas del cerro Cacao, cordillera de Guanacaste, Costa Rica: Codesa: Boletín Geología y de Recursos Minerales, v. 1, p. 268–279.

Cecchi, E., van Wyk de Vries, B., and Lavest, J.-M., 2005, Flank spreading and collapse of weak-cored volcanoes: Bulletin of Volcanology, v. 67, p. 72–91.

Chesner, C.A., and Halsor, S.P., 2006, this volume, The Escuintla and La Democracia debris avalanche deposits, Guatemala: Constraining their sources, in Rose, W.I., Bluth, G.J.S., Carr, M.J., Ewert, J.W., Patino, L., and Vallance, J.W., Volcanic hazards in Central America: Geological Society of America Special Paper 412, doi: 10.1130/2006.2412(06).

Conway, F.M., Vallance, J.W., Rose, W.I., Johns, G.W., and Paniagua, S., 1992, Cerro Quemado, Guatemala: The volcanic history and hazards of an exogenous volcanic dome complex: Journal of Volcanology and Geothermal Research, v. 52, p. 303–323, doi: 10.1016/0377-0273(92)90051-E.

Drummond, M.S., Bordelon, M., de Boer, J.Z., Defant, M.J., Bellon, H., and Feigenson, M.D., 1995, Igneous petrogenesis and tectonic setting of plutonic and volcanic rocks of the Cordillera de Talamanca, Costa Rica-Panama, Central American arc: American Journal of Science, v. 295, p. 875–919.

Duffield, W.A., Heiken, G.H., Wohletz, K.H., Maassen, L.W., Dengo, G., and Mckee, E.H., 1989, Geology and geothermal potential of the Tecuamburro volcano area of Guatemala: Geothermal Resources Council Transactions, v. 13, p. 125–131.

Duffield, W.A., Heiken, G.H., Wohletz, K.H., Maassen, L.W., Dengo, G., and Mckee, E.H., 1991, Geologic map of Tecuamburro volcano and surrounding area, Guatemala: U.S. Geological Survey Miscellaneous Investigations Series, Map I-2197, scale 1:50,000.

Elsworth, D., and Voight, B., 1995, Dike intrusion as a trigger for large earthquakes and the failure of volcano flanks: Journal of Geophysical Research, v. 100, p. 6005–6024, doi: 10.1029/94JB02884.

Escobar, C.D., 2003, San Miguel volcano and its volcanic hazards, El Salvador [M.Sc. thesis]: Houghton, Michigan Technological University, 168 p.

Fairchild, L.H., 1987, The importance of lahar initiation processes: Geological Society of America Reviews in Engineering Geology, v. 7, p. 51–61.

Feldman, L.H., 1993, Mountains of fire, lands that shake: Culver City, California, Labyrinthos, 295 p.

Fernández, E., 1984, Geología y alteración hidrotermal en el Campo Geotérmico Miravalles, Provincia de Guanacaste, Costa Rica [Licensure thesis]: San José, Universidad de Costa Rica, 72 p.

Francis, P.W., and Wells, G.L., 1988, Landsat thematic mapper observation of debris avalanche deposits in the Central Andes: Bulletin of Volcanology, v. 50, p. 258–278, doi: 10.1007/BF01047488.

Gans, P.B., Alvarado-Induni, G., Perez, W., MacMillan, I., and Calvert, A., 2003, Neogene evolution of the Costa Rican arc and development of the Cordillera Central: Geological Society of America Abstracts with Programs, v. 35, no. 4, http://gsa.confex.com/gsa/2003CD/finalprogram/abstract_51871.htm.

García-Palomo, A., Macías, J.L., Arce, J.L., Mora, J.C., Hughes, S., Saucedo, R., Espíndola, J.M., Escobar, R., and Layer, P., 2006, this volume, Geological evolution of the Tacaná Volcanic Complex, México–Guatemala, in Rose, W.I., Bluth, G.J.S., Carr, M.J., Ewert, J.W., Patino, L., and Vallance, J.W., Volcanic hazards in Central America: Geological Society of America Special Paper 412, doi: 10.1130/2006.2412(03).

Glicken, H., 1991, Sedimentary architecture of large volcanic debris avalanches, in Smith, G.A., and Fisher, R.V., eds., Sedimentation in volcanic settings: SEPM (Society for Sedimentary Geology) Special Publication 45, p. 99–106.

Glicken, H., Meyer, W., and Sabol, M., 1989, Geology and ground-water hydrology of Spirit Lake blockage, Mount St. Helens, Washington, with implications for lake retention: U.S. Geological Survey Bulletin, v. 1789, p. 1–33.

Haapala, J.M., Escobar Wolf, R., Vallance, J.W., Rose, W.I., Griswold, J.P., Schilling, S.P., Ewert, J.W., and Mota, M., 2005, Volcanic hazards at Atitlan volcano, Guatemala: U.S. Geological Survey Open-File Report 2005-1403, 13 p., 2 plates.

Hazlett, R.W., 1987, Geology of the San Cristobal volcanic complex, Nicaragua: Journal of Volcanology and Geothermal Research, v. 33, p. 223–230, doi: 10.1016/0377-0273(87)90064-3.

Hidalgo, P.J., Alvarado, G.E., and Linkimer, L., 2004, La *lavina* del Valle Central (Costa Rica): ¿lahar o *debris avalanche*?: Revista Geológica de América Central, no. 30, p. 101–109.

Hurtado de Mendoza, L., and Alvarado, G., 1988, Datos arqueólogicos y vulcanológicos de la region del Volcán Miravalles, Costa Rica: Vínculos (San José), v. 14, p. 77–89.

Incer, J., 1990, Nicaragua: Viajes, rutas y encuentros 1502–1838: San José, Costa Rica, Libro Libre, 638 p.

Instituto de Recursos Hidraulicos y Electrificacion (IRHE), 1987, Final report on the reconnaissance study of geothermal resources in the Republic of Panama: Instituto de Recursos Hidraulicos y Electrificacion–Inter-American Development Bank–Organización Latinoamericana de Energía (IRHE-IDB-OLADE), 72 p.

Iverson, R.M., and Vallance, J.W., 2001, New views of granular mass flows: Geology, v. 29, p. 115–118, doi: 10.1130/0091-7613(2001)029<0115: NVOGMF>2.0.CO;2.

Iverson, R.M., Reid, M.E., and LaHusen, R.G., 1997, Debris-flow mobilization from landslides: Annual Review of Earth and Planetary Sciences, v. 25, p. 85–138, doi: 10.1146/annurev.earth.25.1.85.

Johnson, R.W., 1987, Large-scale volcanic cone collapse: the 1888 slope failure of Ritter volcano, and other examples from Papua New Guinea: Bulletin of Volcanology, v. 49, p. 669–679, doi: 10.1007/BF01080358.

Kerle, N., and van Wyk de Vries, B., 2001, The 1998 debris avalanche at Casita Volcano, Nicaragua; investigation of structural deformation as the cause of slope instability using remote sensing: Journal of Volcanology and Geothermal Research, v. 105, p. 49–63.

Kitamura, S., and Matías, O., 1995, Tephra stratigraphic approach to the eruptive history of Pacaya volcano, Guatemala: Science Reports of the Tohoku University, 7th Series (Geography), v. 45, p. 1–41.

López, D.L., and Williams, S.N., 1993, Catastrophic volcanic collapse: Relation to hydrothermal processes: Science, v. 260, p. 1794–1796.

Macías, J.L., Arce, J.L., Mora, J.C., and García-Palomo, A., 2004, The Agua Caliente debris avalanche deposit: a northwestern sector collapse of Tacaná volcano, Mexico-Guatemala: International Association of Volcanology and Chemistry of the Earth's Interior (IAVCEI) Conference, Pucón, Chile, November 2004, abstract (CD-ROM).

MacLeod, N., 1989, Sector-failure eruptions in Indonesian volcanoes: Geology of Indonesia, v. 12, p. 563–601.

MacMillan, I., Gans, P.B., and Alvarado, G., 2004, Middle Miocene to present plate tectonic history of the southern Central American Volcanic Arc: Tectonophysics, v. 392, p. 325–348, doi: 10.1016/j.tecto.2004.04.014.

Madrigal, R., 1970, Geología del mapa básico Barranca, Costa Rica: Informe Técnicos y Notas Geológicas, 37 p. and 1:50,000 scale geologic map.

Major, J.J., Schilling, S.P., Pullinger, C.R., and Escobar, C.D., 2004, Debris-flow hazards at San Salvador, San Vicente, and San Miguel volcanoes, El Salvador, *in* Rose, W.I., Bommer, J.J., Lopez, D.L., Carr, M.L., and Major, J.J., eds., Natural hazards in El Salvador: Geological Society of America Special Paper 375, p. 89–108.

McGuire, W.J., 1996, Volcano instability: a review of contemporary themes, *in* McGuire, W.J., Jones, A.P., and Neuberg, J., eds., Volcano instability on the Earth and other planets: London, Geological Society Special Publication 110, p. 1–23.

Melekestsev, I.V., and Braitseva, O.A., 1984, Gigantic rockslide avalanches on volcanoes: Volcanology and Seismology, v. 1984(4), p. 14–23 (in Russian: translation in Volcanology and Seismology, 1988, v. 6, p. 495–508).

Melson, W.G., 1988, Major explosive eruptions of Costa Rica volcanoes: Update for Costa Rican Volcanism Workshop: Skyland, Shenandoah National Park, Virginia, November 15–17, 1988, 6 p.

Méndez, J., and Hidalgo, P.J., 2004, Descripción geológica del depósito de *debris avalanche* El Coyol, Formación Barva, Costa Rica: Revista Geológia de América Central, no. 30, p. 199–202.

Mercado, R., Rose, W.I., Matias, O., and Giron, J., 1988, November 1929 dome collapse and pyroclastic flow at Santiaguito dome, Guatemala: Eos (Transactions, American Geophysical Union), v. 69, p. 1487.

Moore, J.G., Clague, D.A., Holcomb, R.T., Lipman, P.W., Normark, W.R., and Torresan, M.E., 1989, Prodigious submarine landslides on the Hawaiian Ridge: Journal of Geophysical Research, v. 94, p. 17,465–17,484.

Mora, J.C., Macías, J.L., García-Palomo, A., Arce, J.L., Espíndola, J.M., Manetti, P., Vaselli, O., and Sánchez, J.M., 2004, Petrology and geochemistry of the Tacaná volcanic complex, Mexico-Guatemala: Evidence for the last 40 000 yr of activity: Geofisica Internacional, v. 43, p. 331–359.

Newhall, C.G., 1987, Geology of the Lake Atitlán region, western Guatemala: Journal of Volcanology and Geothermal Research, v. 33, p. 23–55, doi: 10.1016/0377-0273(87)90053-9.

Oehler, J.-F., van Wyk de Vries, B., and Labazuy, P., 2005, Landslides and spreading of oceanic hot-spot and arc shield volcanoes on Low Strength Layers (LSLs): An analogue modeling approach: Journal of Volcanology and Geothermal Research, v. 144, p. 169–189.

Palmer, B.A., Alloway, B.V., and Neall, V.E., 1991, Volcanic-debris-avalanche-deposits in New Zealand: Lithofacies organization in unconfined, wet avalanche flows, *in* Smith, G.A., and Fisher, R.V., eds., Sedimentation in Volcanic Settings: SEPM (Society for Sedimentary Geology) Special Publication 45, p. 99–106.

Pardo, M., and Suárez, G., 1995, Shape of the subducted Rivera and Cocos plates in southern Mexico: Seismic and tectonic implications: Journal of Geophysical Research, v. 100, p. 12,357–12,373, doi: 10.1029/95JB00919.

Pavanelli, N., Capaccioni, B., Sarocchi, D., Falorni, G., Brenes, J., Vaselli, O., Tassi, F., Duarte, E., and Fernández, E., 2003, Debris avalanche deposits, landslides and related hazards on the southern flank of Irazú volcano (Costa Rica): The Commission on the Chemistry of Volcanic Gases, v. 2003, p. 37–39.

Pavanelli, N., Capaccioni, B., Giacomo, F., Damiano, S., Brenes, J., Vaselli, O., and Tassi, F., 2004, Geology and stratigraphy of the Irazú volcano, Costa Rica [abs.]: 32nd International Geological Congress, Florence, Italy, abstract 129-4, www.32igc.info/igc32/search.

Pullinger, C., 1998, Evolution of the Santa Ana volcanic complex, El Salvador [M.Sc. thesis]: Houghton, Michigan Technological University, 151 p.

Reagan, M., Duarte, E., Soto, G.J., and Fernández, E., 2006, this volume, The eruptive history of Turrialba volcano, Costa Rica, and potential hazards from future eruptions, *in* Rose, W.I., Bluth, G.J.S., Carr, M.J., Ewert, J.W., Patino, L., and Vallance, J.W., Volcanic hazards in Central America: Geological Society of America Special Paper 412, doi: 10.1130/2006.2412(13).

Reiche, P., 1937, The Toreva block, a distinctive landslide type: Journal of Geology, v. 45, p. 538–548.

Reid, M., 2004, Massive collapse of volcano edifices triggered by hydrothermal pressurization: Geology, v. 32, p. 373–376, doi: 10.1130/G20300.1.

Reid, M.E., Sisson, T.W., and Brien, D.L., 2001, Volcano collapse promoted by hydrothermal alteration and edifice shape, Mount Rainier, Washington: Geology, v. 29, p. 779–781, doi: 10.1130/0091-7613(2001)029<0779:VCPBHA>2.0.CO;2.

Reynolds, J.H., 1987, Timing and sources of Neogene and Quaternary volcanism in south-central Guatemala: Journal of Volcanology and Geothermal Research, v. 33, p. 9–22, doi: 10.1016/0377-0273(87)90052-7.

Roobol, M.J., Wright, J.V., and Smith, A.L., 1983, Calderas or gravity-slide structures in the Lesser Antilles island arc?: Journal of Volcanology and Geothermal Research, v. 19, p. 121–134, doi: 10.1016/0377-0273(83)90128-2.

Rose, W.I., Jr., Johnson, D.J., Hahn, G.A., and John, G.W., 1975, Skylab photography applied to geological mapping in northwestern Central America: Proceedings of the National Aeronautics and Space Administration, Earth Resources and Surveying Symposium, v. 1b, p. 861–884.

Rose, W.I., Conway, F.M., Pullinger, C.R., Deino, A., and McIntosh, W.C., 1999, An improved framework for late Quaternary silicic eruptions in northern Central America: Bulletin of Volcanology, v. 61, p. 106–120, doi: 10.1007/s004450050266.

Rüpke, L.H., Morgan, J.P., Hort, M., and Connolly, J.A.D., 2002, Are the regional variations in Central American arc lavas due to differing basaltic versus peridotite slab sources of fluids?: Geology, v. 30, p. 1035–1038, doi: 10.1130/0091-7613(2002)030<1035:ATRVIC>2.0.CO;2.

Scott, K.M., Macías, J.L., Vallance, J.W., Naranjo, J.A., Rodriguez, S., and McGeehin, J.P., 2001, Catastrophic debris flows transformed from landslides in volcanic terrains: mobility, hazard assessment, and mitigation strategies: U.S. Geological Survey Professional Paper 1630, 67 p.

Scott, K.M., Vallance, J.W., Kerle, N., Macías, J.L., Strauch, W., and Devoli, G., 2005, Catastrophic precipitation-triggered lahar at Casita volcano, Nicaragua: Occurrence, bulking and transformation: Earth Surface Processes and Landforms, v. 30, p. 59–79, doi: 10.1002/esp.1127.

Sekiya, S., and Kikuchi, Y., 1889, The eruption of Bandai-san: Tokyo Imperial University College of Science Journal, v. 3, p. 91–172.

Siebert, L., 2002, Landslides resulting from structural failure of volcanoes, *in* Evans, S.G., and DeGraff, J.V., eds., Catastrophic landslides: Effects, occurrence, and mechanisms: Geological Society of America Reviews in Engineering Geology 15, p. 209–235.

Siebert, L., Glicken, H., and Ui, T., 1987, Volcanic hazards from Bezymianny- and Bandai-type eruptions: Bulletin of Volcanology, v. 49, p. 435–459, doi: 10.1007/BF01046635.

Siebert, L., Begét, J.E., and Glicken, H., 1995, The 1883 and late-prehistoric eruptions of Augustine volcano, Alaska: Journal of Volcanology and Geothermal Research, v. 66, p. 367–395, doi: 10.1016/0377-0273(94)00069-S.

Siebert, L., Kimberly, P., and Pullinger, C.R., 2004, The voluminous Acajutla debris avalanche from Santa Ana volcano, western El Salvador, and comparison with other Central American edifice-failure events, *in* Rose, W.I., Bommer, J.J., Lopez, D.L., Carr, M.L., and Major, J.J., eds., Natural hazards in El Salvador: Geological Society of America Special Paper 375, p. 5–23.

Soto, G., 1988, Estructuras volcano-tectónicas del volcán Turrialba, Costa Rica, América Central: Actas V Congreso Geológico Chileno, 8–12 Agosto, 1988, Santiago, v. 3, p. 163–175.

Tournon, J., 1984, Magmatismes du Mesozoique a l'actuel en Amerique Central: l'example de Costa Rica, des ophiolites aux andesites [Ph.D. thesis]: Paris, Memoires Sciences Terre, Université Pierre et Marie Curies, 334 p.

Tournon, J., and Alvarado, G.E., 1997, Carte géologique du Costa Rica: Notice explicative; mapa geológico de Costa Rica: Folleto explicativo, échelle-escala 1:500,000: Edicion Teconológica de Costa Rica, 80 p. + mapa geológico de Costa Rica.

Ui, T., 1972, Recent volcanism in Masaya-Granada area, Nicaragua: Bulletin of Volcanology, v. 36, p. 174–190.

Ui, T., Yamamoto, H., and Suzuki-Kamata, K., 1986, Characterization of debris avalanche deposits in Japan: Journal of Volcanology and Geothermal Research, v. 29, p. 231–243, doi: 10.1016/0377-0273(86)90046-6.

Universidad Tecnológica de Panamá (UTP), 1992, Evaluación de la amenaza, estimación de la vulnerabilidad y del factor costo del riesgo de volcán Barú, Republica de Panamá: Universidad Tecnológica de Panamá, Facultad de Ingeniería Civil, 129 p. and 1:100,000 scale map.

Vallance, J.W., 2000, Lahars, *in* Sigurdsson, H., Houghton, B., McNutt, S., Rymer, H., and Stix, J., eds., Encyclopedia of volcanoes: San Diego, Academic Press, p. 601–616.

Vallance, J.W., and Scott, W.E., 1997, The Osceola Mudflow from Mount Rainier; sedimentology and hazard implications of a huge clay-rich debris flow: Geological Society of America Bulletin, v. 109, p. 143–163, doi: 10.1130/0016-7606(1997)109<0143:TOMFMR>2.3.CO;2.

Vallance, J.W., Siebert, L., Rose, W.I., Jr., Girón, J.R., and Banks, N.G., 1995, Edifice collapse and related hazards in Guatemala, *in* Ida, Y., and Voight, B., eds., Models of magmatic processes and volcanic eruptions: Journal of Volcanology and Geothermal Research., v. 66, p. 337–355.

Vallance, J.W., Schilling, S.P., Matias, O., Rose, W.I., and Howell, M.M., 2001a, Volcano hazards at Fuego and Acatenango, Guatemala: U.S. Geological Survey Open-File Report 01-431, 23 p.

Vallance, J.W., Schilling, S.P., and Devoli, G., 2001b, Lahar hazards at Mombacho volcano, Nicaragua: U.S. Geological Survey Open-File Report 01-455, 14 p.

van Wyk de Vries, B., and Borgia, A., 1996, The role of basement in volcano deformation, *in* McGuire, W.J., Jones, A.P., and Neuberg, J., eds., Volcano instability on the Earth and other planets: The Geological Society of London Special Publication 110, p. 95–110.

van Wyk de Vries, B., and Francis, P.W., 1997, Catastrophic collapse at stratovolcanoes induced by gradual volcano spreading: Nature, v. 387, p. 387–390.

van Wyk de Vries, B., Kerle, N, and Petley, D., 2000, Sector collapse forming at Casita volcano, Nicaragua: Geology, v. 28, p. 167–170.

van Wyk de Vries, B., Shea, T., and Pilato, M., 2005, Mombacho: A volcano falling to bits: Geophysical Research Abstracts, v. 7, abstract 03033, www.cosis.net/abstracts/EGU05/03033/EGU05-J-03033.pdf?PHPSESSID=6d292af3f796f5bd265fa5271f1ed2c6.

Viramonte, J.G., Navarro Collado, M., and Malavasi Rojas, E., 1997, Nicaragua–Costa Rica Quaternary volcanic chain: International Association of Volcanology and Chemistry of the Earth's Interior (IAVCEI) General Assembly, Puerto Vallarta, Mexico, January 19–24, 1997, Fieldtrip Guidebook, 17 p.

Voight, B., Komorowski, J.C., Norton, G., Belousov, A., Belousova, M., Boudon, G., Francis, P., Franz, W., Sparks, S., and Young, S., 2002, The 1997 Boxing Day sector collapse and debris avalanche, Soufrière Hills volcano, Montserrat, B.W.I., *in* Druitt, T., and Kokelaar, B.P., eds., The eruption of Soufrière Hills volcano, Montserrat, from 1995–1999: London, Geological Society Memoir 21, p. 363–407.

Williams, H., and McBirney, A.R., 1969, Volcanic history of Honduras: University of California Publications in Geological Sciences, v. 85, p. 1–101.

Wooller, L., van Wyk de Vries, B., Murray, J.B., Rymer, H., and Meyer S., 2004, Volcano spreading controlled by dipping substrata: Geology, v. 32, p. 573–576.

Wunderman, R.L., and Rose, W.I., 1984, Amatitlan, an actively resurging cauldron 10 km south of Guatemala City: Journal of Geophysical Research, v. 89, p. 8525–8539.

Zamora, N., Mendez, J., Barrahona, M., and Sjóbohm, L., 2004, Volcanoestratigrafía asociage al campo de domos de Canas Dulces, Guanacaste, Costa Rica: Revista Geológia América Central, v. 30, p. 41–58.

Zimbelman, D.R., Watters, R.J., Firth, I.R., Breit, G.N., and Carrasco-Nunez, G., 2004, Stratovolcano stability assessment methods and results from Citlaltépetl, Mexico: Bulletin of Volcanology, v. 66, p. 66–79, doi: 10.1007/s00445-003-0296-8.

MANUSCRIPT ACCEPTED BY THE SOCIETY 19 MARCH 2006

# Origin of the modern Chiapanecan Volcanic arc in southern México inferred from thermal models

**Vlad C. Manea**
**Marina Manea**
*Seismological Laboratory 252-21, Caltech, Pasadena, California 91125, USA*

## ABSTRACT

In southern México, the subducting Cocos slab drastically changes its geometry: from a flat slab in central México to a ~45° dip angle beneath Chiapas. Also, the currently active volcanic arc, the modern Chiapanecan volcanic arc, is oblique and situated far inland from the Middle America trench, where the slab depth is ~200 km. In contrast, the Central America volcanic arc is parallel to the Middle America trench, and the slab depth is ~100 km. A two-dimensional steady-state thermomechanical model explains the calc-alkaline volcanism by high temperature (~1300 °C) in the mantle wedge just beneath the Central America volcanic arc and the strong dehydration (~5 wt%) of the Cocos slab. In contrast, the thermal model for the modern Chiapanecan volcanic arc shows high *P-T* conditions beneath the coast where the extinct Miocene Chiapanecan arc is present, and is therefore unable to offer a reasonable explanation for the origin of the modern Chiapanecan volcanic arc. We propose a model in which the origin of the modern Chiapanecan volcanic arc is related to the space-time evolution of the Cocos slab in central México. The initiation of flat subduction in central México in the middle Miocene would have generated a hot mantle wedge inflow from NW to SE, generating the new modern Chiapanecan volcanic arc. Because of the contact between the hot mantle wedge beneath Chiapas and the proximity of a newly formed cold, flat slab, the previous hot mantle wedge in Chiapas became colder in time, finally leading to the extinction of the Miocene Chiapanecan volcanic arc. The position and the distinct K-alkaline volcanism at El Chichón volcano are proposed to be related to the arrival of the highly serpentinized Tehuantepec Ridge beneath the modern Chiapanecan volcanic arc. The deserpentinization of Tehuantepec Ridge would have released significant amounts of water into the overlying mantle, therefore favoring vigorous melting of the mantle wedge and probably of the slab.

**Keywords:** Chiapanecan volcanic arc, thermal models, alkaline volcanism, flat subduction.

# INTRODUCTION

One of the most interesting volcanic areas in México is located in the southern part, in Chiapas. Here, the volcanism is characterized by two Neogene volcanic arcs: the ancestral Miocene Sierra Madre arc and the modern Chiapanecan volcanic arc (Fig. 1). The Miocene Sierra Madre arc was abandoned between 9 and 3 Ma, when the modern Chiapanecan volcanic arc was formed (Damon and Montesinos, 1978). There are a series of characteristics that make this new modern Chiapanecan volcanic arc special. First, the position of the arc is well inside the continent, at distances of 300–350 km from the Middle America trench. The subducting Cocos slab lies ~200 km below the modern Chiapanecan volcanic arc, whereas the majority of subduction-related volcanic arcs are located where the slab is at ~100 km depth (Gill, 1981). Second, although there is a little variation of the slab dip beneath Chiapas (40–45°) (Rebollar et al., 1999), the modern Chiapanecan volcanic arc trends obliquely (~30°) to the Middle America trench. The modern Chiapanecan volcanic arc holds mainly typical continental arc calc-alkaline magmas (Damon and Montesinos, 1978), but El Chichón volcano presents a transitional magma signature from calc-alkaline to an adakitic-like type (Macías et al., 2003; De Ignacio et al., 2003). This characteristic was associated with melting of the Cocos slab just beneath the volcano at ~200 km by De Ignacio et al. (2003). The triple junction proximity between the North America, Cocos, and Caribbean plates was proposed by Nixon (1982) as a cause for the alkaline volcanism of El Chichón. Other authors suggest that this volcano is associated with the subduction of the Cocos plate (Luhr et al., 1984; Bevis and Isacks, 1984). The northwesternmost closest neighbor to the modern Chiapanecan volcanic arc is the isolated Tuxtla Volcanic Field. On the basis of trace element studies, the Na-alkaline products of Tuxtla Volcanic Field are explained by Nelson et al. (1995) to have been generated from mantle and continental crust melts contaminated by fluids or melts expelled from the subducted lithosphere. To the south, after a gap of ~100 km, the closest neighbor to the modern Chiapanecan volcanic arc is the calc-alkaline volcano Tacaná, which belongs to the Central American volcanic arc (García-Palomo et al., 2004). Although the slab dip, age, and convergence rate of the Cocos slab beneath the Central American volcanic arc in Guatemala are similar to those beneath Chiapas, the two active volcanic arcs are in completely different locations. Therefore, it is likely that an external perturbation of the mantle wedge-flow is responsible for the anomalous position of the modern Chiapanecan volcanic arc. Anisotropy in the form of shear wave splitting has been observed in many subduction zones (Kuril, Japan, Tonga, Izu-Bonin) and appears to be related to the asthenospheric flow produced by present-day plate motion (Fischer et al., 2000). Since thermal and flow models for the mantle wedge can give us advanced insights on the geodynamic processes beneath volcanic arcs, we computed four two-dimensional (2-D) thermal models normal to the Middle America trench, three in southern México and one in Guatemala, using the numerical approach proposed by Manea et al. (2005c). The numerical scheme consists of a system of 2-D Stokes equations and 2-D steady-state heat transfer equation. The subduction slab is considered a rigid body sliding at the convergence rate. The top and bottom limits of the model have fixed temperature of 0 °C and 1450 °C, respectively. The left boundary consists of an oceanic geotherm, which in turn is a function of oceanic plate age at the trench. The right boundary condition is represented by a constant thermal gradient of 20 °C from the surface down to the Moho. All the parameters are independent of time; therefore, the models are steady state.

The relationship between the long-term evolution of the subducting Cocos slab and the origin of modern Chiapanecan volcanic arc has never been addressed. In order to explain the unusual position of the modern Chiapanecan volcanic arc, we propose a geodynamic model mainly related to the initiation of flat subduction in the early-middle Miocene (Ferrari et al., 1999) beneath Central México.

# TECTONIC SETTINGS

Offshore, a large linear feature, the Tehuantepec Ridge, represents a tectonic border between two different regions on the Cocos plate (Manea et al., 2005a). Across the Tehuantepec Ridge, there is an age jump of the Cocos slab at the Middle America trench of ~10 m.y., defining two different thermal gradients across the oceanic lithosphere (Fig. 2). The seafloor age just northwest of the Tehuantepec Ridge is ca. 16 Ma (Kanjorsky, 2003; Klitgord and Mammerickx, 1982), then ca. 26 Ma (Manea et al., 2005b) in front of Chiapas state, and then ca. 23 Ma in Guatemala (Manea et al., 2005b). The rate of convergence between the Cocos plate and the North America and Caribbean plates varies from 6.5 cm/yr in Oaxaca to 7 cm/yr in Chiapas according to a NUVEL 1A model (DeMets et al., 1994).

The geometry of Wadati-Benioff zone shows that the Cocos slab increases its dip from 25° just NW of the Tehuantepec Ridge in Oaxaca to ~40° southeast of the Tehuantepec Ridge in Chiapas (Rebollar et al., 1999; Bravo et al., 2004). Then the slab slightly increases its dip to ~45° beneath the northern edge of the Central American volcanic arc. Also, the maximum depth extent of the intraslab earthquakes increases from 120 km in southern Oaxaca (Bravo et al., 2004) to 240 km in southern Chiapas (Rebollar et al., 1999).

Onshore, an important tectonic feature is represented by the triple junction between three plates: the North America, Cocos, and Caribbean plates. The contact between the North America and Caribbean plates is not clear. Guzmán and Meneses (2000) proposed that this contact is actually a diffuse boundary represented by a fault jog system.

There are two volcanic arcs in Chiapas: the extinct Miocene Sierra Madre arc, which runs parallel to the coast, and the modern Chiapanecan volcanic arc farther inland. The active volcanic arc diverges by ~30° in trend from the Middle America trench. Damon and Montesinos (1978) showed that the Miocene Sierra Madre arc was abandoned between 9 and 3 Ma, then the modern

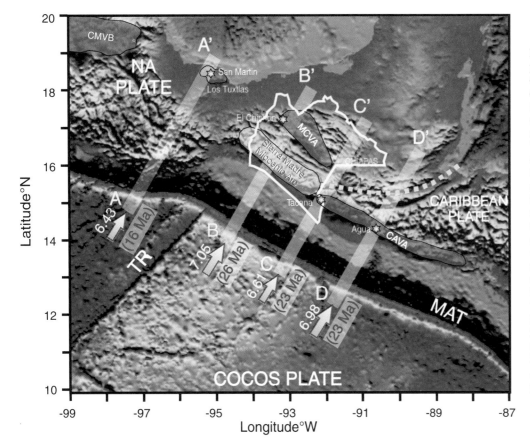

Figure 1. Tectonic setting and position of the four modeled cross sections (straight transparent lines). Transparent red zones show the location of active volcanic belts in México and Guatemala: CMVB—Central Mexican Volcanic Belt; MCVA—modern Chiapanecan volcanic arc; CAVA—Central American volcanic arc. Transparent gray area: the extinct Sierra Madre Miocenic arc. Orange stars are the main active volcanoes on each cross section. Arrows show convergence velocities (noted above the arrows) between the Cocos and North America and Caribbean plates (DeMets et al., 1994). The Cocos plate ages are shown beneath the arrows (Manea at al. 2005c; Kanjorski, 2003; Klitgord and Mammerickx, 1982). NA—North America; MAT—Middle America trench; TR—Tehuantepec Ridge.

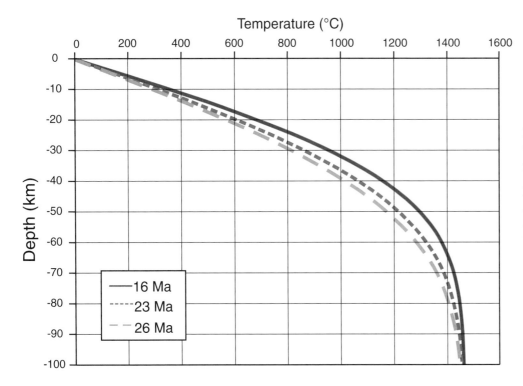

Figure 2. The left (seaward) boundary condition is a one-dimensional geotherm for the oceanic plate at the Middle America trench for ages 16 Ma, 23 Ma, and 26 Ma. The geotherm is calculated using the half-space cooling model (Turcotte and Schubert, 2002). Notice the lower thermal gradient for the older incoming plate.

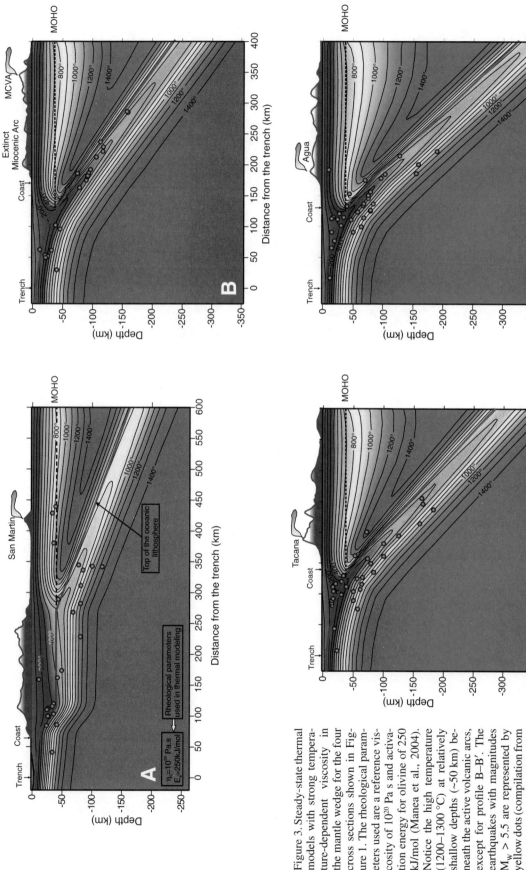

Figure 3. Steady-state thermal models with strong temperature-dependent viscosity in the mantle wedge for the four cross sections shown in Figure 1. The rheological parameters used are a reference viscosity of $10^{20}$ Pa s and activation energy for olivine of 250 kJ/mol (Manea et al., 2004). Notice the high temperature (1200–1300 °C) at relatively shallow depths (~50 km) beneath the active volcanic arcs, except for profile B–B′. The earthquakes with magnitudes $M_w > 5.5$ are represented by yellow dots (compilation from Rebollar et al., 1999; Engdahl and Villaseñor, 2002). Horizontal black dashed line shows the Moho. Topography is represented with 10× vertical exaggeration.

Chiapanecan volcanic arc was born as a consequence of a reorganization of the Cocos plate. Using isotopic ages (Ar-Ar method) for Central México, Ferrari et al. (1999) showed that the position of volcanism in Central México changed during the past 25 m.y., revealing that the Cocos slab became subhorizontal ca. 20 Ma.

## THERMAL MODELS AND MAGMATIC PRODUCTS

In this study, we developed four thermal models constrained by seismic data (Rebollar et al., 1999; Engdahl and Villaseñor, 2002), convergence rates (DeMets et al., 1994), and oceanic plate ages at the Middle America trench (Manea et al., 2005b; Kanjorski, 2003; Klitgord and Mammerickx, 1982) (Fig. 3). The first profile was chosen just northwest of the Tehuantepec Ridge because of the subhorizontal trajectory of the subducting Cocos slab. Because of the temperature-dependent viscosity, this model shows high temperature (1200–1300 °C) at 70–80 km depth beneath the volcanic arc. Using the phase diagrams for basalt and harzburgite (Hacker et al., 2003), we can see that the slab geotherm intersects the dehydration melting solidus at ~75–80 km, implying that melting of the basaltic crust might take place at this depth (Fig. 4). This is in good agreement with the conclusion of Nelson et al. (1995), who showed that a constant ratio of Ba/Nb is an indicator of magma contaminated by fluids or melts derived from the subducted slab. Also, a vertical P-T profile beneath Tuxtla Volcanic Field intersects the wet solidus for peridotite (Wyllie, 1979), thus explaining the calc-alkaline signature of the volcanic material (Fig. 4B). The second profile is situated just southeast of the Tehuantepec Ridge and intersects the modern Chiapanecan volcanic arc in the proximity of El Chichón volcano. The modeling results show high temperatures at ~50 km depth beneath the ancient Miocene volcanic arc rather than beneath the modern Chiapanecan volcanic arc (Fig. 3B). The distant modern Chiapanecan volcanic arc lies ~200 km above the slab, where the entire oceanic slab is completely dehydrated. Also, the slab geotherm does not intersect the melting solidus for basalt (Fig. 4A); therefore, with this model, we cannot explain the K-alkaline products of El Chichón volcano. At this stage, a 2-D steady-state thermal model is inconsistent with the observations, but in the following sections, we will address these issues. The last two models are located at the northern end of the Central American volcanic arc (Figs. 3C and 3D). The results show

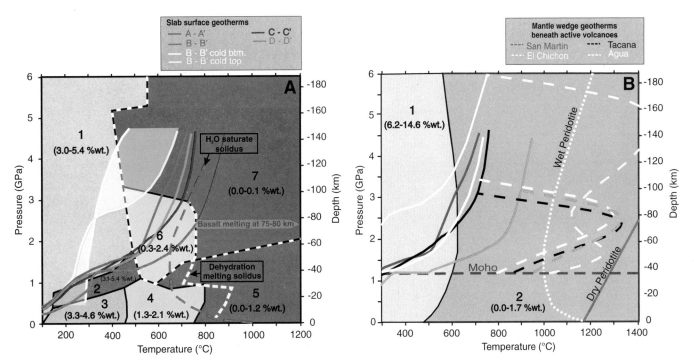

Figure 4. (A) Simplified phase diagram for the hydrous mid-oceanic-ridge basalt (Hacker et al., 2003). Metamorphic facies: 1—jadeite-lawsonite-blueschist-amphibole-talc (3.0–5.4 wt% $H_2O$); 2—lawsonite-epidote-blueschist-jadeite (3.1–5.4 wt% $H_2O$); 3—zeolite-prehnit-pumpellyite-actinolite-greenschist (3.3–4.6 wt% $H_2O$); 4—epidote-amphibolite (1.3–2.1 wt% $H_2O$); 5—garnet-amphibolite-granulite (0.0–1.2 wt% $H_2O$); 6—zoisite-amphibole-eclogite (0.3–2.4 wt% $H_2O$); 7—eclogite-coesite-diamond (0.0–0.1 wt% $H_2O$). The slab geotherms intersect the dehydration melting solidus for basalt only for profile A–A′ at 75–80 km depth. For profiles C–C′ and D–D′ strong oceanic crust dehydration (~5 wt%) occurs just beneath the active volcanic arc. The geotherm B–B′ cold top is for the model with a cold mantle wedge beneath the extinct Sierra Madre Miocene volcanic arc (Fig. 7B). The geotherm B–B′ cold btm. represents the geotherm at the base of the oceanic crust. Note the strong dehydration from phase transformation of lawsonite blueschist into eclogite (~5 wt%). The fluids are released into the mantle wedge at greater depths between 100 and 150 km, close beneath El Chichón volcano. (B) Simplified phase diagram for hydrous harzburgite (Hacker et al., 2003). 1—serpentine-chlorite-brucite-dunite (6.2–14.6 wt% $H_2O$); 2—chlorite-harzburgite-talc-dunite-anthigorite-spinel-garnet-harzburgite (0.0–1.7 wt% $H_2O$).

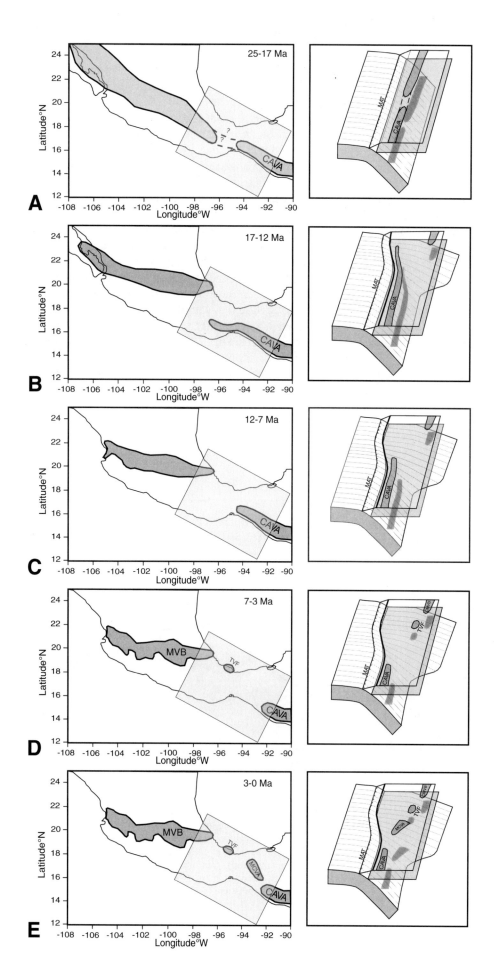

Figure 5. Space-time distribution of the volcanic arcs in central and southern México. (A) 25–17 Ma: the volcanic arc formed an approximately continuous belt. (B) 17–12 Ma: the Central Mexican Volcanic Belt moved inland, suggesting that the subducting slab become subhorizontal. (C) 12–7 Ma: the flattening process of the Cocos slab continued farther SE. The proximity of the new, cold, flat subducting lithosphere to the hot mantle wedge beneath Chiapas would eventually lead to progressive Central American volcanic arc extinction in southern México. (D) 7–3 Ma: the Central American volcanic arc continued to retreat SE, and the volcanic activity close to the Chiapas coast ceased completely. The Tuxtla Volcanic Field is born during this period. (E) 3–0 Ma: The up-bending of the subducting slab for the past 14 m.y. created an asthenospheric inflow from NW to SW, which finally led to the onset of the modern Chiapanecan volcanic arc.

a region with high temperature (~1300 °C) at ~50 km depth beneath the active arc, conditions well above the wet peridotite solidus temperature (Fig. 4B). The phase transformation of the oceanic crust from basalt (low *P-T*) to eclogite (high *P-T*) produces dehydration (4–5 wt%; Fig. 4A). The releasing of fluids into the mantle wedge lowers the melting point of mantle peridotite and favors calc-alkaline volcanism. The slab geotherm does not intersect the basalt solidus, this being consistent with the pure calc-alkaline character of the Central American volcanic arc (no Na-alkaline or K-alkaline products).

## SPACE-TIME EVOLUTION OF THE COCOS SLAB AND CHIAPANECAN VOLCANISM

One of the key parameters for understanding the position through time of a subducting slab is the spatial variation through time of volcanic arcs. Using the distribution of dated rocks from Ferrari et al. (1999) for Central México and from Damon and Montesinos (1978) for Chiapas, we propose the evolution of the subducting slab presented in Figure 5. Between 25 and 17 Ma, the volcanic arc formed an approximately continuous belt in Central México, Chiapas, and Guatemala (Fig. 5A). Then, between 17 and 12 Ma, the Central Mexican Volcanic Belt moved inland, suggesting that the subducting slab become subhorizontal (Fig. 5B). To the south, the rest of the volcanic arc remained parallel to Middle America trench, therefore suggesting that the slab approximately preserved its previous dip. Later, between 12 and 7 Ma, the flattening process of the Cocos slab continued further SE, while the Central American volcanic arc retreated SE (Fig. 5C). Between 7 and 3 Ma, the Central American volcanic arc continued to retreat SE and the volcanic activity close to the Chiapas coast ceased completely (Fig. 5D). The Tuxtla Volcanic Field was born during this time period. The last episode of this scenario took place between 3 and 0 Ma, and is represented by the onset of the modern Chiapanecan volcanic arc (Fig. 5E). Since our thermal models are steady-state, the slab evolution through time is not incorporated in our models.

Since the slab dip, age, and convergence rate of the Cocos slab beneath the Central American volcanic front in Guatemala are similar to those beneath Chiapas, why are the two active volcanic arcs so different? To answer this question, we looked for an external perturbation responsible for the anomalous position of the Chiapanecan volcanism. We propose that an external perturbation resulted from the time evolution of the subducting slab beneath Central México. The onset of the flat-slab northwest of the Tehuantepec Ridge, in Central México, would have produced a strong mantle-wedge inflow in the neighboring region to the SE (Figs. 5B–5E). This intake would have migrated over time, creating the non–trench parallel modern Chiapanecan volcanic arc. The isotopic ages of the modern Chiapanecan volcanic arc decrease from NW to SE (Fig. 6), suggesting that the arc was propagating through time from NW to SE. If such a model can explain the position and orientation of the distant tilted modern Chiapanecan volcanic arc, there is still one question that has to be addressed: why has activity on the Sierra Madre volcanic arc ceased? To answer this question, we look to the implication of the three-dimensional (3-D) temperature distribution of this region. The flat, cold, incoming oceanic lithosphere (profile A–A′ in Fig. 1) is in contact with the hot mantle wedge just ~120 km SE. The flat slab temperature is ~600 °C, while the mantle wedge temperature is ~1300 °C (Fig. 4A and 4B). Therefore, there is a continuous cooling of the mantle wedge beneath the ancient Chiapanecan volcanic arc. If a mantle wedge tip is cooled down, then an interesting phenomenon might appear. From the phase diagram for harzburgite (Fig. 4B), we can see that during the cooling, the mantle wedge enters into the serpentine stability field (~<600 °C). The location of the serpentinized mantle wedge tip is critical because it may control the down-dip coupling between

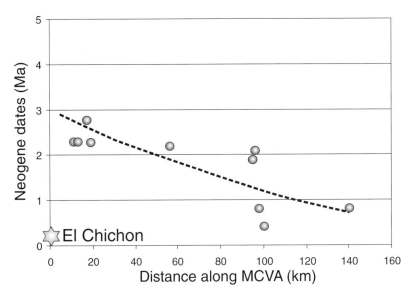

Figure 6. The isotopic ages along the modern Chiapanecan volcanic arc (MCVA) decrease from NW to SE (data from Damon and Montesinos, 1978), suggesting that the arc was propagating through time from NW to SE. Note the anomalous position of El Chichón volcano.

the slab and the overlying material (Manea et al., 2004). Very low shear-wave velocities in the cold forearc mantle have been discovered in the southern Cascadia subduction zone (Bostock et al., 2002). This is evidence of a highly hydrated and serpentinized material in the forearc region. The same conditions should also be expected in the Chiapas subduction zone.

## MODEL CONSTRAINTS

To gather all these assumptions together, we developed a thermal model with a fixed mantle wedge tip, which simulates the decoupling of the subducting slab from the overlying mantle wedge due to serpentinization, and a hot source (1200 °C) beneath the modern Chiapanecan volcanic arc, which represents the NW-SW asthenospheric inflow. The results show a well-developed serpentinized mantle wedge tip (Fig. 7B). From Figure 8, it can be seen that the mantle wedge beneath Chiapas lies just in front of a cold area that corresponds to the flat slab. In order to investigate the existence of a cold and serpentinized mantle wedge beneath Chiapas, we sought external constraints for our model. The serpentinization process has two major effects on the physical properties of peridotite: the density decreases from ~3000 kg/m$^3$ to 2500 kg/m$^3$ (Christensen, 1966; Saad, 1969), and the remnant magnetization increases by at least an order of magnitude (Saad, 1969). A recent study by Blakely et al. (2005) demonstrated that strong magnetic signatures in the mantle are common features in subduction zones like Cascadia, Japan, and southern Alaska. The serpentinized mantle can be detected by the presence of a long-wavelength magnetic anomaly above subduction zones. The availability of a new magnetic map for North America (North American Magnetic Anomaly Group [NAMAG], 2002) allows us to prove the existence of a large amount of serpentine beneath Chiapas. In Figure 9, aeromagnetic data display a distinctive long wavelength positive anomaly (~500 nT) offshore of Chiapas. Unfortunately, it is not possible to infer the maximum extent inland of the serpentinized wedge because magnetic data do not exist for onshore Chiapas. Instead, the contact between the Moho and the subducting Cocos slab is inferred using the onset of the strong positive magnetic anomaly. Since the magnetic anomaly has not been migrated to the pole, the source is not located just beneath the maximum amplitude peak of this anomaly. The reduction to the pole operation is a data processing technique that recalculates total magnetic intensity data as if the inducing magnetic field had a 90° inclination. This transforms dipolar magnetic anomalies to monopolar anomalies centered over the source. Reduction to the pole makes the simplifying assumption that the rocks in the survey area are all magnetized parallel to Earth's magnetic field. Figure 7A shows a 2-D magnetic profile migrated to pole anomaly. The onset of the serpentinized mantle wedge is located ~125 km from the trench and runs parallel to the Middle America trench. This provides a Moho depth beneath the Chiapas coast of ~35 km, which is consistent with the results of Bravo et al. (2004), who give a Moho average depth beneath the Gulf of Tehuantepec of 28.5 ± 3.5 km. Also, the maximum extent of the rupture areas of past subduction earthquakes is in good agreement with the onset of the positive magnetic anomaly and shows that the rupture areas are therefore being controlled by the ductile serpentinized mantle wedge tip rather than by temperature (Fig. 7). On the other hand, the updip seismogenic limit is controlled by temperature, the 100 °C isotherm being in good agreement with the onset of rupture areas (Fig. 9).

## DISCUSSION AND CONCLUSIONS

In this study, we inferred the steady-state thermal structure beneath Chiapas using the numerical scheme proposed by Manea et al. (2005c). The thermal models beneath the northernmost edge of the Central American volcanic arc successfully explain the calc-alkaline character of the magmatic products (Figs. 3C and 3D). Also, the thermal model for the flat slab just northwest of the Tehuantepec Ridge predicts the origin of the calc-alkaline and alkaline magmas of the Tuxtla Volcanic Field (Fig. 3A). The only model that is not consistent with the position of the active volcanic arc is the model located just southeast of the Tehuantepec Ridge beneath the modern Chiapanecan volcanic arc (Fig. 3A). Also, the position and the alkaline character of the El Chichón volcano cannot be explained by this model. To address these issues, we proposed an alternative model in which a mantle wedge flow perturbation is produced by a strong inflow related to the space-time history of the subducting slab beneath Central México (Ferrari et al., 1999). The flattening of the slab ca. 17 Ma would have produced a strong mantle wedge flow from NW to SE, forming the modern Chiapanecan volcanic arc. The migration of the inflow farther southeast can be seen in the age distribution of magmatic rocks through the modern Chiapanecan volcanic arc (Damon and Montesinos, 1978) (Fig. 6). The cessation of the ancient Miocene Sierra Madre volcanic arc suggests that the mantle wedge beneath should be cold enough to prevent mantle-wedge melting. We argue that the proximity of the flat slab is responsible for this cooling. To constrain this model, we used the availability of aeromagnetic data offshore of Chiapas (Fig. 9). A good correlation between the strong positive magnetic anomaly and the hypothesized position of the serpentinized mantle wedge suggests that the mantle wedge beneath the extinct Miocene Sierra Madre arc is sufficiently cold ($T < 600$ °C) to prevent melting (Fig. 7).

Although these models offer a reasonable explanation for the origin of the modern Chiapanecan volcanic arc, there remains an important unknown: El Chichón volcano. De Ignacio et al. (2003) proposed an asthenospheric inflow through major faults in the subducting Cocos slab. This inflow would have melted the oceanic slab and produced adakitic magmas. Unfortunately, this scenario is difficult to prove, because we have to assume that the Cocos slab beneath Chiapas is segmented and decoupled in different blocks along slab-dip faults. If this is the case, then the two parts of the subducting slab, northwest and southeast of the Tehuantepec Ridge, are completely decoupled. But an interesting observation can readily be made in Figure 3: although in front of Chiapas the oceanic plate is older (ca. 29 ± 3 Ma from Manea et al., 2005b)

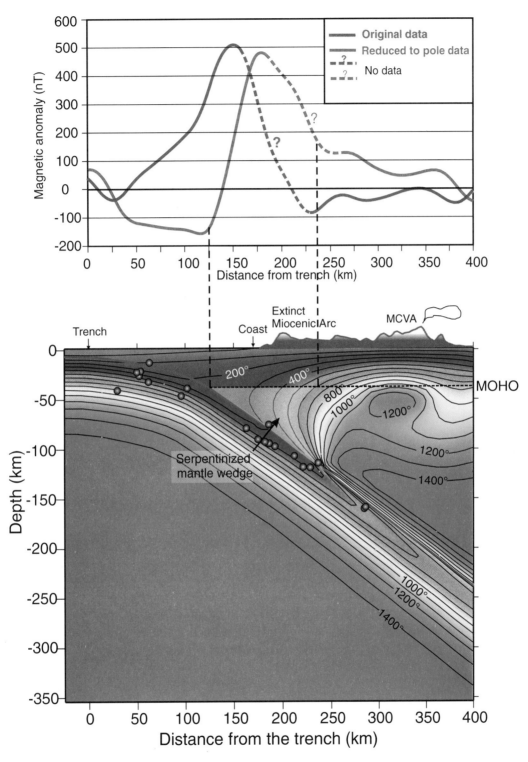

Figure 7. (A) Purple line: Low pass filtered aeromagnetic anomaly along profile B–B′. Blue line is the same aeromagnetic anomaly but migrated to the pole, so the source is just beneath the peak. Dashed lines represent our extrapolation where real data are missing. (B) Steady-state thermal model with a cold mantle wedge tip and a hot asthenospheric region beneath the modern Chiapanecan volcanic arc (MCVA). The foggy triangle represents the serpentinized mantle wedge. Note the good correlation with the position of the pole-migrated aeromagnetic anomaly.

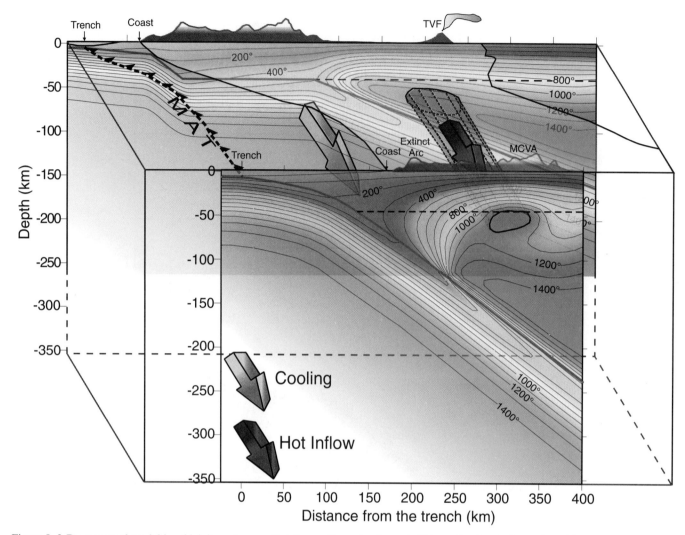

Figure 8. 3-D conceptual model in which it can be seen that the mantle wedge beneath Chiapas lies just in front of a cold area that corresponds to the incoming flat slab. The hot asthenospheric inflow is also shown. MAT—Middle America trench; MCVA—modern Chiapanecan volcanic arc; TVF—Tuxtla Volcanic Field.

and therefore denser, just southeast of the Tehuantepec Ridge, its dip is shallower than the slab dip farther south where the slab is younger (ca. 23 Ma). This paradox can be easily explained by the existence of the flat subduction just northwest of the Tehuantepec Ridge, which actually does not allow the steeper slab to sink freely into the asthenosphere, therefore supporting the nonexistence of such faults in the Cocos slab. Since the model of De Ignacio et al. (2003) looks quite unrealistic, we have to look for an alternative model in order to explain the singularity of El Chichón volcano within the modern Chiapanecan volcanic arc. There is a striking correlation between the Tehuantepec Ridge and the position of El Chichón volcano, also pointed out by Luhr et al. (1984). It is known that the long-offset fracture zones are subject to a double serpentinization process due to bend-faulting at the outer rise (the East Pacific Rise for the Tehuantepec Ridge) and at the trench axis. These faults might be conduits for seawater to react with the slab lithosphere and serpentinize it. If so, then such long fracture zones might produce tears beneath the volcanic arc because they are water rich, therefore favoring more melting just above the slab surface. Indeed, a strong positive magnetic anomaly is observed along the Tehuantepec Ridge (Fig. 8), suggesting that the slab-lithosphere might be highly serpentinized. The dehydration of the oceanic crust (Fig. 4A) and lithosphere (Fig. 4B) for the cold thermal model (Fig. 7B) shows that ~5–10 wt% might be released into the mantle at greater depths between 100 and 150 km, close beneath El Chichón volcano. Since El Chichón volcano is young (ca. 0.2 Ma), we argue that this might reflect the arrival of the serpentinized Tehuantepec Ridge beneath the modern Chiapanecan volcanic arc at least 0.2 Ma ago, resulting from an instantaneous magma rise through the mantle. Assuming a constant convergence rate of 7 cm/yr, the Tehuantepec Ridge would have commenced subduction at least ca. 6 Ma ago.

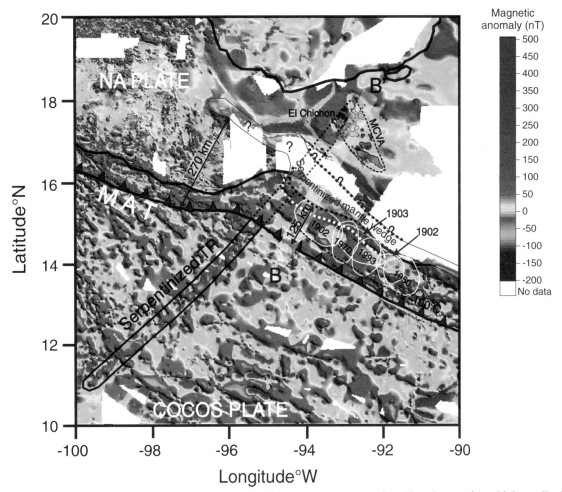

Figure 9. Aeromagnetic map for central and southern México. The white areas represent no data. Note the prominent high-amplitude magnetic anomalies offshore of Chiapas (white dotted line). The reduced to pole anomaly is shifted inland (black dotted line). This anomaly is used to project the serpentized mantle wedge onto the surface (transparent white area). Note the good agreement between the rupture areas and the 100 °C isotherm and the onset of the serpentinized mantle wedge. Also note the high-amplitude magnetic anomalies along the Tehuantepec Ridge, which might be an indicator of a highly serpentinized region. MAT—Middle America trench; MCVA—modern Chiapanecan volcanic arc; NA—North America.

## ACKNOWLEDGMENTS

Very helpful comments and suggestions by Jim Luhr, Mike Willis, and an anonymous reviewer were very important to improving this manuscript. We would also like to acknowledge support from the Gordon and Betty Moore Foundation. This is contribution 9129 of the Division of Geological and Planetary Sciences and contribution 33 of the Tectonics Observatory, California Institute of Technology.

## REFERENCES CITED

Bevis, M., and Isacks, B.L., 1984, Hypocentral trend surface analysis: probing the geometry of Benioff Zones: Journal of Geophysical Research, v. 89, p. 6153–6170.

Blakely, R.J., Brocher, T.M., and Wells, R.E., 2005, Subduction-zone magnetic anomalies and implications for hydrated forearc mantle: Geology, v. 33, p. 445–448, doi: 10.1130/G21447.1.

Bostock, M.G., Hyndman, R.D., Rondenay, S., and Peacock, S.M., 2002, An inverted continental Moho and serpentinization of the forearc mantle: Nature, v. 417, p. 536–538, doi: 10.1038/417536a.

Bravo, H., Rebollar, C.J., Uribe, A., and Jimenez, O., 2004, Geometry and state of stress of the Wadati-Benioff zone in the Gulf of Tehuantepec, México: Journal of Geophysical Research, v. 109, p. B04307, doi: 10.1029/2003JB002854.

Christensen, N.I., 1966, Elasticity of ultrabasic rocks: Journal of Geophysical Research, v. 71, p. 5921–5931.

Damon, P.E., and Montesinos, E., 1978, Late Cenozoic volcanism and metallogenesis over an active Benioff zone in Chiapas, México: Arizona Geological Society Digest, v. X1, p. 155–168.

De Ignacio, C., Castineiras, P., Marquez, A., Oyarzun, R., Lillo, J., and Lopez, I., 2003, El Chichón volcano (Chiapas volcanic belt, México) transitional calc-alkaline to adakitic-like magmatism: Petrologic and tectonic implications: International Geology Review, v. 45, p. 1020–1028.

DeMets, C., Gordon, R., Argus, D., and Stein, S., 1994, Effect of recent revisions to the geomagnetic reversal time scale on estimates of current plate motions: Geophysical Research Letters, v. 21, p. 2191–2194, doi: 10.1029/94GL02118.

Engdahl, E.R., and Villaseñor, A., 2002, Global seismicity: 1900–1999, in Lee, W.H.K. et al., eds., International handbook of earthquake engineering and seismology: International Geophysics Series, v. 81A, p. 665–690.

Ferrari, L., López-Martinez, M., Aquirre-Díaz, G., and Carrasco-Núñez, G., 1999, Space-time patterns of Cenozoic arc volcanism in central México: From the Sierra Madre Occidental to the Mexican Volcanic Belt: Geology, v. 27, no. 4, p. 303–306, doi: 10.1130/0091-7613(1999)027<0303: STPOCA>2.3.CO;2.

Fischer, K.M., Parmentier, E.M., Stine, A.R., and Wolf, E.R., 2000, Modeling anisotropy and plate-driven flow in the Tonga subduction zone backarc: Journal of Geophysical Research, v. 105, p. 16,181–16,191, doi: 10.1029/1999JB900441.

García-Palomo, A., Macías, J.L., and Espíndola, J.M., 2004, Strike-slip faults and K-alkaline volcanism at El Chichón volcano southeastern México: Journal of Volcanology and Geothermal Research, v. 136, p. 247–268, doi: 10.1016/j.jvolgeores.2004.04.001.

Gill, J., 1981, Orogenic andesites and plate tectonics: Springer-Verlag, Berlin, 390 p.

Guzmán-Speziale, M., and Meneses-Rocha, J.J., 2000, The North America–Caribbean plate boundary west of the Motagua-Polochic fault system: A fault jog in Southeastern México: Journal of South American Earth Sciences, v. 13, p. 459–468, doi: 10.1016/S0895-9811(00)00036-5.

Hacker, B.R., Abers, G.A., Peacock, S.M., 2003, Subduction factory 1. Theoretical mineralogy, densities, seismic wave speeds, and $H_2O$ contents: Journal of Geophysical Research, v. 108, doi: 10.1029/2001JB001127.

Kanjorski, M.N., 2003, Cocos Plate structure along the Middle America subduction zone off Oaxaca and Guerrero, México: Influence of subducting plate morphology on tectonics and seismicity [Ph.D. thesis]: San Diego, University of California, 217 p.

Klitgord, K.D., and Mammerickx, J., 1982, Northern east Pacific Rise; magnetic anomaly and bathymetric framework, Journal of Geophysical Research, v. 87, 138, p. 6725–6750.

Luhr, J.F., Carmichael, I.S.E., and Varekamp, J.C., 1984, The 1982 eruptions of El Chichón Volcano, Chiapas, México: Mineralogy and petrology of the anhydrite-bearing pumices: Journal of Volcanology and Geothermal Research, v. 23, p. 69–108, doi: 10.1016/0377-0273(84)90057-X.

Macías, J.L., Arce, J.L., Mora, J.C., Espíndola, J.M., Saucedo, R., Manetti, P., 2003, The ~550 BP Plinian eruption of el Chichón volcano, Chiapas, México: explosive volcanism linked to reheating of a magma chamber: Journal of Geophysical Research, v. 108, ECV3, p. 1–18.

Manea, V.C., Manea, M., Kostoglodov, V., Currie, C.A., and Sewell, G., 2004, Thermal structure, coupling and metamorphism in the Mexican subduction zone beneath Guerrero: Geophysical Journal International, v. 158, p. 775–784, doi: 10.1111/j.1365-246X.2004.02325.x.

Manea, M., Manea, V.C., Kostoglodov, V., and Gúzman-Speziale, M., 2005a, Elastic thickness of the lithosphere below the Tehuantepec Ridge: Geofisica International, v. 44, no. 2, p. 157–168.

Manea, M., Manea, V.C., Ferrari, L., Kostoglodov, V., and Bandy, W.L., 2005b, Structure and origin of the Tehuantepec Ridge: Earth and Planetary Science Letters, v. 238, p. 64–77.

Manea, V.C., Manea, M., Kostoglodov, V., and Sewell, G., 2005c, Thermomechanical model of the mantle wedge in Central Mexican subduction zone and a blob tracing approach for the magma transport: Physics of the Earth and Planetary Interiors, v. 149, p. 165–186, doi: 10.1016/j.pepi.2004.08.024.

Nelson, S.A., Gonzalez-Caver, E., and Kyser, T.K., 1995, Constraints on the origin of alkaline and calc-alkaline magmas from the Tuxtla Volcanic Field, Veracruz, México: Contributions to Mineralogy and Petrology, v. 122, p. 191–211, doi: 10.1007/s004100050121.

Nixon, G.T., 1982, The relationship between Quaternary volcanism in central México and the seismicity and structure of the subducted ocean lithosphere: Geological Society of America Bulletin, v. 93, p. 514–523, doi: 10.1130/0016-7606(1982)93<514:TRBQVI>2.0.CO;2.

North American Magnetic Anomaly Group (NAMAG), 2002, Magnetic anomaly map of North America; Processing, compilation, and geologic mapping applications of the new digital magnetic anomaly database and map of North America: Denver, U.S. Department of the Interior and U.S. Geological Survey, scale 1:10,000,000, plus 31 p. booklet, http://pubs.usgs.gov/sm/mag_map/.

Rebollar, C.J., Espindola, V.H., Uribe, A., Mendoza, A., and Vertti, A.P., 1999, Distributions of stresses and geometry of the Wadati-Benioff zone under Chiapas, México: Geofisica Internacional, v. 38, no. 2, p. 95–106.

Saad, A.F., 1969, Magnetic properties of ultramafic rocks from Red Mountain, California: Geophysics, v. 34, p. 974–987, doi: 10.1190/1.1440067.

Turcotte, D.L., and Schubert, G., 2002, Geodynamics (Second edition): New York, Cambridge University Press, 456 p.

Wyllie, P.J., 1979, Magmas and volatile components: American Mineralogy, v. 654, p. 469–500.

MANUSCRIPT ACCEPTED BY THE SOCIETY 19 MARCH 2006

# Geological evolution of the Tacaná Volcanic Complex, México-Guatemala

**Armando García-Palomo***
*Departamento de Geología Regional, Instituto de Geología, Universidad Nacional Autónoma de México, Coyoacán 04510, México D.F., México*

**José Luis Macías**
*Departamento de Vulcanología, Instituto de Geofísica, Universidad Nacional Autónoma de México, Coyoacán 04510, México D.F., México*

**José Luis Arce**
*Departamento de Geología Regional, Instituto de Geología, Universidad Nacional Autónoma de México, Coyoacán 04510, México D.F., México*

**Juan Carlos Mora**
*Departamento de Vulcanología, Instituto de Geofísica, Universidad Nacional Autónoma de México, Coyoacán 04510, México D.F., México*

**Simon Hughes**
*Department of Geology, State University of New York, 876 Natural Science Complex, Buffalo, New York 14260, USA*

**Ricardo Saucedo**
*Departamento de Vulcanología, Instituto de Geología, Universidad Autónoma de San Luis Potosí, Avenida Dr. Manuel Nava No. 5, 10 Zona Universitaria, San Luis Potosí 78240, México*

**Juan Manuel Espíndola**
*Instituto de Geofísica, Universidad Nacional Autónoma de México, Coyoacán 04510, México D.F., México*

**Rudiger Escobar**
*Coordinadora Nacional para la Reducción de Desastres (CONRED), Avenida Hincapié, 21-72 Zona 1, Ciudad de Guatemala, Guatemala*

**Paul Layer**
*Geophysical Institute, University of Alaska, Fairbanks, Alaska 99775-7320, USA*

## ABSTRACT

The Tacaná Volcanic Complex represents the northernmost active volcano of the Central American Volcanic Arc. The genesis of this volcanic chain is related to the subduction of the Cocos plate beneath the Caribbean plate. The Tacaná Volcanic

---

*E-mail: apalomo@geologia.unam.mx

**Complex is influenced by an important tectonic structure as it lies south of the active left-lateral strike-slip Motozintla fault related to the Motagua-Polochic fault zone. The geological evolution of the Tacaná Volcanic Complex and surrounding areas is grouped into six major sequences dating from the Mesozoic to Recent. The oldest basement rocks are Mesozoic schists and gneisses of low-grade metamorphism. These rocks are intruded by Tertiary granites, granodiorites, and tonalites ranging in age from 12 to 39 Ma, apparently separated by a gap of 9 m.y. The first intrusive phase occurred during late Eocene to early Oligocene, and the second during early to middle Miocene. These rocks are overlain by deposits from the Calderas San Rafael (ca. 2 Ma), Chanjale (ca. 1 Ma), and Sibinal (unknown age), grouped under the name Chanjale–San Rafael Sequence, of late Pliocene–Pleistocene age. The activity of these calderas produced thick block-and-ash flows, ignimbrites, lavas, and debris flows. The Tacaná Volcanic Complex began its formation during the late Pleistocene, nested in the preexisting San Rafael Caldera. The Tacaná Volcanic Complex formed through the emplacement of four volcanic centers. The first, Chichuj volcano, was formed by andesitic lava flows and pyroclastic deposits, after which it was destroyed by the collapse of the edifice. The second, Tacaná volcano, formed through the emission of basaltic-andesite lava flows, as well as andesitic and dacitic domes that produced extensive block-and-ash flows ~38,000, 28,000, and 16,000 yr B.P. The Plan de las Ardillas structure (the third volcanic center) consists of an andesitic dome with two lava flows emplaced on the high slope of the Tacaná ~30,000 yr B.P. Finally, the San Antonio volcanic center was built through the emission of lava flows, andesitic and dacitic domes, and it was destroyed by a Peléan eruption at 1950 yr B.P. that produced a block-and-ash flow deposit. The Tacaná Volcanic Complex was emplaced along a NE-SW trend beginning with Chichuj, followed by Tacaná, Las Ardillas, and San Antonio. This direction is roughly the same as the NE-SW Tacaná graben (as proposed in this work), together with other faults and fractures exposed in the region. The rocks of the Chanjale-San Rafael Sequence and the Tacaná Volcanic Complex have a calc-alkaline signature with medium K contents, negative anomalies of Nb, Ti, and P, and enrichment in light rare earth elements, typical of subduction zones.**

**Keywords:** Southern México, Guatemala, Tacaná, structural geology, volcanic evolution.

## INTRODUCTION

The Tacaná Volcanic Complex is located in the State of Chiapas, in southern México, and in the San Marcos Department of Guatemala. The Tacaná Volcanic Complex represents the northwesternmost volcano of the Central American Volcanic Arc, a WNW-oriented volcanic arc that extends for >1300 km, from the México-Guatemala border to Costa Rica (Fig. 1). The Central American Volcanic Arc is parallel to the trench and consists of several stratovolcanoes that have erupted calc-alkaline magmas from the Eocene to the Recent (Carr et al., 1982; Donnelly et al., 1990). The origin of this volcanic chain is related to the subduction of the Cocos plate beneath the Caribbean plate, which together with the North American plate create a complicated triple point junction in the region (Guzmán-Speziale et al., 1989). The Tacaná Volcanic Complex includes Tacaná volcano, which has shown some recent, though minor, activity as reported by Bergeat (1894) and Sapper (1927). Tacaná reawakened in 1950 and 1986 with small phreatic explosions that reminded villagers and authorities of its potential threat in case of future activity (Müllerried, 1951; De la Cruz-Reyna et al., 1989). Prior to the 1986 eruption, the National Power Company of México (Comisión Federal de Electricidad) conducted a geological survey to evaluate the geothermic potential of Tacaná volcano (De la Cruz and Hernández, 1985). These authors presented the first geological map of the volcano, showing that Tacaná was emplaced on top of Tertiary granodioritic rocks and that it consisted of lava flows and at least three Quaternary fans of pyroclastic flows that they labeled, from youngest to oldest, Qt1, Qt2, and Qt3. De Cserna et al. (1988) presented a geologic map of the volcano based mostly on photo-interpretation. These authors also recognized the existence of three volcanic episodes in the formation of Tacaná. The first radiometric age determinations at the volcano were obtained from charcoal samples extracted from Qt3, which revealed a late Pleistocene eruption dated at 38,000 yr B.P. (Espíndola et al., 1989) that was later confirmed (Espíndola et al., 1993). These latter authors assigned an age of ca. 28,000 yr B.P. to unit Qt2. Mercado and Rose (1992) published a geologic map based mostly on photogeology and presented the first volcanic hazard zonation map of the volcano. This hazard zonation map took into consideration pyroclastic flows, debris avalanches, pyroclastic falls, and lahars. The first study that considered Tacaná as part of a volcanic complex was carried out by Macías et al.

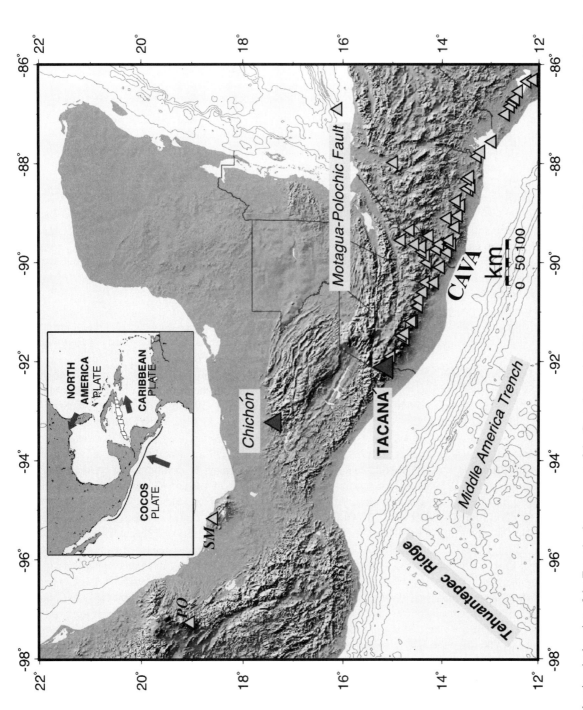

Figure 1. Image showing the location of the Tacaná volcano as part of the Central America Volcanic Arc (CAVA). Other features shown: Motagua Polochic fault zone; El Chichón volcano; SM—San Martin volcano; PO—Pico de Orizaba volcano. Inset shows the three plates that join at the studied area.

(2000). They concluded that Tacaná consisted of three edifices, from oldest to youngest, the Chichuj volcano (3800 m above sea level [asl]), the main summit Tacaná (4060 m asl), and San Antonio Volcano (3700 m asl). They proposed that activity of the Tacaná Volcanic Complex has migrated from the northeast (Chichuj) to the southwest (San Antonio), inside a 9-km-wide caldera hereafter called San Rafael. They studied in detail a Peléan-type eruption originated at San Antonio volcano some 1950 yr B.P. that emplaced the Mixcun flow deposit and associated lahars. These events probably caused the temporary abandonment of Izapa, the main Mayan ceremonial center in the Soconusco region (Macías et al., 2000). Based on these results, these authors proposed a hazard zonation for pyroclastic flows produced by Peléan-type eruptions at Tacaná, associated lahars, and debris avalanche deposits. Mora (2001) and Mora et al. (2004) carried out a petrologic and chemical study of the Tacaná Volcanic Complex for the past 40,000 yr and concluded that the magmas feeding the complex are typical of orogenic zones.

Geological mapping, determination of structural features, and volcanic stratigraphy have been integrated in this work with the purpose of determining the geological evolution of the Tacaná Volcanic Complex and surrounding areas. In particular, we aim to establish the relationship of the Tacaná Volcanic Complex to the tectonic framework, to present the geological units in the area supported by radiometric dating, and finally to synthesize the volcanic evolution of the area.

## TECTONIC SETTING OF SOUTHERN MÉXICO

Southeastern México and northwestern Guatemala are characterized by the Cocos–North America–Caribbean triple point junction delineated by the Middle America Trench and the Motagua-Polochic Fault System (Guzmán-Speziale et al., 1989). The precise location of this triple junction, however, is still a matter of controversy (Guzmán-Speziale et al., 1989, and references therein). In southern México, the Cocos plate subducts toward N45°E at an average rate of 76 mm/yr (De Mets et al., 1990). This process is complicated by the subduction of the Tehuantepec ridge, an aseismic ridge at 95°W (LeFevre and McNally, 1985), where no earthquakes larger than 7.6 $M_s$ (surface wave magnitude) have occurred during the past 190 yr (Singh et al., 1981; McCann et al., 1979) and whose onset occurred ca. 8 Ma (Manea et al., 2005). The Tehuantepec ridge is a narrow linear feature with a maximum vertical relief of 2000 m from the sea bottom that separates shallower sea floor (−3900 m) to the NW from deeper sea floor (−4800 m) toward the SE, that is, the Guatemala basin (Truchan and Larson, 1973; Couch and Woodcock, 1981; LeFevre and McNally, 1985) (Fig. 1). The Tehuantepec ridge also separates oceanic crusts of different ages at the subduction zone; to the NW, the crust has an average age of 12 Ma (Nixon, 1982), dipping 25° (Pardo and Suárez, 1995; Rebollar et al., 1999); whereas, to the SE, the crust has an age of 28 Ma (Nixon, 1982) and dips 40° (Rebollar et al., 1999). However, as summarized by Manea et al. (2005), other authors have estimated a difference in age between these the northwestern and southeastern parts of ca. 12 Ma (Klitgord and Mammerickx, 1982), ca. 8 Ma (Wilson, 1996), and ca. 10 Ma (Couch and Woodcock, 1981). The thickness of the crust at the Tehuantepec ridge is 28.5 ± 3.5 km (Bravo et al., 2004) and dips ~38° according to Ponce et al. (1992) or 30–35° according to Rebollar et al. (1999). Under Chiapas, the thickness of the continental crust is 39 ± 4 km (Rebollar et al., 1999), and thus the Tacaná Volcanic Complex is located ~240 km from Middle America Trench, and the projected Cocos slab underneath would be at a depth of ca. 100 km. Despite the different dipping angles of the subducted Cocos plate at the NW and SE parts of the Tehuantepec ridge, the plate does not break into two segments but rather dips smoothly toward Chiapas and Central America.

The subduction of the Tehuantepec ridge together with the location of the Cocos–North America–Caribbean triple point junction correlate with the apparent truncation of volcanism at the Central American Volcanic Arc in southern Chiapas, the inland migration of scattered volcanism at the Chiapanecan Volcanic Arc (Damon and Montesinos, 1978), and two sites of alkaline volcanism: El Chichón in northern Chiapas (Thorpe, 1977; Luhr et al., 1984; García-Palomo et al., 2004) and Los Tuxtlas Volcanic Field in Veracruz (Nelson et al., 1995).

## Geomorphology

The Tacaná Volcanic Complex consists of four aligned volcanic structures that are described here as Chichuj, Tacaná, Plan de Las Ardillas, and San Antonio (Figs. 2 and 3). The difference in elevation of the Tacaná Volcanic Complex with respect to the surrounding terrain is ~3000 m in the SW part whereas in the NE part (Guatemala) is ~1300 m. These major differences in elevation as well as the complex's asymmetric shape are controlled by tilting of the basement rocks and the array of caldera structures (Fig. 3). These calderas are parallel to the coast as the Cenozoic calderas described in western Guatemala by Newhall (1987), Rose et al. (1987), and Duffield et al. (1993).

Chichuj, the oldest center, consists of a collapsed volcanic structure at a height of 3800 m. Tacaná volcano (4060 m) has a 600-m-wide summit crater breached by a horseshoe-shaped escarpment opened to the NW that contains an andesitic central dome. The escarpment was produced by a flank collapse. The Plan de Las Ardillas dome has an asymmetric shape, and the San Antonio dome has a horse shoe–shaped crater open to the south. The Tacaná Volcanic Complex was built within the remains of a 9-km semicircular structure called San Rafael caldera that borders the northern flank of the volcano, whereas the southern part is characterized by wedges composed of granitic and old volcanic rocks and is bounded by faults. These wedges are named hereafter Desenlace and Agua Caliente–El Aguila. The first is formed by granitic rocks of Tertiary age covered by Pliocene volcanic rocks. The second, located in the western portion of the volcano, is made of Tertiary granitic and Pliocene volcanic rocks tilted toward the west. This wedge is formed by the intersection of NW and NE fault systems and is cut by N-S faults.

Figure 2. View from the south of Chichuj (Ch), Tacaná (T), Plan de Las Ardillas (PA), and San Antonio (SA) edifices forming the Tacaná Volcanic Complex. In the foreground is the town of Santo Domingo located on the southeastern slopes of the volcanic complex.

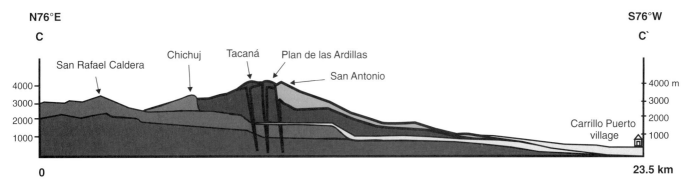

Figure 3. Cross section C–C' shows the geomorphologic features of the Tacaná Volcanic Complex, such as the major difference in elevation, asymmetric shape, and its control by the tilting of the basement rocks and the calderic structures. See location of cross section in Figure 4B.

The southern portion of the Tacaná Volcanic Complex is characterized by large and coalescing pyroclastic and debris fans that reach the coast of the Pacific Ocean, whereas to the north, the pyroclastic fans pinch out against the walls of the rim of the San Rafael caldera.

## Stratigraphy

Based on photogeology and satellite image analysis, field mapping, and radiometric dating, six main stratigraphic units in the Tacaná Volcanic Complex area were recognized (Fig. 4A). From oldest to youngest, they consist of metamorphic rocks intruded by two phases of igneous rocks, the first phase during the late Eocene to early Oligocene and the second phase during the early to middle Miocene. Unconformably, on top of these older units, rests the San Rafael–Chanjale Sequence, constituted by three Pliocene to Pleistocene calderas. On top of these rocks sits the Tacaná Volcanic Complex, which is divided in four sequences ranging in age from Pleistocene to Recent. This stratigraphic sequence is described in detail in the following.

### Mesozoic

Outcrops of this age are randomly exposed in the northwestern portion of the area near the junction of the San Rafael and Coatán Rivers (Fig. 4B). The best exposures of these rocks appear at sites TAC0308a, -0358, -0359, and -0367. These

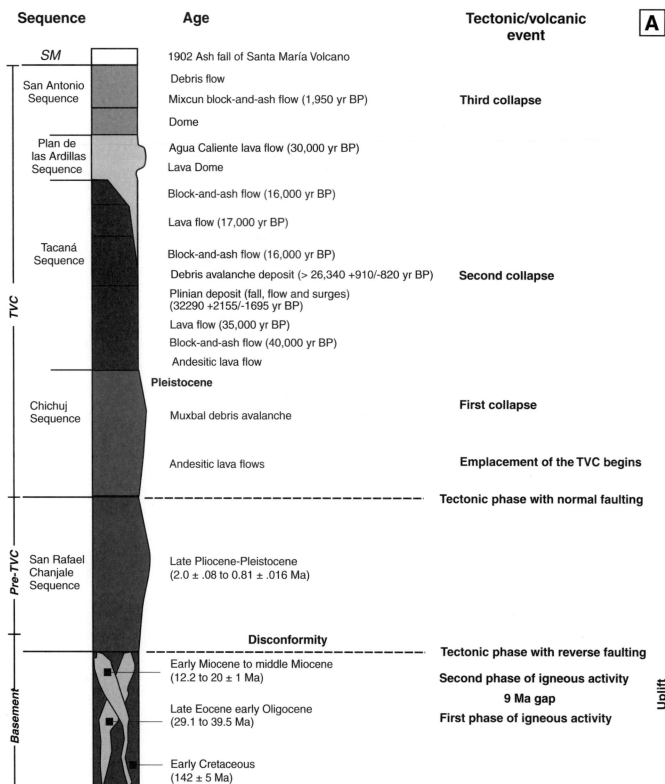

Figure 4 (*on this and following page*). (A) Simplified stratigraphic column of the Tacaná area that displays basement, Pre-Tacaná Volcanic Complex (TVC), and the TVC rocks. (B) Geological map of the TVC displayed upon a digital elevation model of the area. The map shows main units separated on the basis of stratigraphic correlation and radiometric dating as shown in the stratigraphic column. B–B′ and C–C′ represent cross sections shown in Figures 3 and 8, respectively. Asterisks indicate the location of the K-Ar or $^{40}$Ar-$^{39}$Ar dates mentioned in the text. Dots and numbers are the sites of stratigraphic sections.

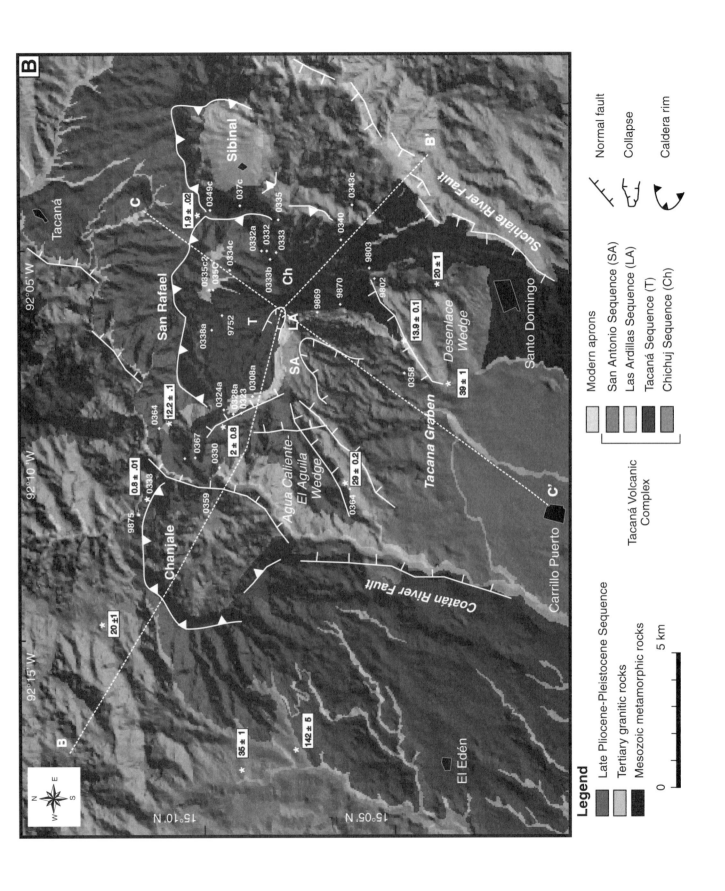

Figure 5. Photographs of the metamorphic basement along the San Rafael River at site TAC0367. Here, the rocks exhibit banding of minerals (A) and crenulation cleavage (B).

outcrops consist of alternating schists and gneisses, which at site TAC0358 have a minimum thickness of 20 m. The schists are light to dark green, forming centimeter-thick layers. The gneisses are composed of alternating green and white centimeter-thick bands (Fig. 5, site TAC0367) with a schistosity and foliation trending N60°W–70°NE. At site TAC0308a, a slightly metamorphosed lava flow, deeply altered and faulted, crops out. A K-Ar determination carried out by Mugica (1987) in similar rocks of the area yielded an age of 142 ± 5 Ma (early Cretaceous) (Table 1).

### Tertiary

These rocks were first described as part of the Coastal Batholith of Chiapas by Mugica (1987). This batholith is 270 km long and 30 km wide and covers an area of 8000 km$^2$. According to Mugica (1987), at outcrop scale, the rocks are light-gray to pink and are petrographically granodiorites composed of Na-plagioclase + quartz + microcline + biotite + hornblende + disseminated oxides. Mugica concluded that these rocks were affected by dynamic metamorphism along major faults. This unit has a late Oligocene–middle Miocene age of 15–29 Ma obtained from biotite concentrates (Mugica, 1987).

*Late Eocene–early Oligocene.* Mugica (1987) described a biotite-hornblende granodiorite exposed southwest of the Santa Rosa village (35 ± 1 Ma), and a hornblende-biotite gneissic diorite exposed in the vicinity of the village of 11 de Abril (39 ± 1 Ma) (Fig. 4B; Table 1). These dates, obtained with the K-Ar method from biotite concentrates, correspond to a late Eocene–early Oligocene event.

Additional granitic rocks are exposed northwest of the El Aguila village at site TAC0364c. Here, the rock is a white granite with equigranular texture (3 mm average length) composed of quartz, plagioclase (up to 0.8 cm), and biotite (up to 1.7 cm). The granite hosts light gray enclaves up to 46 cm in diameter. These are subrounded to elongated, with maximum diameters of ~30 cm. Their boundaries with the host rock are sharp or as a dark-gray aureole. Biotite separates of this rock were dated with the $^{40}$Ar-$^{39}$Ar method, yielding an average age of 29.4 ± 0.2 Ma. This age correlates with a K-Ar date of 29 ± 1 Ma obtained in a gneissic tonalite cropping out along the Huixtla-Motozintla road (Mugica, 1987).

*Middle-late Miocene.* The rocks collected south of the Tacaná Volcanic Complex at Monte Perla (15°03′39″N; 92°05′35″W), a biotite-hornblende gneissic tonalite, and at Finca Zajul, Tapachula (15°13′31″N; 92°15′16″W), a biotite-hornblende granodiorite, yielded ages of 20 ± 1 Ma in biotite (Table 1), respectively (Mugica, 1987).

In this work, we report a new granodioritic stock exposed in the northwestern part of the area in the vicinity of the San Rafael River (site TAC0364). The granodiorite consists of centimeter-sized K-feldspars, plagioclase, biotite, and minor quartz. $^{40}$Ar-$^{39}$Ar analysis of biotite grains yielded ages of 13.3 ± 0.2 and 12.2 ± 0.1 Ma (Table 2). Another small intrusive body is exposed south of San Antonio volcano at site TAC0359c. Here, a granitic rock is intruded by veins rich in centimeter-size biotite; this mineral was dated with the $^{40}$Ar-$^{39}$Ar method at 13.9 ± 0.1 Ma (Table 2). These rocks can be correlated with the youngest part of the Coastal Chiapas Batholith (Mugica, 1987).

*Late Pliocene–Pleistocene.* This volcanic sequence consists of three caldera structures named here San Rafael, Chanjale, and Sibinal, located in the northern portion of the studied area (Fig. 4B).

TABLE 1. SUMMARY OF K-Ar DATES OF TERTIARY INTRUSIVE ROCKS OF THE TACANÁ AREA ACCORDING TO PREVIOUS WORKS

| Sample | Material dated | $^{40*}$Ar (ppm) | $^{40}$K (ppm) | Age (Ma) | Location | References |
|---|---|---|---|---|---|---|
| 2M-26-79 | Bi | 0.00106 | 7.85776 | 20 ± 1 | N15°03'39" W91°05'35" | Mugica, 1987 |
| 2M-28-79 | Bi | 0.001144 | 8.443 | 20 ± 1 | N15°13'32" W92°15'16" | Mugica, 1987 |
| 2M-23-79 | Bi | 0.001603 | 6.6859 | 29 ± 1 | N15°17'33" W92°21'59" | Mugica, 1987 |
| GAP-589 | Bi | 0.001994 | 8.0608 | 35 ± 1 | n.a. | Mugica, 1980 |
| GAP-586 | Bi | 0.002208 | 5.6196 | 39 ± 1 | n.a. | Mugica, 1980 |
| GAP-582 | Hb | 0.0029 | 0.3567 | 142 ± 5 | n.a. | Mugica, 1980 |

Note: $^{40*}$Ar—radiogenic $^{40}$Ar; n.a.—not available.

TABLE 2. SUMMARY OF $^{40}$Ar/$^{39}$Ar ANALYSES OF ROCKS FROM THE TACANÁ AREA

| Sample | Unit | Min. | Integrated Age | % $^{40}$Ar* | Interpreted Age | Age Information |
|---|---|---|---|---|---|---|
| TAC0324a | Plan de las Ardillas | WR#1 | 5 ± 19 ka | 1.1 | 21 ± 19 ka | 3 fraction plateau age: 79% $^{39}$Ar released MSWD = 1.6 |
| | | WR#2 | 22 ± 16 ka | 3.9 | 39 ± 15 ka | 5 fraction plateau age: 90% $^{39}$Ar released MSWD = 1.8 |
| Average | | | | | 32 ± 12 ka | |
| TAC0333 | Chanjale Caldera | WR#2 | 734 ± 38 ka | 46.6 | 811 ± 30 ka | 1 fraction, 56% $^{39}$Ar released |
| | | WR#3 | 769 ± 23 ka | 50.5 | 816 ± 19 ka | 5 fraction plateau age: 94% $^{39}$Ar released MSWD = 0.1 |
| Average | | | | | 815 ± 16 ka | |
| TAC0349c | San Rafael Caldera | WR#1 | 1744 ± 29 ka | 61.7 | 1834 ± 26 ka | 1 fraction 71% $^{39}$Ar released |
| | | WR#2 | 1729 ± 17 ka | 66.4 | 1887 ± 16 ka | 1 fraction 57% $^{39}$Ar released |
| Average | | | | | 1872 ± 24 ka | |
| TAC0323a | San Rafael Caldera | WR#1 | 11.7 ± 5.1 Ma | 5.0 | 1968 ± 147 | 1 precise fraction, 20% $^{39}$Ar released |
| | | WR#2 | 5.8 ± 1.4 Ma | 3.3 | 2007 ± 98 | 8 fraction plateau age: 75% $^{39}$Ar released MSWD = 2.4 |
| Average | | | | | 1995 ± 82 ka | |
| TAC0364 | Tertiary Intrusives | Bi | 13.3 ± 0.2 Ma | 89.1 | 13.3 ± 0.2 Ma | 4 fraction plateau age: 88% $^{39}$Ar released MSWD = 0.4 |
| | | Pl | 11.8 ± 0.1 Ma | 63.4 | 12.2 ± 0.1 Ma | 5 fraction plateau age: 74% $^{39}$Ar released MSWD = 0.9 |
| TAC0359C | Tertiary Intrusives | Bi | 13.7 ± 0.1 Ma | 84.5 | 13.9 ± 0.1 Ma | 9 fraction plateau age: 86% $^{39}$Ar released MSWD = 1.4 |
| TAC0364cgr | Tertiary Intrusives | Bi | 29.1 ± 0.2 Ma | 86.9 | 29.4 ± 0.2 Ma | 10 fraction plateau age: 87% $^{39}$Ar released MSWD = 1.4 |

Note: Whole rock (WR), biotite (Bi), or plagioclase (Pl) samples were laser step-heated at the University of Alaska, Fairbanks Geochronology Laboratory. The standard TCR-2 sanidine with an age of 27.87 Ma was used to calculate the irradiation parameter, except for Tac0359C and Tac0364cgr, where the standard MMhb-1 hornblende with an age of 513.9 Ma was used (Lanphere and Dalrymple, 2000). Calculated ages were corrected for reactor interferences, system blanks, and mass discrimination. Isochrons were used to verify that the initial $^{40}$Ar/$^{36}$Ar composition was, within error, atmospheric (295.5). Interpreted age is either a plateau (if 3 or more fractions), weighted average of 2 best fractions, or best single step depending on heating schedule used. Plateau criteria: 3+ consecutive fractions, mean square of weighted deviates (MSWD) < ~2.5, more than 50% $^{39}$Ar release. Italics: Average age for each sample from multiple runs (errors reported at ± 1 sigma).

## San Rafael Caldera (ca. 2 Ma)

This caldera has a discontinuous structure, with its northern and eastern walls well exposed. By projecting these wall remnants, a 9 km diameter is estimated for this structure. The San Rafael Caldera walls are constituted mainly of a green ignimbrite and capping lava flows.

The basal unit is represented by a 200-m-thick ignimbrite exposed in the caldera rim on top of the Tertiary granites; the ignimbrite starts to crops out at an elevation of 1600 m (site TAC0328a). It appears as a lithified green ignimbrite composed of angular to subrounded lithics, mainly dark gray andesites and rounded pumice fragments embedded in a fine-grained matrix (coarse-ash). Toward the top, this unit becomes enriched in pumice and scoria fragments embedded in a fine ash matrix. This unit crops out at an elevation of 1800 m (TAC0328a) and has juvenile fragments with a composition of 53.81 wt% in silica (Table 3; Fig. 6). A thick sequence of >20-m-thick lava flows covers the

TABLE 3. WHOLE-ROCK CHEMICAL COMPOSITION OF SOME ROCKS OF THE TACANÁ VOLCANIC COMPLEX

| Unit | Caldera San Rafael | | | | Caldera Chanjale | Las Ardillas | |
|---|---|---|---|---|---|---|---|
| Sample | TAC0349C | TAC0323A | TAC0332A | TAC0328A | TAC0333 | TAC0324A | TAC9863 |
| wt% | | | | | | | |
| $SiO_2$ | 57.94 | 54.66 | 59.78 | 53.81 | 58.39 | 59.58 | 62.96 |
| $TiO_2$ | 0.82 | 0.80 | 0.65 | 0.88 | 0.68 | 0.67 | 0.58 |
| $Al_2O_3$ | 17.70 | 17.53 | 17.51 | 17.65 | 17.88 | 17.36 | 16.76 |
| $Fe_2O_3$ | 7.22 | 7.86 | 6.60 | 8.44 | 7.03 | 6.33 | 1.82 |
| FeO | nm | nm | nm | nm | nm | nm | 3 |
| MnO | 0.11 | 0.17 | 0.13 | 0.13 | 0.15 | 0.11 | 0.11 |
| MgO | 2.64 | 3.00 | 2.56 | 2.66 | 2.54 | 2.55 | 2.42 |
| CaO | 6.50 | 7.30 | 6.11 | 9.32 | 7.44 | 5.95 | 5.74 |
| $Na_2O$ | 3.74 | 3.74 | 3.39 | 3.19 | 3.89 | 3.67 | 3.82 |
| $K_2O$ | 2.16 | 1.75 | 2.15 | 1.46 | 1.56 | 2.02 | 2.06 |
| $P_2O_5$ | 0.25 | 0.27 | 0.17 | 0.25 | 0.24 | 0.16 | 0.13 |
| LOI | 0.65 | 1.98 | 1.02 | 2.34 | 0.38 | 0.44 | 0.60 |
| Total | 100 | 99 | 100 | 100 | 100 | 99 | 100 |
| Trace elements (ppm) | | | | | | | |
| Ba | 788 | 788 | 751 | 807 | 821 | 714 | 782 |
| Rb | 70 | 39 | 59 | 26 | 32 | 47 | 89 |
| Sr | 726 | 842 | 515 | 883 | 714 | 507 | 673 |
| Cs | 1.6 | 1.6 | 2.0 | 15.7 | 1.5 | 1.7 | 2.9 |
| Hf | 4.2 | uld | 3.6 | uld | 3.4 | uld | 3.8 |
| Zr | 152 | 155 | 126 | 137 | 135 | 136 | uld |
| Y | 15 | 17 | 15 | 19 | 20 | 17 | 20 |
| Th | 5.6 | 3.9 | 4.0 | 3.6 | 3.2 | 2.7 | 4.7 |
| U | 2.2 | 1.4 | 1.5 | 1.3 | 1.4 | 1.3 | 2.8 |
| Cr | 13.1 | 22.8 | 4.4 | 20.0 | 9.7 | 9.5 | 15.9 |
| Ni | 10 | uld | 7 | uld | uld | uld | uld |
| Co | 16 | 19 | 15 | 23 | 14 | 14 | 13 |
| Sc | 11 | 12 | 13 | 14 | 12 | 12 | 12 |
| V | 139 | 141 | 114 | 170 | 116 | 156 | uld |
| Tb | 0.6 | 0.6 | 0.6 | 0.6 | uld | 0.6 | 0.4 |
| Cu | 14 | 14 | 9 | 20 | 9 | 17 | uld |
| Pb | 5 | uld | 7 | 5 | uld | 4 | uld |
| Zn | 89 | 90 | 76 | 94 | 89 | 77 | 42 |
| Nb | uld | uld | uld | uld | uld | uld | uld |
| Ta | 0.6 | uld | 0.7 | uld | uld | uld | 1.2 |
| La | 22.1 | 20.6 | 17.9 | 19.1 | 17.8 | 15.9 | 20.5 |
| Ce | 40 | 35 | 32 | 33 | 32 | 27 | 39 |
| Nd | 22 | 18 | 17 | 19 | 17 | 13 | 46 |
| Sm | 4.22 | 4.09 | 3.48 | 4.31 | 4.08 | 3.12 | 4.20 |
| Eu | 1.28 | 1.39 | 0.93 | 1.52 | 1.40 | 1.08 | 0.64 |
| Yb | 1.35 | 1.47 | 1.51 | 1.71 | 1.97 | 1.47 | 1.90 |
| Lu | 0.20 | 0.22 | 0.23 | 0.26 | 0.29 | 0.23 | 0.17 |

Note: Major and trace elements analyzed by inductively coupled plasma–mass spectrometry and instrumental neutron activation analysis (<0.01% major elements; Ba—50 ppm; Cs, Co, Cr, Sc, Th, U, La, Tb—0.1 ppm; Pb, Nb, V, Rb—2 ppm; Ce, Nd, Cu, Zn, Ni, Sr, Y, Zr—1 ppm; Hf and Ta—0.2 ppm; Sm, Eu, Yb, and Lu—0.05 ppm; detection limits) at Activation Laboratories, Ancaster, Canada. $Fe_2O_3$* is reported as total iron; nm—not measured; uld—under limit of detection; LOI—loss on ignition.

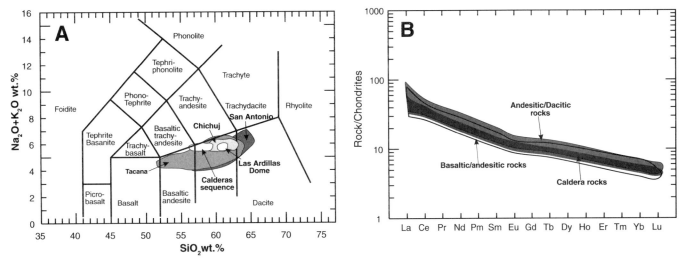

Figure 6. (A) Alkali versus silica classification diagram of all studied rocks (modified after Le Bas et al., 1986). (B) Rare earth element diagram normalized to the chondrite abundances (Nakamura, 1974).

green ignimbrite along the northern rim of the San Rafael caldera. At site TAC0349c, a dark gray lava flow was dated with the $^{40}$Ar-$^{39}$Ar method at 1.87 ± 0.02 Ma (Table 2). This lava has a composition of 57.94 wt% silica and therefore is an andesite (Table 3). We speculate that the green ignimbrite is associated with the caldera formation, while the lava flows represent localized latter emissions at the caldera ring structure.

Inside the Agua Caliente–El Aguila wedge, southwest of the town of Agua Caliente, at site TAC0323a, outcrop several dark gray lava flows that have a porphyritic texture with plagioclase, pyroxene, and rare olivine. Here, the lava flow units are 1 m thick and have fractures and rounded vesicles. An $^{40}$Ar-$^{39}$Ar date obtained from a lava flow yielded an age of ca. 2 Ma (Table 2). This age is similar to those of the lava flows on the northern rim of the caldera. According to this radiometric data, the San Rafael caldera was active from the late Pliocene to the Pleistocene.

At the same TAC0323a site, the lava flows are covered by a massive, matrix supported block-and-ash flow deposit, composed of dark gray andesites and minor red altered andesites set in a medium to coarse ash matrix. The blocks of andesite are subangular to subrounded and contain plagioclase phenocrysts. The dark gray andesites have a silica content of 54.66 wt% (Table 3).

Inside the northern rim of the caldera, near the village of La Vega del Volcán (sites TAC0335c and TAC0337c), a thick debris avalanche deposit is exposed. It consists of megablocks with jigsaw-puzzle structures up to 4 m in diameter. The blocks are banded gray to red porphyritic andesites with hornblende, pyroxene, and plagioclase. The debris avalanche deposit is ponded inside the caldera walls and unconformably covers the granitic rocks (TAC0338c). Its stratigraphic position relative to other units suggests that it is much younger than the green ignimbrite and the lava flows.

### Chanjale Caldera (ca. 1 Ma)

The Chanjale caldera dominates the western portion of the study area. It is a 6.5-km-wide crater opened to the east and cut by the Coatán River (Fig. 4B). The caldera rim consists of several units of lavas, pyroclastic and debris flows (Fig. 7). A gray porphyritic lava flow rich in plagioclase and pyroxene (TAC0333) is exposed within the flanks of the structure near the village of Malacate (Fig. 7). This lava flow has silica compositions of 58.39 wt% and yielded an $^{40}$Ar-$^{39}$Ar date of 0.81 ± 0.02 Ma (Table 2). At site TAC9875, this lava flow has a silica composition of 57.59 wt% (Mora et al., 2004). The total thickness of the unit is ~200 m. A white ignimbrite deeply altered to brown is exposed around the lava flows. It consists of plagioclase, quartz, and altered ferromagnesians embedded in a white ash matrix. On the southern flank of the caldera, in the vicinity of Chespal (TAC0335), a white to light yellow pyroclastic flow is exposed with dense lithics up to 1 m in diameter, angular to subangular, and deeply altered, embedded in a fine ash matrix with millimeter-size pumice fragments. Additionally, xenoliths of biotite-bearing granites were found in this deposit. This pyroclastic flow deposit is related to the caldera stratigraphy; however, with the available data, it is difficult to establish if they are precisely related to caldera formation.

On top of the sequence, several indurated debris flow deposits up to 12 m thick appear, forming a fan toward the southern flank of the caldera. These deposits are heterolithologic, with boulders up to 2 m in diameter embedded in a coarse-sand matrix (TAC0332).

### Sibinal Caldera

The Sibinal caldera is exposed in the northeastern portion of the area with the town of Sibinal, Guatemala, located in its center

Figure 7. View of the vertical walls of the Chanjale Caldera lava flow at site TAC0333 along the Chespal-Pavincul road. Notice columnar jointing of the rocks.

(Fig. 4B). The caldera wall consists of several dark gray, up to 4-m-thick, andesitic lava flows with porphyritic texture, with a total thickness of 60 m. In hand specimens, plagioclase and some ferromagnesians are common (site TAC0340). A lava flow is located on top of the Sibinal caldera, overlying the Tertiary granite. On the northern slopes of the caldera (TAC0341), 2-m-thick lava flows are exposed; they consist of porphyritic gray andesites with plagioclase and altered ferromagnesians. The lava flows encircled the Chamalecón couleé, which is composed of pink to reddish andesites. The Chamalecón rocks contain tabular and subeuhedral phenocrysts of plagioclase and pyroxene embedded in a fine red matrix composed of glass and plagioclase microlites. They are highly altered, spheroidal weathered, and are older than the lavas of Sibinal caldera. The youngest unit of this caldera is a thick volcaniclastic deposit that forms an apron of debris flow deposits interbedded with at least three fall deposits of the Tacaná volcano. One of these fall deposits has been named as the Sibinal Pumice fall and was dated at ca. 32,000 yr B.P. (Arce et al., 2004). According to this stratigraphic relationship, the Sibinal caldera should be younger than the San Rafael and Chanjale calderas. However, the stratigraphy of the Sibinal caldera does not show any evidence of ignimbrite emplacement; therefore, we suggest that its semicircular structure might be related to a series of gravitational collapses.

**Pleistocene-Holocene**

The Tacaná Volcanic Complex is located in the central portion of the area and consists of the Chichuj, Tacaná, Plan de las Ardillas, and San Antonio sequences.

*Chichuj Sequence*

Chichuj volcano has a collapse structure facing west, where it is sealed by Tacaná volcano. Most of the geology of Chichuj is exposed in its eastern flank, where six units were recognized. A unit of gray andesite lava flows crops out near the village of Chocabj (TAC0342). This unit is composed of plagioclase, dark green hornblende, and pyroxene set in a fine gray glassy matrix. These lavas host dark gray to reddish xenoliths that have plagioclase, mafic minerals, and vesicles. The xenoliths are rounded and show reaction rims. The lava flow sequence is 10 m thick, with individual units of ~1 m. This lava covers the granitic rocks at site TAC0336a.

At site TAC9876, a pink debris avalanche deposit at the base of the Muxbal gully is exposed (Espíndola et al., 1993). This deposit is massive, with yellow to orange hydrothermally altered zones and meter-sized jigsaw blocks in a shattered matrix of coarse-grained ash. The Muxbal debris avalanche is also exposed in the eastern flanks of Chocabj. Here, the debris avalanche is covered by a lacustrine sequence and a 28,540 ± 260 yr B.P. block-and-ash flow deposit from Tacaná volcano (Mora et al., 2004) (Table 4). At TAC0334c, a massive block-and-ash flow deposit consists of dark gray andesite and altered scoria clasts embedded in a medium ash matrix. The andesite clasts contain plagioclase + hornblende + clinopyroxene and orthopyroxene ± biotite. The deposit is up to 25 m thick and covers the lava flow unit exposed at site TAC0336a. A light gray porphyritic lava flow (TAC0332a) covers the block-and-ash flow deposit; this lava flow consists of plagioclase, hornblende, and pyroxene embedded in a fine, compact matrix. Atop the Chichuj Sequence, there is a unit composed of pink to dark gray breccias (~2.5 m thick) and laminated lavas with flow banding (50 cm thick) grading to massive lavas (~1 m thick). These rocks consist of plagioclase + clinopyroxene and orthopyroxene + hornblende set in a matrix with rounded, light gray enclaves up to 4 cm in diameter. The phenocrysts display common resorbed rims.

*Tacaná Sequence*

Ten volcanic units can be recognized within the Tacaná Volcano Sequence, most of them ranging in age from late Pleistocene to Holocene. The sequence consists of lava flows and debris avalanche, block-and-ash flow, surge, fall, and lahar deposits. The oldest units of Tacaná are andesitic lava flows exposed only at the base of the gullies surrounding the volcano

TABLE 4. SUMMARY OF RADIOCARBON DETERMINATIONS CARRIED OUT AT ANALYSES OF CHARCOAL SAMPLES

| Sample | Lab. No. | 14C age yr B.P. | δ13PDB (%) | Material dated | Location | References |
|---|---|---|---|---|---|---|
| TAC0343C | A-10442 | 6910 ± 95 | −25.9 | Charcoal | N15°06′10″ W92°04′50″ | This work |
| TAC9752 | n.a. | 16,350 ± 50 | −25.0 | Charcoal | N15°09′29″ W92°06′97″ | Mora et al., 2004 |
| TAC0330a | A-12890 | 26,340 +910/−820 | −26.0 | Wood | N15°09′35″ W92°10′38″ | Macías et al., 2004 |
| TAC9714* | n.a. | 28,540 ± 260 | n.a. | Charcoal | N15°05′03″ W92°04′35″ | Mora et al., 2004 |
| TAC9332 | 6923 | >30,845 | −25.0 | Charcoal | N15°02′34″ W92°05′11″ | Espíndola et al., 1993 |
| TAC0335-C2C | A-13365 | 32,290 +2155/−1695 | −20.8 | Charcoal | N15°09′43″ W92°05′68″ | This work |
| Trinidad | n.a. | 42,000 | n.a. | Charcoal | N15°02′12″ W92°06′51″ | Espíndola et al., 1989 |
| La Trinidad | A-6924 | 38,630 +5100/−3100 | −25.0 | Charcoal | N15°02′12″ W92°06′51″ | Espíndola et al., 1993 |

Note: PDB—Peedee belemnite; n.a.—not available.
*AMS determinations.

(TAC0338a). These lava flows are overlain by block-and-ash flow deposits widely distributed around the volcano, forming two main fans.

The oldest block-and-ash flow deposit is exposed on the southern flank of Tacaná between the villages of Talquian and Santo Domingo. The deposits are massive and consist of light gray dense andesitic blocks and minor red andesites embedded in a coarse ash matrix. The andesitic blocks consist of plagioclase, pyroxene, and dark-brown glass. The deposits contain charred logs that were dated at ca. 42,000 yr B.P. (Espíndola et al., 1989) and 38,630 +5100/−3100 yr B.P. (Espíndola et al., 1993) (Table 4). Another light gray block-and-ash flow deposit overlies the latter deposit in the vicinity of the Cordoban village and over the Muxbal debris avalanche at the Muxbal coffee plantation. At TAC9714, the block-and ash flow consists of glassy juvenile andesites and dense gray andesites embedded in a fine coarse ash matrix. Both types of clasts contain plagioclase, pyroxene, and amphibole. The deposit contains disseminated charcoal that yielded an age of 28,540 ± 260 yr B.P. (Mora et al., 2004) (Table 4).

The northern flank of the volcano is dominated by a block-and ash flow deposit that pinches out against the rim of San Rafael caldera. This block-and ash flow consists of at least four massive gray units, each one composed of andesitic blocks supported by a fine ash matrix. At site TAC9752, near the village of San Rafael, a charcoal sample yielded an age of 16,350 ± 50 yr B.P. (Table 4) (Mora et al., 2004).

The E-NE slopes of the Tacaná Volcanic Complex exhibit complex sequences. The deposits consist of three pumice-rich fall horizons interbedded with pyroclastic flow deposits containing minor pumice and dense clasts set in an ash matrix and several laminated, cross-stratified surge deposits. The juvenile fragments of the oldest pumice-rich fall and flow deposits (Arce et al., 2004) are andesites similar to the most recent products of the volcano, and the mineral assemblage is represented by plagioclase + clinopyroxene + orthopyroxene + hornblende and Fe-Ti oxides, set in a glassy groundmass. A piece of charcoal found at the base of the deposit (TAC0335c) yielded an age of 32,290 +2155/−1695 yr B.P. (Table 4).

The Agua Caliente debris avalanche deposit crops out on the northwestern part of Tacaná (Macías et al., 2004). The deposit is confined by the San Rafael and Tochab rivers up to their junction with the Coatán River. It is 8 km long, covers an area of 6 km$^2$, and has a volume of ~1 km$^3$. It exhibits a block facies with a scarce light brown ash matrix. The deposit has a maximum thickness of 200 m at the Coatán River. The Agua Caliente debris avalanche is covered by block-and-ash flows and debris flow deposits. A wood fragment in the debris flow deposits yielded an age of 26,340 +910/−820 (Table 4).

Within the complex Holocene sequence of Tacaná volcano, there is a clast-supported yellow pyroclastic flow deposit (TAC0343C) rich in pumice. The deposit is 6 m thick and is covered by another 6 m of reworked material. The pumice clasts are white, with pyroxene, amphibole, and plagioclase supported by a glassy matrix. A charcoal sample collected in this unit yielded and age of 6910 ± 95 yr B.P.

### Plan de Las Ardillas Sequence

This sequence is exposed between the San Antonio and Tacaná volcanoes. It consists of a central dome with two lava flows running along the NW and SE flanks of San Antonio and Tacaná volcanoes. The dome consists of porphyritic gray andesitic lavas made up of plagioclase and amphibole and abundant dark gray enclaves set in a glassy matrix (TAC9863; 62.96 wt% SiO$_2$). The two lava flows are dark gray andesites with porphyritic textures of plagioclase and pyroxene phenocrysts in a fine-grained matrix. The lava flows are andesites with a silica content of 59.58 wt% (TAC0324a). At site TAC0324a, the last lava flow developed steep flow fronts and levees with breccias at their base made of meter-size blocks. $^{40}$Ar-$^{39}$Ar analysis of this sample yielded an age of 30 ± 9 ka (Table 2). From field relations, it is clear that the Plan de Las Ardillas domes are younger than the Agua Caliente debris avalanche (<26,000 yr B.P.; Table 4).

### San Antonio Sequence

The San Antonio Sequence is located southwest of the Plan de Las Ardillas dome. Near Santa María La Vega village (site TAC0358c), this sequence is composed of several basaltic-andesitic lava flows. These rocks have a fine-grained matrix with amphibole and plagioclase phenocrysts. Two sites near the village of Talquian (TAC9802 and 9803) were studied. At both localities, the lava flow consists of several units of gray porphyritic basaltic-andesites with plagioclase, hornblende, and olivine set in a medium grain groundmass. Individual flow units are 2 m thick with a total thickness of 5 m.

The youngest product of the San Antonio dome is the Mixcun pyroclastic flow deposit exposed in its southern flank. This is a block-and ash flow deposit that consists of light gray and dark gray, glassy, banded andesites and minor altered red andesites embedded in a coarse ash matrix. The deposit has fumarolic pipes and juvenile lithics with cooling joints. Disseminated charcoal found in this deposit yielded an age of 1950 yr B.P. (Macías et al., 2000). A thick sequence of debris flow, hyperconcentrated flow, and fluvial deposits is exposed at TAC98072 and TAC98073, near the towns of San Salvador Urbina and Union Roja. Each layer is up to 2 m thick and is composed of accidental cobbles up to 20 cm in diameter and pumice embedded in a coarse sand matrix.

## SUMMARY OF WHOLE-ROCK CHEMISTRY

The first study of the chemical composition of rocks from the Tacaná Volcanic Complex was carried out by Mercado and Rose (1992). These authors analyzed 16 samples of lava, ash, and lithics from different deposits of Tacaná Volcano that yielded $SiO_2$ contents between 58 and 64 wt%. Afterward, Macías et al. (2000) analyzed eight samples from the San Antonio volcano and recognized a far larger variation in the composition, from 50 to 64 wt% $SiO_2$. A detailed sampling and analysis of all structures of the Tacaná Volcanic Complex was performed by Mora (2001) and Mora et al. (2004), who analyzed 60 new samples and reported mafic intrusions in granites and mafic enclaves hosted in andesitic rocks. In this work, we present eight new analyses of rocks belonging to the caldera structures and Las Ardillas dome (Table 3).

In the total alkali versus silica (TAS) diagram (Le Bas et al., 1986), the samples plot in the basaltic-andesite (San Rafael) and andesite (Chanjale and Las Ardillas) fields (Fig. 6A). These rocks and those studied by Mora et al. (2004) belong to the subalkaline suite, are calc-alkaline (Irvine and Baragar, 1971), and have low $TiO_2$ (<1 vol%) and high $Al_2O_3$ (14–19 wt%) contents, typical of orogenic volcanic rocks. In summary, all rocks range in composition from andesites to dacites; the most basic compositions are basaltic-andesites found at the Tacaná and San Antonio volcanoes. There is no major variation in the concentration of major elements between the calderas and the Tacaná Volcanic Complex products, which record a continuous emission of magma of intermediate composition.

The pattern of rare earth elements normalized to their chondritic abundances (Nakamura, 1974) shows enrichment in light rare earth elements (LREE) and depletion in heavy rare earth elements (HREE) (Fig. 6B). The concentrations of the rare earth elements in the mafic enclaves show a similar parallel trend as the andesitic rocks, displaying a minor concentration in these elements. These patterns are typical of orogenic environments with a negative anomaly of Eu.

## Structural Geology

Burkart and Self (1985) proposed a structural geometry model of Guatemala and southeastern México (Fig. 8A). They constructed a regional cross section and recognized three volcanotectonic zones. The eastern zone (III) was characterized by horst and graben structures, including the Guatemala and Ipala grabens, with widespread associated monogenetic volcanism; the central zone (II) was characterized by an extreme thinning of the crust, with the Atitlán caldera bounded by two structural highs showing polygenetic volcanism; and the western zone (I) was dominated by volcanoes built upon basement complexes. The Tacaná Volcanic Complex is built upon a structure high in the western zone (Fig. 8B) that is characterized by a negative gravimetric anomaly (Burkart and Self, 1985). Lithologically, the high is made of deeply altered, fractured, and faulted metamorphic, granitic, and volcanic rocks. In the Tacaná Volcanic Complex region, the structural high is affected by three important fault systems; the oldest is located to the west of the complex and consists of fractures and NW-SE faults within the granitic rocks of Mesozoic and Tertiary age. The second fault system is aligned with the Tacaná Volcanic Complex and has a NE-SW trend. The youngest fault system has a N-S trend and crosscuts the other two fault systems. The NE-SW fault system is the most conspicuous structure because it delineates a graben, which is hereafter called the Tacaná graben, inside of which the Tacaná Volcanic Complex has been built (Figs. 8A and 8B).

This graben is 30 km long and 18 km wide with vertical displacements of ~600 m. It is bounded to the NW by a horst on which the Chanjale Caldera is located and to the E by a horst holding the Sibinal caldera. The Coatán and Suchiate rivers follow these major faults, respectively, and act as the main boundaries of the Tacaná graben. Structural analysis on the Coatán and Suchiate rivers (Figs. 9 and 10) indicates that the main fracture systems have a NE-SW trend, which is supported by analyses of satellite images, photogeology, and normal faults (Fig. 11).

The graben became active after the emplacement of the San Rafael (2 Ma) and Chanjale (1 Ma) calderas and before the emplacement of the Chichuj volcano, as the rocks forming this structure are not affected by the NE-SW faults. This fact indicates that the graben has controlled the birth and evolution of the Tacaná Volcanic Complex. This is supported by the alignment of the Chichuj, Tacaná, Plan de Las Ardillas, and San Antonio landforms, which have a general trend of N65°E.

According to our analysis, the region has been affected by a stress field with a NNW minimum principal stress of σ3 since late Pleistocene to the Recent that correlates with the focal mechanisms determined in the region by Guzmán-Speziale et al. (1989). The origin of the stress is related to the direction of movement of the Cocos plate beneath the Caribbean plate (De Cserna et al., 1988).

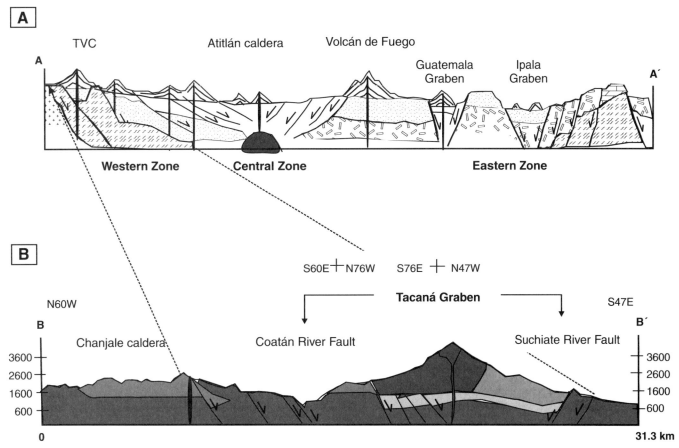

Figure 8. Structural geometry of the Central America region according to Burkart and Self (1985). (A) Regional cross section A–A' that divides Guatemala into three main volcanotectonic zones: The western zone, dominated by volcanoes built upon basement complexes; the central zone, characterized by extreme crustal thinning; and the eastern zone, characterized by horst and graben structures including the Guatemala and Ipala grabens. (B) Section B–B' displays the western zone of Burkart and Self (1985) at the Tacaná Volcanic Complex (TVC).

## DISCUSSION

### Tectonic and Volcanic Evolution

Based upon the stratigraphic relations and radiometric dating, we propose the following tectonic and volcanic evolution of the Tacaná Volcanic Complex and surrounding areas.

The oldest rocks in the region are gneisses, schists, metavolcanics, slates, and granites of Cretaceous age. The origin of these rocks remains uncertain due to the scarcity of absolute dates and other detailed studies. However, the sequence can be correlated with rocks from Central America of the same age (Meschede and Frisch, 1998).

During the Tertiary, the metamorphic rocks were intruded by two main magmatic pulses, including granites and tonalites. The first pulse occurred during the late Eocene to early Oligocene, probably related to the subduction of the Farallon plate under the North America plate (Meschede and Frisch, 1998). The second pulse occurred during the early to middle Miocene, likely associated with the subduction of the Cocos plate under the Caribbean plate. The 9 m.y. gap between these two intrusive pulses could have been related to the reorganization of the Pacific region during the fragmentation of the Farallon plate into smaller microplates (i.e., Cocos plate) (Hey, 1977; Stock and Lee, 1994).

The region was later affected by a Miocene tectonic compressive event accommodated through strike-slip and reverse faults. This tectonic phase has been recorded in other parts of southern México (Campa, 1998), at El Chichón volcano (García-Palomo et al., 2004), and at the Ixtapa Graben in central Chiapas (Meneses-Rocha, 2001). The analysis of these sites indicates a NE-SW trend for the main principal stress σ1 as attested by slickensides, sigmoidal gouge faults, noncohesive breccias, and other sigmoidal structures.

The basement rocks were uplifted and tilted after the middle Miocene and before the Pliocene as a consequence of the subduction process. The basement rocks suffered deep erosion and weathering that created coalescent debris fans widespread in the SE portion of the area.

During the Pliocene and Pleistocene, three caldera structures, San Rafael (ca. 2 Ma), Chanjale (ca. 1 Ma), and Sibinal

Figure 9. Detail of the Coatán River normal fault that bounds two different lithologies: to the left, block-and-ash flow deposits of the Sibinal Caldera; to the right: Tertiary granites. Arrows indicate sense of movement.

(unknown age), were emplaced unconformably on the tilted basement rocks that channeled volcanic products toward the southern portion of the area. This process produced widespread fans of pyroclastic and debris flow deposits. The caldera structures were affected by normal faulting in the late Pliocene to early Pleistocene, forming the NE-SW trending Tacaná graben. The Tacaná Volcanic Complex was emplaced later inside the graben.

The initial episodes of formation of the Tacaná Volcanic Complex began in the late Pleistocene with the emplacement of the Chichuj volcano. It started with the emission of andesitic lava flows that built the volcanic edifice. Intense hydrothermal alteration caused the partial collapse of the SE part of the edifice, producing the Muxbal debris avalanche. The debris avalanche was controlled by the morphology of the region; it was emplaced toward the southeast, abutting the San Rafael caldera walls. The volcano grew through the emission of lavas and minor pyroclastic activity that formed small fans with scoria and block-and-ash flow deposits. Chichuj volcano was destroyed likely by a collapsed event toward the W-SW that left the half-cone morphology we see today. However, there is no evidence of this collapse since subsequent activity buried these deposits.

Tacaná volcano was built west of the remains of Chichuj, initially through emissions of andesitic lava flows followed by Peléan, Plinian, and effusive eruptions. The original edifice of Tacaná was later partly destroyed by a sector collapse directed to the NW of the crater and perpendicular to the NNE-SSW normal faults. The debris avalanche was confined in the San Rafael River valley and was stopped by the western margin of the Coatán River valley. The event continued with a series of

Figure 10. Characteristics of the Suchiate River fault, that affects granitic rocks and forms a fault gouge and shear fracture, indicating a normal fault. Hammer for scale. Arrows indicate sense of movement.

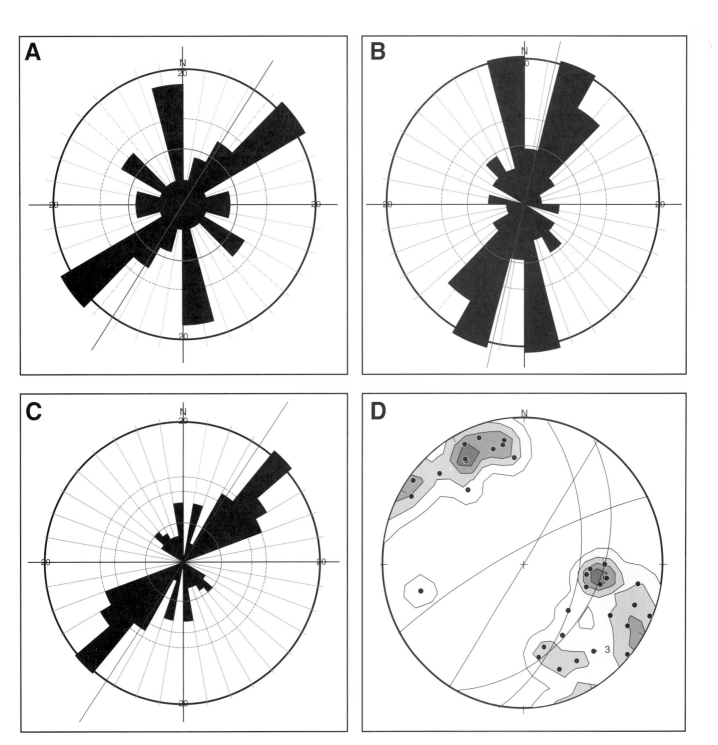

Figure 11. Analysis of faults and fractures in the Tacaná Volcanic Complex area. Rose fracture diagrams of the Suchiate (A) and Coatán (B) river faults, showing N-S and NE-SE trends, respectively. (C) Rose diagram that displays the lineaments obtained from photogeology. (D) Stereographic projection of normal faults (lower hemisphere of this Wulff diagram).

block-and-ash flows that also reached the Coatán River (Macías et al., 2004). These facts suggest that the event was due to the intrusion of a central dome that built overpressure in the volcanic edifice, which collapsed perpendicular to NNE-SSW normal faults.

The activity of the volcano continued with the emplacement of the Plan de Las Ardillas Dome (~30,000 yr), likely coeval with the collapse of Tacaná. This dome was intruded SW of Tacaná following a NW-SE trend that correlates with the NE-SE normal faults. Finally, the San Antonio Volcano was constructed in the SW tip of the NE-SW volcanic alignment, with steep slopes of andesitic lava flows. San Antonio was destroyed by a magma mixing event that produced a Peléan-type eruption with the generation of the Mixcun pyroclastic flow 1950 yr B.P. (Macías et al., 2000).

## CONCLUSIONS

The Tacaná Volcanic Complex is built upon Mesozoic gneisses, schists, metavolcanics, slates, and granites. These rocks were intruded by two episodes of magmatism during the late Eocene to early Oligocene, probably related to the subduction of the Farallon plate underneath the North America plate, and during the early to middle Miocene, likely associated to the subduction of the Cocos plate under the Caribbean plate. These rocks were affected by a tectonic Miocene compressive event that was accommodated through strike-slip and reverse faults. The main principal stress $\sigma 1$ of this event had a NE-SW trend, as supported by slickensides, sigmoidal gouge faults, noncohesive breccias, and sigmoidal structures. These basement rocks were uplifted and tilted after the middle Miocene and before the Pliocene as a consequence of the subduction process. During the Pliocene, three caldera structures (San Rafael, Chanjale, and Sibinal) formed on the basement rocks, and during the Pliocene–early Pleistocene, these calderas were affected by NE-SW normal faults originated by the Tacaná graben, inside of which was emplaced the Tacaná Volcanic Complex during the late Pleistocene. The Tacaná Volcanic Complex was built by the subsequent formation of four NE-SW aligned structures named Chichuj, Tacaná, Plan de las Ardillas, and San Antonio. The area has been affected by a NNW stress field with a minimum principal stress $\sigma 3$ since the late Pleistocene to Recent that correlates with the focal mechanisms determined in the region.

## ACKNOWLEDGMENTS

This project was supported by grants from Consejo Nacional de Ciencia y Tecnología (38586-T to J.L.M. and 48506-F to J.C.M.) and Dirección General de Asuntos del Personal Académico–Universidad Nacional Autónoma de México (IX101404 to J.L.M. and IN103205 to J.L.A.). We are indebted to F. Ortega and M. Alcayde for their review of the first version of this manuscript. Reviews by K. Scott and C. Harpel greatly helped to clarify the ideas stated in this manuscript.

## REFERENCES CITED

Arce, J.L., Macías, J.L., Hughes, S., Saucedo, R., Escobar, R., García-Palomo, A., and Mora, J.C., 2004, Late Pleistocene Plinian activity at the Tacaná Volcanic Complex, México-Guatemala: International Association of Volcanology and Chemistry of the Earth's Interior (IAVCEI) General Assembly 2004, Pucón, Chile, November 14–19, Symposium 03a-07.

Bergeat, A., 1894, Zur Kenntnis der jungen Eruptivgesteine der Republik Guatemala: Zeitschrift der Deutschen Geologischen Gesellschaft, Abhandlungen, p. 131–157.

Bravo, H., Rebollar, C.J., Uribe, A., and Jimenez, O., 2004, Geometry and state of stress of the Wadati-Benioff zone in the Gulf of Tehuantepec, Mexico: Journal of Geophysical Research, v. 109, p. B04307, doi: 10.1029/2003JB002854.

Burkart, B., and Self, S., 1985, Extension and rotation of crustal blocks in northern Central America and effect on the volcanic arc: Geology, v. 13, p. 22–26, doi: 10.1130/0091-7613(1985)13<22:EAROCB>2.0.CO;2.

Campa, M.F., 1998, Una orogenía Miocénica en el sur de México [abs.], in Alaniz-Alvaréz, S., Nieto-Samaniego, A., and Ferrari, L., eds., Primera Reunión Nacional de Ciencias de la Tierra: México, D.F., p. 137.

Carr, M.J., Rose, W.I., and Stoiber, R.E., 1982, Central America, in Thorpe, R.S., ed., Andesites: New York, John Wiley & Sons, p. 149–166.

Couch, R., and Woodcock, S., 1981, Gravity and structure of the continental margins of southwestern Mexico and northwestern Guatemala: Journal of Geophysical Research, v. 86, p. 1829–1840.

Damon, P., and Montesinos, E., 1978, Late Cenozoic volcanism and metallogenesis over an active Benioff Zone in Chiapas, Mexico: Arizona Geological Society Digest, v. 11, p. 155–168.

De Cserna, Z., Aranda-Gómez, J.J., and Mitre-Salazar, L.M., 1988, Mapa fotogeológico preliminar y secciones estructurales del Volcán Tacaná: México, Instituto de Geología, scale 1:50,000.

De la Cruz, V., and Hernández, R., 1985, Estudio geológico a semidetalle de la zona geotérmica del Volcán Tacaná, Chiapas: México, Comisión Federal de Electricidad, Internal Report 41, 30 p.

De la Cruz-Reyna, S., Armienta, M.A., Zamora, V., and Juárez, F., 1989, Chemical changes in spring waters at Tacaná Volcano, Chiapas, México: Journal of Volcanology and Geothermal Research, v. 38, p. 345–353, doi: 10.1016/0377-0273(89)90047-4.

De Mets, C., Gordon, R.G., Argus, D.F., and Stein, S., 1990, Current plate motions: Geophysical Journal International, v. 101, p. 425–478.

Donnelly, T.W., Horne, G.S., Finch, R.C., and López-Ramos, E., 1990, Northern Central America: The Maya and Chortís blocks, in Dengo, G., and Case, J.E., eds., The Caribbean region: Boulder, Colorado, Geological Society of America, Decade of North American Geology, Geology of North America, v. H, p. 37–76.

Duffield, W., Heiken, G., Foley, D., and McEwen, A., 1993, Oblique synoptic images produced from digital data, display strong evidence of a "new" caldera in southwestern Guatemala: Journal of Volcanology and Geothermal Research, v. 55, p. 217–224, doi: 10.1016/0377-0273(93)90038-S.

Espíndola, J.M., Medina, F.M., and De los Ríos, M., 1989, A C-14 age determination in the Tacaná volcano (Chiapas, Mexico): Geofísica Internacional, v. 28, p. 123–128.

Espíndola, J.M., Macías, J.L., and Sheridan, M.F., 1993, El Volcán Tacaná: Un ejemplo de los problemas en la evaluación del Riesgo Volcánico: Simposio Internacional sobre Riesgos Naturales e Inducidos en los Grandes Centros Urbanos de América Latina: Serie Scienza No. 6, Centro Nacional de Prevención de Desastres (CENAPRED), México D.F., p. 62–71.

García-Palomo, A., Macías, J.L., and Espíndola, J.M., 2004, Strike-slip faults and K-Alkaline volcanism at El Chichón volcano, southeastern Mexico: Journal of Volcanology and Geothermal Research, v. 136, p. 247–268, doi: 10.1016/j.jvolgeores.2004.04.001.

Guzmán-Speziale, M., Pennington, W.D., and Matumoto, T., 1989, The triple junction of the North America, Cocos, and Caribbean Plates: Seismicity and tectonics: Tectonics, v. 8, p. 981–999.

Hey, R., 1977, Tectonic evolution of the Cocos-Nazca spreading center: Geological Society of America Bulletin, v. 88, p. 1404–1420, doi: 10.1130/0016-7606(1977)88<1404:TEOTCS>2.0.CO;2.

Irvine, T.N., and Baragar, W.R.A., 1971, A guide to the chemical classification of the common volcanic rocks: Canadian Journal of Earth Sciences, v. 8, p. 523–548.

Klitgord, K.D., and Mammerickx, J., 1982, Northern east Pacific Rise; magnetic anomaly and bathymetry framework: Journal of Geophysical Research, v. 87, p. 6725–6750.

Lanphere, M.A., and Dalrymple, G.B., 2000, First-principles calibration of $^{38}$Ar tracers: Implications for the ages of $^{40}$Ar/$^{39}$Ar fluence monitors: U.S. Geological Survey Professional Paper 1621, 10 p.

Le Bas, M.J., Le Maitre, R.W., Streckeisen, A., and Zanettin, R., 1986, A chemical classification of volcanic rocks based on the total alkali-silica diagram: Journal of Petrology, v. 27, p. 745–750.

LeFevre, L., and McNally, K.C., 1985, Stress distribution and subduction of aseismic ridges in the Middle America subduction zone: Journal of Geophysical Research, v. 90, p. 4495–4510.

Luhr, J.F., Carmichael, I.S.E., and Varekamp, J.C., 1984, The 1982 eruptions of El Chichón Volcano, Chiapas, Mexico: Mineralogy and petrology of the anhydrite-bearing pumices: Journal of Volcanology and Geothermal Research, v. 23, p. 69–108, doi: 10.1016/0377-0273(84)90057-X.

Macías, J.L., Espíndola, J.M., García-Palomo, A., Scott, K.M., Hughes, S., and Mora, J.C., 2000, Late Holocene Peléan style eruption at Tacaná Volcano, Mexico-Guatemala: Past, present, and future hazards: Geological Society of America Bulletin, v. 112, p. 1234–1249, doi: 10.1130/0016-7606(2000)112<1234:LHPASE>2.3.CO;2.

Macías, J.L., Arce, J.L., Mora, J.C., and García-Palomo, A., 2004, The Agua Caliente Debris Avalanche deposit a northwestern sector collapse of Tacaná volcano, México-Guatemala: International Association of Volcanology and Chemistry of the Earth's Interior (IAVCEI) General Assembly 2004, Pucón, Chile, November 14–19, Symposium 11a-03.

Manea, M., Manea, V.C., Kostoglodov, V., and Guzmán-Speziale, M., 2005, Elastic thickness of the oceanic lithosphere beneath the Tehuantepec ridge: Geofísica Internacional, v. 44, no. 2, p. 157–168.

McCann, W., Nishenko, S., Sykes, L., and Krause, J., 1979, Seismic gap and tectonics: Seismic potential for major boundaries: Pure and Applied Geophysics, v. 177, p. 1082–1147.

Meneses-Rocha, J.J., 2001, Tectonic evolution of the Ixtapa Graben, an example of a strike-slip basin of southeastern Mexico: Implications for regional petroleum systems, in Bartolini, C., Buffler, R.T., and Cantú-Chapa, A., eds., The western Gulf of Mexico Basin: Tectonics, Sedimentary Basins, and Petroleum Systems: American Association of Professional Geologists Memoir 75, p. 183–216.

Mercado, R., and Rose, W.I., 1992, Reconocimiento geológico y evaluación preliminar de peligrosidad del Volcán Tacaná, Guatemala/México: Geofísica Internacional, v. 31, p. 205–237.

Meschede, M., and Frisch, W., 1998, A plate-tectonic model for the Mesozoic and early Cenozoic history of the Caribbean plate: Tectonophysics, v. 296, p. 269–291, doi: 10.1016/S0040-1951(98)00157-7.

Mora, J.C., 2001, Studio vulcanologico e geochimico del Vulcano Tacana, Chiapas, Messico [Ph.D. Thesis]: Firenze, Italy, Universita degli Studi di Firenze, 147 p.

Mora, J.C., Macías, J.L., García-Palomo, A., Espíndola, J.M., Manetti, P., and Vaselli, O., 2004, Petrology and geochemistry of the Tacaná Volcanic Complex, Mexico-Guatemala: Evidence for the last 40 000 yr of activity: Geofísica Internacional, v. 43, p. 331–359.

Mugica, M.R., 1980, Estudio radiométrico del prospecto Nizanda, Estados de Oaxaca y Chiapas: Instituto Mexicano del Petróleo, México, C-1084, 69 p.

Mugica, M.R., 1987, Estudio petrogenético de las rocas ígneas y metamórficas en el Macizo de Chiapas: Instituto Mexicano del Petróleo, México, C-2009, 47 p.

Müllerried, F.K.G., 1951, La reciente actividad del Volcán de Tacaná, Estado de Chiapas, a fines de 1949 y principios de 1950: Informe del Instituto de Geología de la Universidad Nacional Autónoma de México, 28 p.

Nakamura, N., 1974, Determination of REE, Ba, Fe, Mg, Na and K in carbonaceous and ordinary chondrites: Geochimica et Cosmochimica Acta, v. 38, p. 757–773, doi: 10.1016/0016-7037(74)90149-5.

Newhall, C.G., 1987, Geology of the Lake Atitlan region, western Guatemala: Journal of Volcanology and Geothermal Research, v. 33, p. 23–55, doi: 10.1016/0377-0273(87)90053-9.

Nelson, S.A., Gonzalez-Caver, E., and Kyser, T.K., 1995, Constrains on the origin of alkaline and calc-alkaline magmas from the Tuxtla Volcanic Field, Veracruz, Mexico: Contributions to Mineralogy and Petrology, v. 122, p. 191–211, doi: 10.1007/s004100050121.

Nixon, G.T., 1982, The relationship between Quaternary volcanism in central Mexico and the seismicity and structure of the subducted ocean lithosphere: Geological Society of America Bulletin, v. 93, p. 514–523, doi: 10.1130/0016-7606(1982)93<514:TRBQVI>2.0.CO;2.

Pardo, M., and Suárez, G., 1995, Shape of the subducted Rivera and Cocos plates in southern Mexico: Seismic and tectonic implications: Journal of Geophysical Research, v. 100, p. 12,357–12,373, doi: 10.1029/95JB00919.

Ponce, L., Gaulon, R., Suárez, G., and Lomas, E., 1992, Geometry and state of stress of the downgoing Cocos plate in the Isthmus of Tehuantepec, Mexico: Geophysical Research Letters, v. 19, p. 773–776.

Rebollar, C.J., Espíndola, V.H., Uribe, A., Mendoza, A., and Pérez-Vertti, A., 1999, Distribution of stress and geometry of the Wadati-Benioff zone under Chiapas, Mexico: Geofísica Internacional, v. 38, p. 95–106.

Rose, W.I., Newhall, C.G., Bornhorst, T.J., and Self, S., 1987, Quaternary silicic pyroclastic deposits of Atitlan caldera, Guatemala: Journal of Volcanology and Geothermal Research, v. 33, p. 57–80, doi: 10.1016/0377-0273(87)90054-0.

Sapper, C., 1927, Vulkankunde: J. Engelhorns Nachf, Stuttgart, p. 1–80.

Singh, S.K., Astiz, L., and Haskov, J., 1981, Seismic gaps and recurrence periods of large earthquakes along the Mexican subduction zone: A reexamination: Bulletin of the Seismological Society of America, v. 71, p. 827–843.

Stock, J.M., and Lee, J., 1994, Do microplates in subduction zones leave a geological record?: Tectonics, v. 13, p. 1472–1487, doi: 10.1029/94TC01808.

Thorpe, R.S., 1977, Tectonic significance of alkaline volcanism in eastern Mexico: Tectonophysics, v. 40, 1926, doi: 10.1016/0040-1951(77)90064-6.

Truchan, M., and Larson, R.L., 1973, Tectonic lineaments on the Cocos plate: Earth and Planetary Science Letters, v. 17, p. 426–432, doi: 10.1016/0012-821X(73)90211-2.

Wilson, D.S., 1996, Fastest known spreading of the Miocene Cocos-Pacific plate boundary: Geophysical Research Letters, v. 23, p. 3003–3006, doi: 10.1029/96GL02893.

MANUSCRIPT ACCEPTED BY THE SOCIETY 19 MARCH 2006

# The chemistry of spring waters and fumarolic gases encircling Santa María volcano, Guatemala: Insights into regional hydrothermal activity and implications for volcano monitoring

**James A. Walker**
Department of Geology and Environmental Geosciences, Northern Illinois University, DeKalb, Illinois 60115, USA

**Sharon Templeton**
Terracon, 2277 W. Spencer Street, Appleton, Wisconsin 54914, USA

**Barry I. Cameron**
Department of Geosciences, University of Wisconsin, Milwaukee, Wisconsin 53201, USA

## ABSTRACT

Springs encircling Santa María volcano in Guatemala generally contain bicarbonate waters. Bicarbonate waters southwest of the persistently active Santiaguito lava dome are characterized by high Mg/Ca. Other springs contain acid sulfate or chloride waters. Most acid sulfate and chloride waters are spatially confined to springs, wells, and streams of the Zunil and Zunil-II (Sulfur Mountain) geothermal fields on the flanks of the Cerro Quemado dome complex, 5 km to the east-northeast of Santa María. Some acid sulfate waters have unusually high S/Cl ratios (20–70). Chloride waters are dilute versions of typical geothermal brines. The $\delta^{13}C$ ratios of all waters fall in a very narrow range (−11.5‰ to −8.5‰). The nonreactive gas compositions from fumaroles encircling Santa María are typical of those sampled at other subduction zone volcanoes. Most fumarolic gases from Santa María have notably lighter $\delta^{13}C$ ratios compared to gases sampled from elsewhere on the Central American volcanic front.

The He and C isotopic values of fumarolic gas samples from the Santa María region indicate significant mantle input. Estimated magmatic $\delta^{13}C$ ratios for Zunil and Zunil-II, however, are lighter than accepted mantle values (−11‰ to −14‰). This is most likely caused by shallow crustal contamination. All of the spring waters from the Santa María region represent variable interactions between magmatic/hydrothermal fluids and meteoric waters. There is, however, only limited mixing between bicarbonate, acid sulfate, and chloride waters. Surface discharges of chloride waters are inhibited by high precipitation rates. The high S/Cl ratios of some of the acid sulfate waters from Zunil/Zunil-II reflect extensive scrubbing by the underlying hydrothermal system. High Mg/Ca bicarbonate waters from springs south-southwest of the Santiaguito dome complex have experienced enhanced water-rock interaction, and their slightly heavier $\delta^{13}C$ ratios (−9.5‰ to −8‰) hint at a small distinction between magmatic $\delta^{13}C$ at Zunil/Zunil-II and Santa María. This supports previous suggestions that the hydrothermal system beneath the Zunil area is independent of Santa María. Gas samples from Zunil/Zunil-II and Cerro Quemado, on the other hand, do share similar $\delta^{13}C$ ratios, strengthening the notion that magmatism at the latter is propelling the hydrothermal system northeast of Santa María. Hence, monitoring of the springs and fumaroles at Zunil/Zunil-II could prove useful in forecasting of future activity at Cerro Quemado.

**Keywords:** Santa María, Guatemala, hydrothermal system, volcanic springs, volcanic gases.

## INTRODUCTION

Hydrothermal systems associated with active, subaerial arc volcanoes can have broad volcanological, environmental, and economic consequences. For instance, circulating thermal waters may react strongly with rocks of the volcanic edifice, leading to alteration, dissolution, and structural weakening, increasing the risks for catastrophic volcanic collapse (López and Williams, 1993; Frank, 1995; Crowley and Zimbelman, 1997; van Wyk de Vries et al., 2000; Reid et al., 2001; Varekamp et al., 2001; Reid, 2004). On the other hand, boiling hydrothermal fluids can precipitate silica in fractures and pore spaces, leading to mineral sealing within a volcanic edifice and increasing the likelihood of a hazardous explosive eruption (Fischer et al., 1994, 1996; Boudon et al., 1998; Stix et al., 1997; Edmonds et al., 2003). Subvolcanic hydrothermal systems can act as natural scrubbing agents for some ascending magmatic gases (Doukas and Gerlach, 1995; Oppenheimer, 1996; Symonds et al., 2001, 2003; Duffell et al., 2003). Leakage of fluids from large subvolcanic hydrothermal systems can adversely affect the quality of regional surface and groundwater resources (Sriwana et al., 1998; Varekamp et al., 2001). Finally, large hydrothermal systems below arc volcanoes often produce prime epithermal ore deposits (Hedenquist and Aoki, 1991; Christenson and Wood, 1993; Hedenquist and Lowenstern, 1994) and can be tapped for electrical power (e.g., Goff and Janik, 2000). Thermal springs, fumaroles, and crater lakes provide windows into the dimensions and nature of volcano-hosted hydrothermal systems (Sturchio et al., 1988, 1993; Shevenell and Goff, 1993; Rowe et al., 1995; Frank, 1995; Hochstein and Browne, 2000) and have also proven useful in volcano monitoring and hazard mitigation efforts (Sturchio et al., 1988; Martini, 1989; Rowe et al., 1992; Tedesco, 1995; Fischer et al., 1996; Martínez et al., 2000; Lewicki et al., 2000; Tassi et al., 2003).

Santa María volcano in Guatemala is historically one of the most active and deadly of the many Quaternary volcanoes along the Central America volcanic arc (Rose, 1987a, 1987b; Sanchez Bennett et al., 1992). Over 300,000 people live within its immediate reach. Santa María has numerous thermal springs and fumaroles within 10 km of its summit, including those within the well-surveyed Zunil geothermal field, just off its northeast flank (A. Adams et al., 1990; M. Adams et al., 1990, 1992). Previous geochemical/hydrogeological studies of the Zunil hydrothermal system have suggested that it may be independent of Santa María (A. Adams et al., 1990; M. Adams et al., 1990, 1992). In this paper, we present new geochemical data on waters and gases collected from springs and fumaroles that girdle Santa María as a baseline for future eruptive monitoring efforts and use this data to obtain further insights into hydrothermal processes below an extremely active portion of the Central American volcanic arc.

## GEOLOGICAL AND VOLCANOLOGICAL MILIEU

Santa María volcano is one of the many polygenetic stratovolcanoes defining the volcanic front of the Central American volcanic arc. This volcanic arc is one of the conspicuous products of the subduction of the Cocos plate beneath the Caribbean plate (Fig. 1). Carr et al. (1982) and Carr (1984), following Stoiber and Carr (1973), have subdivided the Central American volcanic front into eight segments defined by varying geologic, tectonic, or volcanologic parameters, such as the strike of volcanic lineaments. Santa María lies at the northwestern end of the central Guatemalan segment, not far from the northwestern terminus of the arc at the Guatemala–México border. Santa María and the other volcanoes of the central Guatemalan segment form the scenic highlands of Guatemala (Williams, 1960). South-southwest of Santa María is the gently sloping coastal plain, which runs some 60 km to the Pacific Ocean.

Santa María was built on a foundation of Tertiary, subduction-related igneous rocks and remnants of Paleozoic metamorphic rocks that crop out to the north of the volcanic front (Williams, 1960; Rose, 1987b; M. Adams et al., 1990; Foley et al., 1990). Santa María has experienced three stages of development: (1) stratovolcano construction; (2) Plinian eruption; and (3) dome growth (Conway et al., 1994). Paleomagnetic data indicate that the first stage, stratovolcano construction, was completed in perhaps a few thousand years, from 25 to 32 ka (Rose et al., 1977a; Conway et al., 1994). Volcanic rocks from this stage range from basalts to basic andesites, broadly becoming more Si-rich with time (Rose et al., 1977a; Rose, 1987b). The upshot of this activity was the modest volume (~20 km$^3$) Santa María stratovolcano, which today rises to an elevation of ~3800 m. After a lengthy dormancy, Santa María dramatically reawakened in October 1902, explosively ejecting 5–10 km$^3$ dense rock equivalent (DRE) of largely dacitic pyroclastics (Rose, 1972a; Williams and Self, 1983). The 1902 eruption, one of the largest volcanic eruptions of the twentieth century, excavated a huge

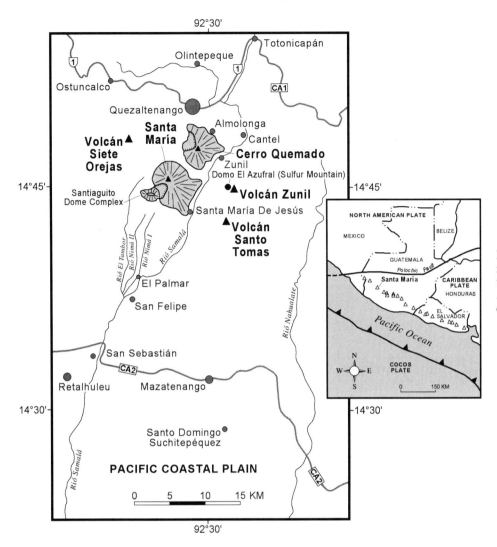

Figure 1. Santa María volcano and environs. Inset shows regional tectonic setting for the northern part of the Central America subduction zone (figure adapted from Sanchez Bennett et al., 1992).

amphitheater/crater on the southwest side of the cone (Fig. 1) and killed ~5000 people (Rose, 1972a; Conway et al., 1994). In 1922, after 20 years of relative quiet, extrusion of dacitic lava began inside the 1902 crater and has continued until the present (Stoiber and Rose, 1969; Rose, 1973, 1987a; Harris et al., 2002). The continuous lava extrusions have constructed the large (~1 km$^3$), intricate Santiaguito dome on Santa María's southwestern flank (Fig. 1; Stoiber and Rose, 1969; Rose, 1972b, 1973, 1987a; Anderson et al., 1995; Harris et al., 2002, 2003; Barmin et al., 2002). Besides continuous lava extrusion, recent activity at Santiaguito has included pyroclastic flows, vertical pyroclastic eruptions, and numerous lahars (Rose, 1973, 1974; Rose et al., 1977b; Rose 1987a). As a result, abundant pyroclastics have been deposited to the south-southwest of Santiaguito, particularly in the Río El Tambor and Río Nimá valleys. Some laharic sequences in these valleys exceed 7 m in thickness.

Santa María is therefore distinguished by a clear, temporal gap in compositions, from early eruption of basic magmas to later eruption of silicic magmas, although it should be noted that the 1902 eruption had a minor basaltic component (Rose, 1987b) that may have triggered the eruption (Self and Williams, 1979). Rose (1987b) has also emphasized the sodium-rich nature of all of Santa María's magmas when compared to other magmas erupted along the Guatemalan volcanic front.

Numerous fumaroles have been active on and near Santiaguito since its birth in 1922 (Stoiber and Rose, 1969, 1970, 1974). Stoiber and Rose (1970) showed that gas condensates from fumaroles at the active Caliente vent typically have lower $Cl/SO_4$ ratios than those collected from more marginal fumaroles on the Santiaguito dome complex. They also indicated that condensate $Cl/SO_4$ ratios at the Sapper fumarole, 0.5 km west of Caliente, decreased before an eruptive period and rebounded during the post-eruptive period, suggesting that the $Cl/SO_4$ of gas condensates might make a useful indicator of an impending eruption.

According to the Guatemalan National Institute of Seismology, Volcanology, Meteorology and Hydrology (INSIVUMEH), the summit of Santa María receives ~1700 mm precipitation per annum. The southern slopes of Santa María receive significantly more rain with decreasing altitude. For example, Finca La Florida, located 9 km south of the summit of Santa María at an elevation of 900 m, gets 4500 mm precipitation per annum. Nearly 80% of the annual rainfall occurs during a six-month period from May to October, with peak amounts in June and September.

Santa María has numerous volcanic neighbors in close proximity to the north and east, some of which have been recently active. To the northwest lies Siete Orejas (Fig. 1), an andesitic stratovolcano that produced a large (>2 km$^3$) Plinian eruption ca. 120–150 ka (Williams, 1960; Rose et al., 1999). Siete Orejas is heavily incised, with a broad south-opening caldera or amphitheater (Williams, 1960; Duffield et al., 1993). Less than 10 km to the northeast of Santa María is Cerro Quemado, an exogenous dome complex that last erupted in 1818 (Conway et al., 1992). Cerro Quemado is the largest in a cluster of domes ringing the southern outskirts of Quetzaltenango, the second largest city in Guatemala (Williams, 1960; Conway et al., 1992). To the east of Santa María are two deeply dissected volcanoes, Santo Tomás (or Pecul) and Zunil, which together coalesce to form a continuous ridge, the Zunil ridge (Williams, 1960). Neither Santo Tomás nor Zunil have had historic activity (Rose, 1987b). According to Foley et al. (1990) and Duffield et al. (1993), many of the volcanoes in this region fall along the southern margin of a large (30 km diameter) caldera that encompasses the Quezaltenango basin. Duffield et al. (1993) indicate that the age of this caldera, which they call the Xela caldera, is anywhere from 84 ka to 12 Ma.

A number of rivers flow from and around Santa María volcano (Fig. 1). To the west lies Río Ocosito, which flows from the deeply incised rim of Siete Orejas southward toward the coastal plain. Several rivers drain the southern flank of Santa María: Río El Tambor, Río Nimá Primero, and Río Nimá Segundo. The extensive Río Samalá runs adjacent to the Zunil geothermal field between Santa María and the Zunil ridge. Río Samalá receives more input from the latter.

The Zunil geothermal field provides clear evidence of significant hydrothermal activity in the vicinity of Santa María. The Zunil geothermal field lies on the southeastern flanks of Cerro Quemado, ~5 km northeast of the summit of Santa María (Fig. 1). The Zunil geothermal field also falls along the northeast-trending Zunil fault zone, which has been linked to the segmentation of the volcanic front proposed by Stoiber and Carr (1973) (Foley et al., 1990; Duffield et al., 1993; Lima Lobato and Palma, 2000). Faults related to the development of Xela caldera could also help control the location of the Zunil geothermal field (Foley et al., 1990). The geochemistry of waters from Zunil has been well characterized (Fournier et al., 1982; A. Adams et al., 1990; M. Adams et al., 1990, 1992; Giggenbach et al., 1992). M. Adams et al. (1990) have also simulated the regional hydrogeologic system around Zunil. Foley et al. (1990) present shallow geophysical surveys of the Zunil area. Lima Lobato and Palma (2000) report on the geology, geochemistry, and geophysics of the Sulfur Mountain region, east of the Zunil geothermal field, which they refer to as Zunil-II.

## SAMPLING AND ANALYTICAL METHODS

A total of 82 water samples were collected from springs, drilled wells, streams, and rain events encircling Santa María during two rainy seasons (August 1994 and August 1995) and one dry season (January 1995). Thirty-five separate locations were sampled (Fig. 2). Water temperature and pH were measured in the field using a thermocouple, an Orion pH meter, or pH strips. Flow rates were estimated visually or with a 1 L bottle. Water samples for anion analysis and alkalinity titrations were filtered using a syringe and a 0.45 μm filter and then stored in high density polyethylene (HDPE) containers. Samples for cation analysis were similarly filtered, then treated with 4N $HNO_3$ and stored in HDPE containers. Samples for stable isotope analyses were collected in separate HDPE containers and sealed against evaporation.

Major anions and cations were determined by ion chromatography and direct-current plasma–multi-element atomic emission spectrometry (DCP-MAES), respectively, at Northern Illinois University. Relative precision for both determinations was <4%. Water samples were prepared for oxygen and hydrogen isotopic analysis via water-$CO_2$ equilibrium (Epstein and Mayeda, 1953) and zinc-reduction (Coleman et al., 1982) methods, respectively. Oxygen, hydrogen, and carbon stable isotope analyses were carried out at Argonne National Laboratories, NASA, and Washington University. Reproducibility of δ values for D is ±1‰, and that for the remaining elements is ±0.2‰.

A total of 16 gas samples were collected from five different fumaroles (Zunil 1, Zunil 2, Sulfur Mountain, Cerro Quemado, and Portales; Fig. 2) during the three field excursions in 1994 and 1995. Gas samples were collected in "Giggenbach" bottles using protocols detailed in Giggenbach and Goguel (1988) and Fahlquist and Janik (1992). Bulk analysis of headspace gases was by mass spectrometry. One aliquot of the oxidized alkaline condensate was titrated for alkalinity. Another aliquot was prepared for carbon isotope analysis using phosphoric acid (McCrea, 1950). Mass spectrometry was done at Argonne National Laboratory (reproducibility as above).

## WATERS

Discharge temperatures of water samples varied from 7 to 92 °C (Table 1). Sampling locations included 15 thermal springs (see definition in Nathenson et al., 2003), 14 nonthermal springs, two thermal wells, and three nonthermal streams. In addition, water from a single rain event was collected 4 km south of the Santiaguito dome. The thermal waters cluster in three general areas: east-northeast of Santa María from the Zunil geothermal field to Sulfur Mountain; in the Siete Orejas valley ~7.5 km northwest of the Santiaguito dome; and in a restricted region ~3 km southwest of Santiaguito (Fig. 2). Except for those in the latter region, the thermal waters all have outlet elevations ~2000 m (Fig. 3). Most of the waters are weakly acidic to weakly basic, except for those from the area of thermal springs east of Santa María, where waters are almost entirely acidic with pH <3.9 (Table 1).

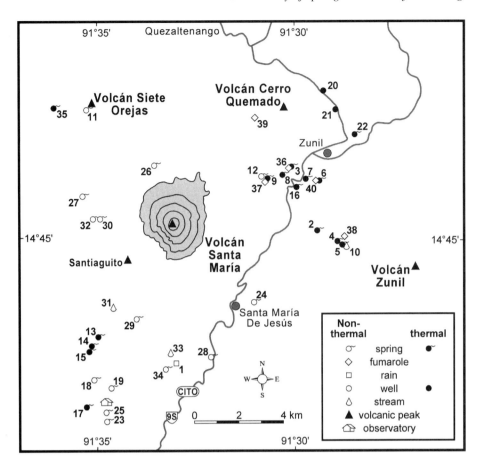

Figure 2. Sampling localities for this study. Numbers correspond to sample locations of specific samples in Tables 1 and 3.

In Figure 4, water samples from this and previous studies of the Zunil geothermal field are classified into sulfate, chloride, and bicarbonate waters based on their major anion chemistry (e.g., Giggenbach et al., 1990). Bicarbonate waters predominate around most of Santa María (Fig. 5). The single rain sample and most of the stream and well waters outside the Zunil geothermal field also fall in the bicarbonate group (Fig. 5). Sulfate and chloride waters, on the other hand, are geographically restricted to springs, wells, and streams in the Zunil–Sulfur Mountain area northeast of Santa María and to the Ocos and Pinabet springs in the Siete Orejas valley to the northwest of Santa María (Fig. 5). The pH of the Zunil and Ocos chloride waters is somewhat variable, ranging from 5.0 to 8.5 (Table 1; A. Adams et al., 1990; Giggenbach et al., 1992). As pointed out previously by A. Adams et al. (1990), the chloride waters are relatively dilute (<2000 ppm Cl) in comparison to average chloride brines from geothermal wells (Goff and Janik, 2000). The overall anion chemistry of Santa María's waters seems to imply considerable mixing between the different groups (Fig. 4A), which contrasts with the more limited mixing observed between water types at some other arc volcanoes (Giggenbach et al., 1990; Shevenell and Goff, 1993). Figure 6, however, suggests there may be only three main mixing (or dilution) trends: one between meteoric water and a sulfate-rich component ($SO_4/Cl > 150$); one between meteoric water and a component with significantly lower sulfate ($SO_4/Cl < 2$) (i.e., the main bicarbonate trend); and a dilution trend amongst the chloride waters. The sulfate waters defining the sulfate-rich trend have unusually high S/Cl ratios (20–70). Most waters from active volcanic areas have much lower (<10) S/Cl ratios (Kiyosu and Kurahashi, 1983; Sturchio et al., 1988, 1993; Giggenbach et al., 1990; Christenson and Wood, 1993; Shevenell and Goff, 1993; Fischer et al., 1997; Taran et al., 1998; Delmelle et al., 2000; Lewicki et al., 2000; Evans et al., 2002) The S/Cl ratios and mixing/dilution trends are discussed further in a following section.

The major cations in the waters from around Santa María are Na, Ca, and Mg (Table 1), likely leached from the basic to silicic rocks of the volcanic edifice (Rowe and Brantley, 1993; Varekamp et al., 2000). On a triangular diagram of these major cations, most of Santa María's waters define a vague trend from the Na apex to the Ca-Mg sideline at a constant Mg/Ca ratio, with the bulk of the waters clustering near the compositions of lavas that have constructed Santa María's cone (Fig. 7). Chloride waters have distinctly low Mg/Ca ratios, while a subset of bicarbonate waters are characterized by elevated Mg/Ca (Fig. 7). These higher Mg/Ca waters are all from thermal and nonthermal springs located south-southwest of the Santiaguito dome or thermal springs located between the Zunil geothermal field and the Río Samalá. Chloride waters have the highest Na (and Ca) concentrations (Fig. 8). In bicarbonate waters, Na concentrations increase with temperature and chloride contents (Fig. 8). Sulfate

TABLE 1. CHEMICAL AND PHYSICAL DATA FOR WATERS FROM THE SANTA MARÍA REGION

| Location Name | Location [Fig.2] | Sample ID | Type | Date mo/dy/yr | Elev. masl | Temp °C | pH | Ca ppm | Mg ppm | Fe ppm | Mn ppm | Sr ppm | K ppm | Na ppm | Cl ppm | SO$_4$ ppm | HCO$_3$ ppm | TDS ppm |
|---|---|---|---|---|---|---|---|---|---|---|---|---|---|---|---|---|---|---|
| Meteoric water | | | | | | | | | | | | | | | | | | |
| Rain | 1 | 94-10 | rain | 08/20/94 | 1395 | 19.8 | 7.78 | 1.08 | 0.19 | 0.05 | 0.03 | 0.01 | 0.47 | 0.1 | 1.4 | 2.372 | 15.86 | 21.60 |
| Acid Sulfate | | | | | | | | | | | | | | | | | | |
| Banos Aguas Amargas | 2 | 94-7 | spring | 08/20/94 | 1999 | 60.5 | 3.6 | 35.32 | 11.92 | 18.40 | 1.09 | 0.23 | 38.03 | 104.2 | 8.8 | 1787.3 | 0.00 | 2007. |
| Banos Aguas Amargas | 2 | 95-13 | spring | 01/05/95 | 1999 | 60.8 | 2.03 | 32.75 | 11.33 | 18.03 | 1.11 | 0.18 | 39.70 | 111.3 | 8.9 | 1615.1 | 0.00 | 1849 |
| Banos Aguas Amargas | 2 | 95-B27 | spring | 08/16/95 | 1999 | 60.1 | 0.88 | 34.77 | 11.12 | 16.58 | 1.13 | 0.19 | 37.09 | 104.3 | 9.8 | 1329.4 | 0.00 | 1677 |
| Azufrales Vent (Z38) | 3 | 95-B1 | spring | 08/11/95 | 2148 | 89.7 | 2.69 | 126.3 | 44.34 | 58.30 | 2.19 | 0.09 | 31.34 | 43.3 | 8.8 | 1235 | 0.00 | 1552 |
| Azufrales Vent (Z38) | 3 | 95-B30 | spring | 08/16/95 | 2148 | 92.3 | 1.17 | 53.54 | 28.34 | 37.63 | 1.17 | 0.09 | 23.37 | 37.0 | 6.1 | 1422 | 0.00 | 1677 |
| Fuentes Georginas #1 | 4 | 94-11 | spring | 08/20/94 | 2470 | 48.3 | 3.5 | 43.82 | 22.76 | 7.47 | 1.06 | 0.12 | 37.26 | 107.8 | 7.8 | 1004 | 0.00 | 1233 |
| Fuentes Georginas #1 | 4 | 95-4 | spring | 01/04/95 | 2470 | 49.9 | 2.24 | 47.80 | 20.19 | 7.30 | 1.08 | 0.14 | 33.61 | 95.5 | 5.2 | 980.54 | 0.00 | 1197 |
| Fuentes Georginas #1 | 4 | 95-B5 | spring | 08/12/95 | 2470 | 43.6 | 2.47 | 43.51 | 22.06 | 9.29 | 1.12 | 0.14 | 35.94 | 104.3 | 5.2 | 917.11 | 0.00 | 1142 |
| Fuentes Georginas #2 | 5 | 94-21 | spring | 08/23/94 | 2386 | 50.6 | 2.43 | 51.52 | 21.89 | 8.83 | 1.19 | 0.14 | 31.83 | 101.8 | 6.8 | 972.4 | 0.00 | 1200 |
| Fuentes Georginas #2 | 5 | 95-6 | spring | 01/04/95 | 2386 | 50.5 | 2.43 | 48.52 | 20.52 | 7.89 | 1.14 | 0.13 | 30.26 | 93.8 | 5.5 | 931.16 | 0.00 | 1143 |
| Fuentes Georginas #2 | 5 | 95-B7 | spring | 08/12/95 | 2386 | 46.9 | 2.93 | 48.31 | 24.78 | 9.17 | 1.23 | 0.14 | 36.60 | 115.3 | 11.8 | 1014.7 | 0.00 | 1263 |
| Portales Mud Pot | 6 | 95-10 | spring | 01/05/95 | 2092 | 93.0 | 3.6 | 61.82 | 24.51 | 2.58 | 0.68 | 0.37 | 14.06 | 57.7 | 4.2 | 392.2 | 0.00 | 558.4 |
| Portales Mud Pot | 6 | 95-B8 | spring | 08/12/95 | 2092 | 92.8 | 3.93 | 54.06 | 24.56 | 0.94 | 0.60 | 0.30 | 13.43 | 57.4 | 4.5 | 376.12 | 0.00 | 532.0 |
| ZMF-12 | 7 | 95-34 | spring | 01/11/95 | 1967 | 91.0 | 2.99 | 21.91 | 26.73 | 1.38 | 2.37 | 0.22 | 7.40 | 16.8 | 3.8 | 407.1 | 0.00 | 488.7 |
| ZMF-12 | 7 | 95-B29 | spring | 08/16/95 | 1967 | 68.0 | 5.88 | 39.42 | 44.08 | 0.00 | 0.21 | 0.27 | 33.97 | 238.3 | 154.9 | 177.2 | 0.00 | 688.4 |
| Zunil (Z47) | 8 | 95-11 | spring | 01/05/95 | 2190 | 85.7 | 3.09 | 32.07 | 16.38 | 10.28 | 0.64 | 0.05 | 7.69 | 22.7 | 4.5 | 328.44 | 0.00 | 423.6 |
| Zunil (Z47) | 8 | 95-B3 | spring | 08/11/95 | 2190 | 85.3 | 2.94 | 44.43 | 24.87 | 20.11 | 0.91 | 0.08 | 8.46 | 27.2 | 4.6 | 441.06 | 0.00 | 572.9 |
| Zunil (Z39) | 9 | 94-1 | spring | 08/18/94 | 2148 | 68.4 | 7.0 | 46.54 | 19.25 | 0.12 | 1.11 | 0.28 | 13.68 | 72.0 | 63.3 | 148.2 | 160.05 | 524.6 |
| Zunil (Z39) | 9 | 94-22 | spring | 08/24/94 | 2148 | 68.1 | 6.8 | 49.84 | 17.54 | 0.14 | 1.09 | 0.26 | 12.16 | 65.1 | 63.4 | 149.74 | 129.50 | 488.8 |
| Zunil (Z39) | 9 | 95-7 | spring | 01/04/95 | 2148 | 70.1 | 6.8 | 45.45 | 15.18 | 0.09 | 1.01 | 0.22 | 11.16 | 58.0 | 50.9 | 137.6 | 132.09 | 451.8 |
| Zunil (Z39) | 9 | 95-B2 | spring | 08/11/95 | 2148 | 71.7 | 7.07 | 46.03 | 17.69 | 0.10 | 1.01 | 0.24 | 13.09 | 68.5 | 56.0 | 209.66 | 106.40 | 518.8 |
| Georginas Stream | 10 | 94-12 | stream | 08/20/94 | 2478 | 12.1 | 5.5 | 11.89 | 4.88 | 0.05 | 0.10 | 0.09 | 3.23 | 9.5 | 5.8 | 41.0 | 27.02 | 103.6 |
| Georginas Stream | 10 | 95-5 | stream | 01/04/95 | 2478 | 10.1 | 7.08 | 9.87 | 3.42 | 0.07 | 0.07 | 0.06 | 2.02 | 4.6 | 3.6 | 3.2 | 49.04 | 75.95 |
| Georginas Stream | 10 | 95-B6 | stream | 08/12/95 | 2478 | 12.9 | 6.44 | 10.74 | 4.52 | 0.01 | 0.07 | 0.06 | 2.75 | 7.3 | 3.3 | 37 | 21.60 | 88.04 |
| Pinabet Cold Spring | 11 | 95-30 | spring | 01/09/95 | 3061 | 7.2 | 7.52 | 14.25 | 4.46 | 0.09 | 0.17 | 0.08 | 2.97 | 6.7 | 19.3 | 32.2 | 28.09 | 108.3 |
| Pinabet Cold Spring | 11 | 95-B21 | spring | 08/14/95 | 3061 | 9.8 | 8.82 | 16.65 | 5.24 | 0.00 | 0.27 | 0.10 | 2.61 | 6.5 | 16.1 | 35.082 | 22.88 | 105.4 |
| Zunil Cold Spring | 12 | 95-B4 | spring | 08/11/95 | 2229 | 25.5 | 6.28 | 18.41 | 12.68 | 0.02 | 0.06 | 0.11 | 3.88 | 10.5 | 4.6 | 15.77 | 0.00 | 68.06 |
| Neutral bicarbonate | | | | | | | | | | | | | | | | | | |
| Santiaguito #1 | 13 | 95-23 | spring | 01/08/95 | 1175 | 53.7 | 6.22 | 66.82 | 104.21 | 0.11 | 2.45 | 0.70 | 40.14 | 206.4 | 129.3 | 224.5 | 821.49 | 1596 |
| Santiaguito #2 | 14 | 95-24 | spring | 01/08/95 | 1165 | 43.3 | 6.36 | 76.76 | 90.64 | 0.08 | 0.05 | 0.73 | 30.62 | 158.7 | 85.4 | 144.38 | 802.04 | 1389 |
| Santiaguito #3 | 15 | 95-25 | spring | 01/08/95 | 1111 | 43.5 | 6.29 | 79.21 | 115.41 | 0.17 | 1.70 | 0.77 | 33.52 | 172.6 | 107.2 | 169.86 | 884.25 | 1565 |
| Rio Samala Spring | 16 | 95-33 | spring | 01/11/95 | 1982 | 71.8 | 6.8 | 40.19 | 35.25 | 0.06 | 0.22 | 0.03 | 35.01 | 274.8 | 151.0 | 158.1 | 594.94 | 1290 |
| Rio Samala Spring | 16 | 95-B28 | spring | 08/16/95 | 1982 | | | | | | | | | | | | | |
| Finca La Florida | 17 | 94-19 | spring | 08/22/94 | 1028 | 25.7 | 6.8 | 57.42 | 57.06 | 0.08 | 0.06 | 0.60 | 10.67 | 63.3 | 47.8 | 55.3 | 456.34 | 748.6 |
| Finca La Florida | 17 | 95-21 | spring | 01/07/95 | 1028 | 25.4 | 6.2 | 53.21 | 52.11 | 0.08 | 0.07 | 0.52 | 10.00 | 59.4 | 47.4 | 53.4 | 491.60 | 767.8 |
| Finca La Florida | 17 | 95-B25 | spring | 08/15/95 | 1028 | 25.5 | 8.34 | 45.82 | 47.21 | 0.00 | 0.08 | 0.46 | 9.13 | 50.4 | 34.6 | 41.203 | 422.58 | 651.5 |
| Agua Sabina | 18 | 94-14 | spring | 08/21/94 | 1040 | 22.5 | 7.0 | 46.05 | 48.75 | 0.07 | 0.08 | 0.41 | 8.65 | 43.2 | 36.8 | 38.1 | 438.27 | 660.4 |
| Agua Sabina | 18 | 95-18 | spring | 01/07/95 | 1040 | 22.1 | 6.42 | 32.96 | 38.13 | 0.09 | 0.08 | 0.28 | 7.32 | 33.3 | 34.4 | 29.1 | 320.41 | 496.1 |
| Agua Sabina | 18 | 95-B23 | spring | 08/15/95 | 1040 | 22.7 | 9.3 | 29.96 | 35.86 | 0.00 | 0.08 | 0.26 | 6.58 | 31 | 20.4 | 21.282 | 307.80 | 453.5 |
| Chichicaste | 19 | 94-13 | spring | 08/21/94 | 1037 | 20.2 | 5.3 | 33.63 | 27.86 | 0.05 | 0.08 | 0.28 | 6.31 | 22.2 | 18.2 | 24.2 | 264.07 | 396.9 |

*Continued*

TABLE 1. CHEMICAL AND PHYSICAL DATA FOR WATERS FROM THE SANTA MARÍA REGION (continued)

| Location Name | Location [Fig.2] | Sample ID | Type | Date mo/dy/yr | Elev. masl | Temp °C | pH | Ca ppm | Mg ppm | Fe ppm | Mn ppm | Sr ppm | K ppm | Na ppm | Cl ppm | SO$_4$ ppm | HCO$_3$ ppm | TDS ppm |
|---|---|---|---|---|---|---|---|---|---|---|---|---|---|---|---|---|---|---|
| Chichicaste | 19 | 95-17 | spring | 01/06/95 | 1037 | 20.2 | 6.49 | 34.92 | 28.56 | 0.09 | 0.08 | 0.26 | 5.86 | 21.6 | 20.1 | 25.8 | 270.46 | 407.7 |
| Chichicaste | 19 | 95-B12 | spring | 08/13/95 | 1037 | 21.1 | 6.59 | 26.14 | 20.4 | 0.02 | 0.06 | 0.2 | 4.87 | 16.7 | 10.3 | 15.747 | 216.73 | 311.2 |
| Almolonga Well | 20 | 95-B2 | well | 01/11/95 | 2288 | 36.9 | 8.23 | 11.26 | 5.74 | 0.19 | 0.03 | 0.07 | 6.23 | 110.1 | 55.1 | 82.5 | 182.90 | 456.5 |
| Banos Cirilo Flores Well | 21 | 94-17 | well | 08/21/94 | 2255 | 46.4 | 6.92 | 12.66 | 3.95 | 0.06 | 0.41 | 0.06 | 6.64 | 90.7 | 51.5 | 75.5 | 147.18 | 388.7 |
| Banos Cirilo Flores Well | 21 | 95-9 | well | 01/04/95 | 2255 | 46.3 | 6.92 | 12.68 | 4.55 | 0.07 | 0.34 | 0.06 | 7.10 | 94.9 | 53.5 | 77.4 | 145.07 | 395.7 |
| Banos Cirilo Flores Well | 21 | 95-B9 | well | 08/12/95 | 2255 | 46.5 | 6.69 | 11.46 | 4.73 | 0.04 | 0.33 | 0.05 | 7.49 | 105.8 | 48.4 | 77.1 | 140.50 | 395.9 |
| Banos Chicovix | 22 | 94-2 | spring | 08/18/94 | 2177 | 48.3 | 5.1 | 9.23 | 4.60 | 0.03 | 0.07 | 0.08 | 12.25 | 80.3 | 58.1 | 26.3 | 146.96 | 337.9 |
| Banos Chicovix | 22 | 95-8 | spring | 01/04/95 | 2177 | 47.4 | 7.58 | 9.69 | 3.69 | 0.08 | 0.06 | 0.06 | 10.86 | 69.7 | 4.3 | 59.6 | 143.67 | 301.8 |
| Banos Chicovix | 22 | 95-B10 | spring | 08/12/95 | 2177 | 47.1 | 7.45 | 9.05 | 4.13 | 0.01 | 0.06 | 0.06 | 12.28 | 82.5 | 54.6 | 22.6 | 141.66 | 327.0 |
| Zunil Cold Spring | 12 | 95-12 | spring | 01/05/95 | 2229 | 24.9 | 6.1 | 17.44 | 11.42 | 0.07 | 0.07 | 0.09 | 3.65 | 11.1 | 7.3 | 19.1 | 120.36 | 190.6 |
| Armadillo | 23 | 94-16 | spring | 08/21/94 | 910 | 22.5 | 5.15 | 7.26 | 9.18 | 0.02 | 0.20 | 0.14 | 1.18 | 4.3 | 5.0 | 4.8 | 74.47 | 106.6 |
| Armadillo | 23 | 95-20 | spring | 01/07/95 | 910 | 22.7 | 5.15 | 8.68 | 11.58 | 0.04 | 0.25 | 0.12 | 1.18 | 6.0 | 3.2 | 2.3 | 109.78 | 145.1 |
| Armadillo | 23 | 95-B11 | spring | 08/13/95 | 910 | 23.0 | 5.1 | 6.55 | 8.31 | 0.01 | 0.19 | 0.11 | 0.51 | 3.4 | 3.0 | 2.263 | 78.06 | 104.5 |
| Inde Dam Spring | 24 | 94-20 | spring | 08/23/94 | 1625 | 18.6 | 6.3 | 13.42 | 8.77 | 0.04 | 0.05 | 0.08 | 3.87 | 19.5 | 8.9 | 30.3 | 81.25 | 166.2 |
| Inde Dam Spring | 24 | 95-14 | spring | 01/06/95 | 1625 | 18.5 | 7.27 | 12.56 | 8.79 | 0.01 | 0.06 | 0.09 | 4.06 | 22.50 | 9.0 | 31.3 | 93.70 | 182.1 |
| Inde Dam Spring | 24 | 95-B26 | spring | 08/16/95 | 1625 | 18.7 | 10.0 | 13.33 | 7.86 | 0.00 | 0.06 | 0.07 | 3.62 | 20.40 | 5.8 | 26.405 | 93.64 | 172.2 |
| Finca El Faro | 25 | 94-15 | spring | 08/21/94 | 975 | 22.0 | 6.7 | 15.98 | 7.01 | 0.02 | 0.07 | 0.15 | 0.26 | 6.7 | 3.7 | 8.2 | 89.05 | 131.1 |
| Finca El Faro | 25 | 95-19 | spring | 01/07/95 | 975 | 22.1 | 6.89 | 12.69 | 6.05 | 0.04 | 0.07 | 0.1 | 0.00 | 5.1 | 1.9 | 3.1 | 81.86 | 113.2 |
| Finca El Faro | 25 | 95-B22 | spring | 08/15/95 | 975 | 22.4 | 8.81 | 14.00 | 6.64 | 0.00 | 0.08 | 0.12 | 0.00 | 6.4 | 2.1 | 3.003 | 80.26 | 115.8 |
| Poligono | 26 | 94-3 | spring | 08/19/94 | 2692 | 12.5 | 6.3 | 29.95 | 13.89 | 0.04 | 0.07 | 0.18 | 5.49 | 12.3 | 31.3 | 40.562 | 80.72 | 216.8 |
| Poligono | 26 | 95-26 | spring | 01/09/95 | 2692 | 11.5 | 7.03 | 27.96 | 15.43 | 0.03 | 0.01 | 0.16 | 6.58 | 11.3 | 34.2 | 45.9 | 98.02 | 239.6 |
| Poligono | 26 | 95-B17 | spring | 08/14/95 | 2692 | 13.3 | 9.27 | 32.30 | 15.78 | 0.00 | 0.08 | 0.17 | 6.76 | 11.9 | 33.8 | 67.052 | 72.46 | 243.0 |
| Finca Parador Los Trece | 27 | 94-6 | spring | 08/19/94 | 2562 | 12.8 | 6.2 | 25.37 | 6.77 | 0.05 | 0.08 | 0.18 | 2.84 | 8.7 | 20.1 | 31.1 | 60.82 | 156.0 |
| Finca Parador Los Trece | 27 | 95-29 | spring | 01/09/95 | 2562 | 12.6 | 7.24 | 24.88 | 7.05 | 0.09 | 0.01 | 0.17 | 2.79 | 9.1 | 22.3 | 30.5 | 65.24 | 162.1 |
| Finca El Canada | 28 | 94-8 | spring | 08/20/94 | 1448 | 18.3 | 5.5 | 7.46 | 2.74 | 0.03 | 0.07 | 0.07 | 0.41 | 4.8 | 2.9 | 3.9 | 46.99 | 69.37 |
| Finca El Canada | 28 | 95-22 | spring | 01/07/95 | 1448 | 18.2 | 7.66 | 8.72 | 3.30 | 0.04 | 0.06 | 0.05 | 0.15 | 5.3 | 1.2 | 1.4 | 63.38 | 83.60 |
| Finca El Canada | 28 | 95-B16 | spring | 08/13/95 | 1448 | 18.5 | 9.05 | 9.12 | 3.15 | 0.00 | 0.06 | 0.05 | 0.00 | 4.7 | 1.1 | 2.256 | 67.92 | 88.50 |
| Agua Cascada | 29 | 95-B24 | spring | 08/15/95 | 1575 | 16.3 | 11.1 | 12.48 | 4.54 | 0.00 | 0.07 | 0.05 | 1.97 | 7.2 | 9.1 | 15.791 | 56.58 | 118.9 |
| Rio Cuache I | 30 | 94-4 | spring | 08/19/94 | 2385 | 12.6 | 6.2 | 13.94 | 4.22 | 0.03 | 0.07 | 0.10 | 2.96 | 9.0 | 19.6 | 25.0 | 32.71 | 107.6 |
| Rio Cuache I | 30 | 95-28 | spring | 01/09/95 | 2385 | 12.7 | 7.23 | 13.21 | 3.80 | 0.04 | 0.02 | 0.07 | 2.66 | 9.2 | 14.7 | 20.3 | 61.51 | 125.5 |
| Rio Cuache I | 30 | 95-B20 | spring | 08/14/95 | 2385 | 13.5 | 9.08 | 12.60 | 3.71 | 0.00 | 0.08 | 0.08 | 2.29 | 8.0 | 15.0 | 19.23 | 44.73 | 105.8 |
| Santiaguito Stream | 31 | 94-18 | stream | 08/22/94 | 1430 | 20.3 | n/a | 15.15 | 6.13 | 0.03 | 0.05 | 0.11 | 2.18 | 5.1 | 5.8 | 41.854 | 43.35 | 119.8 |
| Rio Cuache II | 32 | 94-5 | spring | 08/19/94 | 2399 | 13.2 | 6.0 | 14.04 | 5.08 | 0.03 | 0.08 | 0.11 | 3.16 | 9.4 | 18.9 | 24.2 | 41.27 | 116.3 |
| Rio Cuache II | 32 | 95-27 | spring | 01/09/95 | 2399 | 12.9 | 7.23 | 13.89 | 4.90 | 0.13 | 0.02 | 0.07 | 3.05 | 9.6 | 20.1 | 24.4 | 41.69 | 117.9 |
| Rio Cuache II | 32 | 95-B19 | spring | 08/14/95 | 2399 | 13.6 | 9.43 | 12.79 | 4.31 | 0.00 | 0.08 | 0.08 | 2.36 | 10.6 | 15.8 | 21.2 | 34.78 | 102.8 |
| F. San Juan Patzulin Stream | 33 | 94-9 | waterfall | 08/20/94 | 1401 | 19.1 | 5.7 | 5.76 | 1.93 | 0.03 | 0.07 | 0.07 | 1.37 | 3.1 | 4.7 | 6.0 | 29.16 | 52.19 |
| F. San Juan Patzulin Stream | 33 | 95-15 | stream | 01/06/95 | 1451 | 16.8 | 8.92 | 5.97 | 1.87 | 0.08 | 0.07 | 0.04 | 0.48 | 4.9 | 3.4 | 3.643 | 38.04 | 59 |
| F. San Juan Patzulin Stream | 33 | 95-B13 | stream | 08/13/95 | 1451 | 17.7 | 8.48 | 5.32 | 1.72 | 0.02 | 0.06 | 0.04 | 0.02 | 2.4 | 2.4 | 3.1 | 21.22 | 36.52 |
| Finca La Quina | 34 | 95-16 | spring | 01/06/95 | 1363 | 18.6 | 7.24 | 7.01 | 1.44 | 0.04 | 0.08 | 0.04 | 1.82 | 5.7 | 3.8 | 6.1 | 40.73 | 66.76 |
| Finca La Quina | 34 | 95-B15 | spring | 08/13/95 | 1363 | 18.9 | 9.13 | 6.89 | 1.46 | 0.02 | 0.06 | 0.06 | 1.54 | 3.6 | 1.7 | 3.5 | 30.28 | 49.76 |
| Chloride | | | | | | 67.4 | 6.24 | 135.4 | 13.28 | 0.51 | 0.53 | 1.15 | 35.64 | 505.7 | 614.0 | 381.3 | 391.70 | 2079 |
| Banos De Ocos | 35 | 95-31 | spring | 01/10/95 | 2273 | | | | | | | | | | | | | |

Note: masl—meters above sea level; n/a—not applicable

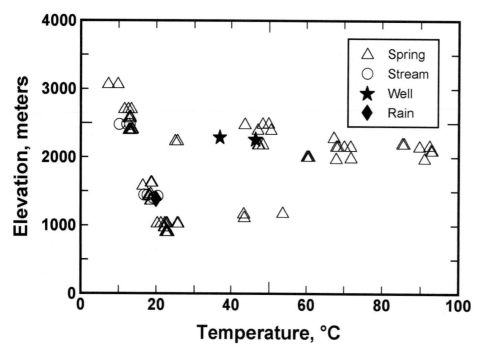

Figure 3. Temperature versus elevation for waters sampled around Santa María volcano.

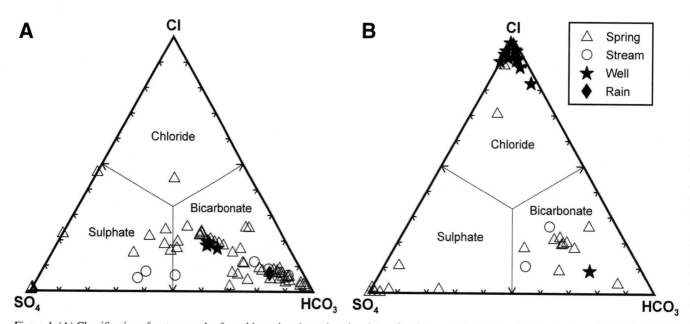

Figure 4. (A) Classification of water samples from this study using anion chemistry after Giggenbach (1988) and Giggenbach et al. (1990). (B) Previously collected anion data on water samples from Zunil and Zunil-II (Sulfur Mountain) (data from Adams et al., 1990; Giggenbach et al., 1992).

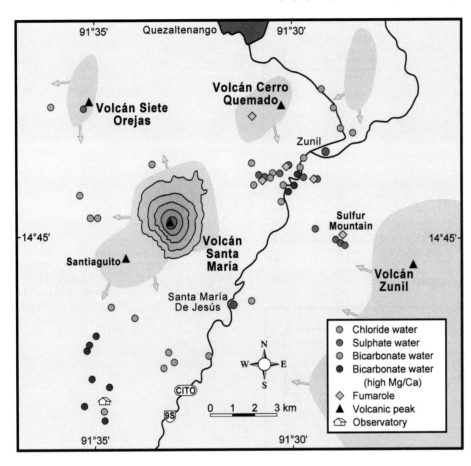

Figure 5. Distribution of water types around Santa María volcano. Includes data from Adams et al. (1990) and Giggenbach et al. (1992). Blue areas are hypothesized recharge areas (see text).

waters above 50–60 °C, on the other hand, are characterized by falling Na contents with increasing temperature (Fig. 8A). Sulfate waters are also characterized by elevated Fe contents (Table 1). Most of Santa María's waters have low (<15 ppm) K concentrations, although a sizable fraction have K > 30 ppm (Table 1). None of the waters from Santa María, with the exception of some of the deep well samples from Zunil (A. Adams et al., 1990; Giggenbach et al., 1992), have reached equilibrium with the altered volcanic rock of the area, classifying them as "immature" waters according to the various geochemical criteria developed by Giggenbach (1988) (Fig. 9).

The stable isotopic compositions of the waters from Santa María are given in Table 2. Their $\delta^{13}C$ values fall in a narrow range from −11.5‰ to −8.5‰ (Table 2). Bicarbonate waters exhibit the largest range in $\delta^{13}C$, with the high Mg/Ca bicarbonate waters from springs south-southwest of Santiaguito being distinctly enriched (Fig. 10).

The $\delta^{18}O$ and $\delta D$ values of rainwater and nonthermal, dilute springs around Santa María define a local meteoric water line nearly identical to the global meteoric line of Craig (1961) (Fig. 11; Templeton, 1999). Only some of the sulfate and chloride waters from the Zunil–Sulfur Mountain area fall significantly off this local meteoric water line (Fig. 11), attesting to the dominance of a meteoric input around most of Santa María. M. Adams et al. (1992) have argued that the $\delta D$ values of waters, largely river waters, from the Zunil-Quetzaltenango area gradually increase with elevation, contrary to the commonly observed isotopic depletion in precipitation with increasing elevation (Dansgaard, 1964). The bicarbonate springs from around Santa María, however, offer strong evidence for stable isotope depletion with increasing elevation for this region, as shown in Figure 12. Sulfate waters, on the other hand, exhibit little to no correlation between $\delta^{18}O$-$\delta D$ and elevation (Fig. 12), implicating other causes in their observed isotopic variability.

**Recharge Elevations**

Correlations between elevation and $\delta^{18}O$-$\delta D$ can be used to determine the elevations of groundwater recharge (Vuataz and Goff, 1986; Janik et al., 1992; Nathenson et al., 2003). The first step is to determine a reference line from springs with known recharge elevations. Since actual recharge elevations are unknown in the study area, they have been estimated from topographic and geologic considerations for six of the samples used to construct the local meteoric water line in Figure 11. These estimated recharge elevations were then used to derive the linear regression lines shown in Figures 12A and 12B. The slopes of these lines yield isotopic gradients of −0.39‰/100 m (O) and −2.7‰/100 m (D), similar to those calculated for other volcanic terrains (Vuataz and Goff, 1986; Janik et al., 1992; Roses et al., 1996; Nathenson

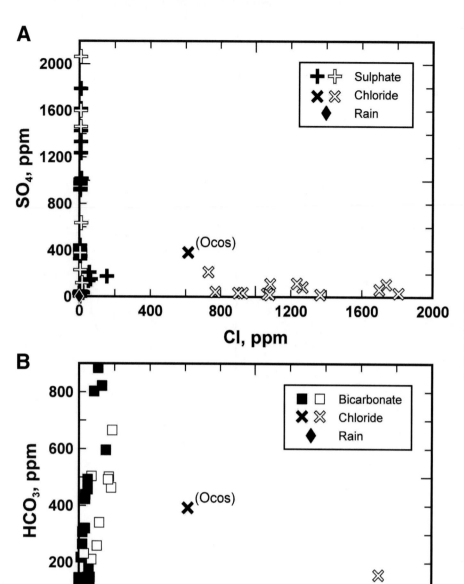

Figure 6. (A) Cl versus $SO_4$ for sulfate and chloride waters. (B) Cl versus $HCO_3$ for bicarbonate and chloride waters. Closed symbols are data from this study. Open symbols are data from Adams et al. (1990) and Giggenbach et al. (1992).

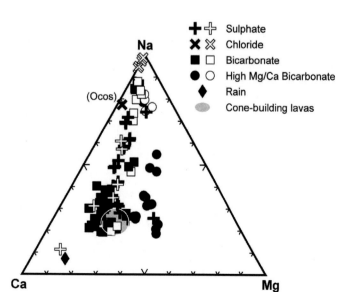

Figure 7. Relative Ca, Na, and Mg contents of waters from the Santa María vicinity. All concentrations are in ppm. Circled shaded area shows relative Ca, Na, and Mg contents of lavas that have constructed Santa María's cone. Closed symbols: data from this study; open symbols: data from Adams et al. (1990) and Giggenbach et al. (1992).

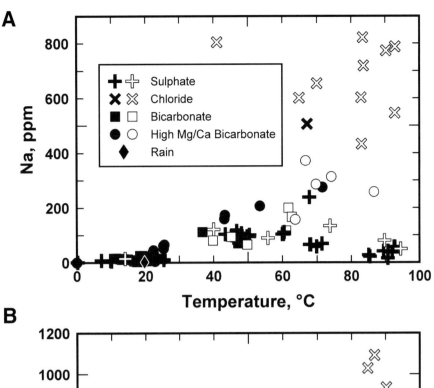

Figure 8. (A) Temperature versus Na concentrations for waters girdling Santa María. Data and symbols as in Figure 7 with the exclusion of three chloride waters from the highest temperature wells at Zunil for better clarity at low Na contents. (B) Cl versus Na concentrations for all waters, as in Figure 7. Closed symbols: data from this study; open symbols: data from Adams et al. (1990) and Giggenbach et al. (1992).

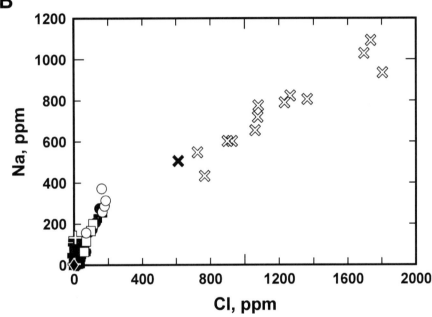

Figure 9. Trilinear diagram of Na/400, K/10, and $(Mg)^{0.5}$ after Shevenell and Goff (1993). Concentrations representing full water-rock equilibrium at temperatures from 20 °C to 300 °C taken from Table II of Giggenbach (1988). Closed symbols: data from this study; open symbols: data from Adams et al. (1990) and Giggenbach et al. (1992).

TABLE 2. STABLE ISOTOPIC COMPOSITIONS OF WATERS FROM THE SANTA MARÍA REGION

| Location | Sample ID | Date mo/dy/yr | Elevation masl | $\delta^{13}C$ | $\delta^{18}O$ | $\delta D$ |
|---|---|---|---|---|---|---|
| Meteoric water | | | | | | |
| Rain | 94-10 | 08/20/94 | 1395 | −9.65* | −3.80* | −18.2[†] |
| Acid Sulfate | | | | | | |
| Banos Aguas Amargas | 94-7 | 08/20/94 | 1999 | −9.52* | −8.32*/[†] | −78.6[†] |
| Banos Aguas Amargas | 95-13 | 01/05/95 | 1999 | −9.48* | −7.90* | − |
| Banos Aguas Amargas | 95-B27 | 08/16/95 | 1999 | − | −8.72[§] | − |
| Azufrales Vent (Z38) | 95-B1 | 08/11/95 | 2148 | − | −6.69[§] | − |
| Azufrales Vent (Z38) | 95-B30 | 08/16/95 | 2148 | − | −6.34[§] | − |
| Fuentes Georginas #1 | 94-11 | 08/20/94 | 2470 | −9.57* | −9.80* | −85.2[†] |
| Fuentes Georginas #1 | 95-4 | 01/04/95 | 2470 | −9.91* | −9.97*/[†] | − |
| Fuentes Georginas #1 | 95-B5 | 08/12/95 | 2470 | − | −10.19[§] | − |
| Fuentes Georginas #2 | 94-21 | 08/23/94 | 2386 | −9.44* | −9.40* | − |
| Fuentes Georginas #2 | 95-6 | 01/04/95 | 2386 | −9.51* | −9.65* | − |
| Fuentes Georginas #2 | 95-B7 | 08/12/95 | 2386 | − | −10.43[§] | − |
| Portales Mud Pot | 95-10 | 01/05/95 | 2092 | −9.69* | −7.95* | − |
| Portales Mud Pot | 95-B8 | 08/12/95 | 2092 | − | −7.49[§] | − |
| ZMF-12 | 95-34 | 01/11/95 | 1967 | − | − | − |
| ZMF-12 | 95-B29 | 08/16/95 | 1967 | − | −11.48[§] | − |
| Zunil (Z47) | 95-11 | 01/05/95 | 2190 | −9.98* | −10.90* | − |
| Zunil (Z47) | 95-B3 | 08/11/95 | 2190 | − | −11.01[§] | − |
| Zunil (Z39) | 94-1 | 08/18/94 | 2148 | − | −12.57[†] | −92.9[†] |
| Zunil (Z39) | 94-22 | 08/24/94 | 2148 | − | −12.69[†] | −96.8[†] |
| Zunil (Z39) | 95-7 | 01/04/95 | 2148 | −9.60* | −11.50* | − |
| Zunil (Z39) | 95-B2 | 08/11/95 | 2148 | − | −12.15[§] | − |
| Georginas Stream | 94-12 | 08/20/94 | 2478 | − | − | − |
| Georginas Stream | 95-5 | 01/04/95 | 2478 | − | − | − |
| Georginas Stream | 95-B6 | 08/12/95 | 2478 | − | −11.65[§] | − |
| Pinabet Cold Spring | 95-30 | 01/09/95 | 3061 | −9.89* | −11.35*/[†] | −83.8[†] |
| Pinabet Cold Spring | 95-B21 | 08/14/95 | 3061 | − | −11.13[§] | − |
| Zunil Cold Spring | 95-B4 | 08/11/95 | 2229 | − | −12.13[§] | − |
| Neutral Bicarbonate | | | | | | |
| Santiaguito #1 | 95-23 | 01/08/95 | 1175 | −8.75* | −7.53*/[†] | −57.7[†] |
| Santiaguito #2 | 95-24 | 01/08/95 | 1165 | −8.89* | −7.80* | − |
| Santiaguito #3 | 95-25 | 01/08/95 | 1111 | −8.69* | −8.10* | − |
| Rio Samala | 95-33 | 01/11/95 | 1982 | −9.82* | −10.80* | − |
| Rio Samala | 95-B28 | 08/16/95 | 1982 | − | −11.09[§] | − |
| Finca La Florida | 94-19 | 08/22/94 | 1028 | − | − | − |
| Finca La Florida | 95-21 | 01/07/95 | 1028 | −8.94* | −7.87*/[†] | −55.3[†] |
| Finca La Florida | 95-B25 | 08/15/95 | 1028 | − | −7.73[§] | − |
| Agua Sabina | 94-14 | 08/21/94 | 1040 | − | − | − |
| Agua Sabina | 95-18 | 01/07/95 | 1040 | − | − | − |
| Agua Sabina | 95-B23 | 08/15/95 | 1040 | − | −7.28[§] | − |
| Chichicaste | 94-13 | 08/21/94 | 1037 | − | −8.79[†] | −60.1[†] |
| Chichicaste | 95-17 | 01/06/95 | 1037 | − | − | − |
| Chichicaste | 95-B12 | 08/13/95 | 1037 | − | −7.54[§] | − |
| Almolonga Well | 95-B2 | 01/11/95 | 2288 | − | − | − |
| Banos Cirilo Flores Well | 94-17 | 08/21/94 | 2255 | − | − | − |
| Banos Cirilo Flores Well | 95-9 | 01/04/95 | 2255 | − | − | − |
| Banos Cirilo Flores Well | 95-B9 | 08/12/95 | 2255 | − | −11.42[§] | − |
| Banos Chicovix | 94-2 | 08/18/94 | 2177 | −10.79* | −11.40* | − |
| Banos Chicovix | 95-8 | 01/04/95 | 2177 | −9.90* | −10.20* | − |
| Banos Chicovix | 95-B10 | 08/12/95 | 2177 | − | −10.79[§] | − |
| Zunil Cold Spring | 95-12 | 01/05/95 | 2229 | −11.50* | −11.50* | − |
| Armadillo | 94-16 | 08/21/94 | 910 | −9.23* | −6.30* | − |
| Armadillo | 95-20 | 01/07/95 | 910 | −8.50* | −6.40* | − |
| Armadillo | 95-B11 | 08/13/95 | 910 | − | −6.44[§] | − |

*Continued*

TABLE 2. STABLE ISOTOPIC COMPOSITIONS OF WATERS FROM THE SANTA MARÍA REGION (continued)

| Location | Sample ID | Date mo/dy/yr | Elevation masl | $\delta^{13}C$ | $\delta^{18}O$ | $\delta D$ |
|---|---|---|---|---|---|---|
| Inde Dam Spring | 94-20 | 08/23/94 | 1625 | −9.69* | −10.93*/† | −83.1† |
| Inde Dam Spring | 95-14 | 01/06/95 | 1625 | −9.90* | −10.40* | − |
| Inde Dam Spring | 95-B26 | 08/16/95 | 1625 | − | −10.80§ | − |
| Finca El Faro | 94-15 | 08/21/94 | 975 | −9.91* | −6.75*/† | −48.2† |
| Finca El Faro | 95-19 | 01/07/95 | 975 | −10.27* | −6.30* | − |
| Finca El Faro | 95-B22 | 08/15/95 | 975 | − | −6.17§ | − |
| Poligono | 94-3 | 08/19/94 | 2692 | −9.84* | −12.12*/† | −93.5† |
| Poligono | 95-26 | 01/09/95 | 2692 | −9.79* | −12.00* | − |
| Poligono | 95-B17 | 08/14/95 | 2692 | − | −12.18§ | − |
| Finca Parador Los Trece | 94-6 | 08/19/94 | 2562 | − | − | − |
| Finca Parador Los Trece | 95-29 | 01/09/95 | 2562 | − | −11.44† | −82.1† |
| Finca Parador Los Trece | 95-B18 | 08/14/95 | 2562 | − | −9.93§ | − |
| Finca El Canada | 94-8 | 08/20/94 | 1448 | −9.99* | −7.40* | − |
| Finca El Canada | 95-22 | 01/07/95 | 1448 | −9.96* | −7.40* | − |
| Finca El Canada | 95-B16 | 08/13/95 | 1448 | − | −7.58§ | − |
| Agua Cascada | 95-B24 | 08/15/95 | 1575 | − | −7.05§ | − |
| Rio Cuache I | 94-4 | 08/19/94 | 2385 | − | − | − |
| Rio Cuache I | 95-28 | 01/09/95 | 2385 | − | − | − |
| Rio Cuache I | 95-B20 | 08/14/95 | 2385 | − | −10.45§ | − |
| Santiaguito Stream | 94-18 | 08/22/94 | 1430 | −10.84* | −7.22*/† | −45.4† |
| Rio Cuache II | 94-5 | 08/19/94 | 2399 | −9.83* | 10.60* | − |
| Rio Cuache II | 95-27 | 01/09/95 | 2399 | −10.25* | −11.10* | − |
| Rio Cuache II | 95-B19 | 08/14/95 | 2399 | − | −10.61§ | − |
| F. San Juan Patzulin Stream | 94-9 | 08/20/94 | 1401 | − | − | − |
| F. San Juan Patzulin Stream | 95-15 | 01/06/95 | 1451 | −9.81* | −7.38*/† | −49.2† |
| F. San Juan Patzulin Stream | 95-B13 | 08/13/95 | 1451 | − | −7.88§ | − |
| Finca La Quina | 95-16 | 01/06/95 | 1363 | −9.86* | −6.99*/† | −46.9† |
| Finca La Quina | 95-B15 | 08/13/95 | 1363 | − | −7.15§ | − |
| **Chloride** | | | | | | |
| Banos De Ocos | 95-31 | 01/10/95 | 2273 | −9.07* | −10.69*/† | −80.7† |

*Note:* masl—meters above sea level.
\*data from Argonne National Laboratory.
†data from the U.S. National Aeronautics and Space Administration.
§data from Washington University.

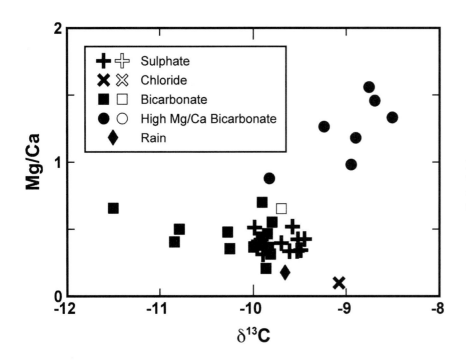

Figure 10. $\delta^{13}C$ versus Mg/Ca for Santa María's waters. Closed symbols: data from this study; open symbols: data from Adams et al. (1990) and Giggenbach et al. (1992).

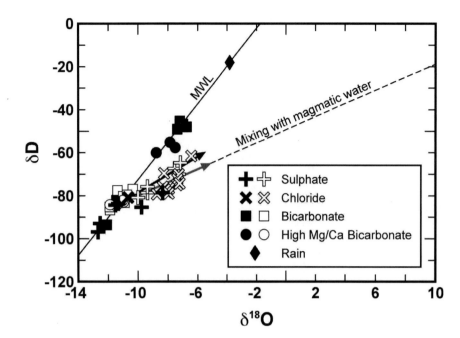

Figure 11. $\delta^{18}O$ versus $\delta D$ for waters from the Santa María region. MWL is the local meteoric water line estimated by Templeton (1999). Dashed line shows hypothesized trend for mixing with magmatic ("andesitic") water (filled box; composition from Giggenbach [1992b]). Gray arrow shows isotopic changes expected in waters for nonequilibrium evaporation at temperatures of ~70–90 °C (Craig, 1963). Black arrow shows isotopic changes calculated to occur with subsurface, single-step boiling at a temperature of 140 °C (from equations given by Fischer et al. [1997] using fractionation factors from Truesdell et al. [1977]). Closed symbols: data from this study; open symbols: data from Adams et al. (1990) and Giggenbach et al. (1992).

et al., 2003). Recharge elevations for the remaining waters are determined by projecting the data in Figure 12 vertically onto the reference lines (e.g., Vuataz and Goff, 1986). Although recharge elevations calculated from the oxygen and deuterium gradients for the waters around Santa María are generally comparable, the latter are preferred because $^{18}O$ is much more susceptible to modification via water-rock interaction in hydrothermal systems (e.g., Vuataz and Goff, 1986). Deuterium-based recharge elevations for Santa María's waters range from 1650–3600 m (Templeton, 1999). The calculated recharge elevations have been used, in conjunction with topographic information, to locate the recharge areas for Santa María's waters (Fig. 5). Not surprisingly, the likely recharge areas are Santa María itself and its volcanic neighbors.

## GASES

Corrected fumarolic gas compositions are reported in Table 3. No temporal and/or seasonal changes are obvious from the data. The relative proportions of nonreactive gases after standard corrections for air contamination are shown in Figure 13. Subduction zone gases generally have high $N_2$ with $N_2/Ar \geq 100$ and $N_2/He \geq 1000$ (Matsuo et al., 1978; Kiyosu, 1986; Hedenquist and Aoki, 1991; Giggenbach, 1992a; Fischer et al., 1998, 2002; Tassi et al., 2003; Zimmer et al., 2004). In contrast, nonsubduction gases have much lower $N_2$, with $N_2/He \leq 200$ (Matsuo et al., 1978; Giggenbach, 1992a; Fischer et al., 1998; Marty and Zimmerman, 1999). The high $N_2$ contents of volcanic gases from subduction zones has commonly been attributed to recycling of nitrogen from subducted oceanic sediments (Matsuo et al., 1978; Kita et al., 1993; Fischer et al., 1998, 2002; Sano et al., 2001; Zimmer et al., 2004). Zimmer et al. (2004) have recently shown that the gases from Poás volcano in the Costa Rican portion of the Central American subduction zone lack high $N_2$ contents and hence may lack a subducted sedimentary component.

Not surprisingly, all but two of the gas samples from the Santa María region fall within the subduction zone field in terms of nonreactive gases (Fig. 13). Most are displaced toward the median to lower portions of the $N_2$-Ar tie line, indicating substantial contributions from air-saturated groundwater or air (Matsuo et al., 1978; Giggenbach, 1992a). A few samples, however, are slightly shifted toward the He apex, suggesting an increased mantle signature. Two problematic samples from a very diffuse vent at Cerro Quemado contain no Ar after our air correction procedure.

The stable isotope composition of gases collected from the Santa María region are given in Table 3. Volcanic gases from many subduction zone volcanoes have $\delta^{13}C$ between –10‰ and –2‰ (Fig. 14A). Volcanic gases from most volcanoes of the Central American volcanic front, by contrast, have characteristically higher $\delta^{13}C$ ratios (> –4‰) (Fig. 14B; Snyder et al., 2001; Shaw et al., 2003). Fumarolic gases from the Santa María region, however, have notably lighter $\delta^{13}C$ ratios, ranging from –14‰ to –4‰ (Fig. 14B). This range in $\delta^{13}C$ is nearly identical to that seen in gas samples taken behind the volcanic front in Honduras (Fig. 14B; Snyder et al., 2001). The fumarolic gases from Santa María also display an intriguing correlation between $\delta^{13}C$ and $CO_2/He$ (Fig. 15).

## DISCUSSION

### Magmatic Signals in Surface Fluids

Giggenbach et al. (1992) and Giggenbach (1992b) have argued that some of the chloride and sulfate waters from the Zunil geothermal field that are enriched in $\delta^{18}O$ and $\delta D$ relative

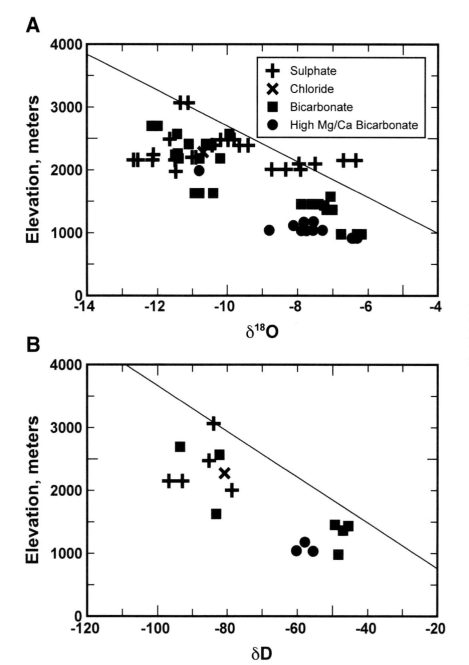

Figure 12. $\delta^{18}O$ (A) and $\delta D$ (B) versus elevation for waters analyzed in this study. Lines are linear regression lines from Templeton (1999) used to estimate recharge elevations (see text for details).

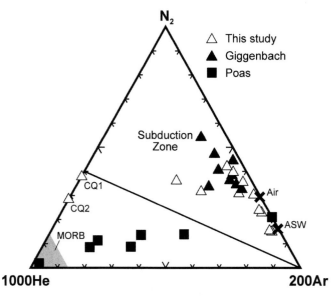

Figure 13. Triangular diagram of the relative amounts of He, $N_2$, and Ar in gas samples from fumaroles around Santa María volcano. Includes data from this study, Giggenbach (1992a), and Giggenbach et al. (1992). Values for air and air saturated water (ASW) from Giggenbach (1992a). Gas data from Poás volcano in Costa Rica from Zimmer et al. (2004). Mid-oceanic-ridge basalt (MORB) field taken from data of Marty and Zimmerman (1999). Field for most subduction zone gases constructed from compilation of data from Kiyosu (1986); Mizutani et al. (1986); Sturchio et al. (1988, 1993); Poorter et al. (1991); Giggenbach (1992a); Janik et al. (1992); Sano and Marty (1995); Sano and Williams (1996); Fischer et al. (1997, 1998); Sano et al. (1998); Lewicki et al. (2000); Tassi et al. (2003); Zimmer et al. (2004). CQ1 and CQ2 are problematic samples from Cerro Quemado.

TABLE 3. FUMAROLIC GAS COMPOSITIONS

| Location | Location [Fig. 2] | Sample ID | Date mo/dy/yr | Temp °C | $H_2$ vol% | He vol% | $CH_4$ vol% | CO vol% | $N_2$ vol% | $C_2H_6$ vol% | $H_2S$ vol% | Ar vol% | $C_3H_8$ vol% | $\delta^{13}C$ ‰ | $CO_2$ umol/mol |
|---|---|---|---|---|---|---|---|---|---|---|---|---|---|---|---|
| Zunil Vent #1 | 36 | Z194 | 08/18/94 | 93.3 | 2.9053 | 0.0635 | 0.0564 | 0.1270 | 95.979 | 0.1270 | 0 | 0.7197 | 0.0225 | −11.17 | 141.1 |
| | | Z195 | 01/04/95 | 93.4 | 2.3611 | 0.0743 | 0.0660 | 0.1486 | 96.721 | 0.1156 | 0 | 0.4804 | 0.0330 | −11.93 | 131.2 |
| | | Z195B | 08/11/95 | 89.7 | 20.18 | 0.015 | 0.2673 | 0.2373 | 77.081 | 0.0367 | 0.0334 | 2.099 | 0.0467 | – | – |
| | | Z195Bd | 08/16/95 | 92.3 | 20.83 | 0.0166 | 0.2294 | 0.1221 | 76.707 | 0.0333 | 0.0370 | 1.9986 | 0.0259 | – | – |
| Zunil Vent #2 | 37 | Z295 | 01/11/95 | 91.8 | 0.2390 | 0.0119 | 1.5315 | 0.033 | 97.72 | 0.0238 | 0 | 0.680 | 0.0087 | −4.59 | 1225.7 |
| | | Z295B | 08/11/95 | 91.8 | 0.2331 | 0.0091 | 1.3484 | 0.0136 | 97.526 | 0.0091 | 0.0212 | 0.8348 | 0.0045 | – | – |
| Sulfur Mountain | 38 | SM94 | 08/23/94 | 91.4 | 0.1601 | 0.0137 | 0.3200 | 0.0365 | 98.846 | 0.003 | 0 | 0.6127 | 0.0076 | −8.55 | 194.3 |
| | | SM95 | 01/04/95 | 91.6 | 0.021 | 0.0063 | 0.0196 | 0.0126 | 98.830 | 0.0126 | 0 | 1.0966 | 0.0011 | −7.62 | 272.5 |
| | | SM95B | 08/12/95 | 90.4 | 0.1557 | 0.017 | 0.5265 | 0.0640 | 98.722 | 0.0106 | 0.0213 | 0.4683 | 0.0149 | – | – |
| Cerro Quemado | 39 | CQ94 | 08/24/94 | 64.5 | 0.5377 | 0.1614 | 0.1434 | 0.3229 | 98.283 | 0.4659 | 0 | 0 | 0.0861 | −12.32 | 166.8 |
| | | CQ94 | 08/24/94 | 64.5 | – | – | – | – | 97.41 | 0.4618 | – | – | – | −12.55 | – |
| | | CQ95 | 01/13/95 | 64.5 | 0.8159 | 0.2448 | 0.2176 | 0.4895 | 97.623 | 0.5440 | 0 | 0 | 0.0653 | −13.86 | 165.2 |
| | | CQ95 | 01/13/95 | 64.5 | – | – | – | – | 96.624 | 0.5384 | – | – | – | −12.53 | – |
| Portales | 40 | P95 | 01/04/95 | 93.3 | 2.9373 | 0.0223 | 0.5467 | 0.0795 | 95.399 | 0.0448 | 0 | 0.9577 | 0.0124 | −9.16 | 119.4 |
| | | P95B | 08/12/95 | 92.8 | 16.547 | 0.0086 | 3.7850 | 0.0228 | 78.326 | 0.0152 | 0.019 | 1.2659 | 0.010 | – | – |
| | | P95Bd | 08/12/95 | 92.8 | 18.550 | 0.010 | 4.3345 | 0.0156 | 75.823 | 0.0832 | 0.0173 | 1.1541 | 0.0121 | – | – |

*Note:* Air corrected headspace results except for $\delta^{13}C$ and $CO_2$. Air corrections were made assuming that all oxygen was from air contamination. Nitrogen and argon corrections were made by subtracting the product of atmospheric proportions and percent air contamination. Totals were normalized to 100 after corrections were made and after removal of $CO_2$.

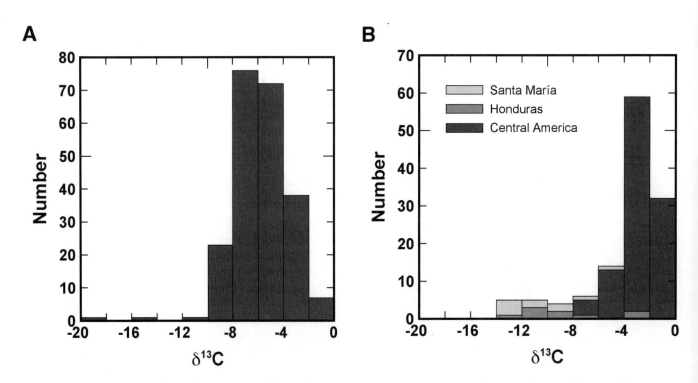

Figure 14. (A) $\delta^{13}C$ values of fumarolic gases from volcanoes at subduction zones other than Central America. Data from Lyon and Hulston (1984); Mizutani et al. (1986); Sturchio et al. (1988); Poorter et al. (1991); Varekamp et al. (1992); Sano and Marty (1995); Sano and Williams (1996); Fischer et al. (1997, 1998); Sano et al. (1997, 1998); van Soest et al. (1998); Lewicki et al. (2000); Tassi et al. (2003). (B) $\delta^{13}C$ of fumarolic gases from Santa María, Honduras, and other Central American volcanoes. Data from Janik et al. (1992); Sano and Marty (1995); Sano and Williams (1996); Snyder et al. (2001); Shaw et al. (2003).

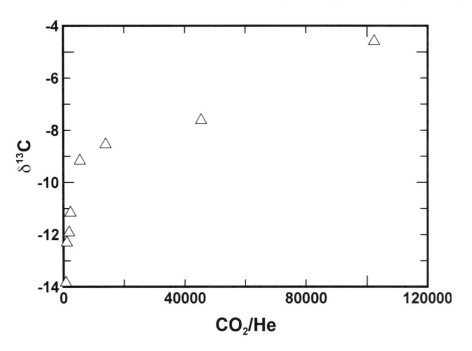

Figure 15. $CO_2$/He versus $\delta^{13}C$ for fumarolic gases from the Santa María vicinity. Data is from this study.

to local meteoric waters have perhaps 20% of an admixed magmatic component. Admixing of Giggenbach's magmatic component (i.e., "andesitic water") with a common meteoric water in the Santa María region is illustrated in Figure 11 and could explain the $\delta^{18}O$ and $\delta D$ enrichments seen in some chloride and sulfate waters. However, surface evaporation, single-step subsurface boiling, and water-rock interaction are other processes that can affect the $\delta^{18}O$ and $\delta D$ values of geothermal waters and need to be individually evaluated (Craig, 1963; Sakai and Matsubaya, 1977; A. Adams et al., 1990; Rowe, 1994). Although water-rock interaction can increase the $\delta^{18}O$ ratios of geothermal waters (Craig, 1963; Rowe 1994), it is unlikely to alter their $\delta D$ ratios because the low hydrogen contents of the rocks results in little leverage on the $\delta D$ of waters passing through them (Craig, 1963; Taylor 1974; Criss and Taylor, 1986; Delmelle et al., 2000). In addition, available data, although restricted, indicates that volcanic rocks of Santa María have similar $\delta D$ values to meteoric waters and thus have even less leverage for altering the latter's hydrogen isotopic ratio (Anderson et al., 1995). Single-step subsurface boiling at temperatures of 140–160 °C can yield increases in $\delta D$ and $\delta^{18}O$ similar to those of "enriched" waters around Santa María (Fig. 11), although it would require extensive subsurface boiling and should produce complimentary isotopic depletions in steam (Janik et al., 1992; Hinkley et al., 1995; Fischer et al., 1997; Goff et al., 2000), which have not been reported from the study area (Giggenbach et al., 1992). Nonequilibrium evaporation at temperatures of ~70–90 °C produces enrichments in both $\delta D$ and $\delta^{18}O$ with a slope of ~3 (Craig, 1963), similar to the trend defined by "enriched" waters from Santa María (Fig. 11). The stable isotope ratios of (acid) sulfate thermal waters, like those from the Zunil–Sulfur Mountain regions, are particularly susceptible to evaporative fractionation (Vuataz and Goff, 1986; Criss, 1995).

Hence, although we can't totally rule out any influence of a magmatic component, we suspect that the small enrichments in $\delta D$ and $\delta^{18}O$ in some chloride and sulfate waters from the Zunil and Sulfur Mountain areas are largely the result of subsurface boiling and nonequilibrium evaporation. M. Adams et al. (1992) also point out that the low Cl concentrations in the chloride waters from Zunil are inconsistent with a substantial magmatic input.

On the other hand, the $\delta^{13}C$ ratios of some gas samples from the Zunil–Sulfur Mountain region fall in the range for mid-oceanic-ridge basalts (MORB) (Fig. 16), generally an indication of substantial mantle magmatic input (Allard, 1983; Sano and Marty, 1995). This is supported by the two available $^3He/^4He$ measurements on gas samples from Zunil (6.32 and 4.69; Sano and Williams, 1996; Fischer et al., 2002), which indicate a significant mantle component (Poreda and Craig, 1989; Hilton et al., 1993; Sano et al., 1998; Fischer et al., 2002). Other gas samples from the Santa María vicinity, including those from Cerro Quemado, have lower $\delta^{13}C$, with some values extending well below the mantle and/or MORB range (Fig. 16). Samples with the lowest $\delta^{13}C$, however, also have the lowest $CO_2$/He ratios (Fig. 15). Since published helium isotopic data has shown the He to be highly magmatic, samples with the lowest $CO_2$/He ratios should have $\delta^{13}C$ values most characteristic of magmatic gas (e.g., Shaw et al., 2004). Hence, for at least portions of the Santa María region, we suggest that magmatic gas actually has $\delta^{13}C$ ratios in the −11‰ to −14‰ range, notably below values typical of the mantle. Low magmatic $\delta^{13}C$ ratios (−9‰ to −11‰) have also been estimated for Mount St. Helens, Lassen, and Three Sisters in the Cascadia subduction zone (Evans et al., 1981, 2004; Janik et al., 1983). Evans et al. (2004) attribute these low ratios to source contamination with isotopically light organic-rich sediments from the subducting plate (e.g., Sano and Marty, 1995). A similar

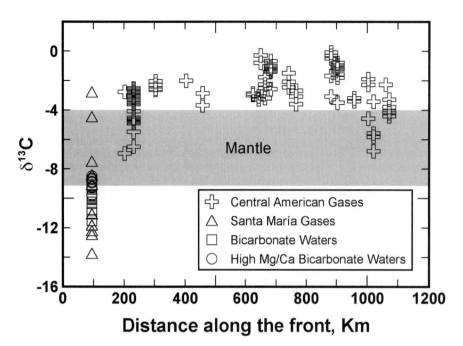

Figure 16. Changes in $\delta^{13}C$ in fumarolic gases with distance along the Central American volcanic front from Guatemala to Costa Rica. Data sources: Janik et al. (1992); Sano and Marty (1995); Sano and Williams (1996); Snyder et al. (2001); Shaw et al. (2003). Also shown are $\delta^{13}C$ in bicarbonate waters from Santa María. Estimated range of $\delta^{13}C$ in mantle rocks from Sano and Marty (1995).

explanation could explain the low magmatic $\delta^{13}C$ around Santa María and would be consistent with the anomalously high $N_2$/He and $\delta^{15}N$ ratios in volcanic gases from this same region (Fischer et al., 2002). However, two other mechanisms could also produce lowered magmatic $\delta^{13}C$. The first is contamination with shallow crustal material containing a substantial organic component (i.e., with $\delta^{13}C \approx -25‰$) (Lyon and Hulston, 1984; Roses et al., 1996; Heiligmann et al., 1997; Sano et al., 1998; Sorey et al., 1998; van Soest et al., 1998; Shaw et al., 2003). This process would also explain the moderately submantle $^3$He/$^4$He ratios (4.69–6.32) in gas samples from Zunil (Hilton et al., 1993; Sano et al., 1998; van Soest et al., 1998; Snyder et al., 2001). A second alternative for lowering magmatic $\delta^{13}C$ ratios is extensive subsurface degassing (Javoy et al., 1978; Gerlach and Taylor, 1990; Macpherson and Mattey, 1994).

The low $\delta^{13}C$ values around Santa María are unusual, as most volcanic gas samples from the Central American volcanic front have $\delta^{13}C$ values greater than MORB (Fig. 16; Snyder et al., 2001; Shaw et al., 2003). Since elevated values are found throughout the arc, at volcanoes built on variable crust and with variable eruptive histories (e.g., Carr et al., 2003), they are most likely the result of higher than normal contributions of $CO_2$ from subducted carbonate sediments, with $\delta^{13}C \approx 0$, in emitted volcanic gases (van Soest et al., 1998; Snyder et al., 2001; Shaw et al., 2003). In Central America, this stems from the unusually high proportions of subducting carbonate sediments (Plank and Langmuir, 1998; Poli and Schmidt, 2002). Thus, if source contamination were the cause of the lowered magmatic $\delta^{13}C$ in the Santa María region as discussed above, it would indicate that northernmost Guatemala is the only portion of the Central American arc where subducted organic-rich hemipelagic sediments, as opposed to carbonate sediments, control carbon recycling. Erupted lava compositions, however, bear no elemental evidence for enhanced hemipelagic contributions in northernmost Guatemala (Patino et al., 2000). Instead, available radiogenic isotopic data suggest that northernmost Guatemala is notable as the only portion of the Central American subduction zone where crustal contamination is significant (Feigenson and Carr 1986; Carr et al., 1990, 2003; Feigenson et al., 2004). This regional perspective suggests that crustal contamination may be the most viable explanation for the lowered magmatic $\delta^{13}C$ in this portion of the arc.

## Origin and Evolution of Waters Encircling Santa María Volcano

### Chloride Waters

As stated earlier, Cl-rich waters, which are the typical reservoir waters in geothermal systems (Henley et al., 1984; Goff and Janik, 2000), are only found in the Zunil geothermal field, west of the Río Samalá, and at the Ocos spring, west of Siete Orejas (Fig. 5). Chloride waters, like these, are believed to form through a series of processes (e.g., White, 1957). Under high-pressure, high-temperature conditions, dense magmatic steam, containing volatile acids and alkali halides, condenses in deeply circulating meteoric waters, forming acidic sodium chloride waters. Chloride concentrations in such parent waters can exceed 5000 ppm (Fournier, 1981; Goff and Janik, 2000). Cl-rich parent waters then experience variable ascent and boiling, further mixing with meteoric waters, and interaction with country rocks, leading to a reduction in acidity. The relatively low Cl contents and variable pH of the Cl-rich waters from Zunil and the Ocos spring suggest substantial dilution of and evolution from parental compositions (A. Adams et al., 1990). Figure 6 suggests the dilutant is meteoric water with little bicarbonate or sulfate. The $\delta^{18}O$-$\delta D$ values of the Cl-rich well waters from Zunil indicate they could possess a magmatic component (Giggenbach et al., 1992), but are equally compatible with extensive subsurface boiling. In addition, the cation concentrations of the Cl-rich well waters from Zunil suggest that they have been "fully equilibrated" with their host rocks at temperatures in excess of 160 °C (Fig. 9; M. Adams et al., 1990; Giggenbach et al., 1992). The low Mg/Ca ratios of the chloride waters from the Santa María region (Fig. 7) are characteristic of such geothermal waters and have been attributed to their high temperatures, at which magnesium minerals have low solubilities (Goff and Janik, 2000).

The chloride water at the Ocos spring is chemically distinct from the Cl-rich well waters from the Zunil geothermal field. Specifically, the Ocos water is distinguished by higher $SO_4$ and $HCO_3$ (Fig. 6), less Na (and K) relative to other cations (Fig. 7), and meteoric $\delta^{18}O$ and $\delta D$ (Fig. 11). Clearly, the Ocos chloride water is more dilute and immature than the Cl-rich well waters from Zunil (Fig. 9).

## Sulfate Waters

Sulfate waters occur in close association with chloride waters in the Zunil geothermal field, in the Sulfur Mountain–Zunil-II region, and near Siete Orejas (Fig. 5). The formation of sulfate waters near active volcanoes is attributed to the condensation and oxidation of hot, sulfur-rich gases in shallow meteoric waters (White, 1957; Ellis and Mahon, 1977; Kiyosu and Kurahashi, 1983; Valentino and Stanzione, 2003). These gases could come directly from degassing magma or from boiling of deep chloride waters. Primary sulfate waters from the Santa María region likely originated by the latter mechanism, given their close association with chloride waters. Concentrated sulfate waters are distinguished by $SO_4$ > 300 ppm (falling at the $SO_4$ apex in Fig. 4), pH < 4.0, high Fe and unusually high S/Cl ratios, and, where analyzed, by $\delta^{18}O$ and $\delta D$ ratios that fall to the right of the meteoric line in Figure 11. The latter characteristic, as discussed above, could reflect a magmatic stamp or originate via surface evaporation.

As mentioned earlier, sulfate-rich sulfate waters have abnormally elevated S/Cl ratios (20–70; e.g., Fischer et al., 1997). One possible explanation of the unusually high S/Cl ratios of the sulfate waters is that they reflect input of magmatic gases with this signature. However, fumarolic condensates from Santiaguito have low (<1), not high, S/Cl ratios (Stoiber and Rose, 1970), and measured $SO_2$ emissions from Santiaguito are not considered excessive (Andres et al., 1993; T. Fischer, 2005, personal commun.). Also, magmatic S/Cl ratios estimated for Guatemalan volcanoes from melt inclusions are exclusively low (<5; Rose et al., 1982; Sisson and Layne, 1993; Roggensack, 2001; K. Roggensack, 2005, personal commun.). The likely cause of the high S/Cl ratios of the sulfate waters is that they reflect extensive scrubbing by a subsurface hydrothermal system (Symonds et al., 2001, 2003; Duffell et al., 2003). Symonds et al. (2003) demonstrate how primary scrubbing by a deep hydrothermal system produces increasing S/Cl ratios. The high precipitation rates around Santa María would enhance the scrubbing process (Doukas and Gerlach, 1995; Oppenheimer, 1996; Symonds et al., 2001). Conversely, significant scrubbing is unlikely below the southwestern flank of Santa María, where decades-long eruptive activity at Santiaguito has dried out subvolcanic pathways (Symonds et al., 2001, 2003).

Sulfate waters with the highest $SO_4$ contents show declining Na contents with increasing temperature (Fig. 8A), suggesting saturation and precipitation of a Na-rich mineral at higher temperatures. One likely candidate is natroalunite, a common mineral in association with acid sulfate waters and mineralization at active subduction zone volcanoes (Slansky, 1975; Graham and Robinson, 1986; Smith et al., 1988; Christenson and Wood, 1993; Wood, 1994). As shown in Figure 6A, concentrated sulfate waters have also experienced variable dilution with a low-$SO_4$, meteoric component.

## Bicarbonate Waters

Bicarbonate waters are found throughout the Santa María region, particularly to the south and west of the volcano (Fig. 5). They occur at thermal wells and springs as well as nonthermal springs (Fig. 8A). The formation of bicarbonate waters near active volcanoes has been attributed to the absorption of $CO_2$-rich vapors, either magmatic or hydrothermal, into shallow meteoric water followed by variable water-rock interaction (White, 1957; A. Adams et al., 1990; Giggenbach and Corrales Soto, 1992; Rowe and Brantley, 1993; Lewicki et al., 2000; Nathenson et al., 2003). Hence, bicarbonate waters around volcanoes can play a significant role in the transport of magmatic carbon (Evans et al., 2002).

As pointed out previously, some of the bicarbonate waters located south-southwest of the Santiaguito dome and just east of the Zunil geothermal field are distinguished by high Mg/Ca ratios (Figs. 5 and 7). These high Mg/Ca bicarbonate waters also have elevated $\delta^{13}C$ (Fig. 10), Na (Fig. 8) and $HCO_3$ (Fig. 17). The higher Mg contents of these waters are likely due to enhanced interaction between meteoric waters and volcanic rocks (Giggenbach, 1974; Giggenbach and Glover, 1975; Giggenbach, 1983; Martin-Del Pozzo et al., 2002). The positive correlation between Mg and bicarbonate (Fig. 17), therefore, implies that the ultimate source of $CO_2$ in these bicarbonate waters is magmatic. This would explain why the most important concentration of high Mg/Ca springs is <10 km down-gradient from the continuously active Santiaguito dome complex (Fig. 5). Hence, the main bicarbonate trend simply represents variable enrichment or dilution with the high Mg/Ca volcanic-magmatic component or meteoric water, respectively (Figs. 8B and 17). The elevated $\delta^{13}C$ ratios of the high Mg/Ca bicarbonate waters (Fig. 10) suggests that magmatic $\delta^{13}C$ ratios extend to slightly heavier values than inferred above from fumarolic gases (to at least −8‰), making the range in magmatic $\delta^{13}C$ similar to that suggested for parts of the Cascades (Evans et al., 1981, 2004; Janik et al., 1983).

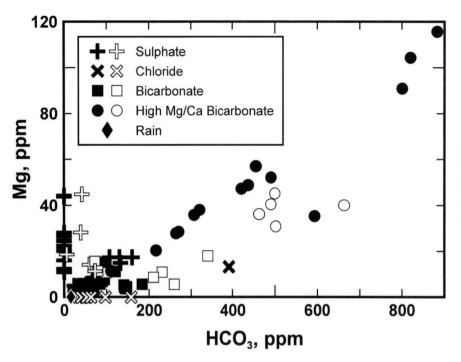

Figure 17. $HCO_3$ versus Mg for waters encircling Santa María volcano. Includes data from Adams et al. (1990) and Giggenbach et al. (1992). Closed symbols: data from this study; open symbols: data from Adams et al. (1990) and Giggenbach et al. (1992).

Giggenbach et al. (1992) hypothesized that chloride and bicarbonate waters shared a common parental water. However, we think the data presented here rule that out (e.g., Figs. 6B, 8B, and 17).

*Summary*

Waters encircling Santa María volcano are similar to hydrothermal waters described at numerous active subduction zone volcanoes. Chloride waters are dilute to very dilute equivalents of typical geothermal reservoir waters. Sulfate waters are closely associated with chloride waters. This and their high S/Cl ratios suggest they originate by boiling of deep chloride waters. Thus, there is extensive primary scrubbing of magmatic gases through a deep hydrothermal system. Sulfate waters also exhibit variable dilution with meteoric waters. Bicarbonate waters are most abundant and show varying degrees of interaction with young igneous rocks of the volcanic edifice. There is only limited mixing between the three water types.

## Hydrothermal Model and Implications

Hydrothermal systems beneath active subduction zone volcanoes are often portrayed as complex, volcano-wide systems with characteristic fluid zonations (Giggenbach et al., 1990; Giggenbach and Corrales Soto, 1992; Goff and Janik, 2000). Shevenell and Goff (1993, 1995), however, have demonstrated that some hydrothermal systems beneath active subduction zone volcanoes can also be quite localized and ephemeral. Clearly, volcano-hydrothermal systems are dynamic systems that can wax and wane in relation to the level of volcanic activity (e.g., McConnell et al., 1997).

Previous studies have together established that an extensive and mature hydrothermal system underlies the Zunil–Sulfur Mountain (or Zunil-II) area, northeast of Santa María (A. Adams et al., 1990; M. Adams et al., 1990, 1992; Foley et al., 1990; Caicedo and Palma, 1990; Giggenbach et al., 1992; Lima Lobato and Palma, 2000). Giggenbach et al. (1992) place the thermal-magmatic core of this system below Volcán Zunil, although the latter has not had historic activity and is deeply dissected. A more credible location of the thermal-magmatic heart of the Zunil–Zunil-II hydrothermal system would be beneath the 1902 crater at Santa María and the long-active Santiaguito dome. According to the modeling of Barmin et al. (2002), the magmatic heart beneath Santiaguito must be large, on the order of 65 km³, in order to sustain the long-lived activity and growth of the dome complex. Such a large and persistent heat and gas source could easily generate an extensive hydrothermal system, particularly in a relatively wet region like the Guatemalan Highlands. The depth to the magmatic core below Santiaguito is perhaps several kilometers since wholesale collapse did not accompany the 1902 eruption (Rose, 1987b). Dry, gas-dominated conditions likely prevail beneath and around the Santiaguito dome because of its continuous activity (Giggenbach et al., 1990; Symonds et al., 2001, 2003). As a result, surface manifestations of a deep and extensive hydrothermal system would be shunted to the periphery of Santa María, preferentially into the Zunil area to the northeast, along faults associated with the Zunil fault zone.

However, arguably the most plausible location of the thermal-magmatic heart of the Zunil–Zunil-II hydrothermal system is below its nearest historically active volcanic neighbor, Cerro Quemado (Fig. 2). This connection is buttressed by the similarity in magmatic gas $\delta^{13}C$ ratios between Cerro Quemado and

Zunil (Fig. 15). Moreover, the slightly lighter $\delta^{13}C$ ratios for high Mg/Ca bicarbonate waters concentrated near the Santiaguito dome hints that magmatic $CO_2$ from Santa María may be isotopically distinct from that at Zunil and Cerro Quemado. Like Zunil, Cerro Quemado and its neighboring dome complexes fall near the southern boundary of the hypothesized Xela caldera. Ring fractures related to caldera formation and other faults of the Zunil fault zone may serve as the loci of upwelling zones within the Zunil–Zunil-II hydrothermal system (Foley et al., 1990). Many studies have stressed the important role played by faults in focusing hydrothermal activity at subduction zone volcanoes (Traineau et al., 1989; Chiodini et al., 1993; Sturchio et al., 1993; Sanford et al., 1995; Finizola et al., 2003). Surface discharges of chloride waters from the Zunil–Zunil-II hydrothermal system are prevented because of the high precipitation (i.e., recharge) rates (Traineau et al., 1989; Rowe et al., 1995).

*Summary*

Hydrothermal activity in the Santa María region of northern Guatemala is concentrated in the Zunil–Zunil-II areas likely associated with shallow magmatic activity at Cerro Quemado. It is also possible that a number of small- to moderate-sized, disconnected, hydrothermal systems exist in the Santa María region (Lima Lobato and Palma, 2000). In particular, the presence of chloride and sulfate waters northwest of Santa María may indicate that a small hydrothermal system underlies Siete Orejas. However, we can't definitively exclude the possibility of one large, regionally connected hydrothermal system centered on Santa María. A comprehensive self-potential survey (Michel and Zlotnicki, 1998; Finizola et al., 2003; Aizawa, 2004) might be the next important step in providing further insights into the extent and connectivity of hydrothermal activity in this region.

If the location of the heart of the Zunil–Zunil-II hydrothermal system is indeed Cerro Quemado, then continued monitoring of the springs and fumaroles at Zunil could be useful in gauging future activity at Cerro Quemado. For monitoring of Santa María, the most useful springs would probably be those currently discharging high Mg/Ca bicarbonate waters south-southwest of Santiaguito.

## CONCLUSIONS

Most springs and wells girdling Santa María volcano in Guatemala contain bicarbonate waters. Sulfate and chloride waters, by contrast, are geographically restricted to the Zunil and Zunil-II geothermal areas east-northeast of Santa María (on both flanks of the Río Samalá) and adjacent to Siete Orejas volcano northwest of Santa María. High Mg/Ca bicarbonate waters have experienced increased reaction with volcanic rocks and have a hint of a greater magmatic component. The high S/Cl ratios of the sulfate waters result from substantial primary scrubbing through a mature hydrothermal system. Geochemical data indicate only limited mixing between bicarbonate, sulfate, and chloride waters. Most waters exhibit extensive meteoric dilution.

Isotopic compositions of gas samples from a number of fumaroles from the Zunil–Zunil-II region provide evidence for significant magmatic input. This magmatic signature is notable for its light $\delta^{13}C$, particularly with respect to the remainder of the Central American volcanic front. Crustal, rather than source, contamination is the suggested cause of the lighter magmatic $\delta^{13}C$ in the Santa María region.

The Zunil–Zunil-II geothermal field is underlain by an extensive hydrothermal system whose magmatic heart is most likely beneath Cerro Quemado, not Santa María. If so, the fumaroles and springs in the Zunil and Zunil-II regions could be quite useful for future monitoring of Cerro Quemado. For monitoring of Santa María, the high Mg/Ca bicarbonate springs south-southwest of the Santiaguito dome complex might prove useful, particularly if a magmatic input can be confirmed. Spring waters northwest of Santa María suggest that hydrothermal activity may also be significant below Siete Orejas.

## ACKNOWLEDGMENTS

Financial support for this research was provided by Northern Illinois University, Sigma Xi, and Michael Realini. Neil Sturchio, Greg Arehart, and Ben Holt provided gracious guidance and assistance at Argonne National Laboratory. Rick Socki of the National Aeronautics and Space Administration is thanked for the additional stable isotope analyses. Eddie Sanchez-Bennett of the Instituto Nacional de Sismología, Vulcanología, Meterorología, e Hidrología (INSIVUMEH) in Guatemala supplied logistical support and allowed INSIVUMEH personnel to accompany us in the field. A million thanks to Rodolfo Morales of INSIVUMEH, who acted as field guide, translator, morale booster, and friend during all of the field work. John Ewert, an anonymous reviewer, and, most particularly, Bill Evans provided criticism and suggestions that greatly improved the final manuscript.

## REFERENCES CITED

Adams, A., Goff, F., Trujillo, P.E., Jr., Counce, D., Medina, V., Archuleta, J., and Dennis, B., 1990, Hydrogeochemical investigations in support of well logging operations at the Zunil geothermal field, Guatemala: Geothermal Resources Council Transactions, v. 14, p. 829–835.

Adams, M.C., Mink, L.L., Moore, J.N., White, L.D., and Caicedo Anchissi, A., 1990, Geochemistry and hydrology of the Zunil geothermal system, Guatemala: Geothermal Resources Council Transactions, v. 14, p. 837–844.

Adams, M.C., Moore, J.N., White, L.D., Mink, L.L., Leiva, O., Ramirez, S., and Caicedo Anchissi, A., 1992, Fluid recharge of the Zunil, Guatemala, geothermal system: Geothermal Resources Council Transactions, v. 16, p. 113–117.

Aizawa, K., 2004, A large self-potential anomaly and its changes on the quiet Mt. Fuji, Japan: Geophysical Research Letters, v. 31, 2004GL019462.

Allard, P., 1983, The origin of hydrogen, carbon, sulphur, nitrogen and rare gases in volcanic exhalations: evidence from isotope geochemistry, *in* Tazieff, H., and Sabroux, J.C., eds., Forecasting Volcanic Events: Amsterdam, Elsevier, p. 337–386.

Anderson, S.W., Fink, J.H., and Rose, W.I., 1995, Mount St. Helens and Santiaguito lava domes: the effect of short-term eruption rate on surface texture and degassing processes: Journal of Volcanology and Geothermal Research, v. 69, p. 105–116, doi: 10.1016/0377-0273(95)00022-4.

Andres, R.J., Rose, W.I., Stoiber, R.E., Williams, S.N., Matías, O., and Morales, R., 1993, A summary of sulfur dioxide emission rate measurements from Guatemalan volcanoes: Bulletin of Volcanology, v. 55, p. 379–388, doi: 10.1007/BF00301150.

Barmin, A., Melnik, O., and Sparks, R.S.J., 2002, Periodic behavior in lava dome eruptions: Earth and Planetary Science Letters, v. 199, p. 173–184, doi: 10.1016/S0012-821X(02)00557-5.

Boudon, G., Villemant, B., Komorowski, J.-C., Ildefonse, P., and Semet, M.P., 1998, The hydrothermal system at Soufriere Hills volcano, Montserrat (West Indies): Characterization and role in the on-going eruption: Geophysical Research Letters, v. 25, p. 3693–3696, doi: 10.1029/98GL00985.

Caicedo, A., and Palma, J., 1990, Present status of exploration and development of the geothermal resources of Guatemala: Geothermal Resources Council Transactions, v. 14, p. 97–105.

Carr, M.J., 1984, Symmetrical and segmented variation of physical and geochemical characteristics of the Central American volcanic front: Journal of Volcanology and Geothermal Research, v. 20, p. 231–252, doi: 10.1016/0377-0273(84)90041-6.

Carr, M.J., Rose, W.I., and Stoiber, R.E., 1982, Central America, in Thorpe, R.S., ed., Andesites: New York, John Wiley, p. 149–166.

Carr, M.J., Feigenson, M.D., and Bennett, E.A., 1990, Incompatible element and isotopic evidence for tectonic control of source mixing and melt extraction along the Central American arc: Contributions to Mineralogy and Petrology, v. 105, p. 369–380, doi: 10.1007/BF00286825.

Carr, M.J., Feigenson, M.D., Patino, L.C., and Walker, J.A., 2003, Volcanism and geochemistry in Central America: Progress and problems, in Eiler, J., ed., Inside the Subduction Factory: American Geophysical Union Geophysical Monograph 138, p. 153–174.

Chiodini, G., Cioni, R., Leonis, C., Marini, L., and Raco, B., 1993, Fluid geochemistry of Nisyros island, Dodecanese, Greece: Journal of Volcanology and Geothermal Research, v. 56, p. 95–112, doi: 10.1016/0377-0273(93)90052-S.

Christenson, B.W., and Wood, C.P., 1993, Evolution of a vent-hosted hydrothermal system beneath Ruapehu crater lake, New Zealand: Bulletin of Volcanology, v. 55, p. 547–565, doi: 10.1007/BF00301808.

Coleman, M.L., Shepherd, T.J., Durham, J.J., Rouse, J.E., and Moore, G.R., 1982, Reduction of water with zinc for hydrogen isotope analysis: Analytical Chemistry, v. 54, p. 2631–2632, doi: 10.1021/ac00251a062.

Conway, F.M., Vallance, J.W., Rose, W.I., Johns, G.W., and Paniagua, S., 1992, Cerro Quemado, Guatemala: The volcanic history and hazards of an exogenous volcanic dome complex: Journal of Volcanology and Geothermal Research, v. 52, p. 303–323, doi: 10.1016/0377-0273(92)90051-E.

Conway, F.M., Diehl, J.F., Rose, W.I., and Matías, O., 1994, Age and magma flux of Santa María volcano, Guatemala: correlation of paleomagnetic waveforms with the 28,000 to 25,000 yr B.P. Mono Lake excursion: Journal of Geology, v. 102, p. 11–24.

Craig, H., 1961, Isotopic variations in meteoric waters: Science, v. 133, p. 1702–1703.

Craig, H., 1963, The isotopic geochemistry of water and carbon in geothermal areas, in Tongiorgi, E., ed., Nuclear Geology in Geothermal Areas: Pisa, Consiglio Nazionale delle Ricerche, p. 17–53.

Criss, R.E., 1995, Stable isotope distribution: variations from temperature, organic and water-rock interactions, in Ahrens, T.J., ed., A Handbook of Physical Constants: Global Earth Physics: American Geophysical Union Reference Shelf, v. 1, p. 292–307.

Criss, R.E., and Taylor, H.P., Jr., 1986, Meteoric-hydrothermal systems, in Valley, J.W., Taylor, H.P., Jr., and O'Neil, J.R., eds., Stable isotopes in high temperature geological processes: Reviews in Mineralogy, v. 16, p. 373–424.

Crowley, J.K., and Zimbelman, D.R., 1997, Mapping hydrothermally altered rocks on Mount Rainier, Washington, with airborne visible/infrared imaging spectrometer (AVIRIS) data: Geology, v. 25, p. 559–562, doi: 10.1130/0091-7613(1997)025<0559:MHAROM>2.3.CO;2.

Dansgaard, W., 1964, Stable isotopes in precipitation: Tellus, v. 16, p. 436–468.

Delmelle, P., Bernard, A., Kusakabe, M., Fischer, T.P., and Takano, B., 2000, Geochemistry of the magmatic-hydrothermal system of Kawah Ijen volcano, east Java, Indonesia: Journal of Volcanology and Geothermal Research, v. 97, p. 31–53, doi: 10.1016/S0377-0273(99)00158-4.

Doukas, M.P., and Gerlach, T.M., 1995, Sulfur dioxide scrubbing during the 1992 eruptions of Crater Peak, Mount Spurr volcano, Alaska, in Keith, T.E.C., ed., The 1992 eruptions of Crater Peak vent, Mount Spurr volcano, Alaska: U.S. Geological Survey Bulletin 2139, p. 47–57.

Duffell, H.J., Oppenheimer, C., Pyle, D.M., Galle, B., McGonigle, A.J.S., and Burton, M.R., 2003, Changes in gas composition prior to a minor explosive eruption at Masaya volcano, Nicaragua: Journal of Volcanology and Geothermal Research, v. 126, p. 327–339, doi: 10.1016/S0377-0273(03)00156-2.

Duffield, W., Heiken, G., Foley, D., and McEwen, A., 1993, Oblique synoptic images, produced from digital data, display strong evidence of a "new" caldera in southwestern Guatemala: Journal of Volcanology and Geothermal Research, v. 55, p. 217–224, doi: 10.1016/0377-0273(93)90038-S.

Edmonds, M., Oppenheimer, C., Pyle, D.M., Herd, R.A., and Thompson, G., 2003, $SO_2$ emissions from Soufrière Hills volcano and their relationship to conduit permeability, hydrothermal interaction and degassing regime: Journal of Volcanology and Geothermal Research, v. 124, p. 23–43, doi: 10.1016/S0377-0273(03)00041-6.

Ellis, A.J., and Mahon, W.A.J., 1977, Chemistry and Geothermal Systems: New York, Academic Press, 392 p.

Epstein, S., and Mayeda, T.K., 1953, Variation of $O^{18}$ content of waters from natural sources: Geochimica et Cosmochimica Acta, v. 4, p. 213–224, doi: 10.1016/0016-7037(53)90051-9.

Evans, W.C., Banks, N.G., and White, L.D., 1981, Analyses of gas samples from the summit crater, in Lipman, P.W., and Mullineaux, D.R., eds., The 1980 eruptions of Mount St. Helens, Washington: U.S. Geological Survey Professional Paper 1250, p. 227–231.

Evans, W.C., Sorey, M.L., Cook, A.C., Kennedy, B.M., Shuster, D.L., Colvard, E.M., White, L.E., and Huebner, M.A., 2002, Tracing and quantifying magmatic carbon discharge in cold groundwaters: lessons learned from Mammoth Mountain, USA: Journal of Volcanology and Geothermal Research, v. 114, p. 291–312, doi: 10.1016/S0377-0273(01)00268-2.

Evans, W.C., van Soest, M.C., Mariner, R.H., Hurwitz, S., Ingebritsen, S.E., Wicks, C.W., Jr., and Schmidt, M.E., 2004, Magmatic intrusion west of Three Sisters, central Oregon, USA: The perspective from spring geochemistry: Geology, v. 32, p. 69–72, doi: 10.1130/G19974.1.

Fahlquist, L., and Janik, C., 1992, Procedures for collecting and analyzing gas samples from geothermal systems. U.S. Geological Survey Open-File Report 92-211, 19 p.

Feigenson, M.D., and Carr, M.J., 1986, Positively correlated Nd and Sr isotope ratios of lavas from the Central American volcanic front: Geology, v. 14, p. 79–82, doi: 10.1130/0091-7613(1986)14<79:PCNASI>2.0.CO;2.

Feigenson, M.D., Carr, M.J., Maharaj, S.V., Juliano, S., and Borge, L.L., 2004, Lead isotope composition of Central American volcanoes: Influence of the Galapagos plume: Geochemistry, Geophysics, Geosystems, v. 5, p. Q06001, doi: 10.1029/2003GC000621.

Finizola, A., Sortino, F., Lénat, J.-F., Aubert, M., Ripepe, M., and Valenza, M., 2003, The summit hydrothermal system of Stromboli. New insights from self-potential, temperature, $CO_2$ and fumarolic fluid measurements, with structural and monitoring implications: Bulletin of Volcanology, v. 65, p. 486–504, doi: 10.1007/s00445-003-0276-z.

Fischer, T.P., Morrissey, M.M., Calvache, M.L., Gómez, D., Torres, R., Stix, J., and Williams, S.N., 1994, Correlations between $SO_2$ flux and long-period seismicity at Galeras volcano: Nature, v. 368, p. 135–137, doi: 10.1038/368135a0.

Fischer, T.P., Arehart, G.B., Sturchio, N.C., and Williams, S.N., 1996, The relationship between fumarole gas composition and eruptive activity at Galeras volcano, Colombia: Geology, v. 24, p. 531–534, doi: 10.1130/0091-7613(1996)024<0531:TRBFGC>2.3.CO;2.

Fischer, T.P., Sturchio, N.C., Stix, J., Arehart, G.B., Counce, D., and Williams, S.N., 1997, The chemical and isotopic composition of fumarolic gases and spring discharges from Galeras volcano, Colombia: Journal of Volcanology and Geothermal Research, v. 77, p. 229–253, doi: 10.1016/S0377-0273(96)00096-0.

Fischer, T.P., Giggenbach, W.F., Sano, Y., and Williams, S.N., 1998, Fluxes and sources of volatiles discharged from Kudryavy, a subduction zone volcano, Kurile Islands: Earth and Planetary Science Letters, v. 160, p. 81–96, doi: 10.1016/S0012-821X(98)00086-7.

Fischer, T.P., Hilton, D.R., Zimmer, M.M., Shaw, A.M., Sharp, Z.D., and Walker, J.A., 2002, Subduction and recycling of nitrogen along the Central American margin: Science, v. 297, p. 1154–1157, doi: 10.1126/science.1073995.

Foley, D., Moore, J.N., Lutz, S.J., Palma, J.C., Ross, H.P., Tobias, E., and Tripp, A.C., 1990, Geology and geophysics of the Zunil geothermal system, Guatemala: Geothermal Resources Council Transactions, v. 14, p. 1405–1412.

Fournier, R.O., 1981, Application of water geochemistry to geothermal exploration and reservoir engineering, in Rybach, L., and Muffler, L.J.P., eds., Geothermal Systems: Principles and Case Histories: New York, John Wiley, p. 109–143.

Fournier, R.O., Hanshaw, B.B., and Urrutia Sole, J.F., 1982, Oxygen and hydrogen isotopes in thermal waters at Zunil, Guatemala: Geothermal Resources Council Transactions, v. 6, p. 89–91.

Frank, D., 1995, Surficial extent and conceptual model of hydrothermal system at Mount Rainier, Washington: Journal of Volcanology and Geothermal Research, v. 65, p. 51–80, doi: 10.1016/0377-0273(94)00081-Q.

Gerlach, T.M., and Taylor, B.E., 1990, Carbon isotope constraints on degassing of carbon dioxide from Kilauea volcano: Geochimica et Cosmochimica Acta, v. 54, p. 2051–2058, doi: 10.1016/0016-7037(90)90270-U.

Giggenbach, W., 1974, The chemistry of Crater Lake, Mt. Ruapehu (New Zealand) during and after the 1971 active period: New Zealand Journal of Science, v. 17, p. 33–45.

Giggenbach, W.F., 1983, Chemical surveillance of active volcanoes in New Zealand, in Tazieff, H., and Sabroux, J.C., eds., Forecasting Volcanic Events: New York, Elsevier, p. 311–322.

Giggenbach, W.F., 1988, Geothermal solute equilibria. Derivation of Na-K-Mg-Ca geoindicators: Geochimica et Cosmochimica Acta, v. 52, p. 2749–2765, doi: 10.1016/0016-7037(88)90143-3.

Giggenbach, W.F., 1992a, The composition of gases in geothermal and volcanic systems as a function of tectonic setting, in Kharaka, Y.K., and Maest, A.S., eds., Proceedings of the 7th International Symposium on Water-Rock Interaction; volume 2, Moderate and High Temperature Environments: Rotterdam, Balkema, p. 873–878.

Giggenbach, W.F., 1992b, Isotopic shifts in waters from geothermal and volcanic systems along convergent plate boundaries and their origin: Earth and Planetary Science Letters, v. 113, p. 495–510, doi: 10.1016/0012-821X(92)90127-H.

Giggenbach, W.F., and Corrales Soto, R., 1992, Isotopic and chemical composition of water and steam discharges from volcanic-magmatic-hydrothermal systems of the Guanacaste geothermal province, Costa Rica: Applied Geochemistry, v. 7, p. 309–332, doi: 10.1016/0883-2927(92)90022-U.

Giggenbach, W.F., and Glover, R.B., 1975, The use of chemical indicators in the surveillance of volcanic activity affecting the crater lake on Mt. Ruapehu, New Zealand: Bulletin Volcanologique, v. 39, p. 70–81.

Giggenbach, W.F., and Goguel, R.L., 1988, Methods of the collection and analysis of geothermal and volcanic water and gas samples: New Zealand Department of Scientific and Industrial Research Report CD-2387, 53 p.

Giggenbach, W.F., Garcia, P.N., Londoño-C., A., Rodriguez-V., L., Rojas-G., N., and Calvache-V., M.L., 1990, The chemistry of fumarolic vapor and thermal-spring discharges from the Nevado del Ruiz volcanic-magmatic-hydrothermal system, Colombia: Journal of Volcanology and Geothermal Research, v. 42, p. 13–39, doi: 10.1016/0377-0273(90)90067-P.

Giggenbach, W.F., Paniagua de Gudiel, D., and Roldan Manzo, A.R., 1992, Isotopic and chemical composition of water and gas discharges from the Zunil geothermal system, Guatemala, in Estudios Geotérmicos con Técnicas Isotópicas y Geoquímicas en América Latina: Vienna, Organismo Internacional de Energía Atómica, p. 245–278.

Goff, F., and Janik, C.J., 2000, Geothermal systems, in Sigurdsson, H., ed., Encyclopedia of Volcanoes: San Diego, Academic Press, p. 817–834.

Goff, F., McMurtry, G.M., Counce, D., Stimac, J.A., Roldán-Manzo, A.R., and Hilton, D.R., 2000, Contrasting hydrothermal activity at Sierra Negra and Alcedo volcanoes, Galapagos Archipelago, Ecuador: Bulletin of Volcanology, v. 62, p. 34–52, doi: 10.1007/s004450050289.

Graham, I.J., and Robinson, B.W., 1986, Natroalunite on Ruapehu volcano, New Zealand: Geochemical Journal, v. 20, p. 249–253.

Harris, A.J.L., Flynn, L.P., Matías, O., and Rose, W.I., 2002, The thermal stealth flows of Santiaguito dome, Guatemala: implications for the cooling and emplacement of dacitic block-lava flows: Geological Society of America Bulletin, v. 114, p. 533–546, doi: 10.1130/0016-7606(2002)114<0533:TTSFOS>2.0.CO;2.

Harris, A.J.L., Rose, W.I., and Flynn, L.P., 2003, Temporal trends in lava dome extrusion at Santiaguito 1922–2000: Bulletin of Volcanology, v. 65, p. 77–89.

Hedenquist, J.W., and Aoki, M., 1991, Meteoric interaction with magmatic discharges in Japan and the significance for mineralization: Geology, v. 19, p. 1041–1044, doi: 10.1130/0091-7613(1991)019<1041:MIWMDI>2.3.CO;2.

Hedenquist, J.W., and Lowenstern, J.B., 1994, The role of magmas in the formation of hydrothermal ore deposits: Nature, v. 370, p. 519–527, doi: 10.1038/370519a0.

Heiligmann, M., Stix, J., Williams-Jones, G., Sherwood Lollar, B., and Garzón, G., 1997, Distal degassing of radon and carbon dioxide on Galeras volcano, Colombia: Journal of Volcanology and Geothermal Research, v. 77, p. 267–283, doi: 10.1016/S0377-0273(96)00099-6.

Henley, R.W., Truesdell, A.H., Barton, P.B., and Whitney, J.A., 1984, Fluid-mineral equilibria in hydrothermal systems: Reviews in Economic Geology, v. 1, p. 1–267.

Hilton, D.R., Hammerschmidt, K., Teufel, S., and Friedrichsen, H., 1993, Helium isotope characteristics of Andean geothermal fluids and lavas: Earth and Planetary Science Letters, v. 120, p. 265–282, doi: 10.1016/0012-821X(93)90244-4.

Hinkley, T.K., Quick, J.E., Gregory, R.E., and Gerlach, T.M., 1995, Hydrogen and oxygen isotopic compositions of waters from fumaroles at Kilauea summit, Hawaii: Bulletin of Volcanology, v. 57, p. 44–51.

Hochstein, M.P., and Browne, P.R.L., 2000, Surface manifestations of geothermal systems with volcanic heat sources, in Sigurdsson, H., ed., Encyclopedia of Volcanoes: San Diego, Academic Press, p. 835–855.

Janik, C.J., Nehring, N.L., and Truesdell, A.H., 1983, Stable isotope geochemistry of thermal fluids from Lassen Volcanic National Park, California: Geothermal Resources Council Transactions, v. 7, p. 295–300.

Janik, C.J., Goff, F., Fahlquist, L., Adams, A.I., Roldán-M., A., Chipera, S.J., Trujillo, P.E., and Counce, D., 1992, Hydrogeochemical exploration of geothermal prospects in the Tecuamburro volcano region, Guatemala: Geothermics, v. 21, p. 447–481, doi: 10.1016/0375-6505(92)90002-Q.

Javoy, M., Pineau, F., and Iiyama, I., 1978, Experimental determination of the isotopic fractionation between gaseous $CO_2$ and carbon dissolved in tholeiitic magma: Contributions to Mineralogy and Petrology, v. 67, p. 35–39, doi: 10.1007/BF00371631.

Kita, I., Nitta, K., Nagao, K., Taguchi, S., and Koga, A., 1993, Difference in $N_2$/Ar ratio of magmatic gases from northeast and southwest Japan: new evidence for different states of plate subduction: Geology, v. 21, p. 391–394, doi: 10.1130/0091-7613(1993)021<0391:DINARO>2.3.CO;2.

Kiyosu, Y., 1986, Variations in $N_2$/Ar and He/Ar ratios of gases from some volcanic areas in northeastern Japan: Geochemical Journal, v. 19, p. 275–281.

Kiyosu, Y., and Kurahashi, M., 1983, Origin of sulfur species in acid sulphate-chloride thermal waters, northeastern Japan: Geochimica et Cosmochimica Acta, v. 47, p. 1237–1245, doi: 10.1016/0016-7037(83)90065-0.

Lewicki, J.L., Fischer, T., and Williams, S.N., 2000, Chemical and isotopic compositions of fluids at Cumbal volcano, Colombia: Evidence for magmatic contribution: Bulletin of Volcanology, v. 62, p. 347–361, doi: 10.1007/s004450000100.

Lima Lobato, E.M., and Palma, J., 2000, The Zunil-II geothermal field, Guatemala, Central America: Proceedings World Geothermal Conference, v. 2000, p. 2133–2138.

López, D.L., and Williams, S.N., 1993, Catastrophic volcanic collapse: relation to hydrothermal processes: Science, v. 260, p. 1794–1796.

Lyon, G.L., and Hulston, J.R., 1984, Carbon and hydrogen isotopic compositions of New Zealand geothermal gases: Geochimica et Cosmochimica Acta, v. 48, p. 1161–1171, doi: 10.1016/0016-7037(84)90052-8.

Macpherson, C., and Mattey, D., 1994, Carbon isotope variations of $CO_2$ in central Lau Basin basalts and ferrobasalts: Earth and Planetary Science Letters, v. 121, p. 263–276, doi: 10.1016/0012-821X(94)90072-8.

Martin-Del Pozzo, A.L., Aceves, F., Espinasa, R., Aguayo, A., Inguaggiato, S., Morales, P., and Cienfuegos, E., 2002, Influence of volcanic activity on spring water chemistry at Popocatepetl volcano, Mexico: Chemical Geology, v. 190, p. 207–229, doi: 10.1016/S0009-2541(02)00117-1.

Martínez, M., Fernández, E., Valdés, J., Barboza, V., Van der Laat, R., Duarte, E., Malavassi, E., Sandoval, L., Barquero, J., and Marino, T., 2000, Chemical evolution and volcanic activity of the active crater lake of Poás volcano, Costa Rica, 1993–1997: Journal of Volcanology and Geothermal Research, v. 97, p. 127–141, doi: 10.1016/S0377-0273(99)00165-1.

Martini, M., 1989, The forecasting significance of chemical indicators in areas of quiescent volcanism: examples from Vulcano and Phlegrean Fields (Italy), in Latter, J.H., ed., Volcanic Hazards: New York, Springer-Verlag, p. 372–383.

Marty, B., and Zimmerman, L., 1999, Volatiles (He, C, N, Ar) in mid-ocean ridge basalts: assessment of shallow-level fractionation and characterization of source composition: Geochimica et Cosmochimica Acta, v. 63, p. 3619–3633, doi: 10.1016/S0016-7037(99)00169-6.

Matsuo, S., Suzuki, M., and Mizutani, Y., 1978, Nitrogen to argon ratio in volcanic gases, in Alexander, E.C., Jr., and Ozima, M., eds., Terrestrial Rare Gases: Tokyo, Japan Science Society Press, p. 17–25.

McConnell, V.S., Valley, J.W., and Eichelberger, J.C., 1997, Oxygen isotope compositions of intracaldera rocks: hydrothermal history of the Long Valley caldera, California: Journal of Volcanology and Geothermal Research, v. 76, p. 83–109, doi: 10.1016/S0377-0273(96)00071-6.

McCrea, J.M., 1950, On the isotope chemistry of carbonates and a paleotemperature scale: The Journal of Chemical Physics, v. 18, p. 849–857, doi: 10.1063/1.1747785.

Michel, S., and Zlotnicki, J., 1998, Self-potential and magnetic surveying of La Fournaise volcano (Réunion Island): Correlations with faulting, fluid circulation, and eruption: Journal of Geophysical Research, v. 103, p. 17,845–17,857, doi: 10.1029/98JB00607.

Mizutani, Y., Hayashi, S., and Sugiura, T., 1986, Chemical and isotopic compositions of fumarolic gases from Kuju-Iwoyama, Kyushu, Japan: Geochemical Journal, v. 20, p. 273–285.

Nathenson, M., Thompson, J.M., and White, L.D., 2003, Slightly thermal springs and non thermal springs at Mount Shasta, California: chemistry and recharge elevation: Journal of Volcanology and Geothermal Research, v. 121, p. 137–153, doi: 10.1016/S0377-0273(02)00426-2.

Oppenheimer, C., 1996, On the role of hydrothermal systems in the transfer of volcanic sulfur to the atmosphere: Geophysical Research Letters, v. 23, p. 2057–2060, doi: 10.1029/96GL02061.

Patino, L.C., Carr, M.J., and Feigenson, M.D., 2000, Local and regional variations in Central American arc lavas controlled by variations in subducted sediment input: Contributions to Mineralogy and Petrology, v. 138, p. 265–283, doi: 10.1007/s004100050562.

Plank, T., and Langmuir, C.H., 1998, The chemical composition of subducting sediment and its consequences for the crust and mantle: Chemical Geology, v. 145, p. 325–394, doi: 10.1016/S0009-2541(97)00150-2.

Poli, S., and Schmidt, M.W., 2002, Petrology of subducted slabs: Annual Review of Earth and Planetary Sciences, v. 30, p. 207–235, doi: 10.1146/annurev.earth.30.091201.140550.

Poorter, R.P.E., Varekamp, J.C., Poreda, R.J., van Bergen, M.J., and Kreulen, R., 1991, Chemical and isotopic compositions of volcanic gases from the east Sunda and Banda arcs, Indonesia: Geochimica et Cosmochimica Acta, v. 55, p. 3795–3807, doi: 10.1016/0016-7037(91)90075-G.

Poreda, R., and Craig, H., 1989, Helium isotope ratios in circum-Pacific volcanic arcs: Nature, v. 338, p. 473–478, doi: 10.1038/338473a0.

Reid, M.E., 2004, Massive collapse of volcano edifices triggered by hydrothermal pressurization: Geology, v. 32, p. 373–376, doi: 10.1130/G20300.1.

Reid, M.E., Sisson, T.W., and Brien, D.L., 2001, Volcano collapse promoted by hydrothermal alteration and edifice shape, Mount Rainier, Washington: Geology, v. 29, p. 779–782, doi: 10.1130/0091-7613(2001)029<0779:VCPBHA>2.0.CO;2.

Roggensack, K., 2001, Unraveling the 1974 eruption of Fuego volcano (Guatemala) with small crystals and their young melt inclusions: Geology, v. 29, p. 911–914, doi: 10.1130/0091-7613(2001)029<0911:UTEOFV>2.0.CO;2.

Rose, W.I., Jr., 1972a, Notes on the 1902 eruption of Santa María volcano, Guatemala: Bulletin Volcanologique, v. 36, p. 29–45.

Rose, W.I., Jr., 1972b, Santiaguito volcanic dome: Geological Society of America Bulletin, v. 83, p. 1413–1434.

Rose, W.I., Jr., 1973, Pattern and mechanism of volcanic activity at the Santiaguito volcanic dome, Guatemala: Bulletin Volcanologique, v. 37, p. 73–94.

Rose, W.I., Jr., 1974, Nuee ardente from Santiaguito volcano April 1973: Bulletin Volcanologique, v. 37, p. 365–371.

Rose, W.I., 1987a, Volcanic activity at Santiaguito volcano, 1976–1984, in Fink, J.H., ed., The Emplacement of Silicic Domes and Lava Flows: Geological Society of America Special Paper 212, p. 17–27.

Rose, W.I., 1987b, Santa María, Guatemala: bimodal soda-rich calc-alkalic stratovolcano: Journal of Volcanology and Geothermal Research, v. 33, p. 109–129, doi: 10.1016/0377-0273(87)90056-4.

Rose, W.I., Jr., Grant, N.K., Hahn, G.A., Lange, I.M., Powell, J.L., Easter, J., and DeGraff, J.M., 1977a, The evolution of Santa María volcano, Guatemala: Journal of Geology, v. 85, p. 63–87.

Rose, W.I., Jr., Pearson, T., and Bonis, S., 1977b, Nuee ardente eruption from the foot of a dacite lava flow: Bulletin Volcanologique, v. 40, p. 23–38.

Rose, W.I., Jr., Stoiber, R.E., and Malinconico, L.L., 1982, Eruptive gas compositions and fluxes of explosive volcanoes: the budget of S and Cl emitted from Fuego volcano, Guatemala, in Thorpe, R.S., ed., Andesites: John Wiley, New York, p. 669–676.

Rose, W.I., Conway, F.M., Pullinger, C.R., Deino, A., and McIntosh, W.C., 1999, An improved age framework for late Quaternary silicic eruptions in northern Central America: Bulletin of Volcanology, v. 61, p. 106–120, doi: 10.1007/s004450050266.

Roses, T.P., Davisson, M.L., and Criss, R.E., 1996, Isotope hydrology of voluminous cold springs in fractured rock from an active volcanic region, northeastern California: Journal of Hydrology, v. 179, p. 207–236, doi: 10.1016/0022-1694(95)02832-3.

Rowe, G.L., Jr., 1994, Oxygen, hydrogen, and sulfur isotope systematics of the crater lake system of Poás volcano, Costa Rica: Geochemical Journal, v. 28, p. 263–287.

Rowe, G.L., Jr., and Brantley, S.L., 1993, Estimation of the dissolution rates of andesitic glass, plagioclase and pyroxene in a flank aquifer of Poás volcano, Costa Rica: Chemical Geology, v. 105, p. 71–87, doi: 10.1016/0009-2541(93)90119-4.

Rowe, G.L., Jr., Ohsawa, S., Takano, B., Brantley, S.L., Fernandez, J.F., and Barquero, J., 1992, Using crater lake chemistry to predict volcanic activity at Poás volcano, Costa Rica: Bulletin of Volcanology, v. 54, p. 494–503, doi: 10.1007/BF00301395.

Rowe, G.L., Jr., Brantley, S.L., Fernandez, J.F., and Borgia, A., 1995, The chemical and hydrologic structure of Poás volcano, Costa Rica: Journal of Volcanology and Geothermal Research, v. 64, p. 233–267, doi: 10.1016/0377-0273(94)00079-V.

Sakai, H., and Matsubaya, O., 1977, Stable isotopic studies of Japanese geothermal systems: Geothermics, v. 5, p. 97–124, doi: 10.1016/0375-6505(77)90014-1.

Sanchez Bennett, E., Rose, W.I., and Conway, F.M., 1992, Santa María, Guatemala: A decade volcano: Eos (Transactions, American Geophysical Union), v. 73, p. 521–522.

Sanford, W.E., Konikow, L.F., Rowe, G.L., Jr., and Brantley, S.L., 1995, Groundwater transport of crater-lake brine at Poás volcano, Costa Rica: Journal of Volcanology and Geothermal Research, v. 64, p. 269–293, doi: 10.1016/0377-0273(94)00080-Z.

Sano, Y., and Marty, B., 1995, Origin of carbon in fumarolic gas from island arcs: Chemical Geology, v. 119, p. 265–274, doi: 10.1016/0009-2541(94)00097-R.

Sano, Y., and Williams, S.N., 1996, Fluxes of mantle and subducted carbon along convergent plate boundaries: Geophysical Research Letters, v. 23, p. 2749–2752, doi: 10.1029/96GL02260.

Sano, Y., Gamo, T., and Williams, S.N., 1997, Secular variations of helium and carbon isotopes at Galeras volcano, Colombia: Journal of Volcanology and Geothermal Research, v. 77, p. 255–265, doi: 10.1016/S0377-0273(96)00098-4.

Sano, Y., Nishio, Y., Sasaki, S., Gamo, T., and Nagao, K., 1998, Helium and carbon isotope systematics at Ontake volcano, Japan: Journal of Geophysical Research, v. 103, p. 23,863–23,873, doi: 10.1029/98JB01666.

Sano, Y., Takahata, N., Nishio, Y., Fischer, T.P., and Williams, S.N., 2001, Volcanic flux of nitrogen from the Earth: Chemical Geology, v. 171, p. 263–271, doi: 10.1016/S0009-2541(00)00252-7.

Self, S., and Williams, S.N., 1979, Triggering of the October 1902 eruption of Santa Maria volcano, Guatemala by a magma mixing event: Eos (Transactions, American Geophysical Union), v. 61, p. 68.

Shaw, A.M., Hilton, D.R., Fischer, T.P., Walker, J.A., and Alvarado, G.E., 2003, Contrasting He-C relationships in Nicaragua and Costa Rica: insights into C cycling through subduction zones: Earth and Planetary Science Letters, v. 214, p. 499–513, doi: 10.1016/S0012-821X(03)00401-1.

Shaw, A.M., Hilton, D.R., Macpherson, C.G., and Sinton, J.M., 2004, The $CO_2$-He-Ar-$H_2O$ systematics of the Manus back-arc basin: resolving source composition from degassing and contamination effects: Geochimica et Cosmochimica Acta, v. 68, p. 1,837–1,856, doi: 10.1016/j.gca.2003.10.015.

Shevenell, L., and Goff, F., 1993, Addition of magmatic volatiles into the hot spring waters of Loowit Canyon, Mount St. Helens, Washington, USA: Bulletin of Volcanology, v. 55, p. 489–503, doi: 10.1007/BF00304592.

Shevenell, L., and Goff, F., 1995, Evolution of hydrothermal waters at Mount St. Helens, Washington, USA: Journal of Volcanology and Geothermal Research, v. 69, p. 73–94, doi: 10.1016/0377-0273(95)00021-6.

Sisson, T.W., and Layne, G.D., 1993, $H_2O$ in basalt and basaltic andesite glass inclusions from four subduction-related volcanoes: Earth and Planetary Science Letters, v. 117, p. 619–635, doi: 10.1016/0012-821X(93)90107-K.

Slansky, E., 1975, Natroalunite and alunite from White Island volcano, Bay of Plenty, New Zealand: New Zealand Journal of Geology and Geophysics, v. 18, p. 285–293.

Smith, I.E.M., Brothers, R.N., Muiruri, F.G., and Browne, P.R.L., 1988, The geochemistry of rock and water samples from Curtis Island volcano, Kermadec group, southwest Pacific: Journal of Volcanology and Geothermal Research, v. 34, p. 233–240, doi: 10.1016/0377-0273(88)90035-2.

Snyder, G., Poreda, R., Hunt, A., and Fehn, U., 2001, Regional variations in volatile composition: isotopic evidence for carbonate recycling in the Central American volcanic arc: Geochemistry Geophysics Geosystems, v. 2, 2001GC000163.

Sorey, M.L., Evans, W.C., Kennedy, B.M., Farrar, C.D., Hainsworth, L.J., and Hausback, B., 1998, Carbon dioxide and helium emissions from a reservoir of magmatic gas beneath Mammoth Mountain, California: Journal of Geophysical Research, v. 103, p. 15,303–15,323, doi: 10.1029/98JB01389.

Sriwana, T., van Bergen, M.J., Sumarti, S., de Hoog, J.C.M., van Os, B.J.H., Wahyuningsih, R., and Dam, M.A.C., 1998, Volcanogenic pollution by acid water discharges along Ciwidey river, west Java (Indonesia): Journal of Geochemical Exploration, v. 62, p. 161–182, doi: 10.1016/S0375-6742(97)00059-9.

Stix, J., Torres, R., Narváez, L., Cortés, G.P., Raigosa, J., Gómez, D., and Castonguay, R., 1997, A model of vulcanian eruptions at Galeras volcano, Colombia: Journal of Volcanology and Geothermal Research, v. 77, p. 285–303, doi: 10.1016/S0377-0273(96)00100-X.

Stoiber, R.E., and Carr, M.J., 1973, Quaternary volcanic and tectonic segmentation of Central America: Bulletin Volcanologique, v. 37, p. 304–325.

Stoiber, R.E., and Rose, W.I., Jr., 1969, Recent volcanic and fumarolic activity at Santiaguito volcano, Guatemala: Bulletin Volcanologique, v. 33, p. 475–502.

Stoiber, R.E., and Rose, W.I., Jr., 1970, The geochemistry of Central American volcanic gas condensates: Geological Society of America Bulletin, v. 81, p. 2891–2912.

Stoiber, R.E., and Rose, W.I., Jr., 1974, Fumarole incrustations at active Central American volcanoes: Geochimica et Cosmochimica Acta, v. 38, p. 495–516, doi: 10.1016/0016-7037(74)90037-4.

Sturchio, N.C., Williams, S.N., Garcia, N.P., and Londono, A.C., 1988, The hydrothermal system of Nevado del Ruiz volcano, Colombia: Bulletin of Volcanology, v. 50, p. 399–412, doi: 10.1007/BF01050639.

Sturchio, N.C., Williams, S.N., and Sano, Y., 1993, The hydrothermal system of Volcan Puracé, Colombia: Bulletin of Volcanology, v. 55, p. 289–296, doi: 10.1007/BF00624356.

Symonds, R.B., Gerlach, T.M., and Reed, M.H., 2001, Magmatic gas scrubbing: implications for volcano monitoring: Journal of Volcanology and Geothermal Research, v. 108, p. 303–341, doi: 10.1016/S0377-0273(00)00292-4.

Symonds, R.B., Janik, C.J., Evans, W.C., Ritchie, B.E., Counce, D., Poreda, R.J., and Iven, M., 2003, Scrubbing masks magmatic degassing during repose at Cascade-Range and Aleutian-Arc volcanoes: U.S. Geological Survey Open-File Report 03-435, 22 p.

Taran, Y., Fischer, T.P., Pokrovsky, B., Sano, Y., Aurora Armienta, M., and Macias, J.L., 1998, Geochemistry of the volcano-hydrothermal system of El Chichón volcano, Chiapas, Mexico: Bulletin of Volcanology, v. 59, p. 436–449, doi: 10.1007/s004450050202.

Tassi, F., Vaselli, O., Capaccioni, B., Macias, J.L., Nencetti, A., Montegrossi, G., and Magro, G., 2003, Chemical composition of fumarolic gases and spring discharges from El Chichòn volcano, Mexico: Causes and implications of the changes detected over the period 1998–2000: Journal of Volcanology and Geothermal Research, v. 123, p. 105–121, doi: 10.1016/S0377-0273(03)00031-3.

Taylor, H.P., Jr., 1974, The application of oxygen and hydrogen isotope studies to problems of hydrothermal alteration and ore deposition: Economic Geology and the Bulletin of the Society of Economic Geologists, v. 69, p. 843–883.

Tedesco, D., 1995, Monitoring fluids and gases at active volcanoes, in McGuire, B., Kilburn, C., and Murray, J., eds., Monitoring Active Volcanoes: London, University College of London Press, p. 315–345.

Templeton, S., 1999, Hydrothermal systems at Volcán Santa María, Guatemala [M.S. thesis]: DeKalb, Illinois, Northern Illinois University, 166 p.

Traineau, H., Westercamp, D., and Benderitter, Y., 1989, Case study of a volcanic geothermal system, Mount Pelée, Martinique: Journal of Volcanology and Geothermal Research, v. 38, p. 49–66, doi: 10.1016/0377-0273(89)90029-2.

Truesdell, A.H., Nathenson, M., and Rye, R.O., 1977, The effects of subsurface boiling and dilution on the isotopic composition of Yellowstone thermal waters: Journal of Geophysical Research, v. 82, p. 3694–3704.

Valentino, G.M., and Stanzione, D., 2003, Source processes of the thermal waters from the Phlegraean Fields (Naples, Italy) by means of the study of selected minor and trace elements distribution: Chemical Geology, v. 194, p. 245–274, doi: 10.1016/S0009-2541(02)00196-1.

van Soest, M.C., Hilton, D.R., and Kreulen, R., 1998, Tracing crustal and slab contributions to arc magmatism in the Lesser Antilles island arc using helium and carbon relationships in geothermal fluids: Geochimica et Cosmochimica Acta, v. 62, p. 3323–3335, doi: 10.1016/S0016-7037(98)00241-5.

van Wyk de Vries, B., Kerle, N., and Petley, D., 2000, Sector collapse forming at Casita volcano, Nicaragua: Geology, v. 28, p. 167–170.

Varekamp, J.C., Kreulen, R., Poorter, R.P.E., and Van Bergen, M.J., 1992, Carbon sources in arc volcanism, with implications for the carbon cycle: Terra Nova, v. 4, p. 363–373.

Varekamp, J.C., Pasternack, G.B., and Rowe, G.L., Jr., 2000, Volcanic lake systematics II, Chemical constraints: Journal of Volcanology and Geothermal Research, v. 97, p. 161–179, doi: 10.1016/S0377-0273(99)00182-1.

Varekamp, J.C., Ouimette, A.P., Herman, S.W., Bermúdez, A., and Delpino, D., 2001, Hydrothermal element fluxes from Copahue, Argentina: A "beehive" volcano in turmoil: Geology, v. 29, p. 1059–1062, doi: 10.1130/0091-7613(2001)029<1059:HEFFCA>2.0.CO;2.

Vuataz, F.D., and Goff, F., 1986, Isotope geochemistry of thermal and nonthermal waters in the Valles Caldera, Jemez Mountains, northern New Mexico: Journal of Geophysical Research, v. 91, p. 1835–1853.

White, D.E., 1957, Thermal waters of volcanic origin: Geological Society of America Bulletin, v. 68, p. 1637–1682.

Williams, H., 1960, Volcanic history of the Guatemalan Highlands: University of California Publications in Geological Sciences 38, 86 p.

Williams, S.N., and Self, S., 1983, The October 1902 Plinian eruption of Santa María volcano, Guatemala: Journal of Volcanology and Geothermal Research, v. 16, p. 33–56, doi: 10.1016/0377-0273(83)90083-5.

Wood, C.P., 1994, Mineralogy at the magma-hydrothermal system interface in andesite volcanoes, New Zealand: Geology, v. 22, p. 75–78, doi: 10.1130/0091-7613(1994)022<0075:MATMHS>2.3.CO;2.

Zimmer, M.M., Fischer, T.P., Hilton, D.R., Alvarado, G.E., Sharp, Z.D., and Walker, J.A., 2004, Nitrogen systematics and volatile fluxes of subduction zones: insights from Costa Rica arc volatiles: Geochemistry Geophysics Geosystems, v. 5, 2003GC000651.

MANUSCRIPT ACCEPTED BY THE SOCIETY 19 MARCH 2006

# Downstream aggradation owing to lava dome extrusion and rainfall runoff at Volcán Santiaguito, Guatemala

**Andrew J.L. Harris***
*Hawaii Institute of Geophysics and Planetology, School of Oceanography and Earth Science Technology, University of Hawai'i, 2525 Correa Road, Honolulu, Hawaii 96822, USA*

**James W. Vallance**
*Cascades Volcano Observatory, U.S. Geological Survey, 1300 SE Cardinal Court, Building 10, Suite 100, Vancouver, Washington 98683, USA*

**Paul Kimberly**
*Smithsonian Institution, Department of Mineral Sciences, Washington, D.C. 20560-0119, USA*

**William I. Rose**
*Department of Geological Engineering and Sciences, Michigan Technological University, Houghton, Michigan 49931, USA*

**Otoniel Matías**
*Instituto Nacional de Sismología, Vulcanología, Meterología y Hidrología (INSIVUMEH), 7a Av. 14-57, Zona 13, Guatemala City, Guatemala*

**Elly Bunzendahl**
*Department of Geological Engineering and Sciences, Michigan Technological University, Houghton, Michigan 49931, USA*

**Luke P. Flynn**

**Harold Garbeil**
*Hawaii Institute of Geophysics and Planetology, School of Oceanography and Earth Science Technology, University of Hawai'i, 2525 Correa Road, Honolulu, Hawaii 96822, USA*

## ABSTRACT

Persistent lava extrusion at the Santiaguito dome complex (Guatemala) results in continuous lahar activity and river bed aggradation downstream of the volcano. We present a simple method that uses vegetation indices extracted from Landsat Thematic Mapper (TM) data to map impacted zones. Application of this technique to a time series of 21 TM images acquired between 1987 and 2000 allow us to map, measure, and track temporal and spatial variations in the area of lahar impact and river aggradation.

---

*E-mail: harris@higp.hawaii.edu

Harris, A.J.L., Vallance, J.W., Kimberly, P., Rose, W.I., Matías, O., Bunzendahl, E., Flynn, L.P., and Garbeil, H., 2006, Downstream aggradation owing to lava dome extrusion and rainfall runoff at Volcán Santiaguito, Guatemala, *in* Rose, W.I., Bluth, G.J.S., Carr, M.J., Ewert, J.W., Patino, L.C., and Vallance, J.W., Volcanic hazards in Central America: Geological Society of America Special Paper 412, p. 85–104, doi: 10.1130/2006.2412(05). For permission to copy, contact editing@geosociety.org. ©2006 Geological Society of America. All rights reserved.

In the proximal zone of the fluvial system, these data show a positive correlation between extrusion rate at Santiaguito (E), aggradation area 12 months later ($A_{prox}$), and rainfall during the intervening 12 months ($Rain_{12}$): $A_{prox} = 3.92 + 0.50\ E + 0.31\ \ln(Rain_{12})$ ($r^2 = 0.79$). This describes a situation in which an increase in sediment supply (extrusion rate) and/or a means to mobilize this sediment (rainfall) results in an increase in lahar activity (aggraded area). Across the medial zone, we find a positive correlation between extrusion rate and/or area of proximal aggradation and medial aggradation area ($A_{med}$): $A_{med} = 18.84 - 0.05\ A_{prox} - 6.15\ Rain_{12}$ ($r^2 = 0.85$). Here the correlation between rainfall and aggradation area is negative. This describes a situation in which increased sediment supply results in an increase in lahar activity but, because it is the zone of transport, an increase in rainfall serves to increase the transport efficiency of rivers flowing through this zone. Thus, increased rainfall flushes the medial zone of sediment.

These quantitative data allow us to empirically define the links between sediment supply and mobilization in this fluvial system and to derive predictive relationships that use rainfall and extrusion rates to estimate aggradation area 12 months hence.

**Keywords:** Santiaguito, lava dome, extrusion, lahar, aggradation.

## INTRODUCTION

Lahar genesis requires water supply, abundant unconsolidated debris, steep slopes, and a triggering mechanism (Vallance, 2000). Pyroclastic fall and flow deposits provide voluminous sources of debris capable of mobilization during rainfall runoff to generate lahars (or hyperconcentrated flow) and more dilute, muddy stream flow. Consequently, frequent lahars and stream sedimentation are common problems in drainage basins blanketed by volcanic fallout or that have their headwaters on active volcanoes. In this regard, sediment loads of $10^3$ to $10^6$ Mg/km$^2$ rank basins affected by volcanic activity among the highest of Earth's sediment producers (Major et al., 2000). Moreover, the rapid and extensive emplacement of lahars means that they can inflict high numbers of fatalities and cause extensive damage (Rodolfo, 2000).

On the basis of temporal variation in sediment flux, we identify two types of volcanic fluvial regimes: transient and persistent. Transient regimes result from a single eruptive event that emplaces a large volume of unconsolidated volcanic material. Remobilization by melt water, lake breaches, or rainfall subsequently triggers a transient increase in lahar volume, frequency, and sediment yield. In time, the sediment source becomes exhausted or stabilized by vegetation growth so that sediment yield shows a decline following an initial peak. Examples of such regimes are the sediment responses to the 1980 eruption of Mount St. Helens (Major et al., 2000) or the 1991 eruption of Pinatubo (Rodolfo et al., 1996).

A regime of persistent lahar activity results when eruptive activity continually supplies unconsolidated volcanic material for remobilization. In such cases, lahars continue as long as volcanic activity persists and sediment yields remain high. Such regimes typically occur in drainages continuously supplied with tephra fall (e.g., Semeru, Indonesia) or unconsolidated pyroclastic debris from episodic collapse of persistently extruded lava domes (e.g., Merapi in Indonesia and Santiaguito in Guatemala).

In 1902, a Plinian eruption on the south flank of Santa María volcano (Guatemala) erupted >8.5 km$^3$ of pumiceous dacite that covered more than $1.2 \times 10^6$ km$^2$ to a thickness of 2 m near the vent (Williams and Self, 1983). Transient lahar activity followed, the dramatic increase in sediment supply causing downstream aggradation of 10–15 m (Kuenzi et al., 1979). By 1922, the sediment supply had waned.

In 1922, after 20 yr of quiescence, a phase of continuous dome extrusion began within the 1902 eruption crater of Santa María (Rose, 1987). Between 1922 and 2000, continuous lava extrusion at an average rate of $0.44 \pm 0.01$ m$^3$ s$^{-1}$ formed a ~1.1 km$^3$ dome complex (Harris et al., 2003). Although the four main units comprising this complex are named Caliente, Monje, Mitad, and Brujo, the complex itself has been named Santiaguito. Although continuous, the rate of extrusion has alternated between 3 and 5 yr periods of above-average extrusion rates, separated by 9–12 yr below-average periods (Rose, 1987; Harris et al., 2003).

The continuous extrusion at Santiaguito has provided a persistent, if variable, sediment supply to the fluvial system. Mobilization of the volcanic material in the wet season triggers lahars and aggradation each rainy season. As a result, river channels become clogged with volcaniclastic material, causing river-beds to aggrade, as well as damming of tributaries and diversion of river channels. Lahar activity and aggradation thus impacts a fluvial system extending 60 km from Santiaguito to the Pacific coast of Guatemala (Fig. 1). This is a heavily populated and farmed zone. The approximate population for the 7 km radius around San Felipe (Fig. 1), for example, is 31,800, giving a population density of ~800 people/km$^2$. Therefore, lahar and aggradation activity has widespread impacts on communities across this zone.

One of the best ways to track such a persistent yet widespread phenomenon is to use the repeat, synoptic coverage afforded by satellite data. The 30-m spatial resolution multispsectral data of the Landsat Thematic Mapper (TM) and Enhanced Thematic Mapper Plus (ETM+) provides data capable of providing regular

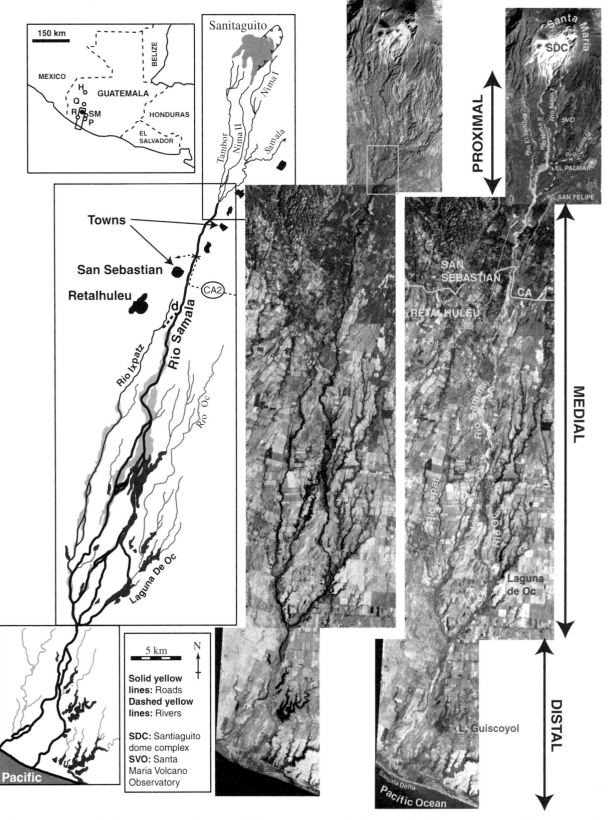

Figure 1. Location maps for the fluvial system fed by volcaniclastic sediment from Santiaguito and a Thematic Mapper (TM) image acquired on 19 January 1990 covering the same area. Guatemala location map inset top left, showing the locations of Santiaguito (black dot), the rain-gauge locations used in this study (H—Huehuetenango; Q—Quezaltenango; R—Retalhuleu; P—Patzulin; SM—Santa Marta), and location of the main maps (rectangle). On the main map, Río Tambor, Río Nimá I and Río Nimá II, and Río Ixpatz are shown by thin black lines, Río Samalá by thick black lines, and all other tributaries not supplied by volcaniclastic sediments by gray lines. Gray zone labeled Santiaguito marks the dome complex, and lighter gray zones along the river courses mark zones of aggradation. The thick dashed line marked "d" indicates the course taken by Río Samalá when diverted, by aggradation, into Río Ixpatz. Two images are displayed on which the proximal, medial, and distal zones of the fluvial system are shown. The first (middle) is a false color composite (TM band 754). This composite best distinguishes areas of aggradation. Vegetation—blue; fallow fields—green/olive; aggraded beds—brown. Yellow box notes location of the sketch map in Figure 2. The second (extreme right) image is a true-color composite (TM band 321) of the same zone. The 754 rendition is given, because, although it represents the vegetation in false color, it is a particularly effective combination for highlighting aggraded zones. SDC—Santiaguito Dome Complex; SVO—Santa María volcano.

maps of impacted zones. Thus, in this paper we use a time series of 21 TM and ETM+ images to document and examine lahar inundation and aggradation in Santiaguito's drainage system between 1987 and 2000. Because we cannot detect individual lahar events using the satellite data, we concentrate on the net effect of the persistent sedimentation regime affecting this fluvial system and focus on the process of aggradation as expressed by changes in the area impacted by fluvial sedimentation.

Our aims are threefold: First, to define techniques that can be applied to the satellite data to allow effective mapping of aggraded zones; second, to use our satellite-image time series to show how the data can be used to measure, map, track, and document aggradation; and third, to use the Landsat data to obtain annual estimates of aggraded area, allowing us to examine the correlations between extrusion rate at the dome complex, rainfall runoff, and annual changes in area aggraded. Our overall aim is to show how satellite data can be used to monitor lahar and aggradation activity in a volcanic setting, while providing a case history of aggradation at Santiaguito for the period 1987–2000.

## BACKGROUND: LAHARS AND AGGRADATION AT SANTIAGUITO

### Lahar-Aggradation Processes

A sequence of processes combines to impact the fluvial system downstream of Santiaguito dome. At Santiaguito, regular ash venting, persistent lava extrusion, and crumbling of silicic lava flow levees and fronts cause avalanches of hot rock and ash (Harris et al., 2002). These provide a continuous source of unconsolidated material that can then be washed into the fluvial system during the rainy season. Heavy rains commonly remobilize these deposits and generate floods and lahars. At Santiaguito, monsoonal rains between May and October (Table 1) commonly trigger lahars. Rarely, heavy rains in February through April have also triggered lahars. Subsequent downstream deposition of rain-mobilized volcaniclastic material causes river channel aggradation such that ravines (barrancas) that were once tens of meters deep are now filled. Ravine-filling, in turn, promotes stream diversion, capture, and channel abandonment.

### *Santiaguito Drainage Basin*

Santiaguito's 825 km$^2$ drainage basin is comprised of a dendritic network of subparallel braided streams that flow across an elongate alluvial fan (Fig. 1). Río Samalá has its headwaters in the volcanic highlands of Guatemala and flows south around the eastern flank of Santa María. Three tributaries that have their headwaters at Santiaguito (Río Tambor, Río Nimá I, and Río Nimá II) join Río Samalá ~11 km south of Santiaguito. Annual rainfall at Patzulin in the headwaters of the Río Nimá II ranges from 3 to 6 m with 85% occurring between June and October (Table 1; Wernstedt, 1961). Although Río Samalá is an ungauged river, calculated discharge indicates a mean value of ~40 m$^3$ s$^{-1}$. Whereas minimum discharge during the dry season is calculated at 2 m$^3$ s$^{-1}$, peak discharge during the wet season is ~2700 m$^3$ s$^{-1}$ (assuming 0.5 m of rain in 12 h) (Kimberly, 1995). Measured minimum and maximum discharges at Río Guacalate, 90 km to the east and on the same coastal slope, are similar. Here, Davies et al. (1979) observed that a minimal flow of 5.5 m$^3$ s$^{-1}$ increased to peak flow of 2200 m$^3$ s$^{-1}$ within 20 min following storms and transported 230-cm boulders.

From ~1975 until 1997, Río Tambor and Río Nimá II headed within the 1902 crater of Santa Maria, and Río Nimá I headed just east of the crater. A lava flow active from 1997 to 1999, however, breached the east crater wall so that Río Nimá I now receives volcaniclastic debris from the 1902 crater (Harris et al., 2003). Each of these three Río Samalá tributaries receives voluminous volcaniclastic debris from the dome complex and each in turn supplies sediment to Río Samalá.

Downstream of its confluence with the three tributaries, Río Samalá flows southward over an alluvial fan 49 km to the Pacific Ocean. Río Ixpatz heads at San Sebastian, 22 km SSW of Santiaguito. Periodically, Río Samalá is diverted, due to aggradation and avulsion of its channel, to flow into Río Ixpatz south of San Sebastian.

### Proximal, Medial, and Distal Zones

We divide the Santiaguito fluvial system into proximal, medial, and distal zones, each defined by characteristic slope, channel features, and environment of deposition (Table 2). Although each zone can experience incision, equilibrium transportation, or aggradation, one process dominates within each zone. These are incision, transport, and deposition in the proximal, medial, and distal sections, respectively (Schumm, 1977).

The proximal zone is the sediment source area and the location of primary sediment production. It comprises Río Tambor, Río Nimá I, and Río Nimá II and extends 11 km from Santiaguito to the confluence with Río Samalá (Fig. 1). Without the continuous sediment flux the dome and its lava flows provide to its upper reaches, this zone would experience erosion and rapid incision. The voluminous flux of debris from the dome complex, in the long run, however, largely balances down-cutting.

The medial zone is the transport zone where, for a graded profile, sediment input approximately equals output. This zone extends 32 km and includes Río Samalá and Río Ixpatz (Fig. 1) and is characterized by sinuous-braided channels. Although sediment transport dominates in this zone, variations in sediment supply can cause aggradation or incision.

The distal zone of the Santiaguito fluvial system is the sediment sink or zone of deposition, where sediment is deposited on alluvial fans, alluvial plains, deltas, or off-shore. This zone extends 17 km from the Samalá-Ixpatz confluence to the Pacific Ocean (Fig. 1). It encompasses the deltaic plain that Kuenzi et al. (1979) describe as arcuate, wave dominated, and in equilibrium with the continual supply of sediment derived from Santiaguito. Floods and hyperconcentrated flow dominate sediment transport and deposition.

TABLE 1. AVERAGE TEMPERATURE AND RAINFALL VALUES FOR 9- TO-27-YEAR-LONG PERIODS PRIOR TO 1961 FROM STATIONS NEAR SANTIAGUITO*

| Month | Jan | Feb | Mar | Apr | May | Jun | Jul | Aug | Sep | Oct | Nov | Dec | Annual |
|---|---|---|---|---|---|---|---|---|---|---|---|---|---|
| (1) Quezaltenango, Elev. 2379.3 m, 22-yr record | | | | | | | | | | | | | |
| Temp (°C) | 11.5 | 12.5 | 14.3 | 16.4 | 17.0 | 16.4 | 16.2 | 16.9 | 15.9 | 15.2 | 14.2 | 12.2 | 14.6 |
| Precip. (cm) | 1.02 | 0.18 | 0.05 | 1.12 | 3.23 | 21.3 | 46.9 | 41.7 | 54.3 | 41.9 | 3.38 | 0.64 | 215 |
| (2) Patzulin, Elev. 899.8 m, 27-yr record | | | | | | | | | | | | | |
| Temp (°C) | 21.5 | 21.7 | 21.9 | 21.9 | 22.2 | 22.3 | 22.4 | 22.1 | 22.1 | 22.0 | 21.6 | 21.6 | 21.9 |
| Precip. (cm) | 3.30 | 4.09 | 9.75 | 20.5 | 53.3 | 70.9 | 59.6 | 63.4 | 84.2 | 64.2 | 25.5 | 9.30 | 468 |
| (3) Retalhuleu, Elev. 239.9 m, 9-yr record | | | | | | | | | | | | | |
| Temp (°C) | 24.9 | 26.8 | 26.4 | 25.0 | 25.5 | 25.3 | 25.0 | 25.0 | 24.1 | 24.6 | 25.3 | 24.9 | 25.2 |
| Precip. (cm) | 1.22 | 1.93 | 10.3 | 18.4 | 32.3 | 43.9 | 29.5 | 34.8 | 55.4 | 48.1 | 12.9 | 1.70 | 290 |

*From Wernstedt, 1961.

TABLE 2. CHARACTERISTICS OF FLUVIAL SYSTEM ZONES FOR THE DRAINAGE THAT HEADS AT SANTIAGUITO*

| | Proximal zone | Medial zone | Distal zone |
|---|---|---|---|
| Elevation range | 500–2770 m | 50–500 m | 0–50 m |
| Distance from volcano | 0–11 km | 11–43 km | 43–60 km |
| Slope (degrees) | 14–3 | 3–0.2 | ~0.2 |
| Major streams | Río Tambor, Río Nimá I, Río Nimá II, Río Samalá | Río Samalá, Ixpatz | Río Samalá |
| Channel form | entrenched–braided | sinuous–braided | sinuous–straight |
| Deposits | gravel, sand | gravel, sand | sand, silt, clay |
| Primary hazards | pyroclastic flows, lava flows, lahars, floods, ashfall | hyperconcentrated mudflows, debris flows, floods | hyperconcentrated mudflows, floods |

*After Kimberly, 1995.

### Aggradation at El Palmar: A Proximal Zone Case Study

We identify El Palmar as our ground truth zone. This village is located within the proximal zone on the banks of the Río Nimá I, ~8 km south of Santiaguito and 3 km above the confluence with Río Samalá. We have visited this location multiple times during 1983–2003, carrying out detailed field surveys annually during 2000–2003. Our aim has been to define and map the nature of the aggradation problem using the remote sensing data.

Aggradation of the Río Nimá II near El Palmar has caused channel wandering and over-bank deposition of a ~1-km-wide wedge of sediment (Fig. 2). From 1957 to 1988, the channel of the Río Nimá II near El Palmar aggraded ~40 m (INSIVUMEH, 1988). During an unusually wet season in 1983, Río Nimá II aggraded up to 10 m and diverted Río Nimá I eastward into Río Samalá (SEAN, 1983). This was achieved when material emplaced on the east bank of Río Nimá II forced the Río Nimá I to take a more easterly course at the point where it joined Río Nimá II (Fig. 2). Lahar inundation from Río Nimá II in 1983 and subsequent flooding from Río Nimá I and II severely damaged El Palmar and eventually forced evacuation. Although the Guatemalan government relocated El Palmar to Las Marias, 2 km east of its original site (SEAN, 1988), people continued to live in less affected parts of the town until 1997.

Growth of the 1997–1999 lava flow into the headwaters of Río Nimá I initiated aggradation in the previously stable Río Nimá I in 1997. As a result, several lahars inundated El Palmar during the 1997 rainy season. During the 1998–1999 rainy seasons, Río Nimá I incised a 20-m-wide, 23–27-m-deep ravine through the center of El Palmar. By 2001, incision had ceased because the Río Nimá I bed had attained the bed level of Río Samalá. Lateral undermining, however, widened the ravine and consumed several more buildings.

Diversion of Río Nimá I into Río Samalá supplied volcaniclastic sediment to the latter and caused dramatic downstream aggradation of Río Samalá from 1997 to 2000. Prior to 1997, Río Samalá received no sediment from Santiaguito above its confluence with Río Nimá II. From 1997 to 2000, however, Río Samalá was supplied by Río Nimá I and hence aggraded to fill its gorge north of San Felipe by up to 20 m. During the 2000 rainy season, erosion removed an average of 3 m of sediment from this location. A 30–50 m deep gorge presently protects San Felipe from lahars and aggradation in Río Samalá (Fig. 2).

Downstream of El Palmar, aggradation has caused the bed level of Río Nimá II to exceed that of surrounding areas. As a result, floods and lahars have spilt westward off of the aggraded Río Nimá II bed into Río Tambor (Fig. 2). Here, Río Tambor has a bed level lower than that of Río Nimá II and now occupies a channel west of its position before 1987 (Fig. 2). In addition, the confluence between the Río Tambor and Río Nimá II has been forced progressively farther south, such that the channel of Río

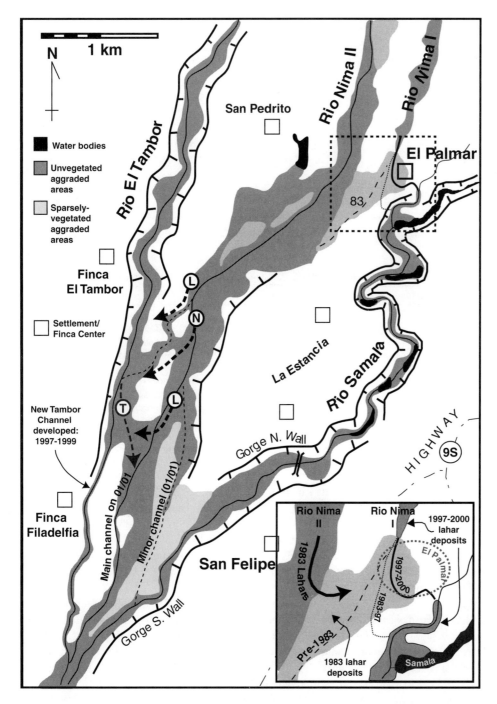

Figure 2. Map of aggraded channels in the vicinity of El Palmar and San Felipe (see Figure 1 for location). Map produced from 15-m pixel, 23 January 2000 ETM+ panchromatic image and field-based observations made during January 2000 and 2001. Thick solid lines with ticks mark ravine rims; thin solid lines mark primary river channels; thin dashed lines mark secondary river channels as of January 2001. "L" identifies paths of lahars flowing off of the aggraded Río Nimá II channel in 2001–2002. "N" locates the course of the Río Nimá II as marked on the 1:50,000 Instituto Geográfico Nacional map sheet for Retalhuleu that conforms to the river course as of 1959. "T" marks the course of the Río El Tambor during 1987–1997, before taking the more westerly route through Finca Filadelfia. Inset: enlarged sketch map of El Palmar; location marked by black, dashed box on main map. On the inset, the dotted circle marks the approximate zone of the town. Sketch shows (1) paths of the 1983 lahars (thick solid arrow); (2) pre-1983 course of Río Nimá I (dashed line); (3) 1983–1997 course of Río Nimá I (dotted line); (4) 1997–2000 course of Río Nimá I (thick solid line); and (5) new (1997–2000) aggradation in Río Nimá I and Río Samalá (gray).

Tambor has been forced along a new path, cutting cultivated land through Finca Filadelfia.

## METHOD

To map, measure, and document aggradation across each zone using our Landsat satellite data, we apply a simple approach that uses a normalized difference vegetation index (NDVI) derived from TM data to determine the extent of vegetation cover in a pixel. Because vegetation reflects strongly in the near-infrared (TM band 4, 0.76–0.90 μm) and weakly in the visible red portion of the spectrum (TM band 3, 0.6–0.69 μm), the band 4/band 3 ratio is sensitive to changes in vegetation health and cover (Barrett and Curtis, 1995; Mather, 1987). The NDVI is commonly defined as

$$\text{NDVI} = (\rho_{nir} - \rho_r)/(\rho_{nir} + \rho_r), \quad (1)$$

where $\rho_{nir}$ and $\rho_r$ are reflectance in the near-infrared and red portions of the spectrum, respectively (e.g., Townshend and Tucker,

1984; Tucker et al., 1984; Holben et al., 1986). The NDVI ranges from −1 to +1, where NDVI < 0 indicates zones with no vegetative cover or water and NDVI > 0 indicates vegetation (Mather, 1987). For our purposes, this is a central concept. Because in inundated areas sediment will remove and/or bury (or partly bury) vegetation, changes in NDVI from positive to negative values will indicate new areas of impact. Furthermore, the tropical climate favors rapid regrowth of vegetation in areas no longer receiving sediment. Thus, the NDVI not only indicates areas newly affected, it also indicates areas that become inactive. We note, however, that the NDVI will also reveal zones that have been cleared of vegetation due to human activities (e.g., plowing, forest clearance) or natural processes (e.g., fires).

We thus use the TM-derived NDVI image to determine whether a pixel contains surfaces subject to sedimentation or not. To achieve this, we first construct an NDVI image and (to save processing time) extract a sub-image containing the Santiaguito fluvial system from the entire 185 × 185 km image. We then apply a simple threshold in which we define pixels with NDVI ≤ 0 as non-vegetated and those with NDVI > 0 as vegetated. We next identify and remove zones of cloud and cloud shadow, fallow fields, and urban areas. This allows us to isolate zones that have NDVI ≤ 0 due to sediment cover. This is achieved by a consideration of location, shape, and association. For example, towns are in fixed and known locations, fallow fields and burn zones have characteristic shapes, and low reflectance cloud shadows are associated with high reflectance clouds. These zones are manually identified, classified, and located on the final map. Thus, the final map shows zones that are apparently vegetation-free due to (1) aggradation and lahar activity (sediment), (2) agricultural practices, (3) presence of urban areas or other man-made structures, and (4) clouds and cloud shadow. In addition, we exclude the vegetation-free zone that covers the dome complex itself. Finally, we multiply the number of pixels identified as sediment covered by the pixel area (900 m$^2$) to produce a quantitative measure of the aggradation area.

Comparing our results with direct observation of the original TM images shows that our approach is reliable and accurately measures areas affected by aggradation and laharic activity (Fig. 3). The method does, however, exclude some pixel areas partially affected by aggradation and partially covered by vegetation. These areas are usually located along the margin of the channel. This may result in an underestimation for the total aggraded area. At worst, assuming that the margin of each aggraded channel comprises a continuous line of unflagged, mixed pixels, we infer that the maximum underestimate would be 35%.

## RESULTS I: SPATIAL CHANGES IN AGGRADATION

### Proximal Zone

The time series of Landsat-generated aggradation maps show that volcaniclastic sedimentation affects all rivers draining the 1902 crater: Río Tambor, Río Nimá I, and Río Nimá II (Fig. 4). We note, however, that the mapped aggradation area waxes and wanes. Aggradation along the Río Tambor, for example, as well as the wedge of sediment at El Palmar, is particularly apparent during 1988–1990, but less so during 1991–1992 (Fig. 4). The second period of enhanced aggradation is evident between 1996 and 2000, particularly along the Río Nimá I and II (Fig. 4). This second period coincides with the emplacement of two new lava flows.

The first lava flow supplied new sediment to Río Nimá I and Río Samalá, such that both of these rivers are mapped as suffering significant aggradation in the proximal zone after 1996. This flow developed during 1996–1999, extended down the east flank of the dome complex, breached the eastern rim of the 1902 crater, and extended into the headwaters of Río Nimá I. As a result, Río Nimá I, a river that had not previously received sediment, began to aggrade and to feed Río Samalá with sediment above its confluence with Río Nimá II. This caused aggradation along Río Samalá above the Río Nimá II confluence. Prior to this period, this section of Río Samalá had not received sediment and we observed a small lake at this location in the Landsat data. Here, water had become dammed above the sediment-clogged confluence with Río Nimá II. However, by 2000 this had been filled.

The second lava flow supplied new sediment to Río Nimá II. As a result, this river is also mapped as suffering increased aggradation by 2000 (Fig. 4). This lava flow began advancing down the south flank of the dome complex during July 1999. By January 2000, the flow had extended ~2.4 km into the headwaters of the Río Nimá II and rejuvenated sediment transport in that river.

### Medial Zone

The map series for the medial zone reveals severe aggradation along both Río Samalá and Río Ixpatz (Fig. 5). Although suffering aggradation in all years, the aggradation area is mapped as remaining broadly stable along the first ~12 km of Río Samalá in this zone during 1988–2000 (Fig. 6). This may partly reflect the success of engineering measures along this section where, since 1997, excavation of ~10 m of sediment from the Río Samalá channel along a 1-km-long stretch of the river has been necessary to preserve the bridge along the coastal highway near San Sebastian (CA2, Fig. 5). Similar efforts are necessary along the entire 3-km stretch of Río Samalá near San Sebastian, where the CA2 highway runs along the river bank. It may also reflect efficient transport of sediment through this section.

However, the time series reveals a highly dynamic system in the lower half of this zone. In this section, constant changes in channel location and over-banking of lahar material to emplace new deposits on the surrounding terrain are apparent from the aggradation maps (Fig. 6). This is evident, for example, to the west of Boxoma, where lahars extending into new land along the eastern margins of the river began in 1993. Thereafter, and as mapped in the Figure 5 summary, continued activity built a lahar fan of increasing size.

Aggradation in the Río Ixpatz shows a different cycle of activity. Aggradation areas are high in 1988–1990, decrease during

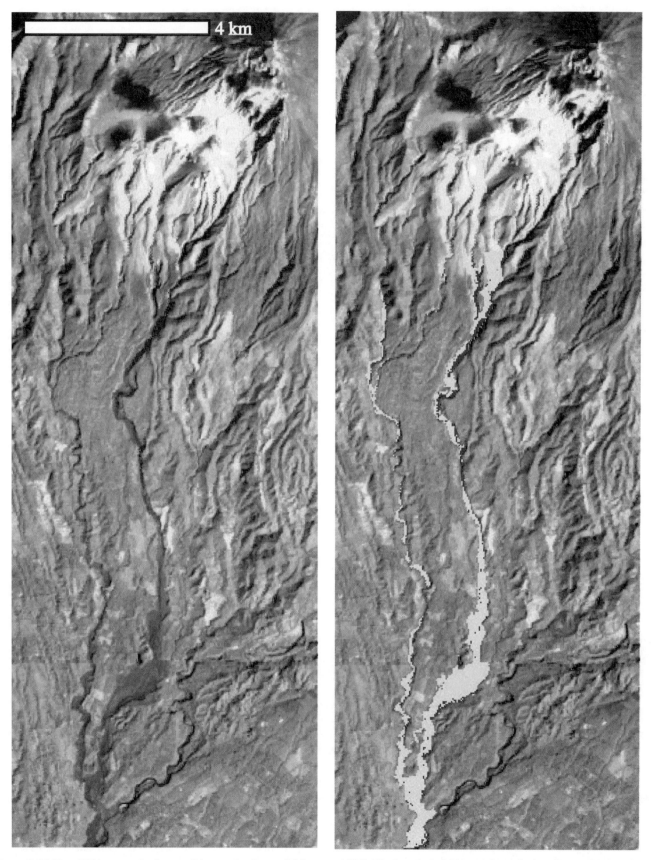

Figure 3. TM band 754 composite image of the proximal zone (19 January 1990). The location of this zone is marked as the proximal section in Figure 1, where the Santiaguito dome complex (volcaniclastic material source) is at the top of the image. In this false color combination, vegetation is blue and aggraded riverbeds are brown. On the right hand image, pixels identified as containing aggraded riverbeds are flagged in yellow. Flagged, vegetation-free pixels of the dome complex and those due to fallow field have been identified and removed, so that only vegetation-free pixels due to aggradation are considered.

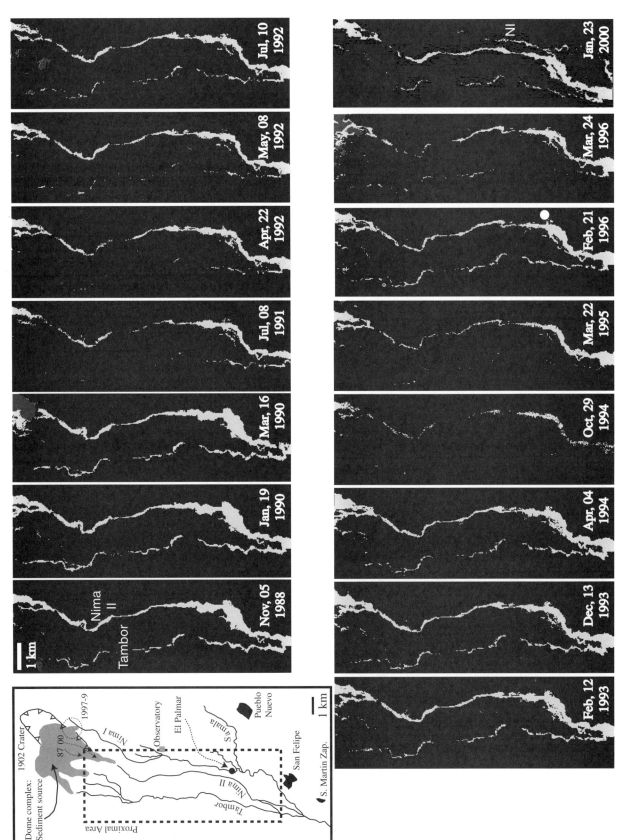

Figure 4. Proximal zone location map produced from 23 January 2000 ETM+ image with TM-derived aggradation map time series. On the location map, arrows on the dome complex marked "87" and "00" indicate the paths of the 1987 and 2000 lava flows. The dashed circle marked 1997-9 marks the location where the 1997–1999 lava flow breached the 1902 crater rim. On the time series, the main rivers are marked (NI—Río Nimá I) along with the location of El Palmar (white circle). Masked clouds are identified in purple, areas of aggradation are yellow, and vegetated zones are black.

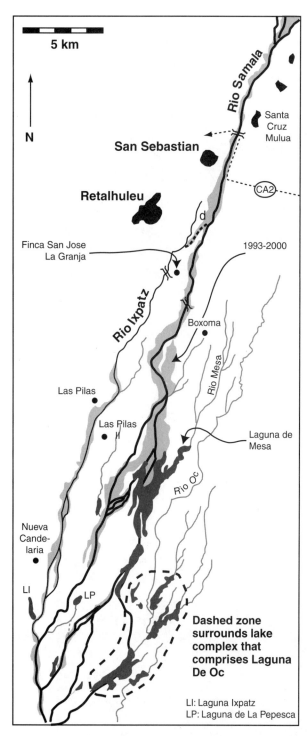

Figure 5. Medial zone location and aggradation summary map compiled from Landsat time series. The location of this zone is marked as the medial section in Figure 1. Aggraded areas—light gray; water bodies—mid-gray; population centers—black. Black lines—Río Samalá and Río Ixpatz (heavy and thin lines, respectively); all other tributaries not supplied by volcaniclastic—gray lines. Lighter gray zones along the river courses mark zones of aggradation; that marked "1993-2000" was a sediment fan that developed during these years. Heavy dashed line marked "d" indicates the course taken by Río Samalá when diverted, by aggradation, into Río Ixpatz; thin dashed line marked "CA2" is the coastal highway.

1991–1994, and are then high again during 1996–2000 (Fig. 6). This probably reflects the success of dyke building to protect Río Ixpatz from sediment supply by Río Samalá, where Río Ixpatz levees did indeed fail in both 1998 and 1999. Río Samalá lahars first overflowed into Río Ixpatz between 1982 and 1984. Boulder dykes constructed along the right bank of Río Samalá now protect Río Ixpatz from further incursions of Río Samalá. Clearly, such measures were more successful in the first half of the 1990s than in the latter half.

### Distal Zone

Our maps for the distal zone show that the margins of this zone are dominated by a series of lakes that owe their origin to the aggradation that followed the 1902 eruption of Santa Maria (Fig. 7). The 1902 eruption resulted in the deposition of ~4 km$^3$ of deltaic sediments (Kuenzi et al., 1979), which created a series of marginal, sediment-dammed lakes. Río Samalá's delta also prograded ~7 km seaward immediately following the 1902 eruption, but after sediment supply was reduced, the delta was eroded and sands were distributed laterally along the shoreline to form the present arcuate delta (Kuenzi et al., 1979). With the onset of persistent extrusive activity at Santiaguito in 1922, however, a new source of sediment was established. Consequently, delta erosion ceased and the position of the shoreline has remained stable since 1947.

Comparison of our maps with the maps of Kuenzi et al. (1979) shows that from 1987 to 1995 the main channel of Río Samalá occupied a channel marked as abandoned during 1954–1967. In contrast, the main channel from 1954 to 1967 was, by 1987, a secondary channel (Fig. 7). Although the main channel suffered constant aggradation, the map time series shows particularly extensive lahar activity during 1987–1988 (Fig. 8). These spread out in a 6 × 3 km fan mostly to the east of the main Río Samalá channel, 4 km inland from the present shoreline. Thereafter, aggradation remained confined to the Río Samalá channel (Fig. 8). However the 1987–1988 activity resulted in the establishment of a new channel that entered the ocean between the main and secondary channels, as well as a new lake (Fig. 7).

## RESULTS II: TEMPORAL CHANGES IN AGGRADATION

On the basis of area affected versus time, we identify five periods of enhanced or diminished aggradation between 1987 and 2000 across the proximal and medial zones (Fig. 9). Because the distal zone is at the image edge, data are not always available. As a result, our time series for the distal zone consists of just eight data points spanning 1987–1995 where, over this period, distal aggradation areas generally declined (Fig. 9).

Within the proximal and medial zones, enhanced aggradation marked the period from 1987 to 1990. Aggradation during this period coincided with emplacement of a 3.6-km-long lava flow that extended southward into a tributary of the Río Nimá II.

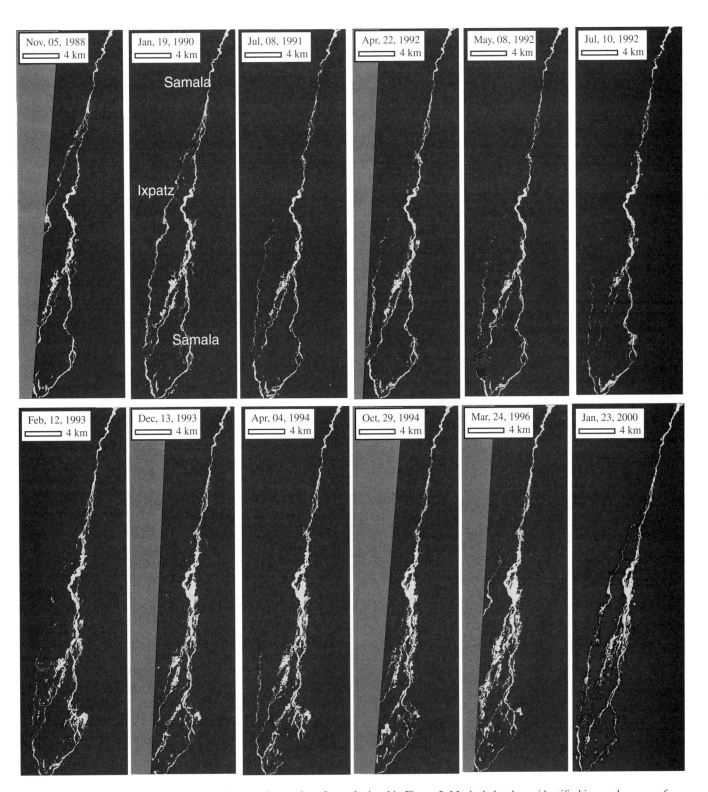

Figure 6. TM-derived medial zone aggradation map time series of area depicted in Figure 5. Masked clouds are identified in purple, areas of aggradation are yellow, and vegetated zones are black. Gray—image edge (no data zone).

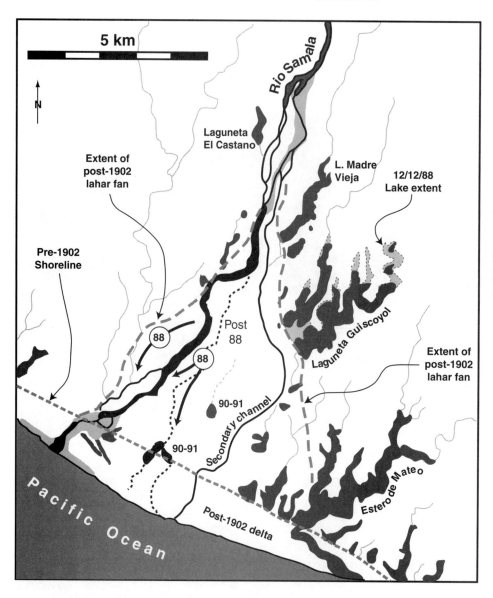

Figure 7. Distal zone location and aggradation summary map produced from Landsat time series. Zone location is marked as the distal section in Figure 1. The post-1902 lahar fan (outlined by thick, dashed gray line) and the 1902 delta are both apparent in this image. Aggraded areas—light gray; water bodies—black. Solid black lines—Río Samalá; thin gray lines—all other tributaries not supplied by volcaniclastic materials. Arrows marked "88" show main lahar paths evident on the December 1988 TM image. Dashed black line marked post-88 indicates the new channel that developed after 1988, and water bodies marked 90-91 identify sediment-dammed lakes that developed during 1990–1991. High lake levels for the sediment-dammed marginal lakes are given for 1988 (light gray) and low levels for 2000 (dark gray).

Flow front and levee collapse at this lava flow presented a new source of unconsolidated material, hence increasing the supply of debris to the system and increasing aggradation.

Diminished aggradation characterized the period from 1991 to 1993. This period followed the emplacement of smaller lava flows in 1990 that extended <500 m. As a result, sediment supply to the system persisted, but at diminished levels. Mapped aggradation areas also diminished in size as plants grew in areas that no longer received sediment. Another period of increased aggradation during 1993–1994 followed the emplacement of longer (~1.25 km) lava flows in 1991–1992.

This apparent relationship between aggradation area and lava flow length and activity is detailed in Table 3 and leads us to next explore the possible correlation between extrusion rate and aggraded area.

## THE RELATIONSHIP BETWEEN AGGRADATION, EXTRUSION RATE, AND RAINFALL

Comparison of the temporal variation in aggradation area and lava extrusion rate at Santiaguito indicates that aggradation in the proximal and medial zones responds to variations in extrusion rate at the dome complex with a lag of 12 months (Fig. 10). That is, an increase in extrusion rate is reflected by an increase in aggradation after a period of ~12 months. To fully examine this correlation, the role of water in mobilizing the supplied material must be considered. We therefore obtained rainfall data from the meteorological station at Huehuetenango, located in the Guatemalan highlands ~63 km north of Santa María (Fig. 1). Although these rainfall data show general (regional-seasonal) trends in precipitation, they do not provide

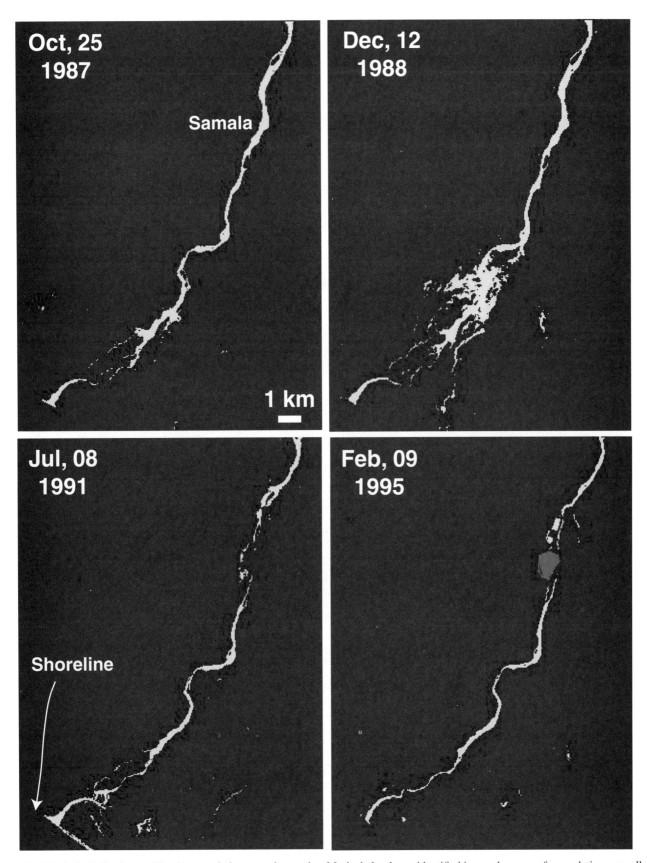

Figure 8. TM-derived distal zone (Fig. 1) aggradation map time series. Masked clouds are identified in purple, areas of aggradation are yellow, and vegetated zones are black.

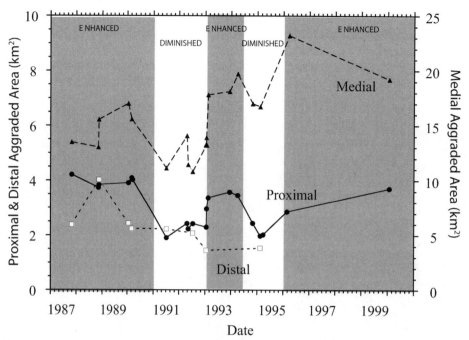

Figure 9. Temporal variation in TM-derived proximal (solid black line), medial (dashed line), and distal (stippled line) aggradation areas (1987–2000). Periods of enhanced and diminished aggradation are marked in gray and white, respectively.

TABLE 3. COMPARISON OF LAVA FLOW AREA, LENGTH AND EXTRUSION RATE*
WITH AGGRADATION AREAS

| Date (day-mo-yr) | Flow area (km²) | Flow length (km) | Extrusion rate (m³ s⁻¹) | Aggradation area (km²)† | | |
|---|---|---|---|---|---|---|
| | | | | proximal | medial | distal |
| 25-Oct-87 | 0.23 | 3.7 | 0.72 | 4.21 | 13.47 | 2.38 |
| 05-Nov-88 | 0.09 | 1.1 | 0.84 | 3.74 | 13.02 | |
| 12-Nov-88 | | | | 3.85 | 15.56 | 4.03 |
| 19-Jan-90 | 0.03 | 0.4 | 0.18 | 3.91 | 16.99 | 2.44 |
| 07-Mar-90 | | | | 4.09 | 15.59 | 2.24 |
| 16-Mar-90 | 0.04 | 0.6 | 0.20 | 4.03 | no data | |
| 08-Jul-91 | 0.01 | 1.2 | 0.09 | 1.91 | 11.11 | 2.23 |
| 22-Apr-92 | 0.03 | 1.25 | 0.24 | 2.43 | 14.06 | |
| 08-May-92 | 0.02 | 1.0 | 0.25 | 2.24 | 11.42 | |
| 10-Jul-92 | 0.02 | 1.1 | 0.17 | 2.42 | 10.80 | 2.08 |
| 18-Jan-93 | | | | 2.30 | 13.23 | 1.45 |
| 19-Jan-93 | 0.05 | 0.2 | 0.35 | 2.97 | 13.90 | |
| 12-Feb-93 | 0.04 | 0.6 | 0.28 | 3.37 | 17.81 | |
| 13-Dec-93 | 0.01 | 0.2 | 0.07 | 3.58 | 18.10 | |
| 04-Apr-94 | 0.01 | 0.4 | 0.09 | 3.46 | 19.71 | |
| 29-Oct-94 | 0.07 | 0.8 | 0.37 | 2.45 | 16.99 | |
| 09-Feb-95 | 0.05 | 0.3 | 0.34 | 1.98 | 16.73 | 1.53 |
| 22-Mar-95 | | | | 2.02 | | |
| 21-Feb-96 | 0.02 | 0.5 | 0.34 | 2.86 | | |
| 24-Mar-96 | 0.05 | 0.25 | 0.25 | | 23.22 | |
| 23-Jan-00 | 0.09 | 2.4 | 0.48 | 3.71 | 19.21 | |

*From Harris et al., 2003.
†Includes sediment and water.

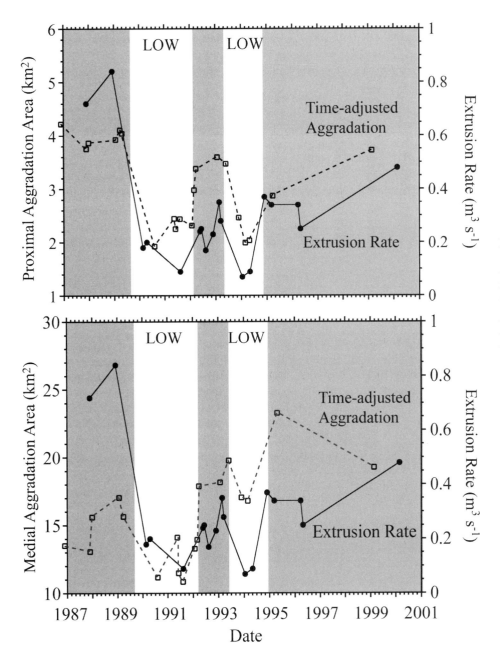

Figure 10. Comparison of proximal (top) and medial aggradation (bottom) areas with extrusion rate at the Santiaguito dome complex. Aggradation, shown as black dashes, has been time-adjusted (offset) by minus one year from the extrusion rate measurement to produce a best fit. Extrusion rates, shown as solid black line, are taken from Harris et al. (2003). Gray and white areas identify periods where extrusion rate and lagged aggradation area were high and low, respectively.

information on localized thunderstorms that are instrumental in triggering individual lahars at Santiaguito.

Although data are available for locations closer to Santiaguito, these records are unavailable after 1979. Pre-1979 precipitation records for stations closer to Santiaguito reveal that seasonal variations in rainfall at Santiaguito are similar to those recorded at more distant locations (Fig. 11). Using the aggradation, extrusion rate, and rainfall data given in Figure 12, we present an analysis of annual-scale relationships between aggraded area, volcaniclastic sediment supply (as expressed by extrusion rate), and rainfall.

**Proximal Zone**

Figure 10 indicates a positive correlation between extrusion rate and aggradation area 12 months later. The relationship is positive and linear:

$$A_{prox} = 2.14 + 2.27\ E_{-12},\ r^2 = 0.69, \quad (2)$$

in which $A_{prox}$ is the aggradation area and $E_{-12}$ is the extrusion rate 12 months previously. This implies that an increase in the rate of supply of material, as expressed by extrusion rate, leads to an

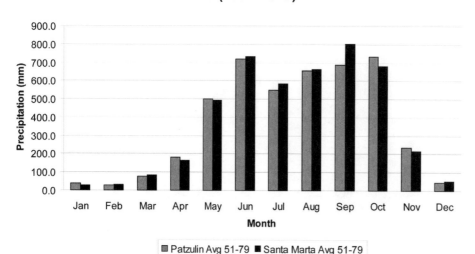

Figure 11. Monthly average rainfall for 1951–1979 for two stations within the Santiaguito fluvial system: Santa Marta and Patzulin (see Figure 1 for locations). The two stations are separated by ~800 m in vertical elevation (1550 m and 720 m above sea level, respectively). Both show differences in absolute values but the same relative trends that mark the wet and dry seasons.

increase in aggradation within 12 months. If we also consider the total rainfall occurring in the intervening 12 months ($Rain_{12}$), the correlation improves:

$$A_{prox} = 3.92 + 0.50\, E_{-12} + 0.31\, \ln(Rain_{12}), \; r^2 = 0.79. \quad (3)$$

The relationship between all variables is thus positive such that any increase in extrusion rate and/or rainfall will result in an increase in aggradation, with increases in sediment supply being transmitted downstream during the rainy season to increase proximal aggradation within 12 months.

If we examine the correlation coefficient ($r$) for total rainfall and proximal aggradation area, we find that $r$ increases as the number of months over which rainfall is summed prior to the area measurement increases, reaching a maximum after ~12 months. In fact, the $r$ value for the correlation between rainfall and proximal aggradation area improves from ~0.25, if rainfall over the preceding month is considered, to ~0.6 for rainfall over the preceding 12 months. This influences our selection of a 12-month interval over which to sum rainfall in our correlations.

We also note, however, that we are somewhat constrained by our sampling frequency. It is difficult to acquire cloud-free satellite images during the rainy season when much of the redistribution of volcaniclastic material no doubt occurs. Hence, our satellite-based extrusion rate and aggradation area estimates for Santiaguito tend to cluster in the dry season, most images having been acquired during November–March (Table 3). Thus, the data are only capable of detecting year-on-year change, a 12-month duration probably being of the order of the minimum temporal change our data are capable of revealing. We add, though, that the relationship between rainfall and aggradation area is very poor for the month immediately following any change in extrusion rate. This, though, is mostly because we are examining a month in the dry season. Thus, the best that we can say is, any new volcaniclastic material supplied into the system is mobilized during the rainy season, such that the full effect is felt by at least the end of the rainy season or after the ~12-month-long period following any measurement. To improve this analysis, further data from the rainy season are required.

**Medial Zone**

Figure 10 also indicates that our best correlation between extrusion rate and medial aggradation area is one that is lagged by 12 months. This relationship is, however, slightly more complex than in the proximal zone, having a general, long-term increase as well as shorter-term variation (Fig. 10). If, however, we normalize our data by considering change in medial aggradation area ($\Delta A_{med}$) and change in extrusion rate ($\Delta E$), we find:

$$\Delta A_{med} = 1.89 + 13.22\, \Delta E, \; r^2 = 0.79. \quad (4)$$

Because the proximal zone can be considered an additional supply source for the medial zone, we also examined the relationship between $\Delta A_{med}$ and the change in proximal aggradation area ($\Delta A_{prox}$). This shows a similar, positive, linear correlation:

$$\Delta A_{med} = 1.69 + 3.67\, \Delta A_{prox}, \; r^2 = 0.77. \quad (5)$$

Considering rainfall in the intervening 12 months improves the correlation further:

$$A_{med} = 18.84 - 0.05\, A_{prox-12} - 6.15\, Rain_{12}, \; r^2 = 0.85, \quad (6)$$

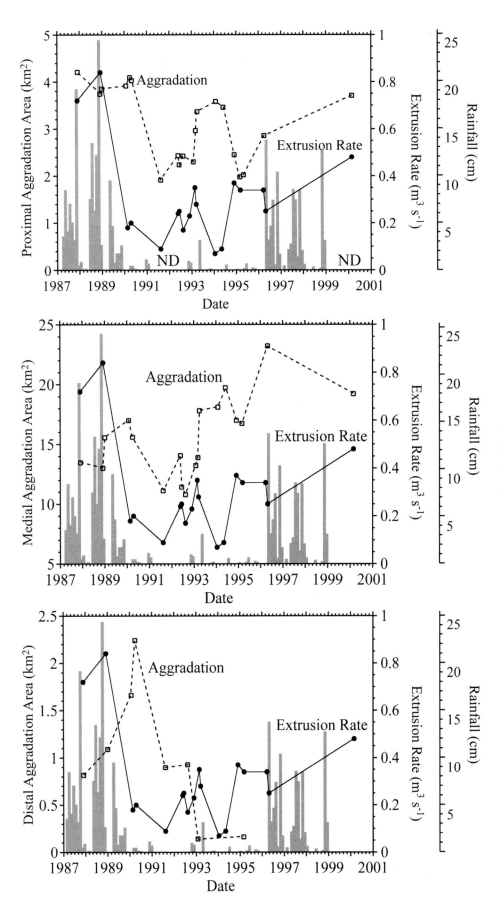

Figure 12. Extrusion rate at Santiaguito (black line), proximal, medial, and distal aggradation areas (black lines), and rainfall (gray bars) at Huehuetenango, 1987–2000 (see Figure 1 for location). ND—data gaps in rainfall records.

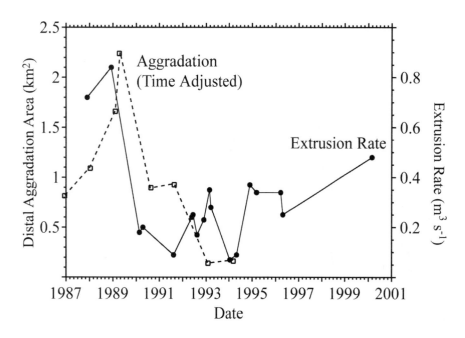

Figure 13. Comparison of distal aggradation (black line) and extrusion rate (black line) at the Santiaguito dome complex. Aggradation has been time-adjusted (offset) by minus one year from the extrusion rate measurement to produce a best fit. Extrusion rates are taken from Harris et al. (2003).

where $A_{prox-12}$ is the area of proximal aggradation 12 months previous to the $A_{med}$ measurement and $Rain_{12}$ is the total rainfall in the intervening 12 months.

We note that the correlation with supply (as expressed by extrusion rate or area of the proximal aggraded zone) is positive, but that with rainfall is negative. Thus, although the medial zone is a zone of transport, deposition does occur, with the amount of deposition increasing if the supply of sediment increases. However, because transport dominates, a negative relationship with rainfall results from the increased ability to wash sediment through and out of this zone during periods of higher rainfall. Within this context, lower rainfall during the 1990s, compared with 1987–1989, may explain the general increase in aggradation area following 1991 (Fig. 12). High rainfall during 1987–1989 would have flushed the medial zone of sediment. With lower rainfall from 1991 onward, however, medial aggradation would have been able to increase, with short-term variation due to variable annual rainfall and sediment supply superimposed on this general increasing trend.

### Distal Zone

Figure 13 indicates a best correlation between extrusion rate and distal aggradation area, if changes in aggradation area are lagged by 12 months (i.e., they follow the change in extrusion rate by 12 months). The best-fit relationship is positive and linear:

$$A_{dist} = 0.26 + 1.76\, E_{-12},\ r^2 = 0.67. \qquad (7)$$

This relationship is consistent with the distal zone being the zone of deposition, where increased supply of material results in increased aggradation area.

## DISCUSSION

Our data constrain a simple working model for volcaniclastic aggradation within the Santiaguito fluvial system. Here, any increase in the supply of material to the system (as expressed by extrusion rate) requires mobilization by rain before the effect of this new input can be felt. Thus, there is a lag before we detect the impact of a supply change, the lag being determined by the timing and severity of the rainy season. In our case, this lag is 12-months and is reflective of the annual cycle of wet and dry periods. Simply, most of the mobilization occurs in the annual rainy season.

In the medial zone, aggradation will increase if sediment supply, either from the dome complex or the reservoir of sediment within the proximal zone, increases. As in the proximal zone, a 12-month lag between a change in sediment supply and aggradation response results from the annual cycle of wet and dry seasons. That is, the rainy season is required if the sediment is to be mobilized. Increases in rainfall, however, will complicate and counter this relationship, where increased transport power during periods of increased stream flow will promote a decrease in aggradation area in the medial zone.

The statistical relationships derived from our 1986–2000 dataset indicate starting points for predictive relationships. Extrusion rate and rainfall data, for example, can be used in our statistical relationships to predict aggradation area 12 months ahead of time. Using, for example, the extrusion rate of 0.58 ± 0.1 m³ s⁻¹ obtained during January 2000 by Harris et al. (2002) with a mean annual rainfall of 18.76 cm gives an empirically predicted increase from a measured proximal aggradation area of 3.7 ± 0.5 km² in January 2000 to a predicted area of 4.6 ± 0.6 km² by January 2001. This predicted area for January 2001 compares with a value of 4.85 ± 0.65 km² measured by us using an ETM+ image acquired on 25 January 2001.

Landsat-7 ETM+ data are available every 16 days and provide quick and easy means to survey aggradation areas across the entire drainage. Between 31 July 1999 and 25 January 2001, for example, a total of 10 cloud-free images were identified as suitable for analysis. Images currently cost ~US$600 per scene, for a total bill of US$6000 for this 18 month period alone. This cost is low when compared to the monetary loss due to farmland inundation in 1995 of US$1.5 to 13.8 million calculated by Kimberly (1995).

Identification of the scale and location of new damage is essential from a hazard assessment point of view. Failure of measures to contain aggradation and lahar activity will pose increased hazard to local communities. For example, at the highway bridge at Finca San Jose La Granja (Fig. 5), Río Ixpatz aggraded ~19 m during 1977–2000 and built a 120-m-wide hyperconcentrated flow and lahar fan. Four-meter-high levees of stream-bed material are constructed each year to channelize Río Ixpatz and protect the bridge. Because the bridge over Río Samalá near Boxoma was washed out in 1983, loss of the Finca San Jose La Granja bridge would effectively isolate communities trapped between Río Ixpatz and Río Samalá for long periods during the wet season (Fig. 5).

## CONCLUSIONS

Our method provides a consistent, repeatable, and easy way to construct aggradation-impact maps and calculate areas of aggradation for large areas from a time series of satellite-based images. The availability of a 21-image sequence collected over a 14-yr period allows us to assemble a time series of maps that indicate spatial and temporal changes in activity. This, for Santiaguito, has permitted construction of a baseline data set of simultaneously acquired extrusion rate and aggraded area measurements that allow us to explore, quantify, and understand the relationships between these parameters.

For the Santiaguito fluvial system, extrusion rate appears to be a good proxy for the rate at which sediment is supplied to the system, where we find reasonable correlations between extrusion rates, rainfall, and aggradation area. We note, however, that the nature of this relationship varies depending on downstream location. Unconsolidated volcanic material from the dome complex is mobilized by rainfall and washed directly into the proximal section of the system. We find that an increase in the supply rate of volcaniclastic material, as expressed by extrusion rate, and/or rainfall will lead to an increase in the area of aggradation. In the Santiaguito case, the lag between a change in supply and downstream impact depends on the timing of the wet season. Here, supply (extrusion) may increase, but the arrival of the rainy season is required to mobilize this new supply and to transport it into the fluvial system.

Our study indicates the potential of remote sensing to monitor lahar activity and map large areas of lahar inundation in a thorough and efficient manner. In rapidly changing systems, the repeat capability of the satellite data provides a considerably more efficient means of mapping aggradation than ground-based methods. To map the changes presented here with the same spatial and temporal resolution from the ground would have required a significant commitment of time and manpower.

## ACKNOWLEDGMENTS

This research was funded by Landsat 7 Science Team National Aeronautics and Space Administration grant NAG5-3451. Travel Support came from the National Science Foundation through INT-9613647 and from the U.S. Geological Survey (USGS) Volcano Disaster Assistance Program. This manuscript was greatly improved following thorough USGS reviews by Jon Major and Larry Mastin, as well as reviews by Barry Cameron and Gail Ashley.

## REFERENCES CITED

Barrett, E.C., and Curtis, L.F., 1995, Introduction to environmental remote sensing: London, Chapman and Hall, 426 p.

Davies, K.R., Vessell, R.K., Miles, R.C., Foley, M.G., and Bonis, S.B., 1979, Fluvial transport and downstream sediment modifications in an active volcanic region: Canadian Society of Petroleum Geologists Memoir 5, p. 61–83.

Harris, A.J.L., Flynn, L.P., Matías, O., and Rose, W.I., 2002, The thermal stealth flows of Santiaguito: Implications for the cooling and emplacement of dacitic block lava flows: Geological Society of America Bulletin, v. 114, p. 533–546, doi: 10.1130/0016-7606(2002)114<0533:TTSFOS>2.0.CO;2.

Harris, A.J.L., Flynn, L.P., and Rose, W.I., 2003, Temporal trends in Lava Dome Extrusion at Santiaguito 1922–2000: Bulletin of Volcanology, v. 65, p. 77–89.

Holben, B., Kimes, D., and Fraser, R.S., 1986, Directional reflectance response in AVHRR red and near-IR bands for three cover types and varying atmospheric conditions: Remote Sensing of Environment, v. 19, p. 213–236, doi: 10.1016/0034-4257(86)90054-4.

INSIVUMEH (Instituto Nacional de Sismología, Vulcanología, Meteorología y Hidrología), 1988, Estudio Preliminar del Problema de "El Palmar," Quezaltenango: Guatemala, Instituto Nacional de Sismología, Vulcanología, Meteorología y Hidrología (INSIVUMEH) Informe Tecnico 3-88, 110 p.

Kimberly, P., 1995, Changing volcaniclastic sedimentary patterns at Santa Maria volcano, Guatemala, detected with sequential Thematic Mapper data, 1987–95 [M.Sc. Thesis]: Houghton, Michigan Technological University, 59 p.

Kuenzi, W.D., Horst, O.H., and McGehee, R.V., 1979, Effect of volcanic activity on fluvial-deltaic sedimentation in a modern arc-trench gap, southwestern Guatemala: Geological Society of America Bulletin, v. 90, p. 827–838, doi: 10.1130/0016-7606(1979)90<827:EOVAOF>2.0.CO;2.

Major, J.J., Pierson, T.C., Dinehart, R.L., and Costa, J.E., 2000, Sediment yield following severe volcanic disturbance—A two-decade perspective from Mount St. Helens: Geology, v. 28, p. 819–822, doi: 10.1130/0091-7613(2000)028<0819:SYFSVD>2.3.CO;2.

Mather, P.M., 1987, Computer processing of remotely-sensed images: Chichester, Wiley, 352 p.

Rodolfo, K.S., 2000, The hazard from lahars and Jokulhlaups, in Sigurdsson, H., et al., eds., Encyclopedia of Volcanoes: San Diego, Academic Press, p. 973–995.

Rodolfo, K.S., Umbal, J.V., Alonso, R.A., Remotigue, C.T., Paladio-Melosantos, M.L., Salvador, J.H.G., Evangelista, D., and Miller, Y., 1996, Two years of lahars on the western flank of Mount Pinatubo: Initiation, flow processes, deposits, and attendant geomorphic and hydraulic changes, in Newhall, C.G., and Punongbayan, R.S., eds., Fire and Mud: Seattle, University of Washington Press, p. 989–1013.

Rose, W.I., 1987, Volcanic Activity at Santiaguito volcano, 1976–1984, *in* Fink, J.H., ed., The emplacement of silicic domes and lava flows: Geological Society of America Special Paper 212, p. 17–27.

Schumm, S.A., 1977, The fluvial system: New York, John Wiley & Sons, p. 2–15.

SEAN, 1983, Santa Maria: Smithsonian Institution Scientific Event Alert Network (SEAN) 8(11).

SEAN, 1988, Santa Maria: Smithsonian Institution Scientific Event Alert Network (SEAN) 13(2).

Townshend, J.R.G., and Tucker, C.J., 1984, Objective assessment of advanced very high resolution radiometer data for land cover mapping: International Journal of Remote Sensing, v. 5, p. 496–504.

Tucker, C.J., Gatlin, J.A., and Schneider, S.R., 1984, Monitoring vegetation in the Nile delta with NOAA-6 and NOAA-7 AVHRR imagery: Photogrammertic Engineering Remote Sensing, v. 50, p. 53–61.

Vallance, J.W., 2000, Lahars, *in* Sigurdsson, H., et al., eds., Encyclopedia of Volcanoes: San Diego, Academic Press, p. 601–616.

Wernstedt, F.L., 1961, World Climatic Data—Latin America and the Caribbean: University Park, Pennsylvania State University, Dept. of Geography, 87 p.

Williams, S.N., and Self, S., 1983, The October 1902 Plinian eruption of Santa Maria Volcano, Guatemala: Journal of Volcanology and Geothermal Research, v. 16, p. 33–56, doi: 10.1016/0377-0273(83)90083-5.

MANUSCRIPT ACCEPTED BY THE SOCIETY 19 MARCH 2006

# The Escuintla and La Democracia debris avalanche deposits, Guatemala: Constraining their sources

**Craig A. Chesner***
*Department of Geology/Geography, Eastern Illinois University, Charleston, Illinois 61920, USA*

**Sid P. Halsor***
*GeoEnvironmental Sciences and Engineering Department, Wilkes University, Wilkes-Barre, Pennsylvania 18766, USA*

## ABSTRACT

The Escuintla and La Democracia debris avalanches are the two largest debris avalanches so far identified in Guatemala, with respective volumes of 9–15 km$^3$ and 2.4–5 km$^3$. Based upon their geographic locations on the Guatemalan coastal plain, both deposits have several possible source volcanoes. The Escuintla debris avalanche could have originated at either the Fuego or Acatenango volcanic complexes, or Agua volcano. Farther to the west, the La Democracia debris avalanche could only have come from the Fuego or Acatenango volcanic complexes. An apparent collapse scar on the east face of the Meseta edifice (the northernmost vent of the Fuego volcanic complex) has been attributed to the formation of the Escuintla debris avalanche. A mostly obscured summit collapse scar on Acatenango and an erosional remnant of a debris avalanche deposit near the base of the cone have been linked to the La Democracia debris avalanche. Petrographic and geochemical analyses of lava blocks collected from the Escuintla debris avalanche suggest that a substantial volume of amphibole-bearing dacitic lavas were present at its source volcano. Examination of rocks from the possible source volcanoes indicate that no dacitic lavas or tephras are known to have erupted from the Fuego volcanic complex and that the rocks exposed in the Meseta scarp bear little resemblance to the Escuintla debris avalanche samples. A few dacitic lavas and tephras are known from the Agua volcano, and several dacitic tephras have erupted from Acatenango. Geochemical comparisons of lavas and tephras from these volcanoes with rocks from the Escuintla debris avalanche showed greater similarities than those from Fuego and Meseta. Even though Acatenango is not known to have erupted dacitic lavas, its geochemistry is the most consistent with that of the Escuintla debris avalanche. Lava blocks from the La Democracia debris avalanche are mostly basaltic, although one andesitic sample contains phenocrystic amphibole. Geochemical analyses of Fuego and Meseta lavas overlap with the La Democracia debris avalanche samples; however, no amphibole-bearing rocks are known from Meseta, and Fuego is presumed to be younger than the La Democracia debris avalanche. Compared to the Acatenango rocks, the geochemistry and mineralogy of the La Democracia debris avalanche are

---

*E-mails: cachesner@eiu.edu; shalsor@wilkes.edu

**quite similar. Furthermore, rocks from the debris avalanche deposit on the flank of Acatenango are also consistent with the chemistry of the La Democracia debris avalanche. Thus, Acatenango produced at least one debris avalanche, the La Democracia debris avalanche, and possibly also generated the Escuintla debris avalanche.**

**Keywords:** debris avalanche, geochemistry, Fuego, Acatenango, Agua.

## INTRODUCTION

Since the eruption of Mount St. Helens in 1980, debris avalanches have been recognized as a common process at stratovolcanoes throughout the world. Many debris avalanche deposits are now easily identified by their conspicuous "hummocky" topography that spreads outward on the gentle lower slopes of their source volcanoes. An arcuate collapse scar, or sometimes a gaping amphitheater, is the best evidence that a volcano has undergone sector collapse and generated a debris avalanche (Siebert, 1984). With time, however, stratovolcanoes rebuild themselves, often obscuring the collapse scars (Clavero et al., 2002; Crandell, 1989) and burying the proximal portions of the deposits. Erosion or burial of the distal deposits can also mask evidence indicative of their source. In Guatemala, five debris avalanche deposits have been identified (Vallance et al., 1995; Duffield et al., 1989, Conway et al., 1992), and four others are inferred from source volcano evidence (Mercado and Rose, 1992; Reynolds, 1987; Siebert et al., 1994; Vallance et al., 1995). The source volcano for a few of these debris avalanche deposits is tentative at best. The two largest debris avalanche deposits in Guatemala, the Escuintla debris avalanche and the La Democracia debris avalanche, each have several possible sources (Fig. 1). The Escuintla debris avalanche lies on the coastal plain downslope from and midway between the Fuego volcanic complex and Agua volcano. The Acatenango volcanic complex, just to the north of the Fuego complex, is also a possible source for this deposit. Farther to the west, also on the coastal plain, the La Democracia debris avalanche could have originated at an edifice in either the Fuego or Acatenango volcanic complexes. This paper will review the physical evidence for the sources of these debris avalanche deposits and further constrain the source for each deposit using petrography and geochemistry.

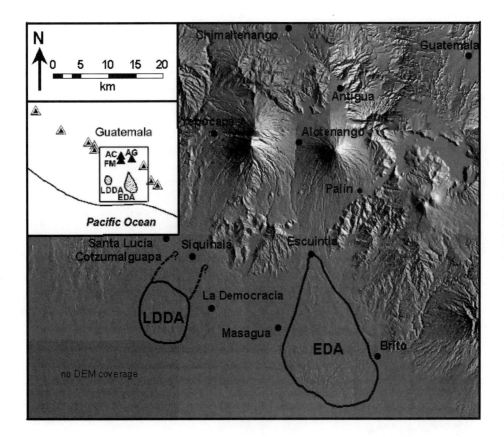

Figure 1. Digital elevation model showing the Escuintla (EDA) and La Democracia (LDDA) debris avalanche deposits and their possible source volcanoes, the Fuego (FM) and Acatenango (AC) volcanic complexes between the towns of Yepocapa and Alotenango, and the Agua Volcano (AG) to the east. Inset map shows the location of the study area.

# DEBRIS AVALANCHE DEPOSITS

## Escuintla Debris Avalanche

Analysis of Sky-Lab photographs of southern Guatemala led Rose et al. (1975) to suggest that a huge fan south of Escuintla might consist of volcaniclastic rocks derived from the stratovolcanoes upslope (Fig. 2). Subsequently, the Escuintla debris avalanche was mapped as Tertiary volcaniclastic and volcanic rocks by Hunter et al. (1984). Vallance et al. (1988, 1995) recognized it as a debris avalanche and described its characteristics within the modern context of debris avalanche deposits. It is 27 km long by 18 km wide, covers 300 km$^2$, and has an estimated volume of 9 km$^3$ (Vallance et al., 1995). Siebert et al. (1994) have inferred a 6 km$^3$ buried proximal portion, increasing the total volume to ~15 km$^3$. Four small outliers of the deposit have been isolated from the main body by fluvial processes. The northern apex of the exposed portion of the Escuintla debris avalanche outcrops just north of Escuintla and the deposit widens toward its southern terminus 45–50 km from its likely source volcanoes. It is bounded more-or-less to the east and west by the Michatoya and Guacalate Rivers, which have eroded small portions of the deposit. An apparent NNW-trending longitudinal axis could point toward its source or simply be an artifact of erosion or paleotopography. Individual hummocks range in height from 5 to 50 m, although most rise ~10 m above a gently sloping surface consisting of the deposit matrix (Figs. 3A and 3B). Their density and size is greatest in a proximal zone extending ~10 km south of Escuintla and in a 3-km-wide zone near the distal and lateral margins. At the terminal margin, the surface of the deposit drops abruptly 20 m to the coastal plain below. Vallance et al. (1995) interpret these hummock zones to represent emplacement of 2 sequential slide blocks from the same collapse event. Retrogressive flank failure would thus deposit lower flank rocks near the distal margin of the deposit and upper flank and summit rocks farther upslope. The relief at the distal end implies that the deposit is at least 20 m thick, and Vallance et al. (1995) have used a 30-m overall thickness as an estimate for their volume calculation. The Escuintla debris avalanche is geomorphologically the younger of the two avalanche deposits and has ~2–5 m of soil development according to Vallance et al. (1995). Because the Los Chocoyos tephra does not overlie the deposit, the Escuintla debris avalanche must be younger than 84 ka (Drexler et al., 1980).

The Puerto Quetzal Autopista transects the center of the deposit N-S from its proximal exposures to its distal end. Highway excavations and active quarries expose the interiors of several hummocks along the axis of the deposit. The interiors of the hummocks display many of the characteristics of debris avalanche deposits, such as large coherent lava blocks, multiple lava blocks of varying mineralogies and textures, "jigsaw" textured blocks (Fig. 3C), intact volcanic stratigraphy, and a multicolored matrix. A total of 26 lava block samples were collected from 16

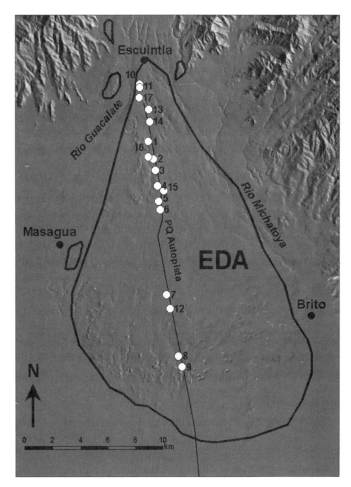

Figure 2. Digital elevation model showing the Escuintla debris avalanche deposit (EDA) and sample locations for this study. Note the four small outliers of EDA to the north and west of the main deposit and the northern and southern hummock concentration zones.

different Escuintla debris avalanche hummocks during field seasons in 1999 and 2003 (Fig. 2).

## La Democracia Debris Avalanche

A portion of the La Democracia debris avalanche was also mapped by Hunter et al. (1984) as Tertiary volcanics, and the entire deposit was recognized and described as a debris avalanche by Vallance et al. (1988, 1995). It is 10 km wide by 15 km long, covers ~120 km$^2$, and has an estimated volume of 2.4 km$^3$ (Vallance et al., 1995). Incorporating an inferred buried proximal portion of the La Democracia debris avalanche, Siebert et al. (1994) estimated a total volume of ~5 km$^3$. Lying south of Siquinalá and Santa Lucía Cotzumalguapa and west of La Democracia, the northern half of the deposit is actively being buried by alluvial fan deposits of the Rio Panteleón drainage system. In this portion of the deposit, scattered hummocks barely protrude through the alluvium. Farther south, numerous hummocks up to

Figure 3. Hummocks from the northern hummock concentration zone of the Escuintla debris avalanche (A and B); fence-posts for scale. (C) A jigsaw block from the Escuintla debris avalanche; pocket-knife for scale in the center of the photograph.

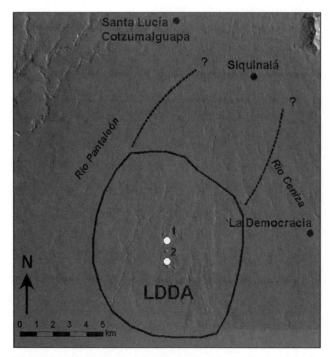

Figure 4. Digital elevation model showing the La Democracia debris avalanche (LDDA) and sample locations for this study.

40 m high are common (Fig. 4). The southern terminus of the deposit is ~40 km from its potential source volcanoes. The La Democracia debris avalanche is more deeply weathered than the Escuintla debris avalanche, with soil developed to a depth of six or more meters (Vallance et al., 1995). Absence of overlying Los Chocoyos tephra constrains its upper age limit to 84 ka also. The Finca El Balsamo road transects this deposit, but intersects few hummocks. Only 10 lava block samples were collected from two of the deeply eroded La Democracia debris avalanche hummocks in 1999 (Fig. 4).

## POSSIBLE SOURCE VOLCANOES

Considering the locations and youthful geomorphologies of the Escuintla and La Democracia debris avalanches, the only possible source volcanoes are the Fuego volcanic complex, Acatenango volcanic complex, and Agua volcano. The Fuego complex consists of the historically active Fuego vent and the inactive prehistoric Meseta vent. These vents are aligned along a N-S trend with the Acatenango and Yepocapa vents of the Acatenango complex farther to the north (Fig. 5A). About 15 km to the east of these paired and aligned volcanoes lies the single vent edifice of Agua volcano (Fig. 5B).

Figure 5. (A) View looking west of the Fuego and Acatenango volcanic complexes. Volcanoes from north to south are Y—Yepocapa, A—Acatenango, M—Meseta, and F—Fuego. (B) View looking west of Agua Volcano with the Fuego and Acatenango complexes in the background.

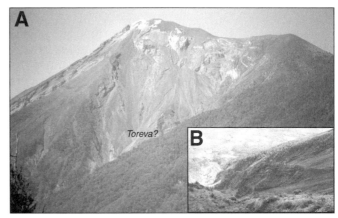

Figure 6. (A) View looking west of the Meseta scarp. The active Fuego vent is the hydrothermally altered summit, and the flat area is the inactive Meseta edifice. A possible toreva block is shown near the base of the scarp. (B) View looking down from Meseta at the possible toreva block in the center of the photograph. (C) Close-up view of sequential lava flows in the Meseta exposure. Lavas FL-2 and FL-27 were dated at 230 ka and 18 ka, respectively. The possible toreva block can also been seen in this view.

## Meseta Volcano

Previous studies have postulated that the source of the Escuintla debris avalanche was Meseta volcano, the northernmost vent of the Fuego volcanic complex (Vallance et al., 1995; Siebert et al., 1994; Chesner and Halsor, 1997). A steep east-facing scarp on Meseta presumably formed by sector collapse, generating the Escuintla debris avalanche (Fig. 6A). The amphitheater shape of the scarp, an apparent displaced slide block (toreva) at the base of the exposure (Fig. 6B), and the absence of a significant portion of the cone all point to a major debris avalanche producing sector collapse. Vallance et al. (1995) also argued that the straightest path for the avalanche would have been from the Fuego and Acatenango complexes, not Agua. The scarp exposes a thick stratigraphic section of lavas and tephras that represents a significant portion of Meseta's eruptive history. A stratigraphic section of 27 consecutive lavas exposed in the upper 75% of the scarp has been characterized petrographically and geochemically by Chesner and Halsor (1997). Marvin Lanphere at the U.S. Geological Survey in Menlo Park has dated two lava block samples from the Meseta exposure by whole-rock $^{40}Ar/^{39}Ar$ dating (Vallance et al., 2001). Sample FL-27, located near the present summit of Meseta (Fig. 6C), yielded an integrated $^{40}Ar/^{39}Ar$ age of 18 ± 11 ka. Near the bottom of the sampled exposure, sample FL-2 yielded an age spectrum plateau of 234 ± 31 ka and an inverse isochron age of 232 ± 93 ka. Thus, if the Escuintla debris avalanche originated at Meseta, it is younger than 18 ka.

## Fuego Volcano

The Fuego vent is the closest edifice to the La Democracia debris avalanche and is only 17 km from the nearest exposures of the Escuintla debris avalanche. However, there is no evidence of a sector collapse scar on the Fuego edifice. Martin and Rose (1981) estimated an age of 13 ka for the Fuego volcanic complex (Fuego and Meseta vents) using a time averaged overall eruption rate. Chesner and Rose (1984) have estimated that the Fuego volcanic complex has a minimum age of 17 ka and that the Fuego vent is ca. 8.5 ka based on projections of its historic eruption rates. Chesner and Halsor (1997) suggested that a shift in activity from the Meseta vent to the Fuego vent was initiated by collapse of the Meseta edifice. Therefore, the Fuego vent is presumed to be younger than both debris avalanches. If this evolutionary model is correct and the Escuintla debris avalanche originated at Meseta, the Escuintla debris avalanche's age is constrained between ca. 18 and 8.5 ka.

## Agua Volcano

Based upon geographic location, the Agua volcano is equally as likely to be the source of the Escuintla debris avalanche as the Fuego vent. The summit of this large, symmetrical stratovolcano also lies ~17 km from the first exposures of the Escuintla debris avalanche. A small debris flow originating at a crater lake that once filled Agua's summit crater destroyed much of the original city of Antigua (the first capital of Guatemala) in 1541, but there is no evidence of a major sector collapse on the Agua cone. A "hump" on the west flank of Agua, possibly thought to represent a former crater wall, was investigated in the field and appears to be a flank vent. Oto Matias (2002, personal commun.) generated a global information system map of the lava flow–dominated cone and found nothing to suggest that Agua produced a debris avalanche. Considering the route from all possible source volcanoes that the Escuintla debris avalanche would have followed, the route from Agua is the most direct. However, an avalanche from Agua would be expected to have a north- or NNE-trending longitudinal axis due to deflection by the ridge of Tertiary volcanics between Escuintla and Palín (Fig. 1). It is impossible for the La Democracia debris avalanche to have originated at Agua because it is located west of the Fuego and Acatenango complexes.

## Acatenango Volcano

Another possible source for the Escuintla debris avalanche and La Democracia debris avalanche is the Acatenango volcanic complex, which overlaps the Fuego volcanic complex to the north. Basset (1996) has demonstrated that at least one debris avalanche has originated from the Acatenango complex. The main evidence for this event is a proximal debris avalanche deposit on the west flank of Acatenango near Yepocapa (Fig. 7). Based upon proximal evidence of a debris avalanche and geochemical arguments, Basset (1996) concluded that the summit region of the ancient Acatenango edifice collapsed between 43 and 70 ka to form the La Democracia debris avalanche. In his evolutionary model, the Yepocapa cone was then built in the ancient Acatenango collapse crater from 43 to 20 ka. Activity then shifted to the modern Acatenango cone during the past 20 k.y.

## PETROGRAPHIC SOURCE CONSTRAINTS

Approximately 75 thin sections from the Fuego and Meseta volcanoes have been examined by Chesner and Rose (1984) and Chesner and Halsor (1997). These samples are predominantly lavas; a few are blocks from block and ash flow deposits. Rocks exclusive to the Fuego vent were collected mostly from historical eruptions and lava flows or block and ash flow deposits outcropping in barrancas that drain the cone. In an effort to assure sampling of the older portion of the Fuego cone, Chesner and Rose (1984) collected 13 lava block samples from alluvial deposits in Quebrada Playa Trinidad, and in January 2006 lava blocks from several nearby drainages and outcrops from Cerro Mongoy were also sampled (Fig. 7). Rocks exclusive to the Meseta vent were collected from lava flows exposed high on the cone within the presumed avalanche scar. Petrographically, both the Fuego and Meseta rocks can be classified as basalts, basaltic andesites, and andesites. Their mineralogy consists of varying proportions of plagioclase, clinopyroxene, orthopyroxene, olivine, and magnetite phenocrysts (Fig. 8A). Amphibole phenocrysts occur in a small proportion of andesitic lava blocks from Quebrada Playa Trinidad and Rio Achiguate and some lavas exposed at Cerro Mongoy. No amphibole-bearing rocks are known at Meseta or in drainages originating in the east-facing Meseta exposure.

A suite of 18 lavas and one light-colored tephra sample (AG-8) were collected mostly from the summit region and the southwest sector of Agua volcano (Fig. 9). Petrographically, the lavas were basaltic andesites and andesites carrying varying proportions of plagioclase, clinopyroxene, orthopyroxene, olivine, and magnetite phenocrysts (Fig. 8B). None of the lavas examined in this study contain amphibole. The tephra sample can be characterized as a hornblende dacite and contains green amphibole phenocrysts.

Petrographic analyses of 48 Acatenango lavas by Basset (1996) indicated a suite of basaltic andesites and andesites. Only

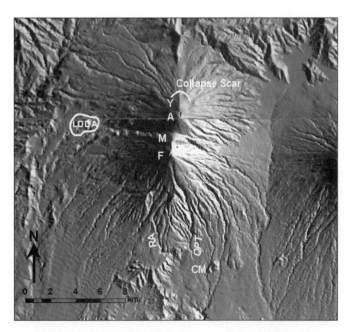

Figure 7. Digital elevation model of the Acatenango and Fuego volcanic complexes. Collapse scar and proximal La Democracia debris avalanche (LDDA) mapped by Basset (1996) are indicated. Yepocapa (Y), Acatenango (A), Meseta (M), and Fuego (F) vents are indicated. The Quebrada Playa Trinidad (QPT) Fuego rock suite and other amphibole bearing rock localities at Rio Achiguate (RA) and Cerro Mongoy (CM) are also shown.

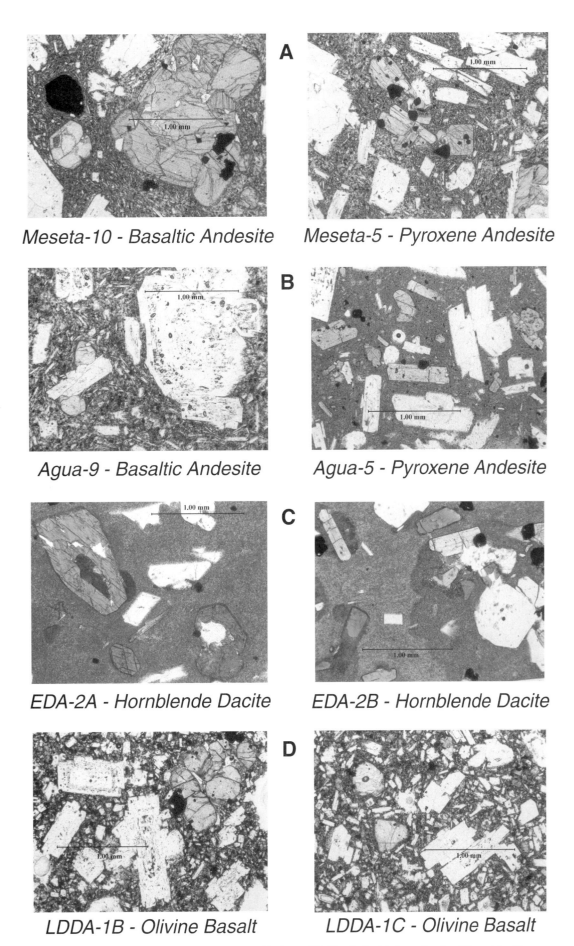

Figure 8. Photomicrographs of typical Meseta (A), Agua (B), Escuintla debris avalanche (C), and La Democracia debris avalanche (D) samples. All images were taken in plane-polarized light.

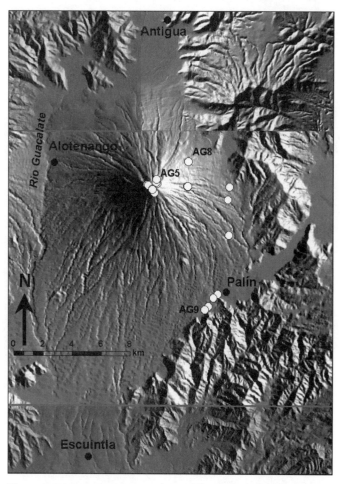

Figure 9. Digital elevation model of Agua volcano showing sample sites for petrographic analyses reported in this study. Lava flows were sampled at all sites except for AG-8, where the amphibole-bearing dacite tephra was collected.

five of these samples showed trace amounts of amphibole phenocrysts. Several andesitic and dacitic tephra samples in this study contained appreciable phenocrystic amphibole, some as high as 10% by volume.

The 26 Escuintla debris avalanche lava blocks that were collected from 16 separate hummocks are mineralogically distinct from the Meseta rocks. They are mostly basaltic andesites, andesites, and dacites. Olivine is less abundant in the Escuintla debris avalanche samples, but the most striking difference between the Escuintla debris avalanche and Meseta samples is that seven lava blocks from five different hummocks contained significant amounts of phenocrystic amphibole (Fig. 8C). The amphibole was distinctly green, except in two samples where it was highly oxidized. All but one of the amphibole-bearing hummocks are located in the northern hummock concentration zone (sites 2/16, 3, 4, and 14, Fig. 2). The other hummock that was found to contain amphibole (Site 7) was in the portion of the deposit between the hummock concentration zones. Highly oxidized amphibole occurred in the two most southern amphibole-bearing hummocks (Sites 4 and 7).

Thin sections of all 10 La Democracia debris avalanche lava blocks collected from two separate hummocks were also examined. These rocks are mostly basalts and basaltic andesites and are generally similar to lavas from the Meseta and Fuego volcanoes (Fig. 8D). One andesite sample, however, did contain abundant, highly oxidized amphibole phenocrysts. The La Democracia debris avalanche samples are not nearly as fresh as the Escuintla debris avalanche samples, often containing highly iddingsitized olivine. This observation is consistent with the greater observed soil development and more subdued topographic expression of the presumably older La Democracia debris avalanche deposit.

These petrographic assessments strongly suggest that the Escuintla debris avalanche source volcano contained a significant volume of amphibole-bearing dacitic lavas or lava domes. Our observations indicate that no such rocks are known to have erupted from the Meseta or Fuego volcanoes. Some debris avalanche deposits have been shown to incorporate significant amounts of underlying basement rocks (Wadge et al., 1995; Francis and Self, 1987). In these cases, the basement rocks tend to concentrate in the distal slide blocks, whereas the proximal slide blocks represent upper flank and summit rocks. Since the dacites in the Escuintla debris avalanche were all collected from the proximal slide block, we do not believe that they have been inherited from the basement. Although the vast majority of La Democracia debris avalanche samples are mineralogically compatible with the Meseta and Fuego vents, the one amphibole andesite precludes its origin from Meseta. Thus, petrographic constraints have reduced the possibility that the Escuintla debris avalanche or La Democracia debris avalanche originated at the Meseta vent as previously postulated.

## GEOCHEMICAL SOURCE CONSTRAINTS

Comparing geochemical analyses of lava blocks from debris avalanche deposits with rocks collected from possible source volcanoes has been shown to help constrain debris avalanche sources, especially when physical evidence is absent or inconclusive (Siebert et al., 2004). Preliminary geochemical studies of eight Escuintla debris avalanche and seven La Democracia debris avalanche lava blocks by Vallance et al. (1995) indicated that the Escuintla debris avalanche's chemistry was consistent with the Meseta, Acatenango, and Agua volcanoes, but not Fuego. They also suggested that the La Democracia debris avalanche lava blocks were more consistent with Fuego than Acatenango or Meseta. However, since Fuego is presumed to be younger than the La Democracia debris avalanche, they invoked an eroded or buried cone as the La Democracia debris avalanche source.

New major and trace element X-ray fluorescence spectrometry (XRF) analyses of 24 Escuintla debris avalanche and 10 La Democracia debris avalanche samples were acquired at Michigan State University using the same laboratory and XRF spectrometer that was used for the Meseta suite (Table 1). These analyses were compared to the Meseta data set previously studied by Chesner and Halsor (1997), assuring internal analytical consistency.

TABLE 1. GEOCHEMICAL ANALYSES OF LAVA BLOCKS FROM THE ESCUINTLA AND LA DEMOCRACIA DEBRIS AVALANCHE DEPOSITS

| Sample: | EDA-1A | EDA-1B | EDA-1C | EDA-2A | EDA-2B | EDA-3A | EDA-4A | EDA-4B | EDA-5A | EDA-5B | EDA-6 | EDA-7 |
|---|---|---|---|---|---|---|---|---|---|---|---|---|
| $SiO_2$ | 61.33 | 59.14 | 58.59 | 66.05 | 66.20 | 64.73 | 61.94 | 64.92 | 56.42 | 54.71 | 57.56 | 65.79 |
| $TiO_2$ | 0.70 | 0.80 | 0.85 | 0.46 | 0.46 | 0.58 | 0.69 | 0.61 | 0.90 | 0.93 | 0.78 | 0.50 |
| $Al_2O_3$ | 17.80 | 17.74 | 18.12 | 17.25 | 17.20 | 16.52 | 17.24 | 18.86 | 18.88 | 19.10 | 18.65 | 17.55 |
| FeO | 6.00 | 6.75 | 6.81 | 3.85 | 3.77 | 5.40 | 5.82 | 5.14 | 7.09 | 7.77 | 6.67 | 4.05 |
| MnO | 0.13 | 0.14 | 0.14 | 0.12 | 0.12 | 0.12 | 0.12 | 0.13 | 0.13 | 0.14 | 0.14 | 0.13 |
| MgO | 2.45 | 3.23 | 3.20 | 1.31 | 1.25 | 1.61 | 2.39 | 1.44 | 3.73 | 4.40 | 3.60 | 1.29 |
| CaO | 5.63 | 6.92 | 7.27 | 4.01 | 4.04 | 3.88 | 5.45 | 2.77 | 8.09 | 8.80 | 8.01 | 3.75 |
| $Na_2O$ | 3.90 | 3.41 | 3.36 | 4.56 | 4.59 | 4.33 | 4.02 | 3.48 | 3.24 | 2.96 | 3.16 | 4.54 |
| $K_2O$ | 1.88 | 1.66 | 1.46 | 2.19 | 2.17 | 2.64 | 2.14 | 2.49 | 1.24 | 0.95 | 1.22 | 2.21 |
| $P_2O_5$ | 0.19 | 0.19 | 0.20 | 0.19 | 0.19 | 0.19 | 0.18 | 0.18 | 0.29 | 0.23 | 0.22 | 0.20 |
| Cr | – | 18 | 2 | – | – | 2589 | – | – | 27 | 36 | 17 | – |
| Ni | 2 | 12 | 11 | 5 | 2 | 214 | 14 | 11 | 24 | 31 | 16 | 6 |
| Cu | 43 | 50 | 94 | 23 | 15 | 33 | 44 | 36 | 71 | 62 | 49 | 21 |
| Zn | 79 | 88 | 88 | 75 | 68 | 70 | 75 | 73 | 89 | 87 | 82 | 71 |
| Rb | 25 | 35 | 26 | 43 | 40 | 75 | 51 | 57 | 25 | 15 | 28 | 41 |
| Sr | 461 | 459 | 532 | 449 | 452 | 402 | 454 | 361 | 624 | 618 | 601 | 453 |
| Y | 23 | 29 | 25 | 26 | 23 | 26 | 27 | 31 | 22 | 21 | 26 | 29 |
| Zr | 173 | 134 | 136 | 179 | 188 | 219 | 168 | 205 | 171 | 117 | 144 | 190 |
| Nb | 8 | – | – | – | 3 | 3 | – | 6 | 3 | – | 5 | 6 |
| La | 23 | – | 26 | 14 | 16 | 41 | 28 | 32 | 13 | – | 17 | 21 |
| Ba | 925 | 607 | 638 | 975 | 1070 | 915 | 753 | 1214 | 559 | 546 | 616 | 882 |

| Sample: | EDA-8 | EDA-9 | EDA-10A | EDA-10B | EDA-10C | EDA-11 | EDA-12A | EDA-12B | EDA-13 | EDA-14A | EDA-14B |
|---|---|---|---|---|---|---|---|---|---|---|---|
| $SiO_2$ | 59.33 | 59.05 | 57.15 | 53.37 | 52.61 | 57.23 | 61.95 | 64.75 | 58.34 | 57.81 | 67.50 |
| $TiO_2$ | 0.73 | 0.73 | 0.95 | 1.01 | 0.97 | 0.81 | 0.73 | 0.58 | 0.68 | 0.83 | 0.54 |
| $Al_2O_3$ | 18.44 | 18.88 | 18.26 | 19.20 | 19.72 | 20.20 | 18.40 | 16.97 | 19.32 | 18.00 | 17.40 |
| FeO | 6.70 | 6.33 | 7.57 | 9.09 | 8.86 | 6.47 | 6.34 | 4.74 | 6.24 | 7.59 | 3.94 |
| MnO | 0.14 | 0.13 | 0.15 | 0.15 | 0.15 | 0.13 | 0.15 | 0.11 | 0.13 | 0.16 | 0.10 |
| MgO | 3.01 | 2.94 | 3.41 | 4.25 | 4.59 | 2.32 | 1.74 | 1.66 | 2.83 | 3.36 | 0.45 |
| CaO | 6.76 | 6.97 | 7.63 | 8.35 | 8.91 | 7.35 | 4.50 | 4.39 | 7.08 | 6.76 | 3.18 |
| $Na_2O$ | 3.40 | 3.45 | 3.25 | 3.35 | 3.14 | 3.85 | 4.09 | 4.22 | 3.75 | 3.53 | 4.43 |
| $K_2O$ | 1.31 | 1.32 | 1.42 | 1.03 | 0.86 | 1.39 | 1.86 | 2.39 | 1.40 | 1.75 | 2.24 |
| $P_2O_5$ | 0.19 | 0.19 | 0.21 | 0.21 | 0.17 | 0.24 | 0.24 | 0.18 | 0.24 | 0.22 | 0.21 |
| Cr | – | – | – | 8 | – | – | – | 12 | 7 | 1745 | – |
| Ni | 9 | 11 | 12 | 12 | 13 | 6 | 9 | 10 | 9 | 139 | 6 |
| Cu | 38 | 48 | 70 | 118 | 81 | 71 | 44 | 42 | 68 | 74 | 30 |
| Zn | 77 | 82 | 90 | 94 | 86 | 83 | 83 | 75 | 82 | 84 | 72 |
| Rb | 24 | 19 | 32 | 27 | 25 | 43 | 52 | 73 | 45 | 55 | 59 |
| Sr | 560 | 608 | 576 | 622 | 633 | 604 | 494 | 440 | 612 | 510 | 408 |
| Y | 29 | 25 | 26 | 18 | 19 | 20 | 20 | 22 | 22 | 21 | 24 |
| Zr | 122 | 116 | 134 | 121 | 109 | 146 | 178 | 215 | 150 | 169 | 197 |
| Nb | 2 | – | – | 4 | 3 | 12 | 8 | 12 | 8 | 9 | 10 |
| La | 12 | 13 | 11 | 32 | 27 | 47 | 50 | 51 | 50 | 43 | 44 |
| Ba | 586 | 551 | 623 | 508 | 454 | 594 | 813 | 925 | 641 | 742 | 994 |

| Sample: | EDA-15 | LDA-1A | LDA-1B | LDA-1C | LDA-1D | LDA-1E | LDA-1F | LDA-1G | LDA-2A | LDA-2B | LDA-2C |
|---|---|---|---|---|---|---|---|---|---|---|---|
| $SiO_2$ | 54.72 | 52.04 | 51.96 | 51.66 | 52.28 | 52.03 | 51.48 | 52.07 | 61.62 | 50.05 | 54.36 |
| $TiO_2$ | 0.97 | 0.95 | 0.93 | 1.01 | 0.96 | 0.96 | 1.02 | 0.99 | 0.58 | 1.01 | 0.90 |
| $Al_2O_3$ | 20.68 | 18.91 | 19.34 | 19.72 | 18.78 | 18.94 | 18.47 | 18.81 | 18.88 | 19.12 | 18.84 |
| FeO | 7.20 | 9.49 | 9.62 | 8.95 | 8.97 | 9.38 | 9.77 | 9.49 | 6.45 | 10.33 | 8.81 |
| MnO | 0.11 | 0.15 | 0.16 | 0.14 | 0.14 | 0.15 | 0.14 | 0.15 | 0.16 | 0.15 | 0.15 |
| MgO | 2.55 | 5.49 | 4.88 | 5.18 | 5.05 | 5.54 | 5.87 | 5.49 | 1.56 | 6.12 | 4.26 |
| CaO | 9.66 | 9.05 | 8.86 | 9.34 | 9.90 | 9.06 | 9.57 | 9.14 | 5.17 | 9.84 | 8.21 |
| $Na_2O$ | 3.09 | 2.94 | 3.04 | 2.99 | 2.98 | 2.94 | 2.83 | 2.93 | 3.84 | 2.68 | 3.32 |
| $K_2O$ | 0.82 | 0.80 | 1.04 | 0.78 | 0.78 | 0.82 | 0.70 | 0.76 | 1.50 | 0.57 | 0.93 |
| $P_2O_5$ | 0.20 | 0.17 | 0.18 | 0.23 | 0.17 | 0.16 | 0.15 | 0.17 | 0.24 | 0.14 | 0.21 |
| Cr | – | 70 | 36 | 45 | 74 | 51 | 76 | 74 | – | 63 | 17 |
| Ni | 3 | 38 | 23 | 26 | 40 | 36 | 42 | 40 | 5 | 36 | 17 |
| Cu | 82 | 99 | 58 | 73 | 56 | 65 | 46 | 92 | 26 | 62 | 44 |
| Zn | 85 | 108 | 83 | 88 | 81 | 86 | 92 | 91 | 98 | 79 | 91 |
| Rb | 16 | 26 | 21 | 7 | 11 | 10 | 5 | 11 | 31 | – | 7 |
| Sr | 629 | 607 | 572 | 624 | 605 | 609 | 601 | 585 | 582 | 594 | 583 |
| Y | 17 | 16 | 22 | 23 | 24 | 24 | 26 | 25 | 26 | 24 | 26 |
| Zr | 115 | 92 | 85 | 112 | 71 | 73 | 66 | 76 | 148 | 57 | 95 |
| Nb | 3 | – | – | – | – | – | – | – | – | – | – |
| La | 32 | 40 | 12 | 18 | – | – | 7 | 34 | 19 | – | 7 |
| Ba | 487 | 419 | 455 | 501 | 506 | 366 | 453 | 426 | 864 | 343 | 476 |

*Note:* Analyses normalized to 100% with all Fe as FeO. Major elements ($SiO_2$ through $P_2O_5$) given in wt%; trace elements (Cr through Ba) given in ppm.

Samples collected from the Escuintla debris avalanche have a wide compositional range (53–67.5 wt% $SiO_2$) and are predominantly andesites and dacites (Fig. 10). Nearly one-third (7 of 24) of the Escuintla debris avalanche lava blocks are dacites. These samples occur in six hummocks (sites 2, 3, 4, 7, 12, and 14, Fig. 2) spanning ~20 km along the N-S axis in the northern part of the deposit. This distribution suggests that dacites were a common rock type at the source volcano. Retrogressive flank failure, eventually capturing a dacitic lava dome complex, is consistent with such a distribution. The Meseta rock suite consists mostly of basaltic andesites with a few andesites and has a more restricted $SiO_2$ range (51–59 wt%) than the Escuintla debris avalanche (Fig. 11). On most geochemical plots, the Escuintla debris avalanche samples form a distinct group with wide linear trends, offset from the tight trends of the Meseta samples. A limited data set of older lavas from Fuego volcano is also plotted in Figure 11. Most of the Fuego rocks are basaltic andesites and andesites with $SiO_2$ contents ranging from 52 wt% to 61.5 wt%. Although the Fuego rocks overlap with Escuintla debris avalanche samples in some plots, in others they are quite distinct. Furthermore, like Meseta, none of the Fuego lavas are dacitic. This data strongly suggests that the Escuintla debris avalanche did not originate from the scarp on Meseta volcano as previously proposed, and its generation from Fuego is also unlikely.

The Escuintla debris avalanche was also compared to a data set consisting of 39 XRF analyses of lava flows collected from Agua volcano by Oto Matias (2002, personal commun.) and analyzed by Barry Cameron (Northern Illinois University). The Agua sample suite ranges from 52%–70% $SiO_2$, and the vast majority of the samples (37 of 39) are basaltic andesites and andesites containing <60 wt% $SiO_2$ (Fig. 12). Only two samples have $SiO_2$ contents between 60% and 70% and can be characterized as dacitic. Agua and Escuintla debris avalanche samples display overlapping trends on most geochemical plots. Although the Agua sample suite does not contain as many evolved rocks as the Escuintla debris avalanche, the Agua dacites plot along or close to the Escuintla debris avalanche trends. This data suggests that Agua volcano is a better candidate as an Escuintla debris avalanche source than the Meseta or Fuego volcanoes.

The Acatenango complex was also evaluated as a potential Escuintla debris avalanche and La Democracia debris avalanche source by utilizing an XRF data set consisting of 94 chemical analyses, 48 of which are lava flows (Basset, 1996). Lava flows from Acatenango are mostly basaltic andesites and andesites with $SiO_2$ contents ranging between 51% and 61.5% $SiO_2$ (Fig. 13). The vast majority of the lavas have $SiO_2$ contents above 55%, but there are no dacitic lavas in the suite. Of the numerous pyroclastic samples in this suite, several (~18) are dacitic tephras; they have been included in Figure 13. The Escuintla debris avalanche samples consistently lie along the Acatenango trends in virtually all chemical variation diagrams. Therefore, of all the data sets, the rocks from the Acatenango suite are the best chemical match to those from the Escuintla debris avalanche.

Samples from the much smaller La Democracia debris avalanche suite are mostly basaltic (50%–55% $SiO_2$) and include only one andesite sample containing ~61.5% $SiO_2$ (Figs. 11 and 13). The majority of La Democracia debris avalanche analyses overlap with those from Meseta and Fuego on most geochemical plots (Fig. 11). Thus, based on geochemical data, neither Fuego nor Meseta can be eliminated as a possible source of the La Democracia debris avalanche. There is considerable overlap of the Acatenango lava flow data and the La Democracia debris

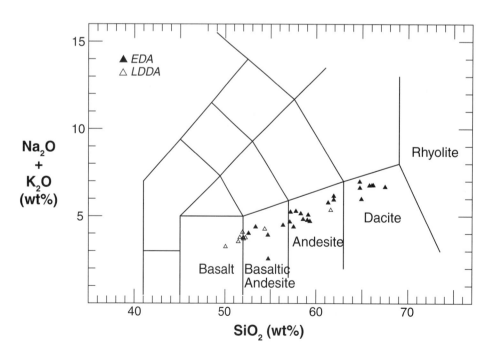

Figure 10. Volcanic rock classification diagram (LeBas et al., 1986) showing the Escuintla debris avalanche (EDA) and La Democracia debris avalanche (LDDA) samples.

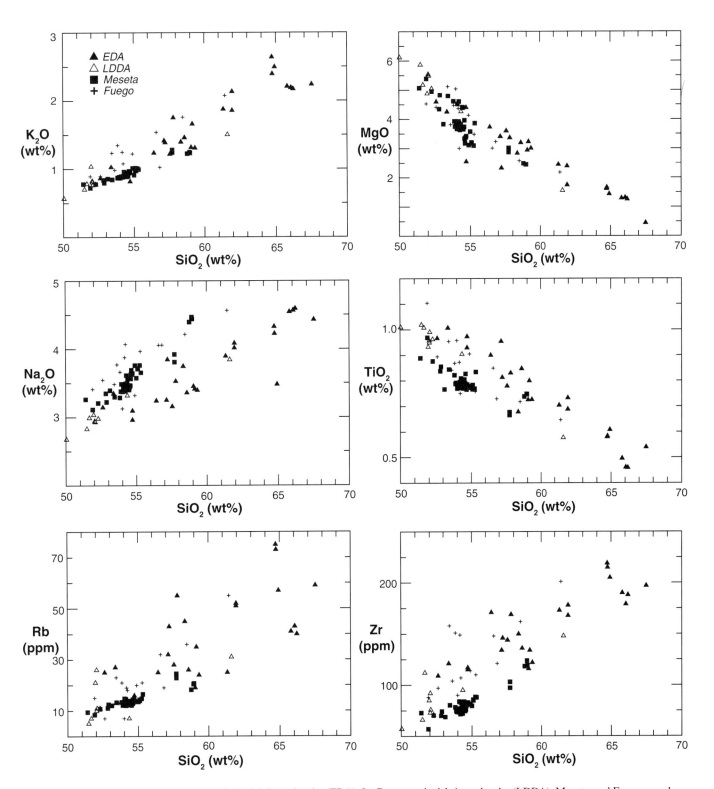

Figure 11. Variation diagrams showing Escuintla debris avalanche (EDA), La Democracia debris avalanche (LDDA), Meseta, and Fuego samples.

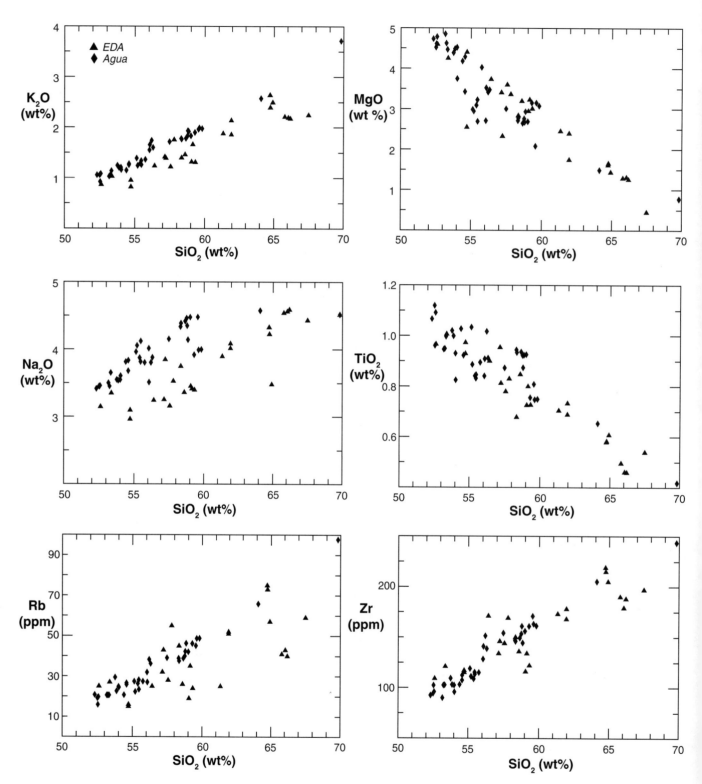

Figure 12. Variation diagrams showing samples from the Escuintla debris avalanche (EDA) and Agua volcano. Agua data is from Oto Matias (2002, personal commun.) and was analyzed by Barry Cameron (Northern Illinois University).

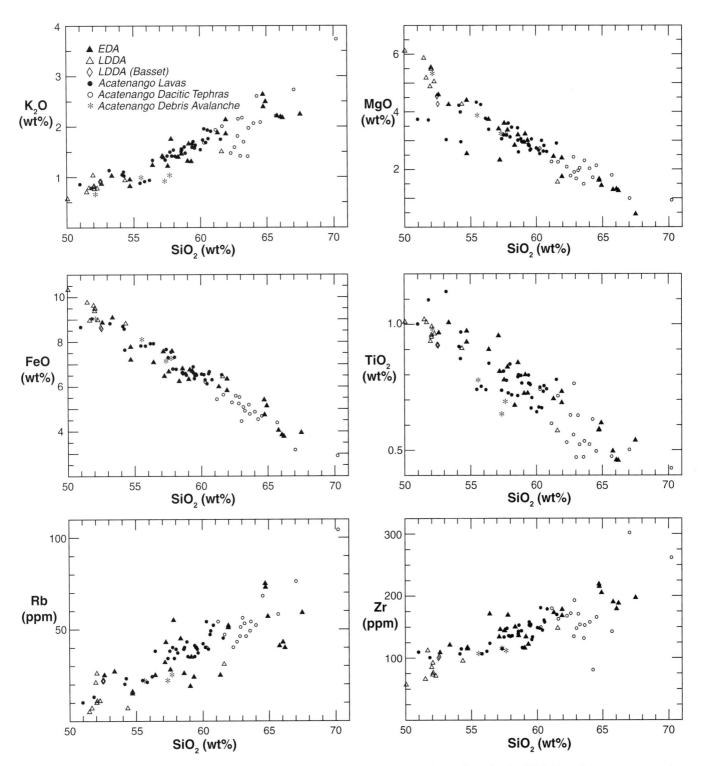

Figure 13. Variation diagrams showing Escuintla debris avalanche (EDA), La Democracia debris avalanche (LDDA), and Acatenango samples. All Acatenango analyses are from Basset (1996). Acatenango lava flows and dacitic tephras are shown with separate symbols. Three LDDA samples and four Acatenango debris avalanche samples analyzed by Basset are also shown.

avalanche analyses (Fig. 13). Two La Democracia debris avalanche lava blocks analyzed by Basset (1996) have also been included in Figure 13 and plot within the mafic cluster of La Democracia debris avalanche samples. Therefore, based on geochemical correlations, it is also possible that the ancient Acatenango edifice was the source of the La Democracia debris avalanche. Analyses of four lava blocks from the debris avalanche identified by Basset on the west flank of Acatenango were also plotted in Figure 13. Three of these samples lie along a trend between the mafic group of La Democracia debris avalanche rocks and the single andesite sample. Consequently, there is nothing in the geochemistry to refute Basset's suggestion that the debris avalanche on the west flank of Acatenango is a proximal remnant of the La Democracia debris avalanche.

## DISCUSSION

### Source Constraints of the Escuintla Debris Avalanche

Several lines of evidence suggest that the Fuego vent was not the source of the Escuintla debris avalanche. Fuego is the youngest and most active of the possible source volcanoes and is generally thought to be younger than the Escuintla debris avalanche. Furthermore, no collapse scars have been identified on the edifice. In terms of petrology, Fuego has erupted mostly basalts and basaltic andesites during its short history, while the Escuintla debris avalanche consists of a more evolved set of rocks containing appreciable dacitic material with $SiO_2$ contents up to 67.5 wt%. No dacitic lavas or tephras are known to have erupted from Fuego. The most silicic Fuego sample is a lava block collected in Quebrada Playa Trinidad containing ~61.5 wt% $SiO_2$. Consistent with this geochemistry, there are relatively more olivine-bearing rocks at Fuego than are found in the Escuintla debris avalanche. However, before Fuego can be eliminated as a potential Escuintla debris avalanche source, further investigation of the amphibole bearing rocks on its southern flank is necessary.

Physical evidence is strongly suggestive that the Meseta edifice was the source of the Escuintla debris avalanche. The steep east-facing exposure at Meseta is reminiscent of a sector collapse amphitheater. Additional evidence consists of a large toreva slide block located at the base of the Meseta exposure that appears to be a portion of the Meseta cone that did not travel far from its original position. According to Vallance et al. (1995), Meseta is also along the most direct path that the Escuintla debris avalanche would have traveled. Petrologic evidence, however, does not support the physical evidence of a Meseta source for the Escuintla debris avalanche. The most silicic rock known at Meseta contains ~59% $SiO_2$. Like Fuego, there are no known dacitic lavas or tephras at Meseta, and none of the Meseta rocks contain amphibole phenocrysts. Conceivably, the summit of Meseta may have consisted of a dacitic lava dome complex prior to a collapse event, but we deem this scenario highly unlikely because even the more mafic rocks at Meseta are somewhat chemically distinct from the Escuintla debris avalanche samples (Fig. 11). Minor variations between geochemical data sets collected at different laboratories might be expected to result in subtle distinctions between two data sets on chemical variation diagrams, such as those noted in the more mafic Escuintla debris avalanche and Meseta samples. However, the Escuintla debris avalanche and Meseta rock suites were both analyzed at Michigan State University by XRF. If the Escuintla debris avalanche did not originate from the Meseta scarp, then an alternate explanation of this exposure is required. Perhaps the exposure does represent a sector collapse and its debris avalanche deposit is buried beneath younger volcanics and volcaniclastics between Meseta and Agua volcanoes or under the alluvial apron to the south. Alternatively, perhaps the Meseta exposure is merely the result of incremental mass wasting, and a debris avalanche never took place.

Although there is no direct physical evidence for a substantial volume sector collapse at Agua volcano, indirect physical evidence and petrologic data do not exclude it as a possible source for the Escuintla debris avalanche. After examining the composite digital elevation model (DEM) image (Fig. 1), it is our opinion that the most direct path for the Escuintla debris avalanche would have been from Agua volcano. Orientation of the longitudinal axis of an Agua debris avalanche is likely to depend upon location of the collapsed sector. Geochemically, the Agua sample suite has greater similarity to the Escuintla debris avalanche than the Fuego and Meseta data sets. Petrographic examination of 18 lava samples collected from Agua did not result in definitive mineralogical consistencies between Agua and the Escuintla debris avalanche. The majority of the Agua samples were similar to the basaltic andesites and andesites of the Escuintla debris avalanche, but none of the Agua lava samples contained phenocrystic amphibole like the Escuintla debris avalanche dacites. Two of the 39 Agua lavas analyzed by Cameron at Northern Illinois University, however, are dacitic, containing ~64 wt% and 70 wt% $SiO_2$. These samples plot within the chemical trend of the Escuintla debris avalanche dacites (thin sections were not available for these samples). A 2-m-thick coarse dacitic tephra unit (AG-8) was collected from the northeast flank of Agua and does contain phenocrysts of amphibole. A charcoal sample from this unit has a $^{14}C$ age of 23,950 ± 270 yr B.P. (B. Rose, 2004, personal commun.). Thus, dacitic amphibole-bearing rocks are known to have erupted from Agua, and the requisite dacitic lava dome complex is not out of the question. If dacitic lavas and tephras were erupting from Agua ca. 24 ka, and Agua collapsed to generate the Escuintla debris avalanche, this age could represent an upper limit for such an event. If the Escuintla debris avalanche did originate from Agua, all physical evidence of a possible collapse and a dacitic lava dome complex are now obscured.

The Acatenango complex provides strong but inconclusive evidence for being the Escuintla debris avalanche source volcano. Basset (1996) identified and mapped the remnant of a former caldera or collapse scar on the northern part of Acatenango's upper cone. He attributed this south-facing feature to a sector collapse that took place between 43 and 70 ka generating the La Democracia debris avalanche. No other collapse scars have been identified on the Acatenango volcanic complex. Petrographic analyses also performed by Basset indicate that some Acatenango lavas contain trace amounts of amphibole. The only Acatenango samples reported to have substantial phenocrystic amphibole contents are dacitic tephras. Several amphibole-bearing tephras were erupted from the Yepocapa and modern Acatenango vents in the past 43 k.y. (Basset, 1996). Thus, amphibole-bearing dacitic tephras have been fairly common since 43 ka at the Acatenango complex. Geochemical analyses indicate that no Acatenango lava flows overlap with the dacites found in the Escuintla debris avalanche. The most silicic lava flow in the Acatenango data set has only 61 wt% $SiO_2$, less silicic than lavas at Agua and Fuego. Acatenango lavas do, however, have overlapping trends with the less silicic Escuintla debris avalanche rocks on variation diagrams. When the chemistry of the dacitic tephras is considered, the overall Acatenango data set is more similar to the Escuintla debris avalanche than that of Meseta, Fuego, or Agua. Determining which data set, Acatenango or Agua, better matched the Escuintla debris avalanche analyses was difficult. Subtle chemical differences between the three data sets, possibly inherited from their respective analytical laboratories, may be responsible for some chemical overlaps or distinctions. Regardless, our evaluations of the data indicated that the best agreement was between the Acatenango and Escuintla debris avalanche data sets. Because Acatenango has produced amphibole-bearing dacitic tephras, and geochemical trends consistently overlap with the Escuintla debris avalanche, Acatenango should be considered a strong candidate as the source of the Escuintla debris avalanche. In order for an Acatenango collapse event to generate the dacite-bearing Escuintla debris avalanche, it would need to have occurred since 43 ka, after eruption and accumulation of dacite from the Yepocapa and modern Acatenango cones. Even though Basset (1996) has attributed the collapse scar on Acatenango to the La Democracia debris avalanche, it is feasible that the scar could be the source of the Escuintla debris avalanche. Basset's calculations, however, suggest that the volume of missing cone associated with this collapse scar is only ~2.7 km$^3$, far too small to have generated the Escuintla debris avalanche.

Based upon the source volcano requirements of significant amounts of amphibole-bearing dacite, we believe that a Fuego-Meseta source of the Escuintla debris avalanche is highly unlikely. Instead, we have demonstrated that both Agua and the Acatenango complex were erupting dacitic magma ca. 24 ka and <43 ka, respectively. One collapse event has already been documented at the Acatenango complex, and several studies have shown that multiple collapses from the same volcano are common (Ponomareva et al., 1998; Tibaldi, 2001; Begét and Kienle, 1992). Furthermore, the N-S alignment of vents in the Acatenango complex and inferred dike-like configuration of the shallow magma chamber could make Acatenango more prone to edifice collapse than the symmetrical Agua cone. An Acatenango collapse from such structural factors would likely occur to the SE or SW, whereas an Agua collapse direction would be more strongly influenced by the regional slope of the basement toward the south (Vallance et al., 1995). Proximity and direct route arguments might favor an Agua source, but the NNW-trending axis of the Escuintla debris avalanche could also imply derivation from Acatenango. Final conclusions concerning the Escuintla debris avalanche source should await dating of the Escuintla debris avalanche dacites and comparison of its amphibole geochemistry with Fuego, Agua, and Acatenango, or discovery and study of proximal Escuintla debris avalanche exposures.

## Source Constraints of the La Democracia Debris Avalanche

Far less mineralogical and geochemical data is available for use in determination of the La Democracia debris avalanche source volcano. Geographical location does, however, preclude its origin from Agua volcano; thus, only the Fuego and Acatenango volcanic complexes need to be considered as possible sources. We do not believe that the Fuego vent generated the La Democracia debris avalanche because the La Democracia debris avalanche is considered to be much older than the Escuintla debris avalanche, and the Fuego vent is presumed to be younger than the Escuintla debris avalanche. Although geochemistry of the La Democracia debris avalanche is quite similar to both Fuego and Meseta, one sample from the small La Democracia debris avalanche suite contained considerable phenocrystic amphibole, precluding its origin from Meseta.

Compelling physical evidence of an Acatenango source for the La Democracia debris avalanche has been presented by Basset (1996) and includes identification of a partial collapse scar on Acatenango's upper cone, a missing volume calculation of 2.7 km$^3$ that is similar to the volume of the La Democracia debris avalanche (2.4–5 km$^3$), and identification of a proximal debris avalanche on the west flank of the Acatenango cone. The age range suggested for this collapse event (40–73 ka) by Basset seems more appropriate for the age of the highly weathered La Democracia debris avalanche than for the less-weathered Escuintla debris avalanche. The majority of our limited suite of La Democracia debris avalanche samples consists of basalts and basaltic andesites, which is consistent with Basset's model of a mafic ancestral Acatenango cone. Our chemical analyses of the La Democracia debris avalanche overlap well with Basset's Acatenango data set. In addition, analyses of four lava blocks from the proximal debris avalanche identified on Acatenango are also consistent with our La Democracia debris avalanche analyses. Thus, we concur with Basset that the La Democracia debris avalanche originated from collapse of the ancestral Acatenango cone.

## ACKNOWLEDGMENTS

Funding for this study was provided by the Eastern Illinois University (EIU) Council for Faculty Research and the Wilkes University Faculty Development. We thank Marvin Lanphere for dating two of the Meseta lavas and Jim Vallance for arranging those analyses. Oto Matias generously shared his map of the Agua lava flows and the Agua geochemical data set. Digital elevation models (DEMs) provided by the Instituto Geográfico Nacional of Guatemala were used in this study. As undergraduate students at EIU, Scott Boroughs prepared and studied the Escuintla debris avalanche and La Democracia debris avalanche thin sections while Cara Schiek assisted with geochemical sample preparation and initial evaluation of the data set. Bill Toothill, director of the Geographic Information Sciences Center at Wilkes University, assisted with the processing of DEMs. Most importantly, we thank Bill Rose for his continuing support and guidance of our efforts to study Central American volcanism. His enthusiasm, wisdom, and friendship continue to sustain us.

## REFERENCES CITED

Basset, T.S., 1996, Histoire éruptive et évaluation des aléas du volcan Acatenango (Guatemala) [Ph.D. dissertation]: Universite de Genève, 240 p. with appendices.

Begét, J.E., and Kienle, J., 1992, Cyclic formation of debris avalanches at Mount St. Augustine volcano: Nature, v. 356, p. 701–704, doi: 10.1038/356701a0.

Chesner, C.A., and Rose, W.I., 1984, Geochemistry and evolution of the Fuego Volcanic Complex, Guatemala: Journal of Volcanology and Geothermal Research, v. 21, p. 25–44, doi: 10.1016/0377-0273(84)90014-3.

Chesner, C.A., and Halsor, S.P., 1997, Geochemical trends of sequential lava flows from Meseta volcano, Guatemala: Journal of Volcanology and Geothermal Research, v. 78, p. 221–237, doi: 10.1016/S0377-0273(97)00014-0.

Clavero, J.E., Sparks, R.S.J., Huppert, H.E., and Dade, W.B., 2002, Geological constraints on the emplacement mechanism of the Parinacota avalanche, northern Chile: Bulletin of Volcanology, v. 64, p. 40–54, doi: 10.1007/s00445-001-0183-0.

Conway, F.M., Vallance, J.W., Rose, W.I., Jr., Johns, G.W., and Paniagua, S., 1992, Cerro Quemado, Guatemala: the volcanic history and hazards of an exogenous dome complex: Journal of Volcanology and Geothermal Research, v. 52, p. 303–323, doi: 10.1016/0377-0273(92)90051-E.

Crandell, D.R., 1989, Gigantic debris avalanche of Pleistocene age from ancestral Mount Shasta volcano, California, and debris avalanche hazard zonation: U.S. Geological Survey Bulletin 1861, 32 p.

Drexler, J.W., Rose, W.I., Sparks, R.S.J., and Ledbetter, M.T., 1980, The Los Chocoyos ash, Guatemala: A major stratigraphic marker in middle America and in 3 ocean basins: Quaternary Research, v. 13, p. 327–340, doi: 10.1016/0033-5894(80)90061-7.

Duffield, W.A., Heiken, G.H., Wohletz, K.H., Maassen, L.W., Dengo, G., and McKee, E.H., 1989, Geology and geothermal potential of the Tecuamburro volcano area of Guatemala: Geothermal Resource Council Transactions, v. 13, p. 125–131.

Francis, P., and Self, S., 1987, Collapsing volcanoes: Scientific American, v. 256, no. 6, p. 91–97.

Hunter, B., Querry, M., Hebbeger, J., and Tharpe, J., 1984, Geologic map of the Escuintla quadrangle, Guatemala: Instituto Geográfico Militar Guatemala, scale 1:50,000.

LeBas, M.J., LeMaitre, R.W., Streckeisen, A., and Zanettin, B., 1986, A chemical classification of volcanic rocks based on the total alkali silica diagram: Journal of Petrology, v. 27, p. 745–750.

Martin, D.P., and Rose, W.I., 1981, Behavior patterns of Fuego volcano, Guatemala: Journal of Volcanology and Geothermal Research, v. 10, p. 67–81, doi: 10.1016/0377-0273(81)90055-X.

Mercado, R., and Rose, W.I., 1992, Reconocimiento geológico y evaluacion preliminar de peligrosidad del volcán Tacaná, Guatemala/Mexico: Geofísica Internacional, v. 31, p. 205–237.

Ponomareva, V.V., Pevzner, M.M., and Melekestsev, I.V., 1998, Large debris avalanches and associated eruptions in the Holocene eruptive history of Shiveluch Volcano, Kamchatka, Russia: Bulletin of Volcanology, v. 59, p. 490–505.

Reynolds, J.H., 1987, Timing and sources of Neogene and Quaternary volcanism in south-central Guatemala: Journal of Volcanology and Geothermal Research, v. 33, p. 9–22, doi: 10.1016/0377-0273(87)90052-7.

Rose, W.I., Jr., Johnson, D.J., Hahn, G.A., and Johns, G.W., 1975, Skylab photography applied to geological mapping in northwestern Central America: Proceedings of National Aeronautics and Space Administration, Earth Resource Survey Symposium, v. 1b, p. 861–884.

Siebert, L., 1984, Large volcanic debris avalanches: Characteristics of source areas, deposits, and associated eruptions: Journal of Volcanology and Geothermal Research, v. 22, p. 163–197, doi: 10.1016/0377-0273(84)90002-7.

Siebert, L., Vallance, J.W., and Rose, W.I., 1994, Quaternary edifice failures at volcanoes in the Guatemalan highlands: Eos (Transactions, American Geophysical Union), v. 75, p. 367.

Siebert, L., Kimberly, P., and Pullinger, C.R., 2004, The voluminous Acajutla debris avalanche from Santa Ana volcano, western El Salvador, and comparison with other Central American edifice-failure events, in Rose, W.I., et al., Natural Hazards in El Salvador: Geological Society of America Special Paper 375, p. 5–23.

Tibaldi, A., 2001, Multiple sector collapses at Stromboli volcano, Italy: How they work: Bulletin of Volcanology, v. 63, p. 112–125, doi: 10.1007/s004450100129.

Vallance, J.W., Giron, J.R., Rose, W.I., Siebert, L., and Banks, N.G., 1988, Volcanic edifice collapse and related hazards in Guatemala, preliminary report: Instituto Nacional de Sismologia, Vulcanología, Meteorología e Hidrologia (INSIVUMEH), Guatemala, Michigan Technological University, U.S. Geological Survey Volcano Crisis Assistance Team, 15 p.

Vallance, J.W., Siebert, L., Rose, W.I., Giron, J.R., and Bands, N.G., 1995, Edifice collapse and related hazards in Guatemala: Journal of Volcanology and Geothermal Research, v. 66, p. 337–355, doi: 10.1016/0377-0273(94)00076-S.

Vallance, J.W., Schilling, S.P., Matias, O., Rose, W.I., and Howell, M.M., 2001, Volcano hazards at Fuego and Acatenango, Guatemala: U.S. Geological Survey Open-File Report 01-431, 23 p., plus plates.

Wadge, G., Francis, P.W., and Ramirez, C.F., 1995, The Socompa collapse and avalanche event: Journal of Volcanology and Geothermal Research, v. 66, p. 309–336, doi: 10.1016/0377-0273(94)00083-S.

Manuscript Accepted by the Society 19 March 2006

# Diverse volcanism in southeastern Guatemala: The role of crustal contamination

**B.I. Cameron***
*Department of Geosciences, University of Wisconsin, P.O. Box 413, Milwaukee, Wisconsin 53040, USA*

**J.A. Walker**
*Department of Geology and Environmental Geosciences, Northern Illinois University, Davis Hall 312, Normal Road, DeKalb, Illinois 60115, USA*

## ABSTRACT

In the Central American arc, southeastern Guatemala hosts the most diverse volcanism. Large stratovolcanoes at the volcanic front (VF) form as a result of subduction of the oceanic Cocos plate beneath the continental Caribbean plate. Behind the volcanic front (BVF) volcanism, however, has undergone a fundamental change in eruptive style during the Quaternary from older, polygenetic central volcanism to younger, monogenetic cinder cone volcanism. Magmas that traverse the 40–45-km-thick crust in southeastern Guatemala are highly susceptible to crustal contamination. Consequently, mineral chemical data, whole-rock oxygen isotope, and light element geochemistry are used to investigate the relationship between edifice type and the magnitude of crustal contamination.

The lack of systematic variation between compositions of phenocryst phases and host rocks strongly suggests that open system processes were operating. Moreover, phenocryst core compositions are generally out of equilibrium with host rock compositions. Olivine from BVF cinder cones deviate only slightly from the equilibrium line in comparison to the older behind the volcanic front (OBVF) central volcanoes and VF stratovolcanoes, suggesting less assimilation of crustal lithologies. Steep arrays on the $\delta^{18}O$-$SiO_2$ diagram cannot be explained by crystal fractionation and favor the incorporation of $^{18}O$-enriched crustal rocks. Higher $\delta^{18}O$ values in the OBVF central volcanoes and VF stratovolcanoes support the idea that larger, shallow magma bodies experienced greater amounts of crustal contamination. Regional extension in the Ipala Graben of southeastern Guatemala likely promoted short residence times in crustal reservoirs and small degrees of crustal assimilation for the BVF cinder cone magmas.

**Keywords:** Arc magmas, geochemistry, stable isotopes, crustal assimilation.

---

*E-mail: bcameron@uwm.edu

# INTRODUCTION

Magmatic systems can operate in an open or closed manner. Closed magmatic systems allow the exchange of heat with the system's surroundings, but not the exchange of mass. Open magmatic systems can exchange both energy and mass with their surroundings. Different volcanic edifice types in continental arcs result from enigmatic subvolcanic processes. In this paper, we try to ascertain whether open or closed systems processes predominated underneath arc volcanoes in southeastern Guatemala, using select geochemical data. For example, open system behavior, such as crustal assimilation, can change the chemical composition of a magma, including its critical volatile content. Volatile content drives explosive volcanic eruptions, which holds significant implications for the type of volcanic hazards expected to be associated with the arc volcanoes.

The extent to which crustal contributions modify primary arc magmas remains a controversial issue amongst petrologists. Studies have convincingly demonstrated that crustal inputs play a significant role in the generation of evolved, high-$SiO_2$ continental arc lavas (Hildreth and Moorbath, 1988; Davidson et al., 1990; Feeley and Sharp, 1995). Greater uncertainty surrounds the contribution of the continental crust in less evolved basaltic arc lavas. In a single arc segment of northern Honshu, Japan, isotopic data for basalts vary systematically with changes in crustal lithology (Kersting et al., 1996). Apparently, even thin crustal lithosphere can modify the compositions of primary melts as they ascend from the mantle.

Although erupted through continental crust, the vast majority of Central American lavas show little evidence for pronounced crustal contamination, based on radiogenic isotopes (Walker et al., 1995; Feigenson et al., 2004). The relatively thin and young continental crust of southern Central America possesses distinct radiogenic isotopic characteristics compared to crust of most other continental margins. Consequently, if crustal assimilation occurs, it does not greatly influence the radiogenic isotopic compositions of the arc magma. The thickness of the continental crust in Central America peaks in western Guatemala at 48 km, from a minimum in Nicaragua of ~32 km (Carr, 1984). Crustal thicknesses decline in central and eastern Guatemala to 45 and 40 km, respectively (Carr et al., 1990). Therefore, Guatemalan lavas hold the most potential for registering crustal geochemical signatures. An unusual trend toward lower $^{206}Pb/^{204}Pb$ and more radiogenic $^{207}Pb/^{204}Pb$ and $^{208}Pb/^{204}Pb$ in southeastern Guatemala cinder cones could result from assimilation of granulitic crust (Walker et al., 1995). Southeastern Guatemala also hosts the most diverse arc volcanism in all of Central America and, consequently, warrants a detailed investigation of the role played by crustal contamination in creating diversity in magma compositions and edifice style.

# GEOLOGICAL CONTEXT

Prominent stratovolcanoes or composite cones form a distinct line of volcanoes called the volcanic front (VF) in Guatemala (Fig. 1). The VF forms as a direct manifestation of subduction of the oceanic Cocos plate beneath the continental Caribbean plate. Common behind the volcanic front (BVF) volcanism adopts contrasting styles in southeastern Guatemala with time. Small composite cones were the site of polygenetic eruptions throughout the late Pliocene and are referred to as older behind the volcanic front (OBVF) central volcanoes. Ubiquitous cinder cones were erupted in proximity to the extensional Ipala Graben (Fig. 1) and locally developed on the flanks of the larger central volcanoes. These monogenetic volcanoes were accordingly assigned Quaternary ages (Walker, 1981). Shield volcanoes such as Las Viboras and calderas such as Retana and Ayarza were also constructed behind the front during the Quaternary (Fig. 1). Overall, however, BVF volcanism in southeastern Guatemala has undergone a fundamental change in eruptive style from older, polygenetic, central volcanism to younger, diffuse, monogenetic, cinder cone volcanism.

Conflicting ideas currently exist concerning the degree of crustal contamination expected to accompany different styles of arc volcanism. A specific plumbing system underlying one of the main edifice types in Guatemala may favor crustal assimilation over the others. In Mexico, as in Guatemala, the cinder cones represent small batches of magma that traverse a separate section of the crust only a single time (Hochstaedter et al., 1996). This eruptive character may make more differentiated members of the cinder cone suites more susceptible to crustal assimilation during their ascent to the surface. In sharp contrast, composite cones may erupt repeatedly through well-established plumbing systems. This feature suggests that older, crustally contaminated lavas may give way to larger volumes of magmas that were shielded in the conduit of the volcano from interacting with crustal rocks. Conversely, crustal contamination might flourish beneath composite cones where magma stalls in large crustal magma reservoirs. In Mexico, magmas erupted from cinder cones seem to experience more crustal contamination (Hochstaedter et al., 1996). In Guatemala, the appropriate data have not existed to test this hypothesis.

Detecting the geochemical signatures of crustal contamination depends not only on the degree of assimilation but also on the compositional contrast between magma and contaminant. Systematic increases in Sr isotopic ratios and decreases in Nd isotopic ratios across the arc in BVF cinder cones of Guatemala reflect either an absolute or apparent increase in crustal contamination (Walker et al., 1995; Feigenson et al., 2004). Absolute increases in crustal contamination associated with cinder cones from the Ipala Graben far behind the VF make geologic sense, if the normal faulting that accompanies extension facilitates assimilation by increasing the surface area of crustal lithologies. Alternatively, the extensional regime of eruption could result in less impedance and more direct access to the surface and, consequently, smaller degrees of contamination. In contrast, assimilation of equal amounts of older, more radiogenic crust in the Ipala Graben could also generate the observed isotopic changes. Low Nd isotopic ratios thought to represent crustal contamination were also measured in mafic lavas from Tegucigalpa in Honduras (Patino et al., 1997).

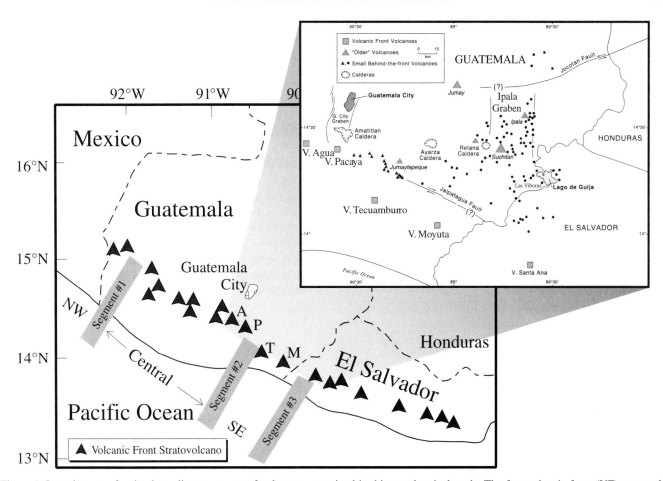

Figure 1. Location map for the three diverse groups of volcanoes examined in this geochemical study. The four volcanic front (VF) stratovolcanoes in southeastern Guatemala are Agua (A), Pacaya (P), Tecuamburro (T), and Moyuta (M). Segment #2 divides Agua and Pacaya from Tecuamburro and Moyuta. Inset: magnified view of the abundant behind the front volcanism (BVF) in southeastern Guatemala. Older BVF volcanism includes central volcanoes such as Ipala and Suchitan and calderas such as Retana and Ayarza. Younger BVF volcanism includes cinder cones that cluster in the Ipala Graben or along the trend of the Jalpatagua fault and shield volcanoes such as Las Viboras near Lago de Guija on the border with El Salvador.

Nd isotopic ratios remain rather uniform throughout the Central American arc except for samples from western Guatemala (Fig. 2A). Here, low Nd isotopic ratios occur in VF lavas that traverse the thickest section of crust (Carr et al., 1990). The erratic behavior of Rb/Nd ratios (Fig. 2B) in volcanic rocks from Guatemala compared to other segments of the arc suggests differing petrogenetic processes. The lower Nd isotopic values and higher Rb/Nd ratios of the VF lavas yield the first order suggestion that the composite cones possess greater degrees of crustal contamination.

## GEOCHEMICAL APPROACH

Examination of relationships between mineral phases and the whole-rock composition lends qualitative information on the petrological evolution of shallow magma chamber systems (Sakuyama, 1981; Hunter and Blake, 1995). Moreover, assimilation of granites and other crustal rocks in Mexico has been shown to enrich differentiated arc lavas in boron (B) (Hochstaedter et al., 1996). In addition, oxygen isotope data on whole-rocks and mineral separates can potentially supply unequivocal evidence of crustal inputs to arc magmas (Ellam and Harmon, 1990; Davidson et al., 1990; Feeley and Sharp, 1995; Pokrovskii and Volynets, 1999) and also provide an excellent means of quantifying the role of the continental crust as a magmatic source (Smith et al., 1996; Harmon and Gerbe, 1992). Thus, in an attempt to augment our limited knowledge on the extent of crustal contamination in Central American lavas, we discuss in this paper mineral composition, whole-rock oxygen isotope, and elemental B, Be, and Li data for lavas erupted in southeastern Guatemala.

## SAMPLES AND METHODS

Representative samples of VF stratovolcanoes, OBVF central volcanoes, and BVF cinder cones were selected for the mineral chemical study. The chemical compositions of groundmass

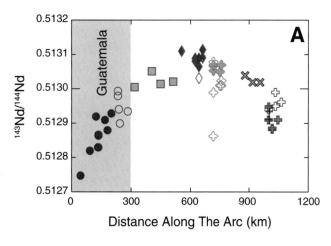
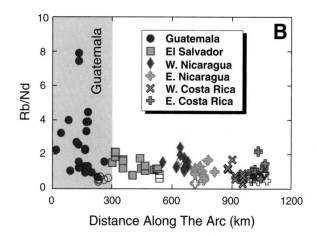

Figure 2. Variations in mafic samples along the Central American arc in geochemical parameters that indicate crustal contamination is most likely in southeastern Guatemala. (A) Low $^{143}$Nd/$^{144}$Nd occurs in Guatemala (circles). Behind the volcanic front (BVF) volcanism is shown by open circles. Crustal contamination or anomalous mantle may explain the complementary low $^{143}$Nd/$^{144}$Nd in Costa Rica (crosses). (B) Guatemalan lavas possess high Rb/Nd ratios compared to the rest of the Central American arc.

and phenocryst phases were measured with a Cameca SX-50 electron microprobe at the University of Chicago, calibrated with natural and synthetic standards. The program audit profile (PAP) correction routine was utilized for all analyses. Typical analytical conditions adopted were a 15 kV accelerating voltage and a 25 nA beam current. Most analyses were run with a focused beam and a beam width of 1–2 µm. Tables 1–3 report representative major element compositions and mineral formulae of olivine, pyroxene, and plagioclase.

Significant crustal inputs of the light elements (i.e., Li, Be, and B) to arc magmas seem especially probable where granitoid rocks serve as the contaminant. Consequently, the light elements were measured on a subset of 23 lavas from southeast Guatemala and two possible crustal contaminants by direct current plasma atomic emission spectrometry (DCP-AES) using an ARL model SpectraSpan (SS) 7 DCP at the University of South Florida. Three standards were run for B (RR-1, NBS-688, and QDF-1), and two standards (STM-1 and QLO-1) were analyzed for Li and Be to

TABLE 1. REPRESENTATIVE OLIVINE MINERAL CHEMICAL DATA

| Sample | GIP.1Pc* | GIP.1Pr | GIP.1GM | GCC837.Pc | GCC837.Pr | GCC837.GM | G201.GM | T302.Pc | T302.Pr |
|---|---|---|---|---|---|---|---|---|---|
| Group | OBVF | OBVF | OBVF | BVF | BVF | BVF | BVF | VF | VF |
| $SiO_2$ | 37.08 | 34.19 | 34.79 | 39.62 | 38.14 | 37.06 | 36.53 | 39.07 | 37.96 |
| MgO | 34.35 | 22.35 | 23.59 | 41.12 | 36.18 | 31.15 | 29.32 | 41.51 | 35.32 |
| FeO | 27.16 | 41.76 | 40.34 | 18.53 | 24.34 | 30.49 | 32.82 | 18.28 | 25.24 |
| MnO | 0.42 | 0.89 | 0.78 | 0.22 | 0.48 | 0.63 | 0.73 | 0.23 | 0.42 |
| CaO | 0.19 | 0.3 | 0.29 | 0.15 | 0.2 | 0.35 | 0.33 | 0.13 | 0.16 |
| NiO | 0.05 | 0.05 | 0 | 0.02 | 0.03 | 0.05 | 0.08 | 0.07 | 0.06 |
| Total | 99.25 | 99.54 | 99.79 | 99.66 | 99.37 | 99.73 | 99.81 | 99.29 | 99.16 |
| Formula based on 4 oxygens | | | | | | | | | |
| Si | 0.998 | 0.993 | 0.998 | 1.013 | 1.009 | 1.008 | 1.006 | 1.004 | 1.011 |
| Mg | 1.378 | 0.968 | 1.009 | 1.568 | 1.427 | 1.264 | 1.204 | 1.59 | 1.402 |
| Fe | 0.611 | 1.014 | 0.968 | 0.396 | 0.538 | 0.694 | 0.756 | 0.393 | 0.562 |
| Mn | 0.01 | 0.022 | 0.019 | 0.005 | 0.011 | 0.015 | 0.017 | 0.005 | 0.009 |
| Ca | 0.005 | 0.009 | 0.009 | 0.004 | 0.006 | 0.01 | 0.01 | 0.004 | 0.005 |
| Ni | 0.001 | 0.001 | 0 | 0 | 0.001 | 0.001 | 0.002 | 0.001 | 0.001 |
| mol% end members | | | | | | | | | |
| Fo | 69.3 | 48.8 | 51.0 | 79.8 | 72.6 | 64.6 | 61.4 | 80.2 | 71.4 |
| Fa | 30.7 | 51.2 | 49.0 | 20.2 | 27.4 | 35.4 | 38.5 | 19.8 | 28.6 |

Note: BVF—behind the volcanic front; OBVF—older behind the volcanic front; VF—volcanic front.
*The prefix of the sample designations stands for Guatemala and the particular volcano, whereas the suffix is composite of texture (eg. P—phenocryst, GM—groundmass) and analysis location (eg. c—core or r—rim).

assess precision and accuracy. Results of other studies quote errors within 10% (relative) of accepted values for both B and Be (Hochstaedter et al., 1996). In this study, precision for B was closer to 20% at the 1 ppm level, but improved to ~1% at the 10 ppm level. Errors on the Be standards were 11 and 16% at the 1 ppm and 10 ppm level, respectively. Li analyses had errors between 0.06% and 4.1%. Table 4 presents the new Li, Be, and B data.

Oxygen-isotopic ratios were determined on ~12 mg samples of 27 separate whole-rock powders at the Stable Isotope Laboratory at Southern Methodist University in Dallas. Oxygen was extracted from powders following reaction with $ClF_3$. Typical analytical precision on whole-rock $\delta^{18}O$ values approximated ±0.2‰ based on standards and repeat analyses. Of the whole-rock powders analyzed, two are crustal rocks, seven are lavas from VF stratovolcanoes, nine are from OBVF central volcanoes, and nine are from BVF cinder cones. Table 4 also reports the analytical results in the standard per mil (‰) notation as deviations relative to the standard mean ocean water (SMOW) standard.

## RESULTS

### Mineral Chemical Data

Olivine occurs as a phenocryst and groundmass phase in basalts and basaltic andesites in southeast Guatemala (Table 1). Normal zoning displayed by backscatter images was confirmed by electron microprobe analyses. Olivine cores (Fig. 3A; mean $Fo_{73.1}$) are generally more enriched in Mg relative to rim compositions (Fig. 3B; mean $Fo_{64.5}$). Groundmass olivine grains (mean $Fo_{56.2}$) are more Mg-poor than both phenocryst cores and rims (Fig. 3C).

A comparison of olivine compositions at the different edifice styles shows that olivine cores at VF composite cones (mean

TABLE 2. REPRESENTATIVE PYROXENE MINERAL CHEMICAL DATA

| Sample | GJM.2Pc* | GJM.2Pr | GJM.2GM | GCC911.GM | GCC837.Pc | T302.Pc | T302.Pr | M210.Pc | M210.Pr |
|---|---|---|---|---|---|---|---|---|---|
| Group | OBVF | OBVF | OBVF | BVF | BVF | VF | VF | VF | VF |
| $SiO_2$ | 52.2 | 51.78 | 49.35 | 52.57 | 48.36 | 50.76 | 51.9 | 51.62 | 51.04 |
| $Al_2O_3$ | 2.15 | 1.92 | 3.68 | 1.51 | 4.29 | 4.09 | 2.94 | 2.03 | 2.1 |
| $Fe_2O_3$ | 2.06 | 2.47 | 2.85 | 0 | 3.46 | 1.5 | 0.43 | 0.11 | 0.1 |
| $TiO_2$ | 0.42 | 0.39 | 1.48 | 0.57 | 1.48 | 0.7 | 0.54 | 0.5 | 0.52 |
| $Cr_2O_3$ | 0.01 | 0 | 0.09 | 0.18 | 0.06 | 0.03 | 0.03 | 0.01 | 0.01 |
| MgO | 24.67 | 24.74 | 14.47 | 17.06 | 12.96 | 14.89 | 15.22 | 13.79 | 13.09 |
| FeO | 15.92 | 15.27 | 7 | 11.85 | 8.4 | 7.32 | 8.53 | 11.66 | 10.72 |
| MnO | 0.38 | 0.44 | 0.24 | 0.4 | 0.29 | 0.23 | 0.29 | 0.4 | 0.4 |
| CaO | 1.84 | 1.75 | 19.87 | 15.06 | 20.02 | 20 | 19.73 | 18.57 | 19.63 |
| $Na_2O$ | 0.03 | 0.03 | 0.37 | 0.2 | 0.38 | 0.34 | 0.27 | 0.37 | 0.41 |
| Total | 99.68 | 98.79 | 99.4 | 99.4 | 99.7 | 99.86 | 99.88 | 99.06 | 98.02 |
| Formula based on 6 oxygens | | | | | | | | | |
| Si | 1.915 | 1.915 | 1.85 | 1.96 | 1.826 | 1.882 | 1.925 | 1.953 | 1.951 |
| Al | 0.085 | 0.084 | 0.15 | 0.04 | 0.174 | 0.118 | 0.075 | 0.047 | 0.049 |
| $Fe^{3+}$ | 0 | 0.001 | 0 | 0 | 0 | 0 | 0 | 0 | 0 |
| Al | 0.008 | 0 | 0.013 | 0.026 | 0.017 | 0.061 | 0.054 | 0.044 | 0.046 |
| $Fe^{3+}$ | 0.057 | 0.068 | 0.08 | 0 | 0.098 | 0.42 | 0.012 | 0.003 | 0.003 |
| Ti | 0.012 | 0.011 | 0.042 | 0.016 | 0.042 | 0.2 | 0.015 | 0.014 | 0.015 |
| Cr | 0 | 0 | 0.003 | 0.005 | 0.002 | 0.001 | 0.001 | 0 | 0 |
| Mg | 0.923 | 0.921 | 0.809 | 0.948 | 0.729 | 0.823 | 0.841 | 0.778 | 0.746 |
| $Fe^{2+}$ | 0 | 0 | 0.053 | 0.005 | 0.112 | 0.053 | 0.077 | 0.161 | 0.19 |
| Mn | 0 | 0 | 0 | 0 | 0 | 0 | 0 | 0 | 0 |
| Mg | 0.426 | 0.443 | 0 | 0 | 0 | 0 | 0 | 0 | 0 |
| $Fe^{2+}$ | 0.488 | 0.472 | 0.166 | 0.365 | 0.153 | 0.174 | 0.188 | 0.208 | 0.153 |
| Mn | 0.012 | 0.014 | 0.008 | 0.013 | 0.009 | 0.007 | 0.009 | 0.013 | 0.013 |
| Ca | 0.072 | 0.069 | 0.798 | 0.602 | 0.81 | 0.795 | 0.784 | 0.753 | 0.804 |
| Na | 0.002 | 0.002 | 0.027 | 0.014 | 0.028 | 0.24 | 0.019 | 0.027 | 0.03 |
| mol% end members | | | | | | | | | |
| Wo | 3.8 | 3.6 | 43.7 | 31.3 | 44.9 | 43.1 | 41.5 | 39.6 | 42.5 |
| En | 70.7 | 71.6 | 44.3 | 49.4 | 40.4 | 44.6 | 44.5 | 41.0 | 39.4 |
| Fs | 25.6 | 24.8 | 12.0 | 19.3 | 14.7 | 12.3 | 14.0 | 19.4 | 18.1 |

*Note:* BVF—behind the volcanic front; OBVF—older behind the volcanic front; VF—volcanic front.
*The prefix of the sample designations stands for Guatemala and the particular volcano, whereas the suffix is composite of texture (eg. P—phenocryst, GM—groundmass) and analysis location (eg. c—core or r—rim).

TABLE 3. REPRESENTATIVE PLAGIOCLASE MINERAL CHEMICAL DATA

| Sample | GJM.2Pc | GJM.2Pr | GJM.2GM | GCH.1GM | GCC837.Pc | GCC837.Pr | GCC837.GM | T302.Pc | T302.Pr |
|---|---|---|---|---|---|---|---|---|---|
| Group | OBVF | OBVF | OBVF | OBVF | BVF | BVF | BVF | VF | VF |
| $SiO_2$ | 52.86 | 51.22 | 54.14 | 52.89 | 46.63 | 49.71 | 49.73 | 45.08 | 49.5 |
| $Al_2O_3$ | 28.78 | 30.24 | 27.7 | 29.08 | 33.48 | 30.69 | 31.37 | 34.35 | 31.3 |
| $Na_2O$ | 4.47 | 3.48 | 5.08 | 4.37 | 1.59 | 3.07 | 2.85 | 1.27 | 3.07 |
| CaO | 12.18 | 13.96 | 11.02 | 12.07 | 17.43 | 14.77 | 15.13 | 17.64 | 14.79 |
| $K_2O$ | 0.25 | 0.18 | 0.34 | 0.34 | 0.05 | 0.15 | 0.13 | 0 | 0.06 |
| BaO | 0 | 0 | 0 | 0.04 | 0 | 0 | 0 | 0 | 0 |
| Total | 98.54 | 99.08 | 98.28 | 98.79 | 99.18 | 98.39 | 99.21 | 98.34 | 98.72 |
| Formula based on 32 oxygens | | | | | | | | | |
| Si | 9.713 | 9.401 | 9.945 | 9.694 | 8.64 | 9.217 | 9.147 | 8.435 | 9.145 |
| Al | 6.233 | 6.542 | 5.998 | 6.282 | 7.312 | 6.707 | 6.801 | 7.576 | 6.816 |
| Na | 1.593 | 1.238 | 1.809 | 1.553 | 0.571 | 1.104 | 1.016 | 0.461 | 1.1 |
| Ca | 2.398 | 2.745 | 2.169 | 2.37 | 3.461 | 2.935 | 2.982 | 3.537 | 2.928 |
| K | 0.059 | 0.042 | 0.08 | 0.08 | 0.012 | 0.035 | 0.031 | 0 | 0.014 |
| Ba | 0 | 0 | 0 | 0.003 | 0 | 0 | 0 | 0 | 0 |
| mol% end members | | | | | | | | | |
| Ab | 39.3 | 30.8 | 44.6 | 38.8 | 14.1 | 27.1 | 25.2 | 11.5 | 27.2 |
| An | 59.2 | 68.2 | 53.4 | 59.2 | 85.6 | 72.0 | 74.0 | 88.5 | 72.4 |

*Note:* BVF—behind the volcanic front; OBVF—older behind the volcanic front; VF—volcanic front.
*The prefix of the sample designations stands for Guatemala and the particular volcano, whereas the suffix is composite of texture (eg. P—phenocryst, GM—groundmass) and analysis location (eg. c—core or r—rim).

TABLE 4. OXYGEN AND Sr ISOTOPIC DATA ALONG WITH SELECT TRACE ELEMENTS

| Sample | Group | $SiO_2$ (wt%) | $\delta^{18}O$ (‰) | $^{87}Sr/^{86}Sr$† | B (ppm) | Be (ppm) | Li (ppm) | La (ppm) | Pb (ppm) | Th (ppm) | Cs (ppm) | U (ppm) | Rb (ppm) | Sb (ppm) | Ba (ppm) |
|---|---|---|---|---|---|---|---|---|---|---|---|---|---|---|---|
| G2 | Crust | 59.65 | 3.71 | 0.7066 | 24.1 | 1.13 | 21.7 | 21.21 | 14.88 | 8.43 | 0.39 | 1.26 | 1.33 | 66.02 | 497.3 |
| M2 | Crust | 81.19 | 14.27 | 0.74277 | 22.0 | 0.45 | 1.9 | 34.1 | 5.76 | 6.78 | 5.612 | 1.773 | | | 1377 |
| GUC-201 | BVF | 50.36 | 6.47 | 0.703547 | | | | 12.53 | 5.17 | 1.25 | 0.53 | 0.53 | 14.12 | 0.17 | 404.8 |
| GUC-911 | BVF | 53.71 | 7.74 | 0.704054 | 9.9 | 1.29 | 9.0 | 28.53 | 7.14 | 1.19 | 0.6 | 0.5 | 25.12 | 0.1 | 1039 |
| GUC-837 | BVF | 49.01 | 6.61 | 0.703975 | 2.3 | 0.87 | 7.8 | 14.05 | 5.25 | 0.96 | 0.39 | 0.41 | 17.41 | 0.09 | 344.4 |
| GUC-839 | BVF | 48.32 | 6.7 | | | | | 11.73 | 3.58 | 0.82 | 0.16 | 0.34 | 7.94 | 0.03 | 263.6 |
| GUC-844 | BVF | 50.39 | 6.63 | 0.703791 | 2.7 | 1.07 | 8.4 | 14.21 | 4.9 | 1.48 | 0.27 | 0.64 | 23.31 | 0.05 | 462.2 |
| GUC-702 | BVF | 49.98 | 6.72 | 0.70353 | 2.7 | 1.33 | 9.2 | 15.18 | 4.1 | 1.07 | 0.24 | 0.5 | 13.6 | 0.02 | 402.2 |
| GUC-800 | BVF | 48.76 | 6.16 | | 1.1 | 0.85 | 7.4 | 10 | 2.59 | 0.95 | 0.15 | 0.22 | 7.33 | | 248.3 |
| GUC-25 | BVF | 53.7 | | 0.70381 | 10.0 | 1.66 | 12.3 | 26 | 6.6 | 2.07 | 0.26 | 0.82 | 20.88 | 0.05 | 624.7 |
| GUC-303 | BVF | 49.9 | | 0.70321 | 4.0 | 0.92 | 7.3 | 9 | 2.25 | 0.67 | 0.04 | 0.17 | 6.64 | | 181.4 |
| GUAT-20 | BVF | | 6.67 | | | | | | | | | | | | |
| GUAT-33 | BVF | | 6.98 | | | | | | | | | | | | |
| GUC-835 | OBVF | 50.33 | 6.87 | 0.70415 | | | | 15.13 | 4.85 | 1.13 | 0.37 | 0.53 | 17.65 | 0.05 | 500.5 |
| GSU-01 | OBVF | 52.56 | 7.5 | 0.704135 | 8.3 | 0.86 | 9.4 | 12 | 6.71 | 1.89 | 1.09 | 0.69 | 36.77 | 0.17 | 470.6 |
| GSU-04 | OBVF | 55.43 | 7.41 | 0.704101 | 4.5 | 1.24 | 9.2 | 22.88 | 8.91 | 3.1 | 1.41 | 1.22 | 50.99 | 0.13 | 668.2 |
| GIP-01 | OBVF | 52 | 7.14 | 0.703877 | 4.1 | 1.36 | 10.0 | 18.6 | 6.45 | 1.93 | 0.83 | 0.78 | 29.92 | 0.17 | 504.8 |
| GIP-04 | OBVF | 51.04 | 7.12 | 0.703884 | 2.6 | 0.87 | 8.6 | 15.13 | 4.43 | 1.47 | 0.35 | 0.63 | 18.96 | 0.07 | 480.8 |
| GJM-02 | OBVF | 56.67 | 7.86 | | 11.5 | 1.08 | 17.3 | 21.74 | 7.64 | 3.9 | 1.38 | 1.26 | 47.37 | 0.15 | 676.9 |
| GJM-03 | OBVF | 51.73 | 7.85 | | 6.8 | 0.76 | 11.5 | 12.64 | 4.97 | 2.04 | 0.89 | 0.68 | 26.93 | 0.11 | 480.7 |
| GCH-01 | OBVF | 51.15 | 6.9 | | 2.6 | 1.01 | 10.0 | 20.08 | 4.47 | 1.47 | 0.67 | 0.61 | 17.33 | 0.08 | 582.5 |
| GUC-812 | OBVF | 49.91 | 6.51 | | 3.7 | 0.98 | 8.4 | 16.04 | 4.39 | 1 | 0.12 | 0.28 | 9.77 | 0.02 | 354.2 |
| LC-2 | OBVF | 54.75 | | 0.704102 | 8.0 | 1.16 | 10.1 | 23.49 | 7.98 | 2.28 | 0.78 | 0.78 | 38.61 | 0.18 | 865.8 |
| FEL-3 | OBVF | 58.25 | | 0.70409 | 18.1 | 1.23 | 11.4 | 23.35 | 12.1 | 3.01 | 0.73 | 1.04 | 54.22 | 0.3 | 935.9 |
| FVH-4 | OBVF | 64.32 | | 0.704056 | 22.8 | 1.33 | 10.4 | 25.61 | 13.6 | 5.02 | 2.32 | 1.83 | 99.48 | 0.39 | 943 |
| GPA-01 | VF | 49.26 | 6.36 | | | | | 8.2 | 3.26 | 0.96 | 0.41 | 0.42 | 10.26 | 0.02 | 341.2 |
| GAG-21 | VF | 51.4 | 6.59 | | 6.0 | 0.75 | | 10.48 | 5.49 | 1.57 | 0.62 | 0.67 | 19.53 | 0.2 | 519.9 |
| GAG-27 | VF | 55.31 | 6.79 | | | | | 13.19 | 5.07 | 2.21 | 0.6 | 0.99 | 39.69 | 0.25 | 638.1 |
| GAG-10 | VF | 59.1 | 6.78 | | | | | 16.09 | 8.12 | 3.17 | 1.93 | 1.46 | 55.92 | 0.4 | 748.7 |
| TCB-302 | VF | 53.87 | 6.73 | | | | | 5.83 | 3.27 | 0.45 | 0.77 | 0.25 | 9.41 | 0.54 | 314.7 |
| E1 | VF | 51.4 | 7.44 | 0.70389 | 3.1 | 0.62 | 9.2 | 10.15 | 5.51 | 1.47 | 0.58 | 0.58 | 13.53 | | 442.3 |
| GMO-201 | VF | 57.46 | 8.41 | 0.703649 | 22.9 | 0.78 | 9.6 | 12.29 | 8.12 | 3.43 | 2.11 | 1.48 | 46.59 | 0.7 | 669 |
| Qam-13 | VF | 49.81 | | | 7.0 | 0.58 | 9.2 | 8.51 | 4.6 | 1.87 | 0.64 | 0.62 | 13.6 | | 486.6 |

*Note:* BVF—behind the volcanic front; OBVF—older behind the volcanic front; VF—volcanic front.
†Sr isotopic ratios were obtained at the mass spectrometry facility at Rutgers University. Sr isotopic ratios are normalized to $^{86}Sr/^{88}Sr$ of 0.1194 and are reported as measured.

$Fo_{75.3}$; Fig. 3A) and BVF cinder cones (mean $Fo_{76.0}$; Fig. 3A) have slight enrichments in Mg compared to the OBVF central volcanoes (mean $Fo_{71.6}$; Fig. 3A). Moreover, VF stratovolcanoes and BVF cinder cones have more narrow ranges of olivine core compositions (Fig. 3). In contrast, OBVF central volcanoes contain olivine that extends the range in forsterite (Fo) content to below 60 mol% (Fig. 3).

Clinopyroxene occurs as both a phenocryst and groundmass phase in most lavas from southeast Guatemala, except for BVF cinder cones, which only contain clinopyroxene as a groundmass phase. In contrast to olivine, backscatter images and electron microprobe analyses showed little chemical zoning in the clinopyroxene phenocrysts. Phenocryst cores (Fig. 4A; mean Mg number = 77.0) have nearly identical ranges and means as phenocryst rims (Fig. 4B; mean Mg number = 77.4). Groundmass clinopyroxene grains possess a similar range in Mg number as phenocrysts (Fig. 4C), but have a slightly lower average value (mean Mg number = 74.2). More evolved samples, such as the dacite (Figs. 4A and 4B), display both normal and reversed zoning, though variations in Mg number are relatively small. Clinopyroxene compositions vary little as whole-rock $SiO_2$ increases.

Clinopyroxene core compositions at VF stratovolcanoes (mean Mg number = 77.0) and OBVF central volcanoes (mean Mg number = 77.0) have similar ranges (Fig. 4A) and an identical mean. Owing to the paucity of clinopyroxene phenocrysts in cinder cone lavas, only one phenocryst core was measured in a BVF cinder cone rock, and its Mg# is 73.3 (Fig. 4A). The range for groundmass clinopyroxene in BVF cinder cones is shifted to lower Mg numbers (62.2–80.8) relative to the phenocrysts in lavas from the other edifice types (Fig. 4). The mean Mg number of the groundmass clinopyroxene at BVF cinder cones (Mg number = 72.7) is lower than the clinopyroxene phenocryst core and rim compositions from the other edifice types, but approximates the mean from the lone phenocryst analysis from a Guatemalan cinder cone (Fig. 4A).

Orthopyroxene occurs in lower modal abundances as phenocryst and groundmass phases than clinopyroxene in most Guatemalan lavas. Groundmass and phenocryst orthopyroxene exhibit a similar range of compositions (Fig. 5). Moreover, orthopyroxene phenocryst cores, rims, and groundmass grains have nearly identical mean Mg numbers (mean Mg number of 71.9, 70.9, and 70.3, respectively). Despite these similar means, the data show both normal and reversely zoned phenocrysts in some of the more evolved magmas. It is noteworthy that Mg-rich cores persist across the compositional spectrum. The lower Mg number of orthopyroxene cores (mean Mg number = 71.9) compared to clinopyroxene (mean Mg number = 77.0) in the lavas from southeast Guatemala may indicate that orthopyroxene occurs later in the crystallization history.

Slight variations in orthopyroxene core composition exist based on the type of volcanic edifice (Fig. 5). Although orthopyroxene phenocrysts from OBVF central volcanoes and VF stratovolcanoes have similar ranges in Mg number (Fig. 5A), orthopyroxene cores in lavas from the former edifice type (mean Mg number = 73.1) have a higher mean Mg number than the latter (mean Mg number = 70.2). No orthopyroxene phenocrysts from BVF cinder cones were analyzed by the electron microprobe.

Plagioclase occurs as a modally abundant phenocryst and groundmass phase in the arc lavas of southeast Guatemala. Both phenocryst cores and rims exhibit a nearly identical, wide range of An contents (Figs. 6A and 6B), yet generally display normal zoning. The mean An values for plagioclase phenocryst cores and rims ($An_{70.5}$ and $An_{61.6}$, respectively) reflects the predominance of normally zoned crystals. However, some more-evolved samples

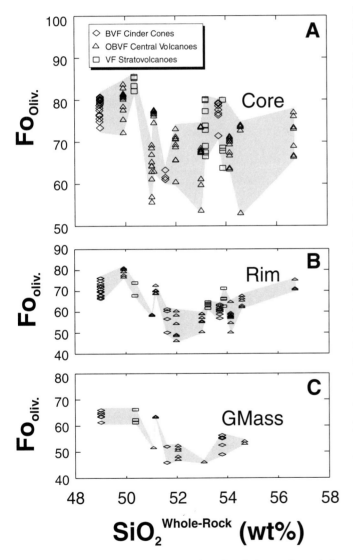

Figure 3. Forsterite (Fo) content variations in olivine against whole rock $SiO_2$ compositions. (A) Core Fo contents in olivine phenocrysts (oliv.). (B) Rim Fo contents in olivine phenocrysts. (C) Groundmass (GMass) olivine Fo contents. Shaded compositional envelopes were drawn to highlight that little systematic variation exists between the Fo contents and the host rock chemistry.

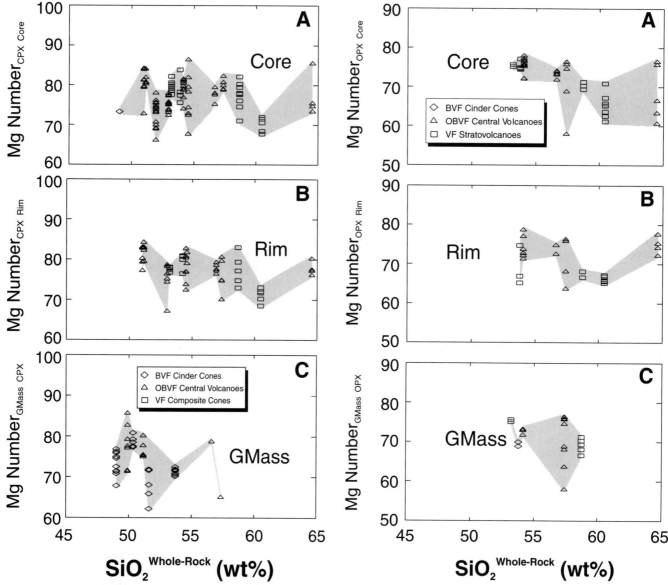

Figure 4. Mg number variations in clinopyroxene grains versus whole-rock $SiO_2$ compositions. (A) Core Mg number values of clinopyroxene phenocrysts (CPX). (B) Rim Mg number values of clinopyroxene phenocrysts. (C) Mg number contents of groundmass (GMass) clinopyroxene grains. Shaded compositional envelopes were drawn to highlight the nonsystematic relationship between Mg number in clinopyroxene versus host rock composition.

Figure 5. Mg number variations in orthopyroxene grains versus whole-rock $SiO_2$ compositions. (A) Mg number values for orthopyroxene phenocrysts (OPX). (B) Rim Mg number values of orthopyroxene phenocrysts. (C) Mg number values for groundmass (GMass) orthopyroxene. Shaded compositional envelopes were drawn to highlight that Mg number in orthopyroxene remains relatively constant over a range of host rock $SiO_2$ compositions.

do display both normal and reversed zoning. A lower mean An content in groundmass plagioclase (mean $An_{58.6}$) relative to phenocryst rims (mean $An_{61.6}$) continues the overall shift in the range of An contents to lower values with inferred crystallization order. An content in plagioclase crystals varies considerably at whole-rock $SiO_2$ contents <57 wt%, but generally decreases at $SiO_2$ contents >57% wt% (Figs. 6A and 6B).

Plagioclase crystals of varying composition occur in lavas erupted from the different edifice types (Fig. 6). Plagioclase cores from polygenetic VF stratovolcanoes and OBVF central volcanoes exhibit large ranges in An content (Fig. 6A), whereas those from monogenetic BVF cinder cones display a more narrow range toward the high end of the An spectrum (Fig. 6A). The mean An content of plagioclase phenocryst cores from BVF cinder cones is substantially higher (mean $An_{74.9}$) than those from VF stratovolcanoes (mean $An_{64.5}$) or OBVF central volcanoes (mean $An_{70.8}$). The same pattern of An variation occurs amongst plagioclase rim compositions (Fig. 6B).

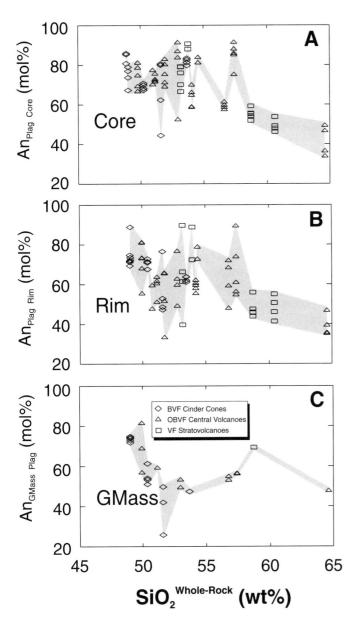

Figure 6. An content variations in plagioclase crystals versus whole-rock SiO$_2$ compositions. (A) Core An contents of plagioclase phenocrysts (Plag). (B) Rim An contents of plagioclase phenocrysts. (C) An contents of groundmass (GMass) plagioclase. Shaded compositional envelopes show no relationship between An content and host rock composition.

## Oxygen Isotopic Compositions

The whole-rock oxygen isotope data from southeastern Guatemala span a wide range for young, fresh arc lavas (6.16–8.46‰). Basalts associated with subduction at convergent margins show enrichments in whole-rock δ$^{18}$O relative to mid-oceanic-ridge basalt (MORB) (δ$^{18}$O = +5.7‰). Forearc trough and backarc basin basalts have a mean δ$^{18}$O of 5.9‰; oceanic arc basalts have a mean δ$^{18}$O of 6.0‰; and continental arc basalts average 6.2‰

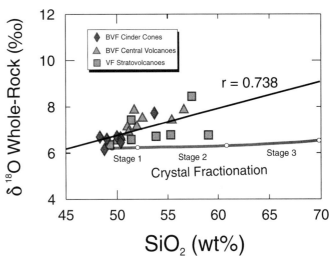

Figure 7. δ$^{18}$O$_{whole-rock}$ versus SiO$_2$ diagram. δ$^{18}$O correlated positively with SiO$_2$, with a Spearman rank correlation coefficient of +0.738. The relatively steep positive trend of the Guatemalan arc lavas contrasts with the near-horizontal crystal fractionation trend. Owing to the large range of compositions found at subduction zone volcanoes (i.e., from rhyolite to basalt), the crystal fractionation trend was modeled in three different stages. Stage 1, 2, and 3 refers to least squares mixing crystal fractionation models. See text for details. Behind the volcanic front (BVF) cinder cone lavas have lower δ$^{18}$O values than both volcanic front (VF) stratovolcanoes and older behind the volcanic front (OBVF) central volcanoes.

(Harmon and Hoefs 1995). Oxygen isotope analyses of olivines in oceanic arc lavas by laser fluorination are significantly less variable than the whole-rock data (Eiler et al., 2000) and might highlight the role of subsolidus alteration on whole-rock samples. The progressive enrichment in whole-rock δ$^{18}$O in arc settings may reflect interaction of parental magmas with volcanic arc basement and continental crust. Notably, δ$^{18}$O values in the arc lavas from southeastern Guatemala correlate positively with SiO$_2$ so that the most evolved lavas have the highest δ$^{18}$O values (Fig. 7). Among the three different edifice types in southeastern Guatemala, BVF cinder cones exhibit the best correlation ($r$ = 0.85) and fall closest to MORB in oxygen isotopic composition. The mean δ$^{18}$O of the BVF cinder cones (+6.8‰) still exceeds the average oxygen isotopic composition of continental arc basalts. VF stratovolcanoes display a relatively poor correlation ($r$ = 0.41), and the subhorizontal trend roughly parallels the crystal fractionation trend (Fig. 7). Lava from VF stratovolcanoes have a mean δ$^{18}$O value of +7.0‰, whereas the OBVF central volcanoes have a higher mean at δ$^{18}$O = +7.2‰. A quartzite crustal sample has the highest δ$^{18}$O value measured in this study at 14.27‰. The granodiorite crustal sample has a δ$^{18}$O value of 3.71‰, suggesting alteration by a high temperature, meteoric fluid.

## Light Element Contents

Boron (B) ranges from 1.1 to 22.9 ppm in the arc lavas from southeast Guatemala. The twenty-fold increase in abundance of

B underscores its highly incompatible chemical character. These analyses extend the narrow range of B concentration (2.4–15.2 ppm) found in an along-arc study of Central America (Leeman et al., 1994). The highest B values were recorded in more evolved lavas, and B contents correlate positively with $SiO_2$ (Fig. 8A).

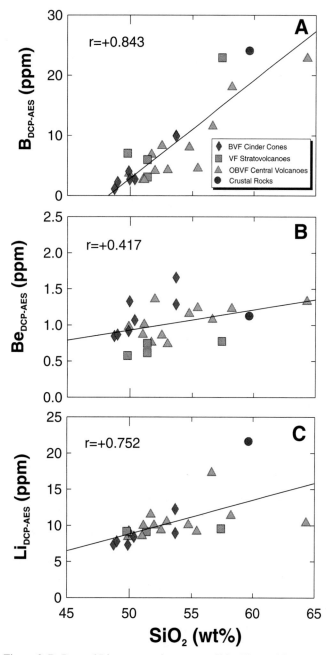

Figure 8. B, Be, and Li concentrations versus $SiO_2$. The positive correlations with $SiO_2$ reflect the incompatible nature of the light elements. (A) B. (B) Be. (C) Li. Only the granodiorite crustal sample is plotted to represent a possible contaminant because quartzite plots far off the $SiO_2$ scale. BVF—behind the volcanic front; OBVF—older behind the volcanic front; VF—volcanic front; DCP-AES—direct current plasma–atomic emission spectrometer.

VF stratovolcanoes and OBVF central volcanoes generally have higher B contents than the BVF cinder cones. Indeed, at a given $SiO_2$ content, VF stratovolcanoes tend to have slightly elevated B abundances relative to the central volcanoes and cinder cones. Granodiorite and quartzite samples taken to represent possible crustal contaminants contain B abundances comparable to the most enriched lavas (mean B of 23.1 ppm).

The 23 analyses reported here represent the first comprehensive study of Be in Guatemala. DCP-AES analyses of Be contents in rocks of this study display a limited range compared to B (0.6–1.7 ppm). The nearly threefold increase in Be affirms its less incompatible chemical character. Four other analyses record a limited range in Be of 0.5–0.8 ppm (Morris et al., 1990; Patino, 1997). In contrast to the behavior of B, the arc lavas exhibit a poor positive correlation between Be and $SiO_2$ and less systematic variation with respect to edifice type (Fig. 8B). Lavas from VF stratovolcanoes display the lowest Be contents, whereas BVF cinder cones and OBVF central volcanoes have a similar range of Be abundances and overall higher concentrations. At a given $SiO_2$ content, VF stratovolcanoes have lower Be abundances compared to the other edifice types. The two crustal samples analyzed cover a range of Be values (0.4–1.1 ppm) comparable to the lavas, with the granodiorite more enriched than the quartzite.

The Li analyses in this study represent the first compositional data of their type for arc lavas from Guatemala. Li contents analyzed by DCP-AES range between 7.3 and 17.3 ppm. The approximate twofold increase in Li concentrations suggests an incompatibility sequence of B >Be >Li. As for B and Be, a positive correlation exists between Li and $SiO_2$ ($r = +0.75$; Fig. 8C). BVF cinder cones generally have the lowest Li contents except for a sample far behind the volcanic front (GUC-25). VF stratovolcanoes exhibit uniform Li abundances of ~10 ppm and OBVF central volcanoes display the greatest variation from 8.4 to 17.3 ppm Li. The crustal samples span a larger range of Li composition than all the lavas with the granodiorite far exceeding the quartzite (21.7 versus 1.9 ppm Li).

## DISCUSSION

### Equilibrium in Minerals

Relations between bulk rock and phenocryst compositions provide a key perspective on whether a magma evolved in an open or closed system (Sakuyama, 1981; Hunter and Blake, 1995). Variations between mineral compositions and whole-rock compositions and assessment of whether equilibrium existed between mineral-mineral and mineral–whole-rock compositions lend insight into petrologic processes that influenced magma chemistry at depth. Obviously, these Guatemalan samples do not represent a cogenetic suite. The samples were collected from numerous locations at diverse edifice types. Indeed, the monogenetic nature of cinder cones essentially dictates sampling a single flow. Accordingly, this section places emphasis on the equilibrium approach. In fact, one might not expect systematic variation in mineral chemistry with whole-rock compositions that do not represent a cogenetic suite. This caveat aside, the smooth major element trends suggest broadly common petrogenetic processes at Guatemalan volcanoes (Cameron et al., 2003).

Closed system evolution of a single magma chamber favors smooth systematic relationships between mineral and whole-rock chemistry (Davidson and de Silva, 1995; Hunter and Blake, 1995). However, the Guatemalan olivine phenocrysts show a range of compositions from $Fo_{86}$ to $Fo_{46}$ with no clear systematic variation with host rock composition (Figs. 3A and 3B). Groundmass olivine crystals exhibit similar compositional trends to the phenocryst rims (Fig. 3C). Compositional envelopes were drawn in Figure 3 to visually aid recognition of abrupt mineral chemical variations with changing whole-rock compositions. Mg numbers in clinopyroxene show an even more saw-toothed pattern of mineral variation with whole-rock composition (Fig. 4). A clinopyroxene phenocryst core from the most evolved sample analyzed records the maximum Mg number (Fig. 4A). Akin to clinopyroxene, Mg-rich orthopyroxene phenocrysts occur throughout the entire range of whole-rock compositions (Figs. 5A and 5B). Plagioclase phenocrysts exhibit many excursions to more anorthitic compositions at the basalt–basaltic andesite end of the whole-rock compositional spectrum but shift to a smooth declining trend at the andesite-dacite end (Figs. 6A and 6B). In summary, the overall lack of correlation between mineral chemistry and host lava composition hints strongly at the operation of open system processes. The existence of both normally and reversely zoned minerals in the same sample of some of the more evolved rocks suggests that magma mixing may be an important petrologic process. The lack of detailed compositional traverses across phenocryst phases prevents an extensive evaluation of this important petrologic process.

A more rigorous attempt to evaluate open system processes involves assessment of the degree of equilibrium attained by minerals with their host lavas. Under closed system fractional crystallization, equilibrium should exist between phenocryst cores and the whole-rock composition, between the groundmass melt and groundmass mineral phases, and between contiguous phenocrysts (Hunter and Blake, 1995). Consequently, plots of elemental ratios in minerals against identical ratios for bulk rocks should yield trends consistent with experimentally determined distribution coefficients.

Olivine-liquid $K_D^{Fe-Mg}$ values (where $K_D^{Fe-Mg}$ = (FeO/MgO)$_{olivine}$/(FeO/MgO)$_{whole-rock}$) decrease with increasing alkali content of coexisting liquid (Falloon et al., 1997). Based on the experiments of Falloon et al. (1997), a $K_D^{Fe-Mg}$ value of 0.32 would characterize a basaltic melt with an alkali content ($Na_2O + K_2O$) of ~5 wt%. This value falls in the intermediate range determined by a series of experiments run by Sisson and Grove (1993). Core compositions of olivine phenocrysts scatter well off the equilibrium line constructed using a $K_D^{Fe-Mg}$ of 0.32 (Fig. 9). A majority of the microprobe analyses plot above the equilibrium line in the field of olivine crystals too evolved for the whole-rock composition. Visual inspection of the data suggests that olivines from VF composite cones and OBVF central volcanoes stray farther from the equilibrium line than olivine from BVF cinder cones.

An equilibrium $K_D^{Fe-Mg}$ value of 0.27 for orthopyroxene was determined by Gerlach and Grove (1982). In experimental melts, clinopyroxene has a slightly lower $K_D^{Fe-Mg}$ value of 0.23 (Grove and Bryan, 1983). Electron microprobe analyses of orthopyroxene phenocryst cores plot predominantly in the upper field of crystals too evolved for the whole-rock composition (Fig. 10A). The VF stratovolcano and OBVF central volcano orthopyroxene grains deviate equally from the equilibrium line into the upper field (Fig. 10A); no orthopyroxene phenocrysts were measured in the BVF cinder cones. At higher FeO/MgO$_{whole-rock}$ compositions, the orthopyroxene cores extend to higher FeO/MgO$_{orthopyroxene}$ contents. The core compositions of clinopyroxene phenocrysts plot in the upper evolved crystal field, although more data points fall below the equilibrium line than for orthopyroxene (Fig. 10B). Clinopyroxene cores from OBVF central volcanoes exhibit the largest deviations below the equilibrium line, whereas OBVF central volcano and VF stratovolcano clinopyroxene cores show equal departures above the equilibrium line over the entire range of FeO/MgO$_{whole-rock}$. One core analysis of a clinopyroxene microphenocryst from a BVF cinder cone lava plots far into the field for evolved crystals.

Melting experiments using basalts and basaltic andesites suggest that $K_D^{Ca-Na}$ values for plagioclase-melt equilibrium vary depending on the water content of the liquid (Sisson and Grove, 1993). Although independent of pressure, the $K_D^{Ca-Na}$ value increases from 1.1 under anhydrous conditions to 5.5 within $H_2O$-saturated magmas at 2 kbar (Sisson and Grove,

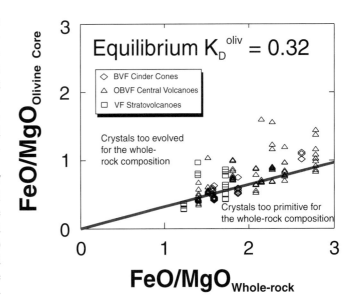

Figure 9. FeO/MgO$_{olivine}$ versus FeO/MgO$_{whole-rock}$. Core compositions of olivine phenocrysts largely fall above the equilibrium line ($K_D$ = 0.32). BVF—behind the volcanic front; OBVF—older behind the volcanic front; VF—volcanic front.

1993). Variable compositions of the melt created by magmatic differentiation exert no control on the $K_D^{Ca-Na}$ value at constant water contents. Without an idea of the water content of the melt, interpreting the plagioclase equilibrium plot becomes a serious challenge. Water contents in melt inclusions hosted by olivines from both BVF cinder cones and VF stratovolcanoes are ~2.0 wt% (Walker et al., 2003). However, regardless of any reasonable water content and associated $K_D^{Ca-Na}$ value chosen, the plagioclase core compositional data exhibit disequilibrium. Isolated VF stratovolcano and OBVF central volcano samples plot well above the equilibrium lines in the primitive crystal field (Fig. 11). Plagioclase from BVF cinder cones show smaller deviations from the equilibrium lines, with more uniform intermediate compositions, suggestive of moderate water contents in agreement with the melt inclusion data.

In a very general sense, any mineral core that falls in the evolved crystal field almost certainly did not crystallize from a liquid similar in composition to the bulk rock. Instead, these minerals likely formed from a more evolved liquid. Geologic explanations for the mineral cores that are too primitive for the whole-rock composition include early-formed phases that failed to react with the surrounding melt or crystals inherited from a more primitive liquid during magma mixing. These latter types definitely constitute the minority in this study.

Owing to the lack of pyroxene phenocrysts in the BVF cinder cones and the uncertainty in melt water contents with respect to plagioclase, we use olivine compositional data to try to determine whether closed or open system behavior was taking place. An argument for closed system behavior can be made if the electron microprobe did not analyze true olivine cores, but instead zones rimward of these points. True cores would be strictly defined as the first solid crystallizing from the bulk liquid. Thin sections are two-dimensional slices through a three-dimensional crystal, so there is no guarantee that the true core is analyzed when the electron beam is positioned at the center of a phenocryst. Thus, the numerous analyses that plot above the equilibrium line in the evolved field represent parts of the mineral that crystallized from interstitial liquid slightly more evolved than the bulk rock composition but still under equilibrium conditions. Although there is a coherent positive correlation between minimum FeO/MgO olivine core compositions with whole-rock FeO/MgO, the lack of true core compositions relative to rims plotting on the equilibrium line is troublesome.

If deviations from the mineral equilibrium line do indeed reflect disequilibrium and open system processes in the magma chamber, then the magnitude of the deviation logically holds important information on the extent of the process. It becomes a trivial matter to calculate the mean deviation from the olivine equilibrium line for each volcano type. Olivine cores from BVF cinder cones have a range in deviations from the equilibrium line of −0.12 to +0.29 with a mean of 0.01. OBVF central volcanoes have a range of deviations from −0.14 to +0.90 with a mean of 0.10, and VF stratovolcanoes exhibit a range of −0.11 to +0.56 with a mean deviation of 0.12. Based on the magnitudes of these

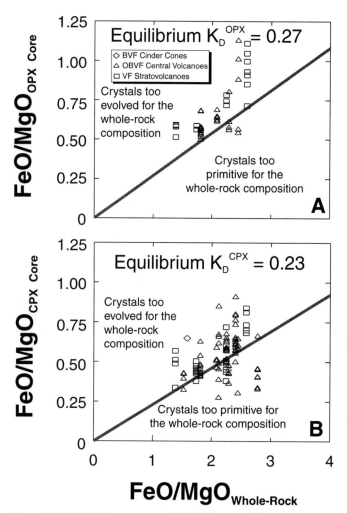

Figure 10. FeO/MgO$_{pyroxene}$ versus FeO/MgO$_{whole-rock}$ diagrams. (A) Orthopyroxene (OPX) phenocryst core data plot largely above an equilibrium defined by $K_D = 0.27$. (B) Clinopyroxene (CPX) phenocryst core data also plot largely above the equilibrium line defined by $K_D = 0.23$. BVF—behind the volcanic front; OBVF—older behind the volcanic front; VF—volcanic front.

deviations, the open system process seems more prevalent at the VF stratovolcanoes and OBVF central volcanoes.

Assuming that true core regions of phenocrysts were analyzed more often than not by the electron microprobe, the nonsystematic variation of mineral chemistry with whole-rock composition and the nonequilibrium mineral core compositions together lead to the conclusion that open system processes played a significant role in creating compositional diversity in the Guatemalan lavas.

**Oxygen Isotope Variations**

Two features of the whole-rock $\delta^{18}O$ data warrant comment. First, the arc lavas of southeastern Guatemala have elevated values of $\delta^{18}O$ compared to their respective global counterparts. The

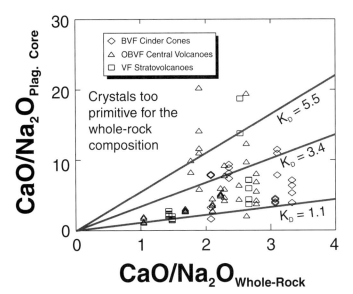

Figure 11. CaO/Na$_2$O$_{plagioclase}$ versus CaO/Na$_2$O$_{whole-rock}$ diagram. The equilibrium $K_D$ value depends on the water content of the magma. $K_D$ values of 1.1, 3.4, and 5.5 are shown. The scatter exhibited by the plagioclase core data suggests disequilibrium behavior. BVF—behind the volcanic front; OBVF—older behind the volcanic front; VF—volcanic front.

range in BVF cinder cones from 6.16 to 7.74‰ falls above the global mean for whole-rock backarc basins of 6.0‰ (Harmon and Hoefs, 1995). Likewise, the range for VF stratovolcanoes of 6.36–8.46‰ lies above the mean for continental arc basalts of 6.2‰ (Harmon and Hoefs, 1995). Second, relatively large ranges in δ$^{18}$O characterize the basalts and evolved lavas from southeastern Guatemala.

### Crystal Fractionation

The positive correlation between δ$^{18}$O and SiO$_2$ might suggest that crystal fractionation played an important role in the chemical variation (Fig. 7). However, the small isotopic fractionations that occur between melt and crystals at high temperature suggest that crystal differentiation has little influence in changing the δ$^{18}$O of evolving magmas. Oxygen isotope studies of closed system magmatic suites confirm this general observation (Muehlenbachs and Byerly, 1982; Chivas et al., 1982; Kalamarides, 1984). At the Galapagos Spreading Center, 90% fractionation from low K-tholeiite to rhyodacite only enriched the residual melt by ~1.2‰ (Muehlenbachs and Byerly, 1982). Crystal fractionation in the Koloula Igneous Complex on the island of Guadalcanal generated δ$^{18}$O values that range from 5.7 to 7.2‰ (Chivas et al., 1982). δ$^{18}$O values of the Kiglapait layered intrusion vary from 6.0‰ for the first liquids in the Lower Zone to 6.3‰ for the final liquids in the Upper Zone (Kalamarides, 1984). Thus, empirical knowledge suggests that the range of δ$^{18}$O values in the Guatemalan arc lavas exceeds the normal fractionation spread. In fractionating suites of magmas with demonstrable closed system behavior, the slope of the δ$^{18}$O versus SiO$_2$ diagram varies according to temperature.

Further insight on the δ$^{18}$O variations can be gleaned from quantitative modeling (Sheppard and Harris, 1985; Grunder, 1987; Woodhead et al., 1987; Harmon and Gerbe, 1992; Singer et al., 1992). Closed system Raleigh fractional crystallization enriches the residual liquid in δ$^{18}$O according to the equation:

$$\delta^{18}O_{melt}^{\,t} = (\delta^{18}O_{melt}^{\,i} + 1000)f^{(\alpha-1)} - 1000, \qquad (1)$$

where $\delta^{18}O_{melt}^{\,t}$ is the $^{18}O/^{16}O$ ratio of the magma after some finite amount of crystal fractionation, $\delta^{18}O_{melt}^{\,i}$ is the initial $^{18}O/^{16}O$ ratio of the magma prior to fractionation processes, $f$ is the fraction of melt remaining at time $t$, and α is the mean crystal-melt fractionation factor. In high temperature magmatic systems, α values approximate unity, so the relative enrichment of the crystal-melt phases in $^{18}O$ is given simply by

$$\Delta_{A-B}(‰) = 10^3 \ln \alpha_{A-B} = \delta_A - \delta_B. \qquad (2)$$

The mean fractionation factor can be uniquely calculated for distinct stages in the evolutionary history of the Guatemalan magmas. The calculation involves multiplying the molar proportion of each phase by the mole fraction of oxygen in each phase and the mean δ$^{18}$O fractionations between crystal and melts taken from the literature (e.g., Sheppard and Harris, 1985; Kalamarides, 1986). Substitution of the mean fractionation factor into the general Raleigh fractionation formula allows calculation of the expected δ$^{18}$O value of residual melts.

Table 5 provides the details of least squares crystal fractionation model developed to support interpretation of the oxygen isotope data. In Stage 1, 28.6% crystallization of plagioclase, clinopyroxene, and olivine from parent magma GUC-800 (recalculated SiO$_2$ = 49.53%) yields the daughter magma GCH-01 (recalculated SiO$_2$ = 52.04%) with an acceptable $r^2$ = 0.805. The mean fractionation factor ($\Delta_{xl-melt}$) is −0.06 or α = 0.99994. The calculated δ$^{18}$O$_{melt}$ after 28.6% crystallization using a δ$^{18}$O$_{melt}^{\,i}$ of 6.21‰ is 6.23‰.

Plagioclase, clinopyroxene, olivine, and titanomagnetite could crystallize from GCH-01 in Stage 2 to produce a nearly equal volume of daughter melt similar in composition to FEL-3 (recalculated SiO$_2$ = 60.83%) with $r^2$ = 0.880. The resulting $\Delta_{xl-melt}$ is −0.121 or α = 0.99988. The Raleigh fractionation formula yields a calculated δ$^{18}$O of 6.31‰ for the melt.

In Stage 3, 46.2% crystallization of plagioclase, clinopyroxene, olivine, and titanomagnetite from a FEL-3 parental melt produces a daughter magma GAG-13A (recalculated SiO$_2$ = 69.81%) with a very low $r^2$ of 0.095. More significant titanomagnetite fractionation increased the mean fractionation factor (Δ) to −0.492 or α = 0.99951. The calculated δ$^{18}$O$_{melt}^{\,t}$ at the end of the fractionating sequence is 6.53‰. Large shifts in δ$^{18}$O of magma (i.e., >1‰) result only through the removal of significant amounts of either an $^{18}$O-depleted phase such as olivine or an oxide mineral or an $^{18}$O-enriched phase such as quartz (Harmon and Gerbe, 1992). Otherwise, closed system differentiation of magma from a basalt to rhyolite generates much

TABLE 5. LEAST SQUARES MIXING CRYSTAL FRACTIONATION MODEL

| Stage | Parent | Proportion | Percentage | Mineral | Fit |
|---|---|---|---|---|---|
| 1 | GUC-800 | 0.123 | 43.30% | Plagioclase ($An_{92}$) | $r^2 = 0.805$ |
| | | 0.12 | 42.30% | Cpx ($Wo_{46}$-$En_{37}$-$Fs_{17}$) | |
| | | 0.041 | 14.30% | Olivine ($Fo_{72}$) | |
| | | 0.714 | | GCH-01 (daughter) | |
| 2 | GCH-01 | 0.029 | 6.00% | Ti-magnetite ($Ti_{10}$) | $r^2 = 0.880$ |
| | | 0.263 | 56.00% | Plagioclase ($An_{67}$) | |
| | | 0.085 | 18.00% | Cpx ($Wo_{44}$-$En_{47}$-$Fs_9$) | |
| | | 0.093 | 19.80% | Olivine ($Fo_{68}$) | |
| | | 0.517 | | FEL-3 (daughter) | |
| 3 | FEL-3 | 0.04 | 11.00% | Ti-magnetite ($Ti_{12}$) | $r^2 = 0.095$ |
| | | 0.256 | 71.00% | Plagioclase ($An_{59}$) | |
| | | 0.023 | 6.30% | Cpx ($Wo_{42}$-$En_{44}$-$Fs_{14}$) | |
| | | 0.042 | 11.70% | Olivine ($Fo_{60}$) | |
| | | 0.638 | | GAG-13A (daughter) | |

smaller variations in $\delta^{18}O$. The predominance of plagioclase in the fractionating assemblage of calc-alkaline magmatic systems often results in $\delta^{18}O$ changes of <0.5. Table 6 presents a summary of the calculated variations in $\delta^{18}O$ of a melt dictated by crystal fractionation only. The oxygen isotope variations traced through the three stages define a near-horizontal crystal fractionation line on the $\delta^{18}O$ versus $SiO_2$ diagram (Fig. 7). Open-system processes produce deviant $\delta^{18}O$-$SiO_2$ arrays compared with the near-horizontal crystal fractionation trend when the contaminant has an isotopic signature distinct from the melt (Harmon and Gerbe, 1992). Mafic to intermediate magmas that assimilate $^{18}O$-rich crust during differentiation form steep, positive $\delta^{18}O$-$SiO_2$ trends (Davidson and Harmon, 1989; Ellam and Harmon, 1990), whereas assimilation of $^{18}O$-depleted crustal rocks modified by meteoric hydrothermal alteration yield negative $\delta^{18}O$-$SiO_2$ arrays (Grunder, 1987; Harmon and Gerbe, 1992). The positive array with a Spearman rank correlation coefficient of +0.738 defined by the Guatemalan arc lavas (Fig. 7) favors involvement of an enriched $^{18}O$ contaminant. The groundmass of lava flows readily exchange oxygen during low temperature weathering or interaction with hydrothermal fluids, and conceivably could account for some of the positive shift in the $\delta^{18}O$ data. The random nature of alteration argues against it being a significant variable in light of the respectable correlation between $\delta^{18}O$ and $SiO_2$. Trace element mobility has been documented in spheroidally weathered Guatemalan lavas (Patino et al., 2003), but this study avoided visibly weathered samples and collected only the freshest samples from the volcanoes in question. Ideally, complementary measurements of the oxygen isotopic composition of refractory minerals such as olivine or clinopyroxene by laser fluorination or ion microprobe techniques would more accurately determine magmatic values (Singer et al., 1992; Macpherson and Mattey, 1998; Eiler et al., 1998, 2000; Dorendorf et al., 2000; Vroon et al., 2001). Correction procedures based on assumptions of primary $H_2O$ contents generally result in small adjustments of $\delta^{18}O$ values. Lavas from Martinique in the Lesser Antilles typically had measured $\delta^{18}O$ reduced by <0.5‰ (Davidson and Harmon, 1989). Owing to the young age and fresh nature of the Guatemalan lavas in thin section and the small modifications attributed to water uptake, crustal contamination plays a more prominent role in enriching the magmas in $\delta^{18}O$.

## Source versus Crustal Contamination

Oxygen isotopic compositions in combination with radiogenic isotopes can distinguish, in principle, between crustal and source contamination (James, 1981). Sr/O ratios of the mantle and crustal end members define mixing trajectories on a plot of $\delta^{18}O$ versus $^{87}Sr/^{86}Sr$ (Fig. 12). The high Sr content of mafic magmas compared to that of the potential crustal contaminants accounts for a convex crustal contamination mixing curve. Conversely, the low Sr content of the bulk mantle relative to subducted crustal material accounts for a concave source contamination trend. Partial melting of mantle peridotite produces primary magma enriched in incompatible elements such as Sr and Nd relative to lithologies of the continental crust. In sharp contrast, the bulk mantle has low incompatible element contents relative to enriched sediments and thus represents the depleted end member in the source contamination case. These differing chemical characteristics produce convex trajectories for crustally contaminated arc magmas and concave trends after recycling sediments into a MORB-like subarc mantle source.

Whole-rock oxygen isotopic data have poor but positive correlations with the Sr isotopic composition of the Guatemalan

TABLE 6. CALCULATED VARIATIONS IN $\delta^{18}O$ OF A MELT CONTROLLED BY CRYSTAL FRACTIONATION

| Model | $\delta^{18}O$ Values | | Mean $\Delta$ (‰) | $\Delta = \delta^{18}O_{xl} - \delta^{18}O_{magma}$ | | | |
|---|---|---|---|---|---|---|---|
| | Parent | Daughter | | Plag | Cpx | Oliv | Mt |
| Stage 1 | 6.21 | 6.23 | −0.06 | 0.2 | −0.2 | −0.5 | 0 |
| Stage 2 | 6.23 | 6.31 | −0.121 | 0.2 | −0.2 | −0.5 | −1.9 |
| Stage 3 | 6.31 | 6.53 | −0.492 | 0 | −1 | −1.8 | −3.6 |

arc lavas. Mixing curves can be calculated using the binary mixing equation:

$$R^X_M = [R^X_A X_A f + R^X_B X_B (1-f)]/[X_A f + X_B (1-f)], \quad (3)$$

where $R^X_M$ is the isotope ratio of Sr or O in a mixture of components $A$ and $B$, $X_A$ and $X_B$ are concentrations of Sr or O in $A$ and $B$, and $f$ is the weight fraction of $A$. Each curve was calculated using a different ratio of Sr in the mantle versus Sr in the crust (i.e., $Sr_M:Sr_C$). Five different concentrations (100, 40, 20, 10, and 2 ppm) were assumed for the mantle, whereas the enriched end member was taken to have a Sr concentration of 20 ppm. The model calculations assume equal concentrations of O in the two end members but different $\delta^{18}O$ values of 6.0‰ for the mantle and 12.0‰ for the crust. Varying values of $f$ and substitution of these concentrations into the general two-component binary mixing equation generates the five separate curves. The data plot within or in close proximity to the field representing magmas contaminated by continental crust (Fig. 12A). The spread of the data suggests that between 5% and 35% crustal material is added to the parental arc magma.

A more detailed, two-stage model can be constructed based on the distribution of data in the simple model using more geochemically relevant parameters. Two-stage models have been successful in explaining Sr and O isotopic relationships in the Marianas (Woodhead et al., 1987) and the Aeolian Islands (Ellam and Harmon, 1990). Source contamination shifts the data away from the mantle source in the direction of higher $^{87}Sr/^{86}Sr$, whereas crustal contamination profoundly increases the $\delta^{18}O$ contents of the magma compared to only moderate increases in $^{87}Sr/^{86}Sr$. Again, oxygen concentrations were assumed to be constant in all components. The mantle Sr concentration of 15 ppm (Chen and Frey, 1985), $^{87}Sr/^{86}Sr$ ratio of 0.7027 (Zindler and Hart, 1986), and $\delta^{18}O$ of 5.7 (Woodhead et al., 1987) characterize a MORB source. The mean hemipelagic sediment from Deep Sea Drilling Project (DSDP) Site 495 off Guatemala has ~300 ppm Sr and a $^{87}Sr/^{86}Sr$ ratio of 0.708 (Patino, 1997). The sediments from DSDP 495 were not analyzed for $\delta^{18}O$. Unaltered hemipelagic sediment from Middle Valley in the northern Pacific Ocean has $\delta^{18}O$ values that approach 12.0‰ (Goodfellow et al., 1993). Extrapolation of the Guatemalan Sr-SiO$_2$ trend back to a SiO$_2$ content of 46% approximates the composition of a hypothetical primitive arc magma in Guatemala and yields the model Sr abundance for the magma of 750 ppm. The magma $^{87}Sr/^{86}Sr$ ratio of 0.7032 and $\delta^{18}O$ value of 5.73 were products of the Stage 1 model calculation of 0.5% sediment contamination in the source. Finally, quartzite in southeastern Guatemala has Sr contents of ~30 ppm, a $^{87}Sr/^{86}Sr$ ratio of 0.740, and a $\delta^{18}O$ value of 14.0‰ (Walker et al., 1995; this study). The quartzite sample was incorporated into the model simply because of the altered nature and unrealistically low $\delta^{18}O$ value of 3.71‰ determined for the granodiorite sample. The high $\delta^{18}O$ value of the quartzite used in this model would approximate other more likely crustal contaminants.

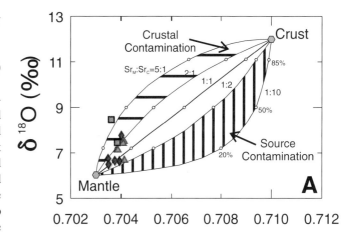

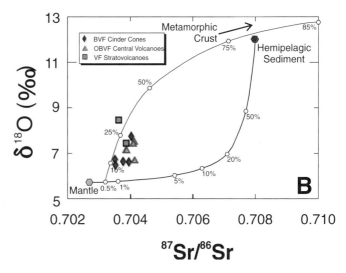

Figure 12. (A) $\delta^{18}O$ versus $^{87}Sr/^{86}Sr$ diagram (James, 1981). The Guatemalan arc lavas plot within the field of magmas contaminated by continental crust as opposed to mantle contaminated by sediments. (B) Diagram of $\delta^{18}O$ versus $^{87}Sr/^{86}Sr$ showing a two-stage evolution model for arc lavas from southeastern Guatemala. Crustal contamination of primary magmas by metamorphic crustal samples follows limited contamination of the mantle by hemipelagic sediments. A series of mixing hyperbolas between sediment-modified mantle and crustal contaminants would best fit the spread of data points. BVF—behind the volcanic front; OBVF—older behind the volcanic front; VF—volcanic front.

The two-stage model illustrated in Figure 12B represents a more realistic explanation of the contamination history in Guatemala. A combination of between 0.5% and 2.0% hemipelagic sediment addition to a MORB-source mantle and subsequent mixing of magma with between 10% and 30% metamorphic crustal material could explain the Sr-O isotopic data. The proposed two-stage process conforms to the general conceptual model for arc magmatism in which material transfer at the slab-mantle wedge interface induces partial melting and ascending magmas digest crustal materials during stagnation in reservoirs at various depths.

## Effects of Crustal Contamination on Elemental Concentrations

Plots of various elements versus $\delta^{18}O$ should convey information about elemental inputs to magmas during crustal contamination assuming that assimilation accounts for increases in $\delta^{18}O$. Two statistical quantities contribute valuable insight into the nature of the relationship between the two participating variables. The Spearman rank correlation coefficient $r$ essentially measures how well the data fit a straight line. A high value of $r$ represents a high degree of correlation. The threshold $r$ value that indicates a statistically significant correlation depends on the number of samples. On the other hand, the slope of the linear regression line relates to the dependence or independence of one variable relative to the other. Simple examples illustrate their interpretative value. A data set with a high $r$ value near one and a slope near zero suggests an excellent correlation, but also that $y$ is independent of $x$. A second data set with $r = 0.75$ and $m = 1$ indicates a moderate to good correlation and that $y$ depends on $x$, or in geochemical phrasing, $y$ and $x$ behave similarly. In the current discussion, $\delta^{18}O$ is used to evaluate the effect of crustal contamination on critical incompatible elements. Spearman rank correlation coefficients screen combinations for statistical significance, whereas slope measures the magnitude of the behavioral similarity.

Overall, incompatible element versus $\delta^{18}O$ plots feature relatively high Spearman rank correlation coefficients. The key threshold value for correlation coefficients with a 0.1% level of significance are 0.618 at $n = 23$ (LeMaitre, 1982). Be, Sb, and La have correlation coefficients below the threshold value and consequently appear unrelated to $\delta^{18}O$. Compilation of the slope data suggests that incompatible elements exhibit a complete spectrum of coherency with $\delta^{18}O$ directly and crustal contamination indirectly (Fig. 13). Crustal contamination more significantly affects those incompatible elements with a slope of ~1 or greater (i.e., B, La, Pb, Li, and Th). Assimilation of crust exerts less control on incompatible element concentrations of Rb, Sb, Ba, and Be based on slopes <0.3. An intermediate group of elements with slopes between 0.3 and 0.9 consists of Cs and U. Thus, the sequential order of the incompatible elements on the $x$-axis from left to right in Figure 13 serves as a relative measure of the chemical influence of crustal contamination.

Igneous petrologists rely heavily on incompatible trace element ratios to provide important petrogenetic information. Common differentiation processes such as crystal fractionation and partial melting have little effect on ratios involving trace elements with similar degrees of incompatibility. Thus, any information on the nature of elemental modification by crustal contamination helps petrologists make more intelligent decisions when assembling these key trace element ratios. Crustal contamination will influence a ratio formed from elements with disparate slopes (Fig. 13). B/Be ratios in Guatemalan arc lavas correlate positively with $SiO_2$ and trend toward but do not exceed the field of crustal rocks (Fig. 14A). Many studies of continental arc lavas indiscriminately use B/Be as a measure of the slab signature without thorough assessment of the control exercised by crustal contamination. The effects of crustal contamination will be canceled out in ratios involving elements with similar slopes. Not only does a poor correlation exist between Ba/La versus $SiO_2$, but many of the lavas possess higher Ba/La ratios than the analyzed crustal rocks (Fig. 14B). Accordingly, Ba/La monitors slab signatures much better than a ratio such as B/Be in an arc with thick continental crust.

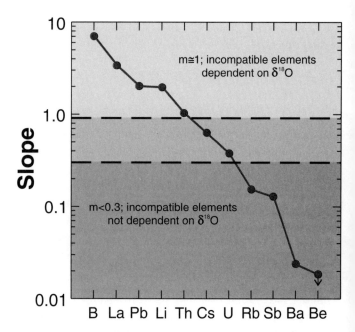

Figure 13. Magnitude of the slope for select trace elements and $\delta^{18}O$. Given that slope measures the interdependency of variables on variation diagrams, elements with slopes greater than 0.9 have been most affected by crustal contamination.

## Edifice Style and Crustal Contamination

Porphyritic lavas with low-pressure phase assemblages develop at the large stratovolcanoes of the VF owing to extensive crystallization in shallow magma chambers. Analogous rocks with similar phenocryst assemblages occur at OBVF central volcanoes. In sharp contrast, the mafic BVF cinder cone lavas have aphyric textures and were produced by small volume eruptions associated with the extensional regime of the Ipala Graben. These diffuse vents often coincide with systematic lineaments interpreted as deep crustal structures. The whole-rock oxygen isotope data support this volcanological scenario. The highly elevated $\delta^{18}O$ values measured at the VF stratovolcanoes and OBVF central volcanoes result from crustal assimilation accompanying fractional crystallization in high-level magma chambers. Erupted lavas have common low-pressure phenocryst phases and evolved whole-rock compositions altered by the assimilation of crustal contaminants. The slightly elevated $\delta^{18}O$ values observed in the BVF cinder cone lavas indicate that the magmas ascended

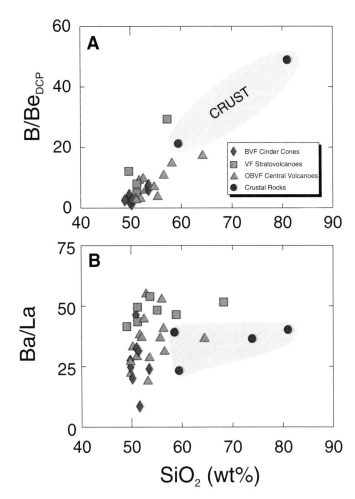

Figure 14. Critical trace element ratios versus $SiO_2$. (A) B/Be versus $SiO_2$. Guatemalan arc lavas essentially plot on a mixing line between mantle and crust. (B) Ba/La versus $SiO_2$. Ba/La in Guatemalan arc lavas exceeds that recorded in possible crustal contaminants. Ba/La ratio was less affected by crustal contamination than B/Be. BVF—behind the volcanic front; OBVF—older behind the volcanic front; VF—volcanic front; DCP—direct current plasma.

more rapidly from deep crustal magma chambers underlying the extended crust. These eruptive products have more mafic compositions, an aphyric to sparsely olivine-phyric texture, and little chemical crustal signature. Short residence times in shallow reservoirs suggest that the small enrichment in $\delta^{18}O$ probably records deeper crustal contamination processes. Unusual trends on Pb isotope plots for the BVF cinder cones identified granulitic crust as a possible contaminant (Walker et al., 1995). Curiously, cinder cone lavas in Mexico seemed vulnerable to crustal assimilation based on their B/Be-$SiO_2$ relationships (Hochstaedter et al., 1996). Small batches of magma from monogenetic centers in Mexico traversed separate sections of crust, and therefore may be more susceptible to crustal contamination. A supplementary oxygen isotope study in the Mexican volcanic belt might help resolve this discrepancy.

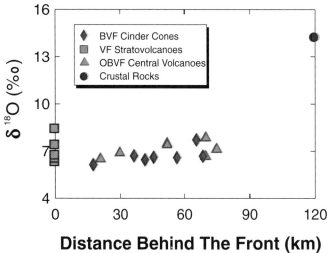

Figure 15. $\delta^{18}O$ versus distance behind the volcanic front in southeastern Guatemala. $\delta^{18}O$ remains essentially constant at behind the volcanic front (BVF) cinder cones across the arc, implying no absolute increase in crustal contamination. BVF cinder cones have $\delta^{18}O$ values only slightly above normal mid-oceanic ridge basalt. OBVF—older behind the volcanic front; VF—volcanic front.

## Across-Arc Changes in $\delta^{18}O$

Nd and Pb isotopic ratios in BVF cinder cones display relatively consistent and continuous across-arc changes: both $^{207}Pb/^{204}Pb$ and $^{208}Pb/^{204}Pb$ generally increase, whereas $^{143}Nd/^{144}Nd$ declines with distance behind the front (Walker et al., 1995). Either an absolute or apparent increase in crustal contamination across the arc can explain the regular radiogenic isotopic variations. Walker et al. (1995) discounted an absolute increase in the amount of contamination as unreasonable owing to the likelihood of rapid magma ascent caused by tectonic extension behind the front. Instead, assimilation of older, more isotopically evolved crust farther behind the front only creates the illusion of increasing extents of crustal contamination. Oxygen-isotopic compositions do not evolve with age like the radiogenic isotopes, and therefore offer another means of evaluating this question. The $\delta^{18}O$ values of the BVF cinder cones remain relatively constant across the arc (Fig. 15), ruling out the possibility that the overall amount of contamination increases behind the front. The highest $\delta^{18}O$ values occur at VF stratovolcanoes and at OBVF central volcanoes.

## CONCLUSIONS

Geochemical studies of Guatemalan arc lavas demonstrate the importance of open system processes in modifying mantle melts. Mineral chemistry varies nonsystematically with whole-rock compositions and together with disequilibrium core compositions in olivine, pyroxene, and plagioclase indicate that the crustal magma chambers behave as an open system. Whole-rock oxygen isotope data are compatible with crustal contamination being an important open system process in southeastern Guatemala. Crystal fractionation cannot explain the large ranges in $\delta^{18}O$ measured in the erupted lavas. Rather, steep arrays on the $\delta^{18}O$-$SiO_2$ diagram favor incorporation of $^{18}O$-rich crustal rocks during ascent of primitive magmas. Enrichments in $\delta^{18}O$ imposed by low-temperature weathering or alteration likely prove insignificant compared to the effects of crustal contamination. Combined Sr and O isotope models indicate minor additions of 0.5%–1% sediment to the mantle wedge followed by variable degrees of crustal contamination between 10% and 30%. Lower $\delta^{18}O$ values from BVF cinder cones compared to the stratovolcanoes of the VF and OBVF suggest that relatively lower amounts of crustal contamination accompany smaller scale, deeper magmatic systems. Extension associated with the Ipala Graben promoted short crustal residence times for the mafic cinder cone magmas and contributed to the lower $\delta^{18}O$ values. Variations in $\delta^{18}O$ better reflect the magnitude of the crustal contamination process than the age-controlled radiogenic isotopes. Open system processes such as crustal contamination and magma mixing might control eruption style, leading to the construction of different edifice types. More significant volcanic hazards would therefore be associated with the eruption of VF stratovolcanoes and OBVF central volcanoes.

## ACKNOWLEDGMENTS

The authors wish to thank Ian Richards for help collecting the oxygen isotopic data at Southern Methodist University. Jeff Ryan provided access to his direct current plasma laboratory at University of South Florida for the light element chemistry. Ian Steele assisted with the mineral chemical data at the University of Chicago. Mark Feigenson at Rutgers opened his lab to collect the Sr isotopic data. Detailed and constructive reviews by Richard Conrey, Todd Feeley, and Gregg Bluth were extremely helpful in improving the manuscript.

## REFERENCES CITED

Cameron, B.I., Walker, J.A., Carr, M.J., Patino, L.C., Matías, O., and Feigenson, M.D., 2003, Flux versus decompression melting at stratovolcanoes in southeastern Guatemala: Journal of Volcanology and Geothermal Research, v. 119, p. 21–50, doi: 10.1016/S0377-0273(02)00304-9.

Carr, M.J., 1984, Symmetrical and segmented variation of physical and geochemical characteristics of the Central American volcanic front: Journal of Volcanology and Geothermal Research, v. 20, p. 231–252, doi: 10.1016/0377-0273(84)90041-6.

Carr, M.J., Feigenson, M.D., and Bennett, E.A., 1990, Incompatible element and isotopic evidence for tectonic control of source mixing and melt extraction along the Central American arc: Contribution to Mineralogy and Petrology, v. 105, p. 369–380, doi: 10.1007/BF00286825.

Chen, C.Y., and Frey, F.A., 1985, Trace element and isotopic geochemistry of lavas from Haleakala volcano, east Maui, Hawaii: Implications for the origin of Hawaiian basalts: Journal of Geophysical Research, v. 90, p. 8743–8768.

Chivas, A.R., Andrew, A.S., Sinha, A.K., and O'Neil, J.R., 1982, Geochemistry of a Pliocene-Pleistocene oceanic-arc plutonic complex, Guadalcanal: Nature, v. 300, p. 139–143, doi: 10.1038/300139a0.

Davidson, J.P., and Harmon, R.S., 1989, Oxygen isotope constraints on the petrogenesis of volcanic arc magmas from Martinique, Lesser Antilles: Earth and Planetary Science Letters, v. 95, p. 255–270, doi: 10.1016/0012-821X(89)90101-5.

Davidson, J.P., and de Silva, S., 1995, Late Cenozoic magmatism of the Bolivian Altiplano: Contributions to Mineralogy and Petrology, v. 119, p. 387–408.

Davidson, J.P., McMillan, N.J., Moorbath, S., Worner, G., Harmon, R.S., and Lopez-Escobar, L., 1990, The Nevados de Pachachata volcanic region (18 S/69 W, N. Chile) II. Evidence for widespread crustal involvement in Andean magmatism: Contributions to Mineralogy and Petrology, v. 105, p. 412–432, doi: 10.1007/BF00286829.

Dorendorf, F., Wiechert, U., and Wörner, G., 2000, Hydrated sub-arc mantle: a source for the Kluchevskoy volcano, Kamchatka/Russia: Earth and Planetary Science Letters, v. 175, p. 69–86, doi: 10.1016/S0012-821X(99)00288-5.

Eiler, J.M., McInnes, B., Valley, J.W., Graham, C.M., and Stolper, E.M., 1998, Oxygen isotope evidence for slab-derived fluids in the sub-arc mantle: Nature, v. 393, p. 777–781, doi: 10.1038/31679.

Eiler, J.M., Crawford, A., Elliot, T., Farley, K.A., Valley, J.W., and Stolper, E.M., 2000, Oxygen isotope geochemistry of oceanic-arc lavas: Journal of Petrology, v. 41, p. 229–256, doi: 10.1093/petrology/41.2.229.

Ellam, R.M., and Harmon, R.S., 1990, Oxygen isotope constraints on the crustal contribution to the subduction-related magmatism of the Aeolian Islands, southern Italy: Journal of Volcanology and Geothermal Research, v. 44, p. 105–122, doi: 10.1016/0377-0273(90)90014-7.

Fallon, T.J., Green, D.H., O'Neill, H., St, C., and Hibberson, W.O., 1997, Experimental tests of low degree peridotite partial melt compositions: Implications for the nature of anhydrous near-solidus peridotite melts at 1 GPa: Earth and Planetary Science Letters, v. 152, p. 149–162, doi: 10.1016/S0012-821X(97)00155-6.

Feeley, T.C., and Sharp, Z.D., 1995, $^{18}O$/$^{16}O$ isotope geochemistry of silicic lava flows erupted from Volcan Ollague, Andean Central Volcanic Zone: Earth and Planetary Science Letters, v. 133, p. 239–254, doi: 10.1016/0012-821X(95)00094-S.

Feigenson, M.D., Carr, M.J., Maharaj, S.V., Juliano, S., and Bolge, L.L., 2004, Lead isotope composition of Central American volcanoes: Influence of the Galapagos plume: Geochemistry, Geophysics and Geosystems, v. 5, 14 p., doi: 10.1029/2003GC000621.

Gerlach, D.C., and Grove, T.L., 1982, Petrology of Medicine Lake Highland Volcanics: Characterization of end members of magma mixing: Contributions to Mineralogy and Petrology, v. 80, p. 147–159.

Goodfellow, W.D., Grapes, K., Cameron, B.I., and Franklin, J.M., 1993, Hydrothermal alteration associated with massive sulfide deposits, Middle Valley, northern Juan de Fuca Ridge: Canadian Mineralogist, v. 31, p. 1025–1060.

Grove, T.L., and Bryan, W.B., 1983, Fractionation of pyroxene-phyric MORB at low pressure: An experimental study: Contributions to Mineralogy and Petrology, v. 84, p. 293–309, doi: 10.1007/BF01160283.

Grunder, A.L., 1987, Low $^{18}O$ silicic volcanic rocks at the Calabozos caldera complex, southern Andes: Contributions to Mineralogy and Petrology, v. 95, p. 71–81, doi: 10.1007/BF00518031.

Harmon, R.S., and Gerbe, M.-C., 1992, The 1982–83 eruption at Galunggung volcano, Java (Indonesia): Oxygen isotope geochemistry of a chemically zoned magma chamber: Journal of Petrology, v. 33, p. 585–609.

Harmon, R.S., and Hoefs, J., 1995, Oxygen isotope heterogeneity of the mantle deduced from global $^{18}O$ systematics of basalts from different geotectonic settings: Contributions to Mineralogy and Petrology, v. 120, p. 95–114.

Hildreth, W., and Moorbath, S., 1988, Crustal contributions to arc magmatism in the Andes of Central Chile: Contributions to Mineralogy and Petrology, v. 98, p. 455–489, doi: 10.1007/BF00372365.

Hochstaedter, A.G., Ryan, J.G., Luhr, J.F., and Hasenaka, T., 1996, On B/Be ratios in the Mexican Volcanic Belt: Geochimica et Cosmochimica Acta, v. 60, p. 613–628, doi: 10.1016/0016-7037(95)00415-7.

Hunter, A.G., and Blake, S., 1995, Petrogenetic evolution of a transitional tholeiitic-calc-alkaline series: Towada volcano, Japan: Journal of Petrology, v. 36, p. 1579–1605.

James, D.E., 1981, The combined use of oxygen and radiogenic isotopes as indicators of crustal contamination: Annual Reviews of Earth and Planetary Science, v. 9, p. 311–344, doi: 10.1146/annurev.ea.09.050181.001523.

Kalamarides, R.I., 1984, Kiglapait geochemistry, VI. Oxygen isotopes: Geochimica et Cosmochimica Acta, v. 48, p. 1827–1836, doi: 10.1016/0016-7037(84)90036-X.

Kalamarides, R.I., 1986, High-temperature oxygen isotope fractionation among the phases of the Kiglapait intrusion, Labrador, Canada: Chemical Geology, v. 58, p. 303–310.

Kersting, A.B., Arculus, R.J., and Gust, D.A., 1996, Lithospheric contributions to arc magmatism: Isotope variations along strike in volcanoes of Honshu, Japan: Science, v. 272, p. 1464–1468.

LeMaitre, R.W., 1982, Numerical Petrology, Statistical Interpretation of Geochemical Data. Developments in Petrology 8: Amsterdam, Elsevier Scientific Publishing Company, 281 p.

Leeman, W.P., Carr, M.J., and Morris, J.D., 1994, Boron geochemistry of the Central American arc: Constraints on the genesis of subduction-related magmas: Geochimica et Cosmochimica Acta, v. 58, p. 149–168, doi: 10.1016/0016-7037(94)90453-7.

Macpherson, C.G., and Mattey, D.P., 1998, Oxygen isotope variations in Lau Basin lavas: Chemical Geology, v. 144, p. 177–194, doi: 10.1016/S0009-2541(97)00130-7.

Morris, J.D., Leeman, W.P., and Tera, F., 1990, The subducted component in island arc lavas: Constraints from Be isotopes and B/Be systematics: Nature, v. 344, p. 31–36.

Muehlenbachs, K., and Byerly, G., 1982, $^{18}$O-enrichment of silicic magmas caused by crystal fractionation at the Galapagos spreading center: Contributions to Mineralogy and Petrology, v. 79, p. 76–79, doi: 10.1007/BF00376963.

Patino, L.C., 1997. Geochemical characterization of Central American subduction zone system: Slab input, output from the volcanoes, and slab-mantle interactions [Ph.D. thesis]: Piscataway, New Jersey, Rutgers University, 183 p.

Patino, L.C., Carr, M.J., and Feigenson, M.D., 1997, Cross-arc geochemical variations in volcanic fields in Honduras, C.A.: Progressive changes in source with distance from the volcanic front: Contributions to Mineralogy and Petrology, v. 129, p. 341–351, doi: 10.1007/s004100050341.

Patino, L.C., Velbel, M.A., Price, J.R., and Wade, J.A., 2003, Trace element mobility during spheroidal weathering of basalts and andesites in Hawaii and Guatemala: Chemical Geology, v. 202, p. 343–364, doi: 10.1016/j.chemgeo.2003.01.002.

Pokrovskii, B.G., and Volynets, O.N., 1999, Oxygen isotope geochemistry of extrusive rocks of the Kurile-Kamchatka arc: Petrology, v. 7, p. 223–245.

Sakuyama, M., 1981, Petrological study of the Myoko and Kurohime volcanoes, Japan: crystallization sequence and evidence for magma mixing: Journal of Petrology, v. 22, p. 553–583.

Sheppard, S.M.F., and Harris, C., 1985, Hydrogen and oxygen isotope geochemistry of Ascension Island lavas and granites: Variation with crystal fractionation and interaction with sea water: Contributions to Mineralogy and Petrology, v. 91, p. 74–81, doi: 10.1007/BF00429429.

Singer, B.S., O'Neil, J.R., and Brophy, J.G., 1992, Oxygen isotope constraints on the petrogenesis of Aleutian arc magmas: Geology, v. 20, p. 367–370, doi: 10.1130/0091-7613(1992)020<0367:OICOTP>2.3.CO;2.

Sisson, T.W., and Grove, T.L., 1993, Experimental investigations of the role of $H_2O$ in calc-alkaline differentiation and subduction zone magmatism: Contributions to Mineralogy and Petrology, v. 113, p. 143–166, doi: 10.1007/BF00283225.

Smith, T.E., Thirlwall, M.F., and Macpherson, C.M., 1996, Trace element and isotope geochemistry of the volcanic rocks of Bequia, Grenadine Islands, Lesser Antilles arc: A study of subduction enrichment and intra-crustal contamination: Journal of Petrology, v. 37, p. 117–143.

Vroon, P.Z., Lowry, D., van Bergen, M.J., Boyce, A.J., and Mattey, D.P., 2001, Oxygen isotope systematics of the Banda arc; low $\delta^{18}O$ despite involvement of subducted continental material in magma genesis: Geochimica et Cosmochimica Acta, v. 65, p. 589–609, doi: 10.1016/S0016-7037(00)00554-8.

Walker, J.A., 1981, Petrogenesis of lavas from cinder cone fields behind the volcanic front of Central America: Journal of Geology, v. 89, p. 721–739.

Walker, J.A., Carr, M.J., Patino, L.C., Johnson, C.M., Feigenson, M.D., and Ward, R.L., 1995, Abrupt change in magma generation processes across the Central American arc in southeastern Guatemala: Flux-dominated melting near the base of the wedge to decompression melting near the top of the wedge: Contributions to Mineralogy and Petrology, v. 120, p. 378–390.

Walker, J.A., Roggensack, K., Patino, L.C., Cameron, B.I., and Matías, O., 2003, The water and trace element contents of melt inclusions across an active subduction zone: Contributions to Mineralogy and Petrology, v. 146, p. 62–77, doi: 10.1007/s00410-003-0482-x.

Woodhead, J.D., Harmon, R.S., and Fraser, D.G., 1987, O, S, Sr, and Pb isotope variations in volcanic rocks from the northern Mariana Islands: Implications for crustal recycling in intra-oceanic arcs: Earth and Planetary Science Letters, v. 83, p. 39–52, doi: 10.1016/0012-821X(87)90049-5.

Zindler, A., and Hart, S., 1986, Chemical geodynamics: Annual Reviews of Earth and Planetary Science, v. 14, p. 493–571, doi: 10.1146/annurev.ea.14.050186.002425.

MANUSCRIPT ACCEPTED BY THE SOCIETY 19 MARCH 2006

# Volcanic hazards in Nicaragua: Past, present, and future

**Armin Freundt***

*SFB 574 at Kiel University, Wischhofstr. 1-3, D-24148 Kiel, Germany, and IFM-GEOMAR (Leibniz Institute of Marine Sciences), Wischhofstr. 1-3, D-24148 Kiel, Germany*

**Steffen Kutterolf**

*SFB 574 at Kiel University, Wischhofstr. 1-3, D-24148 Kiel, Germany*

**Hans-Ulrich Schmincke**
**Thor Hansteen**

*SFB 574 at Kiel University, Wischhofstr. 1-3, D-24148 Kiel, Germany, and IFM-GEOMAR (Leibniz Institute of Marine Sciences), Wischhofstr. 1-3, D-24148 Kiel, Germany*

**Heidi Wehrmann**
**Wendy Pérez**

*SFB 574 at Kiel University, Wischhofstr. 1-3, D-24148 Kiel, Germany*

**Wilfried Strauch**
**Martha Navarro**

*Instituto Nicaragüense de Estudios Territoriales (INETER), Apdo. 2110, Managua, Nicaragua*

## ABSTRACT

We review the most important types of volcanic hazards that have occurred in Nicaragua during the past ~40,000 yr and that are expected to occur in the future. Population density within the potential hazard area is clearly essential in defining and understanding volcanic hazard and risk. There are three main groups of volcanic events that pose major hazards: Group 1 comprises several types of explosive volcanic eruptions that impact society (people and infrastructure) directly. The most hazardous types are pyroclastic surges, particularly those generated by water-magma interaction, pyroclastic fallout, and pyroclastic flows, as well as tsunamis generated by volcanic eruptions within and close to Nicaragua's large lakes. Group 2 includes nonexplosive volcanic activity such as lava flows and the permanent or episodic emission of volcanic gases from open vents. Group 3 comprises chiefly lahars generated by mixing of volcanic debris with water and volcano flank collapses (landslides) sometimes unrelated to synchronous volcanic eruptions but being conditioned chiefly by the stability of a volcanic edifice. We discuss the present database on the age and type of the most recent eruptions emphasizing those that potentially pose major hazards to the populated areas. These include volcanogenic tsunamis in Lake Managua and

---

*E-mail: afreundt@ifm-geomar.de

Lake Nicaragua, scoria cone and maar formation chiefly in the western part of Managua, and major explosive eruptions of Chiltepe and Masaya volcanoes, a large eruption from Masaya volcano having devastated the entire area of present Managua only ~2000 yr ago. We discuss the most important techniques for monitoring volcanoes to detect unrest and predict the time and magnitude of upcoming eruptions, emphasizing techniques presently employed in Nicaragua. Finally, we address the subjects of risk assessment, including hazard and risk maps, and the importance of long-term development plans to reduce vulnerability.

**Keywords:** debris flows, eruption forecasting, monitoring, tephrostratigraphy, tsunamis.

## INTRODUCTION

The subduction of the oceanic Cocos plate beneath Nicaragua (Fig. 1) is the common cause of the earthquakes, tsunamis, and volcanic eruptions that have repeatedly affected the country in the past (Carr et al., 1982; Carr, 1984; Protti et al., 1995; Ranero et al., 2000; Stoiber and Carr, 1973; Walker et al., 2001; Walther et al., 2000). The history of Nicaraguan arc volcanism has been studied by McBirney and Williams (1965), Weyl (1980), and Bice (1985). Carr et al. (1982) and Carr (1984) related the magmatic evolution of the arc to the structure and composition of the subducted slab. Weinberg (1992) studied the relationship between volcanism and subduction-related tectonics.

Every few years, one or more of the roughly 10 active volcanoes erupt along the volcanic front in Nicaragua. Major eruptions, or collapse of volcano flanks unrelated to synchronous eruptions, may also occur at volcanoes that are not presently active. Loss of life through volcanic activity has been rare in Nicaragua in the brief span of recorded history (~400 yr). Nevertheless, many volcanic eruptions that occurred during the past 6000 yr would be disastrous if they were to occur today. We emphasize that the frequency of potentially disastrous large volcanic eruptions is more than one per 1000 yr. Major disruption is inevitable in inhabited areas close to active volcanoes, especially where various kinds of volcanic flows are potentially major types of eruptive activity. The main factors determining volcanic risk in Nicaragua include:

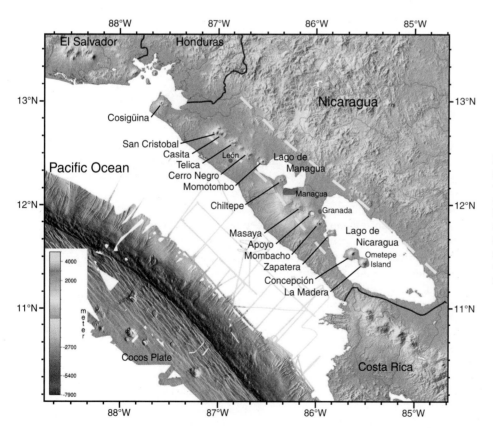

Figure 1. Map of western Nicaragua showing lakes Managua and Nicaragua within the NW-SE trending Nicaraguan depression (yellow dashed lines; Weinberg 1992), the deep-sea trench (violet colors) where the oceanic Cocos plate subducts beneath Central America, and the arc volcanoes parallel to, and ~200 km from, the trench. Volcanoes mentioned in the text are indicated.

- Forty percent of Nicaragua's population lives within, and close to, areas of many active volcanoes.
- The low elevation of the Nicaraguan depression favors extremely common interaction between rising magma and (ground)water. This can generate especially explosive and thus hazardous eruptions.
- Large Lake Managua, which forms the northern boundary of the city of Managua with several eruptive centers lying within the lake, as well as Lake Nicaragua to the south with the volcanoes Mombacho, Concepción, and Maderas, may be sites of highly explosive eruptions and tsunamis.
- The large spectrum of locations, magnitudes, and mechanisms of possible future eruptions presents a challenge for emergency planning. This is even more of a challenge because Nicaragua has also been victim to other disastrous natural events. Examples are the earthquakes that destroyed the capital, Managua, in 1931 and 1972, a tsunami that devastated the Pacific coast in 1992, and major storms, such as Hurricane Mitch in 1998, which triggered the disastrous lahar at Casita volcano.

Sometimes the eruption magnitude (e.g., the volume of material erupted) is equated with the degree of hazard or risk—in other words, the larger the eruption, the greater the hazard or risk. This correlation commonly does not hold, however, since other factors are more important. Increasing investment in communication, power lines, pipelines, transportation lines, etc., as well as the increasing complexity of such networks, have greatly increased vulnerability. The rising population density close to active volcanoes, not only in sprawling urban areas such as Managua and the nearby regions, but also in the agricultural areas in the Nicaraguan depression (Fig. 1), is a particularly acute factor of rising vulnerability. Volcanic and other types of natural disasters are certain to increase in the future because the degree of vulnerability of Nicaragua, as well as of societies worldwide, is rising rapidly.

In this review, we describe major types of volcanic hazards in general terms and illustrate each by a brief summary of case histories of well-studied volcanic events in Nicaragua and their impact on the environment. Discussions of large eruptions include our results from on-land and off-shore mapping of widespread pyroclastic deposits erupted during the past ~40,000 yr in west-central Nicaragua. Some results are included in Kutterolf et al. (2006), Freundt et al. (2006), Wehrmann et al. (this volume), and Pérez and Freundt (this volume). We group the volcanic hazards into three categories: (1) hazards from explosive eruptions; (2) hazards from nonexplosive activity; and (3) hazards from mass flows at volcanoes. We conclude by discussing several aspects of volcanic risk mitigation.

## VOLCANOES OF NICARAGUA

We will discuss types of volcanic eruptions and their hazards only briefly since these are treated more fully in current textbooks (Francis and Oppenheimer, 2004; Schmincke, 2004). There are basically four main types of volcanoes in Nicaragua, defined by their morphology, internal structure, eruption mechanism, and dominant magma composition.

### Stratocones

Most common and most impressive are large volcanic edifices called composite volcanoes or stratocones (Fig. 2B). They form the main backbone of the country from Cosigüina in the north to Concepción and Maderas on Ometepe island in the south. These volcanoes are complex in that they consist of intrusions, such as dikes and sills; extrusive lava domes and flows; several types of pyroclastic deposits (also called tephras), such as fallout, agglutinates, and deposits of pyroclastic density currents; and epiclastic deposits, including various types of breccias and mass-flow deposits, such as lahars (mud flows), extending far into the surrounding plains. The structural complexity of stratocones reflects their evolution by a wide range of eruptive mechanisms, intensities, and magnitudes, including highly explosive Plinian eruptions, and by volcanotectonic and mass-wasting processes. The dominant composition of these volcanoes is andesite. Studies of Nicaraguan stratocones include San Cristóbal (Hazlett, 1987) and Concepción (Borgia and van Wyk de Vries, 2003).

### Caldera Volcanoes

The second type of volcano forms broad, low edifices with shallow-dipping outer flanks and with a major, steep-walled subsided basin in the center called a caldera. Calderas are commonly filled with a lake and may be several kilometers in diameter, such as Apoyo (7 km in diameter, Fig. 2A; Sussman, 1985). Caldera volcanoes are the sites of particularly voluminous, highly explosive Plinian eruptions resulting in thick, widespread sheets of both pumice-fall deposits (Fig. 3A) and massive, ash-rich deposits of pyroclastic flows called ignimbrites. Eruptions of major pyroclastic flows are not common, but when they occur, they can cause widespread havoc because they are hot (often >500 °C), travel rapidly, and destroy everything in their paths. At subduction zones, the large-volume Plinian eruptions are commonly of silicic composition, mainly rhyolite and dacite. In Nicaragua, however, Masaya Caldera has repeatedly produced such large eruptions from mafic magmas (Williams 1983; Pérez and Freundt, this volume; Wehrmann et al., this volume).

### Scoria Cones

Scoria cones, rarely more than 200 m high, are especially common. They are generally of mafic composition, olivine-bearing basalt being most common. Scoria cones consist dominantly of fallout scoria blocks and lapilli and dikes and lava flows produced by Strombolian eruptions (Figs. 2D and 3B). Particularly vigorous Strombolian eruptions can form generally black layers of fallout lapilli and ash transported up to many kilometers from the scoria cone. The most prominent and active scoria cone in

Figure 2. (A) Apoyo caldera with Mombacho stratocone in the background. (B) Concepción stratocone. (C) 15–20-km-high eruption column advected sideways by the wind and producing a widespread fallout deposit. Kliuchevskoy, Kamchatka, 1994. U.S. National Aeronautics and Space Administration Space Shuttle STS-068 photograph. (D) Small Strombolian eruption of Cerro Negro in 1995. Photograph from Instituto Nicaragüense de Estudios Territoriales. (E) Asososca maar in western Managua with superimposed image of phreatomagmatic eruption ejecting dark jets of pyroclastic debris (from Fukutoku-Okanoba volcano 1986, Japan; Hydrographic and Oceanographic Dept., Japan coast guard, www1.kaiho.mlit.go.jp.). Note refinery near crater.

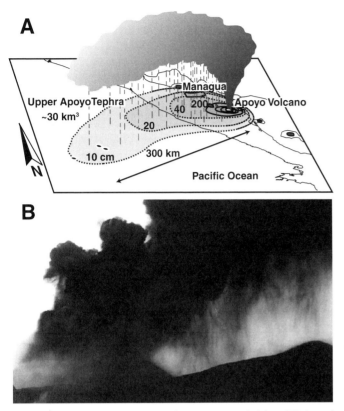

Figure 3. (A) Illustration of fallout from a westward-driven Plinian ash cloud from an eruption of Apoyo volcano ~25,000 yr ago, generating a tephra blanket covering an area of 50,000 km² with >10 cm thickness (Upper Apoyo Tephra). Contours (isopachs) of tephra thickness (cm) are indicated. (B) Ash raining out of the ash cloud of the 1995 Cerro Negro Strombolian eruption. Photograph from Instituto Nicaragüense de Estudios Territoriales, Managua.

Nicaragua is Cerro Negro (McKnight and Williams, 1997; Hill et al., 1998; La Femina et al., 2004). Young scoria cones also occur in western Managua and near Granada (Ui, 1972; Walker, 1984).

## Maar Volcanoes

When magma interacts with groundwater or surface water, the magma is quickly quenched, while the water is explosively vaporized, resulting in phreatomagmatic eruptions. The volcanic landform so generated is a maar, a crater with steep to vertical walls, generally <1 km in diameter, and often filled with a lake (Fig. 2E). A maar is surrounded by a ring of well-bedded deposits that may largely consist of fragments of country rock shattered by the explosive vaporization of the external water. Most maars form from basaltic magma, such as the prominent Asososca and Nejapa maars along the Nejapa-Miraflores alignment and Tiscapa maar in Managua city. Laguna Xiloá on Chiltepe peninsula is a maar that erupted dacitic magma. Masaya Caldera is a complex basaltic volcano in that it also comprises maar-type structures, its eruptions alternating between pyroclastic eruptions and lava flows and major phreatomagmatic eruptions (McBirney, 1956; Walker et al., 1993; Pérez and Freundt, this volume). Dominantly Plinian eruptions from stratocones or calderas often also include phreatomagmatic eruption phases.

## HAZARDS FROM EXPLOSIVE ERUPTIONS

The extent of hazards from volcanic eruptions depends on the eruption style and the mass of magma discharged. Explosive volcanic eruptions disperse pyroclastic material as fallout from eruption columns spreading with the wind and/or as deposits from pyroclastic density currents moving along the ground. Non-explosive eruptions produce basaltic lava flows (e.g., Masaya) or highly viscous dacitic lava domes (e.g., Santiaguito, Guatemala), which episodically collapse to form hazardous pyroclastic flows. Fortunately, such flows have not occurred at Nicaraguan volcanoes during their recent history.

### Pyroclastic Fallout

Most of the magma volume erupted during explosive volcanic eruptions in Nicaragua during the past ~40,000 yr was deposited as fallout from eruption columns that were injected ~15–40 km up into the atmosphere and then were transported laterally by the prevailing wind (Fig. 2C). Figure 3A illustrates the ash dispersal to >300 km during a Plinian eruption from Apoyo caldera 25,000 yr ago, the biggest eruption in Nicaragua's recent volcanic history. The area affected by fallout depends primarily on the height of the eruption column, which in turn depends on the thermal mass flux of the eruption. Compared to the Apoyo eruption, Cerro Negro volcano has produced numerous (at least 22) but small eruptions since its birth in 1850 (McKnight and Williams, 1997; Hill et al., 1998). The Strombolian eruptions of Cerro Negro formed lava flows, lava fountains of coarse spatter building the cone (Fig. 2D), and higher ash plumes, such as the one in Figure 3B of the 1995 eruption. Though relatively small, a similar but larger eruption in 1992 covered the city of León, some 20 km west of the volcano (Fig. 1), with an ash blanket 4 cm thick that caused some roofs to collapse. At least two people were killed and 146 injured; crops and infrastructure worth $19,000,000 were destroyed (Connor et al., 2001).

Wind directions change with height in the atmosphere and with the season of the year (Fig. 4). The main direction of fallout dispersal thus depends on the height of the eruption column and the season. Ash from either <5-km-high Strombolian or >20-km-high Plinian eruptions will most probably be dispersed westward from the volcano. Dispersal directions can be more variable for eruption columns of intermediate height, with westerly directions prevailing from June to September but northeasterly directions for the remaining year (Fig. 4B). Southerly regions a few kilometers from the vent (depending on eruption size) are least affected by fallout.

Here we are mainly concerned with Plinian eruptions that are characterized by large thermal mass flux and produce

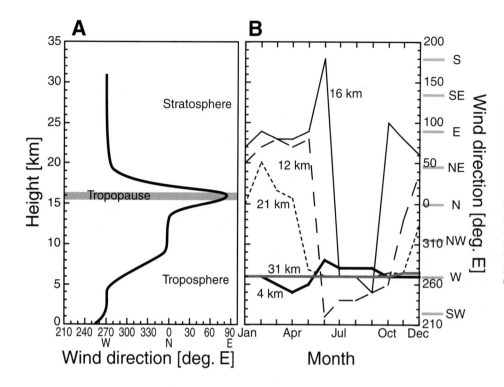

Figure 4. Present-day wind directions in Nicaragua. (A) Annual average wind direction versus atmospheric height above ground. (B) Monthly averages for selected heights. In meteorology, wind direction is noted as where the wind is blowing from; here we show the direction it is blowing to for easier comparison with tephra dispersal directions. Data from the U.S. National Oceanic & Atmospheric Administration–Cooperative Institute for Research in Environmental Sciences Climate Diagnostics Center, www.cdc.noaa.gov.

eruption columns reaching high into the stratosphere (i.e., more than 15 km). While such large-magnitude eruptions occur much less frequently, areas affected by their associated hazards are many orders of magnitude larger than for Strombolian eruptions. Almost all of west-central Nicaragua would, in fact, be affected by a Plinian eruption from any of the potential vents in the area. Plinian eruptions are typically produced by magmas of silicic compositions, but in Nicaragua basaltic magmas from the Masaya volcanic system also generated such eruptions (Wehrmann et al., this volume; Pérez and Freundt, this volume).

We mapped the areal thickness and grain-size distributions of the deposits from past Plinian eruptions at Apoyo and Masaya calderas and the Chiltepe volcanic complex (Fig. 1), including Apoyeque stratocone and Xiloá maar. Data from such maps allows us to estimate erupted masses and column heights. The 15–30-km-high ash clouds from these eruptions were driven by the wind mostly to the west or northwest, gradually losing their load of ash and lapilli to form fan-shaped blankets of tephra on the ground, covering areas on the order of $10^3$ to $10^4$ km$^2$ and >10 cm in thickness (Fig. 3A). The magnitudes of these eruptions, $M = \log_{10}(m_e) - 7$, where $m_e$ is the erupted mass in kilograms (Pyle, 2000), range from 4.3 to 6.3 (i.e., over two orders of magnitude). For comparison, magnitudes of the well-known Mount St. Helens 1980 (USA) and Mount Pinatubo 1991 (Philippines) eruptions are 4.8 and 6.0, respectively (Pyle, 2000). The largest magnitude, $M = 7.3$, of a historical eruption was reached by Tambora volcano, Indonesia, in 1815.

The impacts of large blocks ejected ballistically from a crater or falling out of the lower part of an eruption column are limited to a few kilometers from the vent. A major hazard from pyroclastic fallout that can extend to great distances from the volcano is the partial to total collapse of buildings that become heavily loaded with ash. At least partial building collapse is to be expected when the load pressure exceeds ~1 kPa (Pyle, 2000). Ash that is wet (from eruption or soaked with rain) exerts 1.5 times the load stress of dry ash of the same thickness. Collapse may occur at even lower loads where load stress is focused at the fixings of pole-supported roofs rather than distributed along walls. Probabilities of roof collapse under given ash load for a range of designs have been determined for some volcanic regions (Spence et al., 2005) but are not yet available for Nicaragua. The destructiveness index $D = \log_{10}(A)$, where $A$ is the area [m$^2$] over which ash accumulation exceeds 1 kPa (~10 cm thickness), ranges from $D = 2.8$ to 4.0 for the Nicaraguan Plinian eruptions. For comparison, $D = 2.9$ and 3.4 for the Mount St. Helens and Pinatubo cases, and the largest recorded value of $D = 4.5$ is from the Taupo 186 A.D. eruption, New Zealand (Pyle, 2000).

One way to estimate the possible impact of future fallout on west-central Nicaragua is to consider the average effect of past eruptions. The average thickness of tephra deposited at a particular locality from a past large eruption is obtained by dividing the total tephra thickness locally accumulated (Fig. 5B) by the number of eruptions that contributed to each locality over a period of time. Figure 5C shows the average thickness based on the past ~40,000 yr, while the average thickness in Figure 5D is based on the past 6000 yr. The resulting distribution pattern is quite different from that of a single eruption (Fig. 3A) because the three potential source volcanoes (Chiltepe, Masaya, Apoyo) are considered

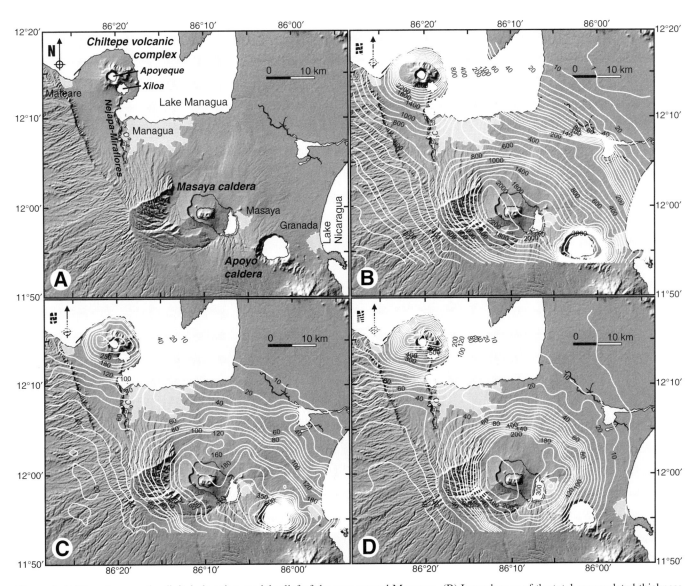

Figure 5. (A) Locations on the digital elevation model relief of the area around Managua. (B) Isopach map of the total accumulated thickness (cm) of all widespread fallout deposits formed during the past ~40,000 yr (see Fig. 7A). (C) Average fallout thickness per eruption (total accumulated thickness divided by number of eruptions represented at each locality) emplaced by eruptions during the past ~40,000 yr. (D) Average thickness of fallout per eruption based on the eruptions of the past 6000 yr only.

simultaneously without taking into account the probability of a future eruption at any of them, a topic that is discussed below in the section on eruption recurrence. The map shows, however, that all of west-central Nicaragua is potentially threatened by destructive fallout accumulating to a thickness of more than 10 cm.

Volcanic ash fallout also creates hazards other than structural collapse. The inhalation of very fine (<10 µm) ash particles can cause respiratory problems for days after an eruption (Baxter, 2000). Ground traffic will be inhibited by very low visibility, and engine air filters will be clogged with ash. Air traffic will also be inhibited. In the 1990s, 12 eruptions worldwide caused damage to or failure of turbines on aircraft even several hundred kilometers away from the volcano (Miller and Casadevall, 2000).

## Pyroclastic Density Currents

Pyroclastic density currents are hot mixtures of pyroclastic debris and gas that move along the ground. They form during explosive eruptions by partial or total collapse of an eruption column too dense to ascend higher into the atmosphere, or by direct ejection from the crater. They also form by partial collapse of extruding lava domes, but such flows (called block-and-ash flows) are not considered here since they have not been recorded in Nicaragua. The volume, runout distance, and flow behavior of pyroclastic density currents can vary widely. Two types of pyroclastic density currents are commonly distinguished: (1) pyroclastic flows consisting of a high concentration of ash form massive deposits (ignimbrites), and (2) gas-rich, highly turbulent pyroclastic surges that typically leave wavy, stratified deposits with dune structures (Freundt et al., 2000; Valentine and Fisher, 2000). Pyroclastic flows are more strongly confined to valleys, while pyroclastic surges can spread across valley shoulders. Large-volume pyroclastic density currents of any form, however, can spread across the entire landscape.

All types of flows pose the greatest hazards to communities. People tend to live in valleys fertilized by volcanic soil with nearby water resources. These valleys are also the favored pathways of volcanic flows. The main loss of lives (47% in 1900–1986; Blong, 1996) and the largest economic losses in the past century from volcanism resulted from pyroclastic flows and surges. Pyroclastic density currents are particularly dangerous phenomena of explosive volcanic eruptions for three reasons: (1) their occurrence during an eruption is sudden and unpredictable, and they spread very rapidly (10–150 m/s), allowing for extremely short warning times; (2) their high temperatures (several hundred degrees Celsius) and magmatic gases are lethal; and (3) they have enormous destructive power, virtually destroying everything in their path. The dynamic pressure such currents exert on the upstream face of buildings is $P = \frac{1}{2}\rho u^2$, where $\rho$ is the bulk density of the pyroclast-gas mixture and $u$ is the velocity of the current. Hence, a pyroclastic flow with 40 vol% pyroclasts moving at 15 m/s has the same destructive potential ($P \approx 50$ kPa) as a pyroclastic surge with only 1 vol% pyroclasts but racing at a speed of 100 m/s. Pyroclastic density currents typically develop dynamic pressures of 1–1000 kPa, and 20 kPa is already sufficient to inflict severe to total damage on most buildings (Valentine, 1998).

In Nicaragua, pyroclastic flows formed during a Plinian eruption from Apoyo caldera ~25,000 yr ago inundated the low-lying areas between Apoyo and Mombacho to the south and around Granada to the east (where they entered Lake Nicaragua), leaving 8 km³ of massive ignimbrite reaching >20 m thickness (Sussman, 1985). Several pyroclastic flow deposits of yet unknown age form widespread sheets on the plains west and northeast of Momotombo volcano (Hradecky, 2001). A ~20 m thick ignimbrite, erupted only 34,000 yr ago from an unknown source at the volcanic arc, covers the plains descending toward the Pacific west of Jinotepe (our observation and $^{14}$C dating). The most recent Nicaraguan eruption producing voluminous pyroclastic flows occurred in the very north at Cosigüina in 1835 (Self et al., 1989; Scott et al., this volume); the deposits cover large parts of the peninsula, and large amounts of ash were shed into the surrounding sea. All these pyroclastic flows filled valleys but were voluminous enough to also flood large plains.

Pyroclastic surges were associated with virtually each of the Plinian eruptions, although most did not reach farther than ~3 km from vent. They formed either by collapse of the Plinian eruption columns or by phreatomagmatic phases of these eruptions. A spectacular example is the coarse-pumice surge deposits exposed on the rim of Apoyo caldera, which were produced during the terminal phase of the last Apoyo eruption ~25,000 yr ago. The Plinian Chiltepe eruption from Apoyeque stratocone less than 2000 yr ago generated surges reaching >5 km from the crater, and phreatic explosions (driven by magmatic heat but without ejection of new magma), producing smaller surges, probably continued for months after the main Plinian eruption, building a tuff-ring on the rim of Apoyeque crater.

Pyroclastic surges are an integral part of phreatomagmatic eruptions, which are discussed in the next section. Such surges spread radially around the vent and can initially have supersonic velocities. The largest phreatomagmatic surge deposit known in Nicaragua, the <2000 yr old Masaya Tuff (Pérez and Freundt, this volume), has spread near-radially up to 40 km around Masaya caldera.

Pyroclastic surge deposits commonly are the accumulated sediment from numerous successive surge flows, each leaving a deposit that may be only a few dm to cm thick. These change from dune-bedded to laminated ash deposits while exponentially thinning away from vent. Distal laminated ashes mainly represent fallout from drifting ash clouds remaining after a surge has lost most of its pyroclast load and kinetic energy. The dune-bedded deposits, however, mark the minimum area over which the surge moved rapidly and was highly destructive. These areas are indicated in Figure 6 and form an almost continuous ~20-km-wide zone along the 80 km length of the volcanic arc from Granada through Masaya and west-Managua to north of Mateare.

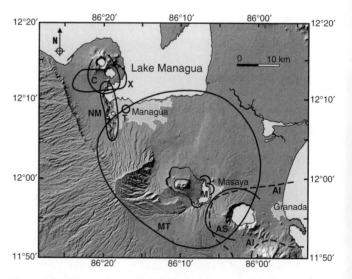

Figure 6. Outline of areas that suffered severe to total destruction by pyroclastic density currents during past eruptions. C—surges of Chiltepe Tephra eruption; X—Xiloá surges; NM—Nejapa-Miraflores maars; T—Tiscapa maar; MT—Masaya Tuff (25-cm isopach from Pérez and Freundt, this volume); M—small phreatomagmatic eruptions in SE part of Masaya caldera; AS—Apoyo surges; AI—Apoyo ignimbrite (dashed line; after Sussman 1985) of Upper Apoyo Tephra. Solid lines approximately outline areas where surge deposits are cross-bedded; lower-energy flow occurred beyond these limits. See Figure 5A for localities.

## Recurrence of Large Eruptions

Periods of quiescence, particularly between large-volume eruptions, last much longer than the few hundreds of years of recorded human history in Nicaragua. Assessment of the probability of occurrence of another such eruption has to rely on extrapolation of the past record of activity of the volcanoes, assuming that the past behavior continues into the future. This approach requires careful mapping, stratigraphic subdivision, dating, and volcanological interpretation of the older volcanic deposits. Here we focus on Masaya caldera and the Chiltepe peninsula—a volcanic complex hosting Apoyeque stratocone, Xiloá maar, and other now hidden vents—for which we have the most complete data on their eruptive histories. Apoyo caldera experienced two Plinian eruptions ~25,000 yr ago separated by a relatively short period of time (perhaps a few hundred years; Fig. 7A), but little is yet known of its earlier history.

A simple way to estimate the time frame for a future eruption is to assume that a volcano follows a periodic pattern; the period is obtained by dividing the time over which a volcano was active by the number of eruptions that occurred. Masaya caldera produced four large eruptions during the past <6000 yr,

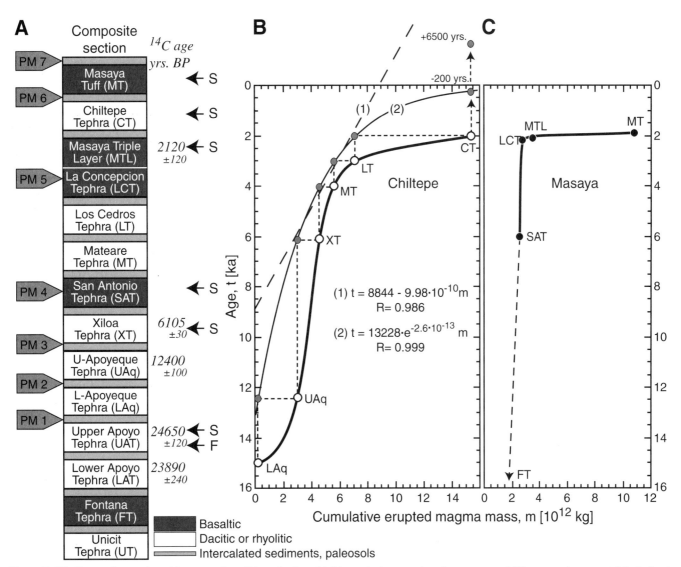

Figure 7. (A) Composite stratigraphic succession of deposits from highly explosive eruptions in west-central Nicaragua. Arrows at right indicate the production of pyroclastic surges (S) and pyroclastic flows (F) during these eruptions. Gray arrows at left mark the stratigraphic position of major phreatomagmatic (PM) eruption phases along the Nejapa-Miraflores lineament, in Managua city, and along the southwestern shore of Lake Managua. Our age data were determined by the Leibniz Laboratory for Radiometric Dating and Stable Isotope Research at Kiel University. (B) Cumulative mass of erupted magma versus eruption age for the Chiltepe volcanic complex. Ages of undated tephras are guessed considering the nature of intercalated deposits. (1) and (2) are linear and exponential regressions, respectively, through the gray dots as explained in the text. (C) Cumulative mass of erupted magma versus eruption age for the Masaya caldera.

indicating an average recurrence period of <1500 yr. Over a similar period of time of 6100 yr, four eruptions were also produced at the Chiltepe volcanic complex, giving a similar average recurrence period. Another simple assumption is that a volcano is characterized by an average eruption magnitude. Dividing the mean ejected magma mass per eruption at Chiltepe by the long-term average mass production rate yields an average recurrence interval of ~1500–2500 yr, considering the past 6100 or 15,000 yr, respectively. For the past 6000 yr at Masaya caldera, the analogous calculation yields ~1200 yr. The last events at both volcanoes, the large phreatomagmatic Masaya Tuff eruption and the Plinian Chiltepe Tephra eruption (Fig. 7A), occurred less than 2000 yr ago. However, both these simple assumptions are poorly constrained here, although they seem quite reasonable in other cases (e.g., at Cerro Negro; Hill et al., 1998).

A more detailed impression of how a volcanic system evolved can be gained from the pattern in which erupted magma mass accumulated with time through a series of eruptions. Using the Chiltepe complex as an example, the white dots in Figure 7B are the cumulative erupted magma mass (added up starting from the Lower Apoyeque Tephra, Fig. 7A) plotted at the age of each respective eruption (we guessed some ages since absolute age data are incomplete). The gray dots are the same cumulatives masses projected upward to the age of the respective next eruption. If eruptive activity at Chiltepe is controlled by a steady process (e.g., steady extension of the Nicaraguan depression), then the gray dots should lie on or close to a straight line (Bacon, 1982; Wadge, 1982). Line (1) in Figure 7B is a linear regression through the Xiloá Tephra (XT) to Los Cedros Tephra (LT) data. A better fit, however, is obtained for an exponential regression using all data (Lower Apoyeque Tephra [LAq] to LT), yielding line (2) in Figure 7B. This would imply that whatever process controls eruptive behavior at Chiltepe is not linear but exponentially accelerating. A forecast of the next future eruption can be made by projecting the cumulative erupted mass of the last eruption (Chiltepe Tephra [CT]) up to either line (1) or (2). For the exponential regression, the next eruption should already have occurred ~200 yr ago, whereas it would be expected in 6500 yr for the linear regression (Fig. 7B). Clearly, a more useful forecast requires a better understanding of the processes that control the evolution of the Chiltepe volcanic complex, which has only produced eruptions of magnitude M > 4 during the past few thousand years and is likely to continue doing so.

The data for Masaya caldera (Fig. 7C) comprise three moderately large eruptions (San Antonio Tephra [SAT], La Concepción Tephra [LCT], Masaya Triple Layer [MTL]; Fig. 7A), the unusually large Masaya Tuff eruption (MT), and a long repose period of ~30,000 yr back to the Fontana eruption (which actually occurred outside the caldera; Wehrmann et al., this volume). The large variations in magnitudes and recurrence periods within this small amount of data do not allow for extrapolation analogous to Chiltepe. Post-caldera volcanic activity since the last large eruption from the Masaya caldera has been cone-forming Strombolian and phreatomagmatic activity and lava effusion (Walker et al., 1993), none of which was very hazardous outside the caldera. This may indicate that the Masaya magmatic system has been reduced to a state favoring less hazardous volcanism, but it does not exclude the possibility that the system may return to a state of much higher activity.

## Magma-Water Interaction

Most of Managua is built on volcanic deposits whose mode of eruption was largely determined by explosive interaction of magma with groundwater or surface water, as clearly demonstrated by the maars of Tiscapa, Asososca, and Nejapa in the city (Fig. 8). The abundance of phreatomagmatic eruptions is not surprising since Managua is situated in the Nicaragua depression, a large catchment area of water as evidenced by the two large lakes (Fig. 1). This environmental setting facilitates the access of water to the magmatic systems in this depression. Phreatomagmatic eruptions from vents along the Nejapa-Miraflores lineament west of Managua and reaching north up to Chiltepe peninsula discharged juvenile magma precooled by mixing with water and lithic material, and hence had a low thermal mass flux and produced ~3 km diameter tuff rings around the vents with only little ash fallout to greater distance. The well-stratified tuff-ring deposits (Fig. 9) were formed by fallout, surges, and debris jets (Fig. 2E) discharged by numerous very powerful explosions.

Little is yet known about the chronology of these phreatomagmatic eruptions. Our preliminary results, based on some of the widespread Plinian pumice layers being intercalated with the tuff-ring deposits (Fig. 9), show that at least seven phases of phreatomagmatic activity occurred from more than 13,000 yr to less than 2000 yr ago (Fig. 7A) and are thus likely to continue in the future. Eruptions from this zone, although small in a global context, would be extremely disastrous because of the high population density and industrialization in western Managua.

## Eruption-Triggered Tsunamis

Tsunamis, waves formed by sudden displacement of a large water volume, are infamous for their occurrence in the oceans. Notable examples are the disastrous events in 2004 in the Indian Ocean or the 1992 tsunami that struck the Pacific coast of Nicaragua. Less well known is the fact that tsunamis also occur in lakes. Lake tsunamis can be triggered by volcanoes in several ways (Beget, 2000; Freundt, 2003): by volcanic earthquakes; by underwater eruption and phreatomagmatic explosions; and by entrance of pyroclastic flows, lahars, or debris avalanches. The debris avalanche of 18 May 1980 at Mount St. Helens entered Spirit Lake and caused a tsunami that ran up the shore to 260 m above the lake level (Voight et al., 1981). Flank-collapse avalanches from Mombacho entered Lake Nicaragua and probably triggered tsunamis. The pyroclastic flows generated during the Upper-Apoyo eruption ~25,000 yr ago also entered Lake Nicaragua and may have triggered tsunamis, as did the flows of the 1883 Krakatau eruption entering the Java Sea at Indonesia (Carey et al., 2001).

Phreatomagmatic explosions that occurred in Xiloá maar, which lay under shallow water at the time of eruption (Freundt et al., 2006), and at vents hidden in Lake Managua, which produced the thick phreatomagmatic deposits along the southwestern lake shore, also had the potential to trigger tsunamis like those witnessed during explosive eruptions at Lake Karymskoye, Kamchatka, in 1996 (Belousov et al., 2000). Another place in Nicaragua where volcanogenic tsunamis can be generated is Cosigüina peninsula at the Gulf of Fonseca, an inlet of the Pacific Ocean. A tsunami generated there could spread over the gulf within minutes and affect not only the villages on the Nicaraguan and Honduran coasts but also the relatively large town of La Union in El Salvador.

Fortunately, in Nicaragua, the size of lake tsunamis would be limited by the shallow depths of Lake Managua (<26 m) and Lake Nicaragua (<70 m). Tsunami deposits related to the above-mentioned events have not yet been identified. However, we have recently documented a tsunami deposit intercalated with the Mateare Tephra (Fig. 7A) that was erupted from a vent near the northwestern shore of Chiltepe peninsula between 3000 and 6000 yr ago (Freundt et al., 2006). The tsunami deposit is a sand layer (MS in Fig. 10) that was emplaced during the initial phase of the eruption, which consisted of discrete explosive eruption pulses (unit A in Fig. 10). The detailed mechanism of tsunami formation remains unknown, but tsunami formation apparently terminated when the eruption assumed a steady Plinian character (unit B in Fig. 10). The areal distribution of the sand layer suggests that the tsunami waves flooded inland to elevations of ~20 m above the lake level at the time of eruption.

Once formed, tsunamis travel away from their source at a velocity $c = \sqrt{gD}$ only depending on water depth, $D$, and acceleration due to gravity, $g$. Tsunami speed in the ocean is typically several hundred kilometers per hour. In the shallow Nicaraguan lakes, the velocity would be ~25–50 km/h. Considering the shore-to-shore distances, however, this means there would be less than an hour—and in most situations only minutes—of warning time. It is important to realize that volcanogenic lake tsunamis are a serious hazard along the shores of the two Nicaraguan lakes.

## HAZARDS FROM NONEXPLOSIVE ACTIVITY

### Lava Flows

Lava flows do not generally pose a direct hazard to people because they advance relatively slowly. Nevertheless, lava flows can cover major stretches of cultivated or inhabited land and destroy buildings and infrastructure. Long lava flows were erupted from Masaya volcano ~10 times during the past 500 yr. These include the 1772 lava, the longest lava flow (16 km) erupted in Central America during historic times (Pérez, 1993). Long lava flows also formed between 1527 and 1529 from the Maribios Range, comprising the volcanoes San Cristobal, Telica and Las Pilas (de Oviedo, 1855). The lava flow of the 1995 eruption of Cerro Negro volcano reached a distance of 2.5 km within a few days. Other major lava flows extruded from Momotombo

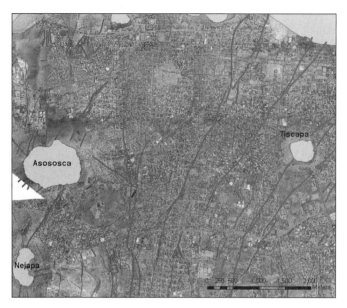

Figure 8. Aerial photograph of western Managua including the maars of Tiscapa, Asososca, and Nejapa. Thick gray lines are tectonic faults. From Instituto Nicaragüense de Estudios Territoriales, Managua.

Figure 9. Two gray stratified phreatomagmatic units (PM2 and PM3 of Fig. 7A) separated by the intercalated white Upper Apoyeque Tephra (UAq; 12,400 yr B.P.) and an erosional unconformity (dashed line). Quarry in northwestern Managua city.

volcano in 1886 and 1905 (Sapper, 1925; Gaceta Oficial de Nicaragua, 1886). Such lava flows are a risk for the geothermal power plant at the foot of this volcano.

### Volcanic Gases

Hot and toxic gases are an integral part of hazards associated with volcanic activity. Gases emitted by large-magnitude eruptions are known to affect local and global climate and atmospheric

Figure 10. (A) Section of the Mateare Tephra in the western low-lands of Chiltepe peninsula showing the stratified fallout unit A, the Mateare Sand (MS) above wavy contact, comprising lower reworked pumice and upper sand bed, and the massive pumice fallout unit B. (B) Detail of MS with hole from driftwood pole (now rotten away) that affected emplacement. This outcrop is the highest elevation at which MS was found.

chemistry for years. However, large amounts of volcanic gases are not only released by magma degassing at depth during eruptions but also from open vents and fumaroles during repose periods between eruptions, which may last several decades. Steam mixed with sulfur dioxide ($SO_2$), carbon dioxide ($CO_2$), and many other components such as fluorine, chlorine, and bromine are continuously emitted from open vents and fumaroles, and, in sufficient quantities, may pollute air, water, and soil over hundreds to thousands of km$^2$ downwind from a volcano. Such gas plumes can affect human and animal health and destroy cultivated and natural crops, particularly in areas suffering long-term exposure due to steady wind direction. Permanent open vent and fumarole degassing in Nicaragua releases some 1500 metric tonnes/day (t/d) of sulfur dioxide that pollute the environment downwind from volcanoes (Andres and Kasgnoc, 1998; our data). In the long run, this is one to two orders of magnitude larger than the long-term average sulfur flux from explosive eruptions. The permanent gas plumes of Masaya and San Cristóbal volcanoes (Fig. 11) are the most prominent examples.

The three largest continuous emitters of volcanic gases in Nicaragua are Masaya, San Cristóbal, and Telica volcanoes (Duffell et al., 2003; Andres and Kasgnoc, 1998; our data). Each currently releases several hundred t/d $SO_2$. The large amount of $SO_2$ released into the atmosphere by these volcanoes produces volcanic air pollution, manifested by poor air quality, hazy atmospheric conditions, and acid rain. During the 1990s, the $SO_2$ emission rates for two of the three volcanoes fluctuated significantly and included periods with anomalously high gas emission rates ("gas crisis"). During such gas crises, one single volcano may produce more $SO_2$ than all other Nicaraguan volcanoes combined. $SO_2$ emission rates exceeding 3000 t/d were measured at Masaya on a few days in February–April 1998 and February–March 1999 (Duffell et al., 2003), while emission rates above 1000 t/d have been measured several times at San Cristóbal. Thus, episodes of strong magma degassing occur repeatedly at both volcanoes.

The plumes of San Cristóbal and Masaya seriously affect areas >20 km downwind, with atmospheric $SO_2$ concentrations that compare to, or exceed, those reported from large cities and areas downwind of industrial point sources. The low-altitude gas plume at Masaya has destroyed crops and other vegetation over several hundred km$^2$, and probably affects the health of thousands of people (Delmelle et al., 2002). San Cristóbal's plume affects the entire area between the volcano and the city of Chinandega, 20 km away. The gas plume normally passes at the height of the crater rim, but under certain meteorological conditions, the gas flows down the volcano flank, and very high $SO_2$ concentrations are periodically observed in populated areas (data from Instituto Nicaragüense de Estudios Territoriales [INETER], www.ineter.gob.ni/). In the past 10 yr, Nicaraguan authorities repeatedly evacuated people from villages near San Cristóbal to prevent

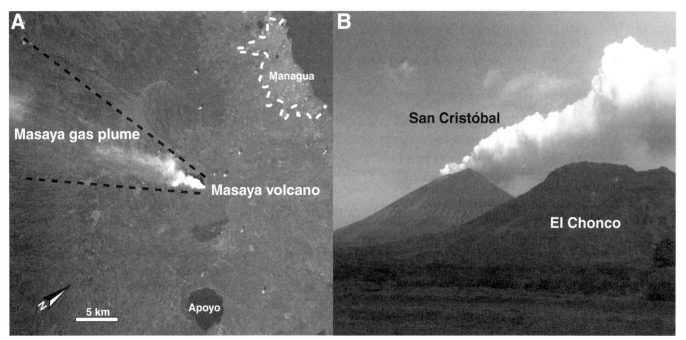

Figure 11. (A) Satellite image of the gas plume (between dashed lines) drifting from Masaya volcano. Satellite image from www.volcano.si.edu. (B) Gas plume emanating from San Cristóbal's crater.

injuries due to high gas concentrations. The U.S. National Air Quality Standard over the year is 30 ppbv of $SO_2$ (data from U.S. Environmental Protection Agency [EPA], www.epa.gov/oar/airtrends/sulfur2.html). Concentrations above this threshold value occur within several hundred km² at Masaya, San Cristóbal, and Telica volcanoes (Delmelle et al., 2002, 2001). Short-term (hours to days) concentration peaks can be much higher than long-term averages, thereby increasing the hazard.

Volcanic gases, particularly $CO_2$, emitted into deep crater lakes dissolve in the water until saturation is reached, which causes overturn of the water body and massive release of $CO_2$ clouds that flow along the ground. Such an invisible $CO_2$-cloud from Lake Nyos (Cameroon) killed 1700 people by asphyxiation in 1985 and raised the awareness that crater lakes need to be monitored at volcanoes with significant $CO_2$ degassing (Rice, 2000). Although no data is presently available on $CO_2$ degassing into Nicaraguan lakes, $CO_2$ degassing does occur along the volcanic arc (e.g., Lewicki et al., 2003). Deep lakes, such as Laguna de Apoyo (200 m) or Laguna Asososca near Managua (95 m), would be capable of generating large reservoirs of $CO_2$-saturated deep water.

## HAZARDS FROM MASS FLOWS AT VOLCANOES

### Lahars

Debris or mud flows that form at or near volcanoes are called lahars (Vallance, 2000) and are among the most dangerous volcanic phenomena, accounting for 10% of all volcanic fatalities (Schmincke, 2004). Lahars can originate by partial edifice collapse, by a variety of other mechanisms destabilizing sediment masses such as earthquakes or volcanic explosions, or by breaching and overspilling of a crater lake or melting of snow and ice by volcanic heat (Scott et al., 2005). Heavy rainfall onto loose ash is a particularly frequent cause of lahar formation (Hodgson and Manville, 1999; Lavigne and Thouret, 2003; Scott et al., 2005). Hence, they occur both related and unrelated to volcanic activity. Lahars are highly mobile and destructive flows that often reach tens of kilometers from a volcano, particularly when they ingest more sediment and water during runout. Mothes et al. (1998) reported a lahar deposit of almost 4 km³ volume, reaching 326 km from Cotopaxi volcano in Ecuador.

The physical nature of lahars ranges from highly concentrated, coarse-grained debris flows of considerable strength through hyperconcentrated flows of intermediate sediment concentration (20–60 vol%) and viscosity to more dilute floods of muddy water (Pierson and Costa, 1987). The character and flow behavior of lahars typically change during runout. They may begin as concentrated mass flows that progressively dilute by loss of sediment or by mixing with stream water, or they may start as watery floods that may entrain up to four times their own volume in sediment (Cronin et al., 1997, 1999). Such bulking of lahars is facilitated where vegetation is absent or has been removed over large areas. Lahars flow through valleys but may spill over shallow channels and become wide on the flat plains around stratovolcanoes (Scott et al., 2005).

Lahars triggered by rainfall occurred repeatedly in Nicaragua during recent years (Scott et al., 2005). On 27 September

1996, a lahar from Maderas volcano destroyed houses in the village of El Corozal (600 inhabitants) and killed six people. Lahars descend yearly from Concepción volcano in the rainy season and reach up to 5 km away. Strong lahars occurred during Hurricane Mitch in 1998. Fresh ash erupted from San Cristóbal volcano from November 1999 on was remobilized as lahars with the beginning of the rainy season in May 2000 and flowed up to 15 km from the volcano.

Geologic mapping of historic and prehistoric deposits is essential to identifying lahar pathways, although numerical tools are available that analyze digitized topography for possible lahar channels (e.g., Iverson et al., 1998). Such models, however, do not yet fully capture the effects of bulking and de-bulking or the amount of sediment that can be mobilized, and predictions of strength and runout of lahars thus remain speculative.

Before the disaster of Hurricane Mitch, the hazard of landslides and lahars in Nicaragua was largely unknown, ignored, or underestimated, both by the general population and the scientific community. After 1998, an intense process started that embraced many projects carried out by foreign and national technical institutions and development aid agencies, improving the technical capacity at INETER, training local geoscientists in techniques for landslide identification and mapping, carrying out field investigations and air photo mapping of landslide occurrence, building up knowledge databases and geographic information systems (GIS) on these phenomena. In the last years, the INETER geophysical department has created a large GIS-based landslide inventory that contains georeferenced data of ~17,000 separate landslide events known in Nicaragua, including lahars and debris flows at the volcanoes. Landslide susceptibility and hazard maps were elaborated. Both conventional and interactive maps are accessible at INETER's Web site, www.ineter.gob.ni/geofisica/desliza/desliza.html. INETER worked also on the early warning system for possible lahars and landslides. A pilot project was carried out on the installation of telemetric meteorological stations at the volcanoes San Cristóbal, Casita, Mombacho, and Concepción to obtain data on the level of intensity and duration of intense precipitations that trigger lahars. Meteorological alerts in case of hurricanes or tropical depressions now always contain more detailed information about possible landslide occurrence. Also, public awareness of landslide and lahar hazards is much higher than before Hurricane Mitch. Local authorities report occurrences of lahars and landslides to INETER and request technical support for risk estimation and proposals for mitigation measures. The Nicaraguan Emergency Commission and Civil Defense actively support this process.

**Sector Collapse and Debris Avalanches**

Processes leading to loss of structural integrity of volcanic edifices are various and occur on a range of timescales, from short-lived eruptive activity to gravitational settling over millennia (Borgia et al., 2000). Acidic waters reduce the competence of rocks that disintegrate to clay in the hydrothermal system inside a large volcano. Elevated pore pressures make volcanic edifices susceptible to failures. Edifices constructed with slopes at the angle of repose of volcaniclastic materials, or even with gentler slopes, that are internally weakened can suddenly collapse in response to some trigger (Cecchi et al., 2005). Such triggers include the inflation of the edifice by new magma, regional or volcanotectonic earthquakes, or massive intrusion of water during heavy rainfalls such as at Casita volcano in 1998. Most large volcanic edifices experience one or more sudden collapses, producing highly mobile gravity-controlled debris avalanches. These accelerate downhill and can flow at up to 100 km/h to distances of tens of kilometers. Debris avalanches destroy everything in their path and can leave breccia deposits thicker than 10 m. All stratocones in Nicaragua that possess active hydrothermal systems, and particularly those also affected by tectonic faults, can potentially experience flank collapse in the future. Below, we briefly summarize two collapse events that have occurred in the recent and historic past.

On 30 October 1998, a disastrous avalanche and lahar occurred after partial collapse of the southern summit of Casita (Sheridan et al., 1999), an older stratocone with flanks deeply incised by erosion. The collapse volume has been determined to be 1,600,000 m$^3$ (Kerle, 2002). The collapse was triggered by the exceptionally high precipitation (reaching a peak value of 59 mm/h prior to collapse; Scott et al., 2005) from Hurricane Mitch, which occurred late in the rainy season when the ground was already soaked with water. This trigger could operate because the Casita edifice was already structurally weakened by strong hydrothermal and meteoric alteration and fracturing of the rocks (Vallance et al., 2001a) with a high capacity for water infiltration and buildup of pore pressure (Kerle et al., 2003; Scott et al., 2005). Van Wyk de Vries et al. (2000) and Kerle et al. (2003) argue that the edifice was weakened by gravitational spreading and that the principal rupture of the collapse occurred along an ~400-m-long segment of a NE-trending fault intersecting the summit of Casita in an area of fumarolic and hydrothermal activity. Kerle et al. (2003) demonstrate that slow deformation had been going on at least since 1954 before the 1998 collapse and that there is geologic evidence of earlier landslide events, and van Wyk de Vries et al. (2000) argue that the tectonic, morphological, and hydrothermal state of Casita favors continuing deformation and the likelihood of another future collapse event.

The collapsing portion of the flank consisted of fractured andesitic lava and clay formed from hydrothermally altered scoria and was soaked with water. The avalanche descended the flank of the volcano through a deep erosional gully (Fig. 12), where it eroded sediment down to the bedrock. When spreading onto the plain at the base of Casita, the rock avalanche had transformed to a watery debris flow and a separate, highly concentrated flow that did not extend far onto the plain (Scott et al., 2005). The volume of the debris flow increased by entrainment of loose sediment as it surged across the plain, facilitated by deforestation (Kerle et

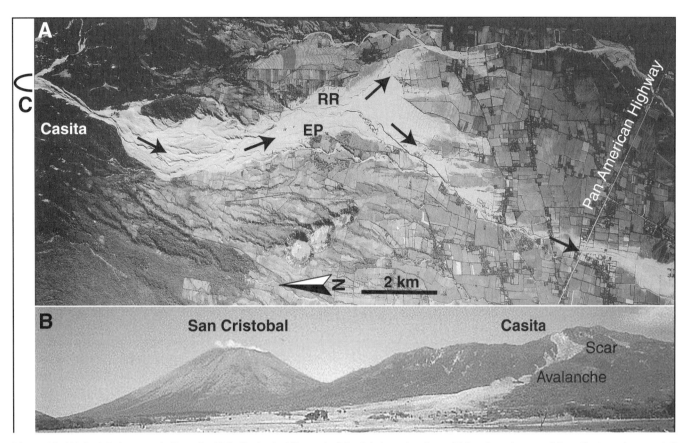

Figure 12. (A) Aerial photograph (from the U.S. Geological Survey) of the debris avalanche and lahar deposits spread from Casita volcano at left into cultivated and inhabited areas at right. C at left indicates source of collapse just outside picture. The avalanche descended through the narrow gorge on the flank, and the lahar spread widely onto the plain (arrows indicate flow directions). El Porvenir (EP) and Rolando Rodriguez (RR) were communities overrun by the lahar. (B) Surface of lahar deposits extending from the light-colored avalanche on the flank of Casita volcano in the background. Photograph from Instituto Nicaragüense de Estudios Territoriales, Managua.

al., 2003). When the lahar reached the villages of El Porvenir and Rolando Rodriguez, ~6 km from the collapse site (Fig. 12), only ~3 min after the collapse started (indicating an average speed of 120 km/h), it was 3–6 m deep, 1 km wide, and had bulked to ~4 times the initial volume (Scott et al., 2001, 2005). The two villages, which had been built only 15 yr before the disaster on deposits of prehistoric lahars, were completely destroyed, and more than 2000 people perished.

Similar to Casita, deeply dissected Mombacho volcano has a hydrothermally altered core. In 1570, an avalanche and debris flow on the south side of the volcano obliterated a town named Mombacho and killed more than 400 people. Historical accounts indicate that the collapse was preceded by earthquakes and accompanied by heavy rainfall, both potential triggers (Vallance et al., 2001b). Geologic evidence reveals three older, much larger, debris avalanches at the volcano that extend ~10 km from the summit to the southeast, northeast, and south and cover areas of 20–30 km$^2$. Two of these avalanches left thick, hummocky deposits that form a peninsula with a group of small islands, Las Isletas, along the shore of Lake Nicaragua (Vallance et al., 2001b). Entrance into the lake by these avalanches is likely to have triggered tsunamis.

## FROM HAZARD TO RISK

Volcanic risk is the product of volcanic hazard and vulnerability. Volcanic hazards are commonly described as the physical processes generating them and the areas that are potentially affected. Ideally, the hazards should be quantified in terms of the probability that they occur within a certain time and affect a given area, based on the past records of volcanoes, in order to facilitate long-term planning. This has been done, for example, for the Cerro Negro eruptions by Hill et al. (1998). For most other Nicaraguan volcanoes, however, extensive volcanological studies are still needed before such probabilistic quantifications can be made. Warning of an impending eruption on a shorter time scale (typically weeks to months) is achieved by a variety of techniques for monitoring unrest at volcanoes.

Vulnerability includes the threat to life and health and the degree of destruction of buildings, lifelines, industry, agriculture, and other economic activities by any kind of volcanic hazard. Vulnerability can be reduced by preparedness, which includes monitoring, mitigation, and emergency planning based on hazard maps and risk scenarios and public awareness and education.

**Forecast and Prediction**

The ability to predict eruptions is especially important in densely populated areas to enable timely evacuation. Forecasts of volcanic activity are usually based on the past record of activity of a volcano and, therefore, are a relatively imprecise statement that can only be reasonably quantified if the volcano evolved in a fairly systematic fashion (cf. Fig. 7). However, apart from such empirical assessment, forecasting becomes increasingly supported by an improved understanding and modeling of the nonlinear dynamics that control the evolution of volcanoes (Sparks, 2003). Prediction is a relatively precise statement of the place, time, and ideally the nature of an impending eruption (Swanson et al., 1985) that is best derived from a combination of different monitoring techniques for detecting pre-eruptive unrest, as discussed in the next section.

Forecasts are hampered by periods of quiescence between eruptions varying greatly among, but also within, volcanoes. Generally, the longer a volcano has been dormant, the larger and more explosive a new eruption is (Simkin, 1993; Blong, 1996). Frequently active volcanoes such as Cerro Negro (22 eruptions since 1850), Masaya (23 since 1524), and Concepción (25 since 1800) have average quiescent periods on the order of 10 yr and produce eruptions with a volcanic explosivity index (VEI; Newhall and Self, 1982) of 0–3 on a logarithmic scale from 0 to 8 (data from Smithsonian Global Volcanism Program, www.volcano.si.edu). Cosigüina had seven eruptions since 1500, an average repose time near 100 yr, and eruptions ranged at VEI = 2–5. Eruptions from Chiltepe volcano, separated by periods on the order of $10^3$ years, had VEI = 4–5, and Apoyo caldera had two cataclysmic eruptions (VEI = 5–6) after an estimated quiescence on the order of $10^4$ years. Hence, the most hazardous volcanoes are commonly those for which eruptive phases are separated by hundreds or thousands of years of quiescence. There are no precise or generally accepted definitions for the terms *active volcano*, *dormant volcano*, and *extinct volcano*, but volcanoes that have erupted within the past 10,000 yr are commonly regarded as active by volcanologists (Simkin and Siebert, 2000), emphasizing the need for analysis of the deposits to arrive at reasonable estimations of the probability of future eruptions.

Volcanic eruptions typically comprise several episodes interrupted by quiescent periods. Total durations range from less than a day to tens of years; the median duration is ~50 days (Simkin and Siebert, 2000). Almost half the recorded eruptions reached their paroxysmal phase on the first day (Simkin and Siebert, 2000), followed by a slowly declining late phase. Obviously, emergency plans need to be ready in order to initiate countermeasures during the relatively short warning time prior to eruption provided by volcano monitoring.

**Monitoring**

Monitoring is the continuous or frequently repeated measurement of geophysical, geological, and geochemical data that reflect processes operating at volcanoes. The four most important precursors to an impending eruption are (1) type, strength, and frequency of volcanic earthquakes; (2) varying magmatic pressure leading to surface deformation; (3) changes in volcanic gas flux and composition; and (4) increased heat flux. Other, less clear signs of unrest include changes in the electrical, magnetic, and gravity fields at a volcano. Most important is the analysis of seismic events and deformation of the surface that in many cases may allow short-term prediction of eruption site, onset (days to hours), but also of the type of eruption. Apart from eruption forecasting and hazard assessment, monitoring also provides essential data for understanding volcanic processes. With all monitoring techniques, possible precursors for an imminent volcanic eruption can only be estimated realistically when background behavior is known. In other words, a volcano must be observed over some time to recognize significant deviations from its "normal state."

*Volcanic Earthquakes and Magma Ascent*

Volcanic earthquakes can be defined as seismic events generated in, below, and close to a volcano, as well as prior to, during, and after eruptions. Earthquakes from sudden stress release generate short seismic pulses, whereas magmatic and hydrothermal flow processes and related phenomena generate seismic waves that last from minutes to days, the so-called tremor. Permanent tremor recorded at Masaya volcano in 1992–1993, for example, originated from continuous degassing of magma residing in the conduit of Santiago crater (Métaxian et al., 1997).

All presently active volcanoes in Nicaragua (Cerro Negro, Concepción, Masaya, Momotombo, San Cristóbal, and Telica) are seismically monitored by INETER. At each of these volcanoes, a set of several short-period and two broad-band stations are installed. Also, all other important volcanic centers are monitored by at least one seismic short-period station. The data processing software employed by INETER's geophysical department (www.ineter.gob.ni/geofisica/geofisica.html) facilitates real-time analysis of the recordings. Earthquakes are located and their magnitudes are determined in near real time 24 hours daily by a qualified technician working with a monitoring and early warning system. Thus, it is sometimes possible to observe how a volcano builds up to an imminent eruption.

Although each volcano behaves individually, some common features of earthquake activity observed at different volcanoes suggest a general model (McNutt, 1996, 2000b). An initial swarm of high-frequency (5–15 Hz) earthquakes, occurring over a significant period of time, reflects fracturing of country rock at depth in response to increasing magma pressure. As magma reaches shallower levels, magmatic degassing and heat-induced hydrothermal flow and boiling induce lower-frequency (1–5 Hz) earthquakes and volcanic tremor (McNutt, 2000a, 2000b). In about a quarter of the investigated cases, the eruption was actually preceded by a period of relative seismic quiescence, reflecting stress release possibly by pressure venting through open cracks or heated wall rocks responding in ductile rather than brittle fashion (McNutt, 1996). During eruption, explosion earthquakes and eruption tremor result from magma flow and fragmentation. Magma withdrawal to depth and stress relaxation can return seismicity to deep, high-frequency earthquakes.

Real-time seismic amplitude measurements (RSAM), as made by INETER at San Cristóbal, Telica, Cerro Negro, Momotombo, Masaya, and Concepción volcanoes (www.ineter.gob.ni/geofisica/datrsam/), determine the seismic energy released, which is a proxy of strain rate (Sparks, 2003) if the seismic events are of tectonic type. Application of laws of materials failure provides a physical framework to interpret such data to predict the time of failure (i.e., the onset of eruption; Voight and Cornelius, 1991). Detailed seismic behavior at each volcano depends on numerous factors, such as geometry of the plumbing system, flow rate and volatile content of magma, and distribution of groundwater (McNutt, 2000b). Also, rather than magma ascent determining volcanic earthquakes, regional tectonic stresses and their associated earthquakes can control magma ascent and trigger an eruption, such as the one at Cerro Negro in 1999 (La Femina et al., 2004). Elevated seismicity is commonly the first sign of volcanic unrest, beginning hours to weeks before an eruption, but in many cases it is not followed by an eruption (Sparks, 2003). In order to judge whether earthquakes herald a forthcoming eruption or not, it is important to monitor the background seismicity in a volcanic area for several years (McNutt, 2000b).

During the last decade, INETER gathered experience in the monitoring and early warning of volcanic eruptions. Though not claiming the capacity for making a real prediction of the hour, magnitude, and duration of an eruption, INETER in several cases emitted warning messages hours, days, or weeks before the beginning of an eruption (Lesage et al., 2006). This was the case on 4 August 1995 at Cerro Negro. At 10 p.m. local time, very strong seismic activity with magnitudes up to 5.0 (Richter scale) was detected near the volcano. INETER immediately informed Civil Defense and the public. Civil Defense officials from their office in the city of León traveled to the volcano and stayed there during the night. The seismic activity turned into a continuous tremor, and INETER published, in the morning of 5 August, a warning about a pending volcanic eruption. Twelve hours after the seismic activity started, the Civil Defense officials observed the opening of a long fissure on the southern part of the volcano, and lava fountaining started. This and other examples are reported at www.ineter.gob.ni/geofisica/vol/alerta-temprana.html.

### *Volcano Deformation and Subsurface Mass Movements*

Beginning with the pioneering work of Eaton and Murata (1960), measuring tilt on the surface of volcanoes is the premier method for detecting an accumulation of magma in the interior of a volcano. The vertical and lateral expansion of the surface of a volcano can be measured very precisely and in real time with modern geodetic instruments, both conventional and using the global positioning system (GPS) (Van der Laat, 1996; Murray et al., 2000). A dramatic and accelerating radial tilt (>2000 μrad) was observed over two months at Mount St. Helens prior to its 1982 eruption (Dzurisin et al., 1983), whereas a much smaller tilt (<20 μrad) over about three weeks preceded the Pu'u 'O'o eruptions in Hawaii (Tilling and Dvorak, 1993). Eruption typically induces deflation of the edifice. Measurable ground deformation can occur even before seismic unrest begins, but then there are also volcanoes that do not deform prior to eruption (Sparks, 2003). Ground deformation is a measure of strain rates, which can also be measured directly using strainmeters (commonly installed in boreholes). Strain analysis in terms of failure criteria can thus principally be applied analogous to seismic data to determine the time of failure and eruption. The precision of such prediction, however, suffers from the complex mechanical response of volcanic edifices to magmatic pressure (Sparks, 2003). Next to predicting eruption time, horizontal and vertical displacements and tilt vectors measured by geodetic networks across a volcano can be used to determine the volume, shape, depth, and pressure of a magmatic intrusion and their variation with time (Van der Laat, 1996). Moreover, deformation of volcanic edifices not related to magma intrusion but to gravity, hydrothermal alteration, or regional tectonics (e.g., at Casita volcano; Van Wyk de Vries et al., 2000) can be geodetically monitored to identify possible future collapse sites. INETER employs GPS-based surface monitoring at all active Nicaraguan volcanoes, but does not have the capacity for real-time measurements and data processing. So, even when unrest has been detected, such as during the present crisis at Concepción volcano on Ometepe island (since July 2005), GPS equipment has to be transported to the field to set up the GPS sites, get the data to the office, send that data to cooperating scientific groups abroad for analysis, and get back the results. This process needs several days or weeks and is not actually useful for short-time prediction or warning.

Subsurface changes in density due to magma movement or degassing at depth are detected by gravity measurements, a technique that is best employed in conjunction with geodetic measurements, since changes in surface elevation also change the local gravity field. The combination of both techniques, called microgravity monitoring, is best applied to volcanoes where intruding magma has a significant density contrast to the host rock or where large volumes of exsolved gas may accumulate (Rymer, 1996). Applications of microgravity surveys have been most successful at andesitic stratovolcanoes, particularly when combined with seismic monitoring (e.g., see Voight et al., 1998). Gravity and deformation measurements made at Masaya volcano identified that the degassing crisis beginning in 1993 was not related to intrusion of new magma at depth (heralding a new phase of eruption) but to convective exchange of degassed with gas-rich resident magma at shallow levels (Rymer et al., 1998).

### *Volcanic Gas Flux and Composition*

The efflux and/or composition of gases that are released from a volcano can change prior to an eruption. $SO_2$ emissions may increase months or years prior to an eruption when new gas-rich magma intrudes the volcano (Kazahaya et al., 2004). Volcanic gases are mixtures of components (e.g., $H_2O$, $CO_2$, $SO_2$, F, Cl) released from magma at different depths and rates. Due to differences in pressure-dependent solubilities in magmas between such volatile species, changes in their proportions may indicate magma movement at depth potentially leading to eruption. Alternatively, the role of groundwater in the shallow magma plumbing system can be monitored using $SO_2$/halogen ratios in the volcanic plume (Edmonds et al., 2001). At Masaya volcano, the $SO_2$/HCl ratios changed abruptly from 1.8 to 4.6 prior to a phreatic eruption in 2001 (Duffell et al., 2003). This change could have reflected hydrothermal scavenging of HCl. Increased magma-water interaction may have played a role in triggering the eruption, showing the importance of chemical monitoring of volcanic gases. Thus, coupled changes in emission style and rate and gas composition are expected, and are highly relevant for monitoring purposes. Electrical self-potential measurements have been tested at Masaya volcano as a promising tool for investigating interactions between magmatic and groundwater systems (Lewicki et al., 2003). Routine sampling and analysis of fumarolic gases from assumed dormant volcanoes is also considered important, because reawakening of volcanoes should be visible through changes in the gas chemistry (e.g., Garofalo et al., 2006).

$SO_2$/HCl ratios can be measured using infrared spectroscopy (Fourier Transformation Infra-Red [FTIR] spectroscopy; e.g., Andres and Rose, 1995) or by direct analysis of fumarolic gases (Giggenbach, 1996). Due to the high solubility in water of both HCl and HBr, $SO_2$/BrO ratios in the plume measured by the mini-differential optical absorption–UV spectrometer (Galle et al., 2003) may similarly be used for monitoring the state of a volcano, though more needs to be understood concerning the chemistry of bromine in volcanic plumes (Bobrowski et al., 2003). Continuous ground-based remote sensing of $SO_2$ and BrO fluxes in volcanic plumes using the Mini-DOAS is a very promising tool for the chemical monitoring of permanently degassing volcanoes.

Through the past two decades, measurements of gas composition and fluxes have been made at Nicaraguan volcanoes (Mombacho, Masaya, Momotombo, Cerro Negro, Telica, San Cristóbal). INETER, in cooperation with the Instituto de Nacional de Sismología, Vulcanología, Meteorología, e Hidrología de Guatemala (INSIVUMEH), uses a correlation spectrometer to measure sulfur dioxide fluxes during volcanic crises. This was employed in August 2005 at the erupting Concepción volcano, where an emission of 400 t/d of $SO_2$ was recorded. Nevertheless, such measurements are limited to periods of only a few days per year and thus cannot be considered to reflect the real behavior of the gas emission of the volcanoes both in quantity and chemical composition. Therefore, application of the above mentioned Mini-DOAS technique to San Cristóbal and Masaya volcanoes is planned to begin in 2006 as part of a European Union program. The plan is to carry out a continuous automatic sampling during the day and to correlate the measurements with seismic and other continuous data streams.

### Satellite Remote Sensing

The rapid development in remote sensing by satellites during the past two decades has greatly increased the potential for monitoring volcanoes in the entire electromagnetic spectrum (Francis et al., 1996). Remote sensing is simply detection and measurement of electromagnetic radiation emitted from the surface of a volcano or from the atmosphere above. The radiation can be reflected sunlight, suitable for measuring changes in $SO_2$ emission (e.g., Krueger et al., 2000); infrared radiation, indicating changes in surface heat flux (e.g., Flynn et al., 2002); or the reflected pulse from a synthetic aperture radar (SAR) system that bombards a volcano with microwave energy and allows measurement of changes in ground elevation. Observations in the visible part of the electromagnetic spectrum are limited by cloud cover and daylight. Radar penetrates any type of weather but is insensitive to temperature. Modern satellites, therefore, carry multispectral sensors (Francis et al., 1996). Temporal resolution of satellite remote sensing is determined by their orbital path and revolution. Polar-orbiting satellites used for environmental monitoring (e.g., LANDSAT–thematic mapping instrument, National Oceanic and Atmospheric Administration–advanced very high resolution radiometer) revolve in ~90 min, but, since Earth rotates underneath them, the repeat cycle for a given area is ~12 h.

An advantage of space-borne surveillance is total areal coverage of a volcano. This is particularly interesting for surface deformation measurements using radar interferometry (Francis et al., 1996). This technique uses radar images from successive overflights in conjunction with topographic data to construct an interferogram that shows the temporal changes in distance between satellite and surface spots with a high vertical and lateral resolution. The high precision and dense areal coverage make SAR interferometry an interesting alternative to ground-based deformation measurements.

Measuring temperature is a standard monitoring technique that is applied by INETER during monthly visits to active volcanoes. Direct measurements of temperature in fumaroles, springs, and soil are complemented by satellite-based infrared sensors that can detect a 1000 °C hot spot of only 0.1 $m^2$ area, or detect temperature changes of a few tenths of a degree. Such thermal mapping has mostly been used to monitor ongoing eruptions. It proved useful, for example, in combination with seismic measurements to monitor the evolution of the 1997–2000 Colima eruption in México (Galindo and Dominguez, 2002). Weak ground heating as a possible precursor of eruption, however, is difficult to distinguish from other (mostly meteorological) effects (Francis et al., 1996). Twenty-three active volcanoes in Nicaragua and the whole of Central America are thermally monitored by INETER in real time using satellite images (AVHRR; e.g., Flynn et al., 2000, 2002). Thermal anomalies in the large open-vent volcanoes Telica and San Cristóbal can be detected by satellite (http://sat-server.ineter.gob.ni/page03/countries.htm).

Another important application of space-borne sensors is to monitor the height, topography, and aerosol content of spreading volcanic plumes (e.g., Carn et al., 2003). Moreover, satellite images are a valuable tool to help create hazard maps. After a disaster has struck, satellite images can provide an overview of the situation to adjust management strategies (Francis et al., 1996).

### Risk Assessment

Volcanic eruptions cannot be influenced by man now or in the foreseeable future. The risk, however, can be reduced by preparedness, which includes both plans for immediate crisis response as well as long-term plans to rearrange infrastructure in a way that reduces risk. The first step is risk identification by recognizing the volcanic hazards that may affect a given area and the factors that determine the vulnerability of that area (Blong, 2000). A convenient way to document the areal distribution of volcanic hazards is through hazard maps. Such maps outline areas that may be affected by volcanic phenomena, such as fallout, pyroclastic flows, or lahars, with a certain probability, typically distinguishing low, medium, and high risk zones. Hazard maps are traditionally based on volcanological studies of the eruptive history of a volcano. Each volcano behaves differently, requiring careful analysis of the local factors. In addition, advanced numerical models are increasingly used to define hazardous areas (Sparks, 2003). Tephra fallout dispersal can be modeled for given eruption conditions considering the local meteorological conditions (Carey, 1996; Rosi, 1998; Bonadonna et al., 2005). Other models, such as LAHARZ (Schilling, 1998) or TITAN-2D (Pitman et al., 2003), determine the spreading of various types of volcanic flows over digitized terrain. The risk for each area is then determined by the vulnerability derived from all information on infrastructure, such as population, buildings, lifelines, and industrial and agricultural use (Blong, 2000). Volcanic hazard and risk maps for Nicaraguan volcanoes were constructed by INETER (Delgado et al., 2002a, 2002b). Other hazard mapping projects are recently on the way for the volcanoes Telica, Cerro Negro, and El Hoyo. Figure 13 shows risk maps for some volcanic hazards at Concepción volcano, which was erupting (beginning 28 July 2005) at the time of this writing.

A useful tool in risk analysis is risk scenarios (Blong, 1996). These assume an eruption of a certain magnitude and intensity and evaluate its impacts on the various sectors of society. Such scenarios serve to develop crisis management plans but also to detect flaws in such plans. Another purpose is to define aims and priorities of long-term development planning both in urban and rural areas to reduce vulnerability (Blong, 1996). Hazard maps must be available and consulted by developers in order to avoid mistakes, such as the construction of the towns of El Porvenir and Rolando Rodriguez on the young lahar deposits at Casita volcano.

A critical question in creating risk scenarios is what kind of probable maximum eruption to use for a volcano (Blong, 1996). In the case of Cerro Negro, for example, the 1992 eruption is probably a reasonable choice of a maximum event (Hill et al., 1998). Masaya volcano has only produced small and mostly effusive eruptions largely limited to the caldera basin during the past <2000 yr (Walker et al., 1993), but large-magnitude eruptions as in earlier times (Pérez and Freundt, this volume) could occur again. Hence, it seems reasonable to create a set of risk scenarios considering a range of eruption types and magnitudes.

When creating risk scenarios or communicating with non-volcanologists, it is useful to quantify hazards in terms of probability. Probabilities can be estimated with the help of probability trees. A simple example, based on the six youngest eruptions from the Chiltepe volcanic complex, is shown in Figure 14. In column 1, $P$ is the probability that an eruption will occur within a given time frame, which is not given a value here because of the uncertainties discussed in the context of Figure 7. Column 2 is the height of the eruption column, showing that 50% of the eruptions reached $H > 21$ km. Column 3 evaluates the sectorial direction of fallout dispersal, which depends on the column height and the present-day wind conditions as shown in Figure 4. For example, intermediate-height (14–21 km) columns have a 75% chance of being driven to the west. Column 4 is the thickness of fallout at a downwind distance of 15 km, where all (100%) past eruptions of $H > 21$ km produced deposits >1 m thick. Application to the town of Mateare indicates a 50% probability that $H > 21$ km, in which case, there is a 100% probability that the tephra thickness is >1 m. Hence, the probability that Mateare would get covered by >1 m tephra by the next eruption at Chiltepe is $0.5 \times 1$, or 50%. The probability that this will occur within a given time span would be $P \times 50\%$ (if, say, $P = 20\%$, then total probability is 10%). Column 4 may be replaced by column 5, which gives the thickness at 15 km distance in the cross-wind direction. For a westward dispersing eruption, this would apply to Managua. Again, there is a 50% probability that the eruption column would be higher than 21 km where the dispersal is always to the west. Since 66% of such eruptions deposited tephra thicker than 1 m in western Managua, the probability is $0.5 \times 1 \times 0.66$, or 33%, that western Managua would be covered by such thickness during the next Chiltepe eruption. The factors considered in such probability trees depend on the kind of hazard studied. An analogous scheme could, for example, be designed to estimate the probability that a given area would be affected by pyroclastic surges. The reliability of the estimated probabilities, of course, increases with the number of past eruptions that have been recorded. The probability tree for fallout in León from Cerro Negro constructed by Connor et al. (2001) is based on 24 past events.

### Reducing Vulnerability

Preparedness of official authorities and individual people to manage an actual volcanic crisis through its advance, occurrence, and aftermath is an essential element in reducing vulnerability. Apart from the technical aspects discussed above, this includes efficient transfer of information between all parties involved and in particular, timely and appropriate information to the public, in a form that is understandable to nonspecialists (Johnston and Ronan, 2000). INETER publishes the actual conditions at the Nicaraguan volcanoes on its Web page (www.ineter.gob.

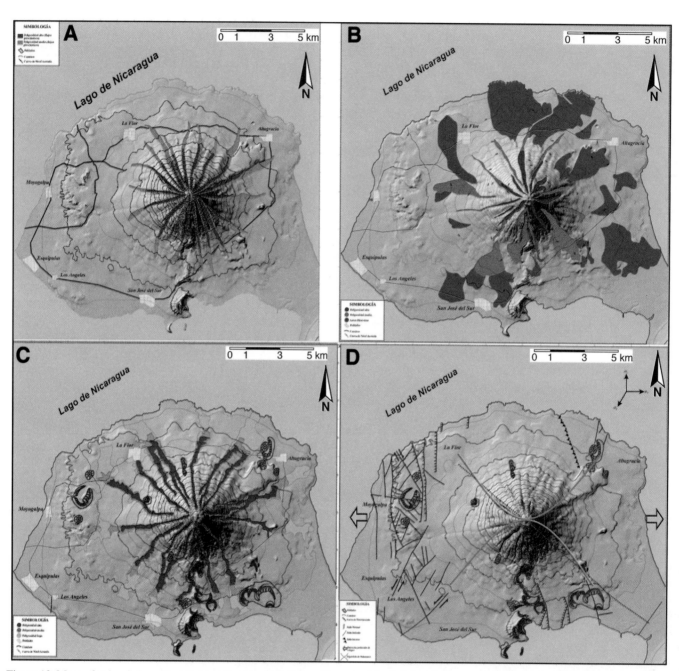

Figure 13. Maps of volcanic hazards and risk at Concepción volcano (Ometepe island, Lake Nicaragua) constructed by the Instituto Nicaragüense de Estudios Territoriales. (A) Pyroclastic-flow hazard zones. Red—high; orange—medium risk. (B) Lava-flow hazard zones. Red—high; orange—medium risk; dark gray—historic lava flows. (C) Lahar hazard zones. Red—high; orange—medium; yellow—low risk. (D) Probable direction of flank collapse to either east or west (arrows) determined from tectonic structures (blue—faults). Flow paths in (A) and (C) were determined by the Flow3D numerical model (see Delgado et al., 2002a).

ni/geofisica/vol/dep-vol.html) and in monthly bulletins (www.ineter.gob.ni/geofisica/sis/bolsis/bolsis.html). Receptivity of the public to information is best when awareness of volcanic hazards has been created by public education. Education programs have been set up in many volcanic areas around the world (Johnston and Ronan, 2000), including community disaster training. The International Association of Volcanology and Geochemistry of the Earth's Interior (IAVCEI; www.iavcei.org) produced video tapes, "Understanding volcanic hazards" and "Reducing volcanic risk," that may be used in such programs. One step in this direction in Nicaragua, for example, is the book *Amenazas Naturales de Nicaragua (Natural Hazards in Nicaragua)*, published and distributed by INETER (www.ineter.gob.ni/geofisica/amenazas/libro-amenazas.html).

In the long run, adjusting rural and urban development to avoid high-risk areas is probably the most efficient way to reduce vulnerability. For one, when a new settlement is planned somewhere, the ground must be checked for previous hazardous volcanic events. Secondly, based on hazard and risk maps, areas of high risk must be delineated far in advance in order to avoid the pressure of individuals, communities, or commercial interest to open up new land for developments. In the last few years, several projects have been carried out in Nicaragua to achieve these goals. A World Bank–financed project directed by the Nicaraguan Emergency Commission (SINAPRED) and technically supervised by INETER combined hazard and vulnerability maps with proposals for land use planning in 30 high-risk municipalities in the country, most of them located in or near the volcanic chain. Since 2004, INETER, by request of the National House Construction Authority (INVUR) and local authorities, elaborated hazard studies at 90 sites all over the country (www.ineter.gob.ni/geofisica/proyectos/INVUR/index.html), some of them in rural areas that are affected by volcanic hazards. For these sites, the construction of new houses in high risk areas is avoided or only permitted with accompanying mitigation or prevention measures.

In times of crisis, the efficiency of monitoring and early warning systems is certainly a crucial factor in reducing vulnerability. The instrumental base of the volcano monitoring and early warning system developed and maintained by INETER has dramatically improved over the last decade by international help, support from the Nicaraguan government, and engagement of the Nicaraguan scientists and technicians. During this period, no disastrous volcanic emergency has occurred in Nicaragua (the 1998 Mitch disaster had no direct volcanic cause). However, as discussed above, a dramatic volcanic emergency might arise, for example, at Apoyeque volcano, very close to the Managua-Masaya area, the largest Nicaraguan population center. The capacity of technical personnel is certainly decisive as to whether optimal use is made of the monitored data in case of an emergency. INETER's volcanology group is very small, but it is permanently assisted by seismology, applied geology, and GIS groups comprising INETER's geophysical department, which is the multidisciplinary body necessary to understand and manage the technical aspects of a volcanic crisis. General knowledge, academic level,

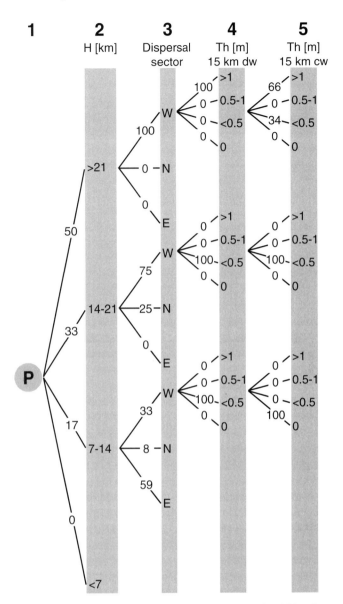

Figure 14. Probability tree for fallout tephra accumulation 15 km from vent by an eruption at Chiltepe peninsula. Only some of the branches are continued from column 1 through 4; column 5 (thickness at 15 km cross-wind-distance) may replace column 4 (thickness at 15 km down-wind-distance). P in column 1 is the probability that an eruption will occur within a given time frame. Numbers on connecting lines give probabilities in percent based on past six eruptions since ~15,000 yr ago (columns 2, 4, 5) and on present-day wind conditions (column 3). See text for discussion.

and scientific capacity are steadily improving in the geophysical department. However, as in other institutions important for the handling of volcanic crises, many deficiencies remain and need to be coped with. Both progress and problems must be seen in the light of the fact that Nicaragua has only recently passed through a deep socioeconomic crisis and is slowly recovering but is still considered the second poorest country in the Americas. Thus, permanent relations with foreign expert groups such as the U.S. Geological Survey–Volcano Disaster Program (VDAP) and other aid institutions must be, and are considered by Nicaragua to be, essential in the case of a volcano emergency with its dramatic human, social, economic, and political implications.

## SUMMARY AND CONCLUSIONS

The flanks and foothills around the active volcanoes of Nicaragua are endangered in many different ways. Highly destructive flank collapses of volcanoes, producing huge debris avalanches and debris flows, are comparatively rare. Particularly hazard-prone areas are the valleys that radiate from volcano flanks, because they channelize high-velocity pyroclastic flows and watery debris flows (lahars), which carry destruction for many tens of kilometers into the foreland (rarely >100 km) and can be generated many years to decades after a volcanic eruption. Over the past 200 yr, most people worldwide were killed during volcanic eruptions by pyroclastic flows and surges, lahars, and volcanogenic tsunamis (Blong, 1996). Fallout of ash and coarser rock fragments cause the overloading and collapse of roofs, particularly relatively close downwind of a volcano. The hazards of fine ash clouds to air traffic have also become apparent in the past few decades from the more than a dozen near catastrophes, which occurred when large jets crossed ash clouds and their turbines became clogged. The entry of debris avalanches or large pyroclastic flows into Lake Managua or Lake Nicaragua, as well as explosive eruptions within or close to these lakes, can generate destructive and lethal tsunamis. The explosive interaction between rising magma and groundwater or surface water is especially dangerous. In many volcanic eruptions, several of these hazard phenomena can occur almost simultaneously. Toxic gases permanently emitted from open vents are a health hazard to people and animals and pollute water and soil downwind of active volcanoes.

The greatest volcanic fatalities in the twentieth century resulted from relatively small eruptions in areas with high population density. The catastrophic effect was due to their sudden onset, which prevented a timely evacuation, or to political reasons, which delayed response planning. Predicting the type of an eruption, its location, and if possible its onset requires very careful analysis of the development of the history of a volcano. Successful predictions require long-term monitoring to understand the characteristic behavior of volcanoes prior to large eruptions. The most important monitoring methods are to record magnitudes and rate changes of volcanic earthquakes, swelling of the volcanic edifice, and gas emissions, especially of $SO_2$. The rapid developments in remote sensing via satellites have significantly increased the possibility for monitoring volcanoes using the entire electromagnetic spectrum and providing complete areal coverage.

Volcanic catastrophes have developed when people did not protect themselves early enough in the face of an impending volcanic eruption. This problem is magnified by the pressure of growing populations who settle the fertile slopes of active volcanoes. This makes it increasingly difficult to prevent potentially hazardous areas from being occupied by people. Clearly, societies today are much more vulnerable vis-à-vis volcanic and other natural hazards than in the past, particularly through the rapid growth of megacities and the increasingly intricate networks of traffic, communication, and supply lines.

After the disaster of Hurricane Mitch in 1998, Nicaragua has significantly increased its technical and organizational capacity for disaster management, the extent and quality of databases on volcanic and other hazards, and the availability of hazard, vulnerability, and risk maps for scientific purposes and land use planning by local and central governmental administrations and the general public. Nicaragua's monitoring and early warning system on volcanic phenomena has developed greatly over the last few years. Real-time GPS measurements, continuous sampling of amount and composition of volcanic gas emissions, and densification of seismic networks around the active volcanoes and densely populated areas were important steps for the improvement of the system. Cooperation with international expert groups is essential for volcano hazard assessment and adequate management of volcanic crisis. Challenges for the future include the training of volcanologists to be able to correctly interpret the signals of the complex volcano monitoring system, the education of the population for adequate behavior when an eruption occurs, and the training of the planners and decision makers to optimize land use while taking into account volcano risks.

## APPENDIX: GLOSSARY OF MAJOR TECHNICAL TERMS

Definitions are from Schmincke, 2004.

**Volcanic hazard** to people and property is a threatening event that is defined by a particular type, magnitude, and probability of occurrence of volcanic activity. Volcanic hazards in Nicaragua take many forms, as discussed in this chapter.

**Vulnerability** is defined as the degree to which the entire community is subject to impacts of hazards. This includes lives, property, essential environmental resources, economy, and standards of living.

**Volcanic risk** is applied to volcanic hazard in relation to life or property in a particular spot and at a particular time. Risk is then a combination of hazard features (which are specific to each volcano) and local valuable elements (including population, resources and infrastructure). People, property and natural resources are vulnerable to hazards when situated too close to a volcano, or, more specifically, to potential processes within a particular form of eruption.

**Disasters** result from the interaction of natural events with political, economic, social, and technological processes. Disasters occur when people do not recognize the potential hazards of a volcano or particular types of volcanic eruptions, warnings from the volcano itself, or from scientists and hence do not protect themselves through mitigation and emergency response plans. In addition, political issues and insufficient communication between scientists, community leaders, and emergency managers often lead to the worst types of volcanic disaster. A volcanic disaster or catastrophe strikes when the scale or particular nature of a volcanic event exceeds the capacity of a community to cope. The ability to cope depends on the economic potential of a community or country and the strength of private or public institutions capable of quickly and effectively responding to a hazardous natural event.

**Disaster mitigation** includes all activities toward reducing risk, either the hazards (via physical interventions; e.g., sabo dams) or vulnerability (via hazard mapping, land use planning). It also includes preparation by responsible administrative bodies and civil protection authorities as well as a public that is fully informed about all possibilities, including evacuation measures. This is a formidable task, because the spectrum of volcanic eruptions in Nicaragua and, therefore, hazard types is large.

## ACKNOWLEDGMENTS

We gratefully acknowledge the support by the Instituto Nicaragüense de Estudios Territoriales, Managua, during field work in Nicaragua. Constructive reviews by Crystal Mann and Willie Scott helped to improve this paper. Radiocarbon ages were determined at the Leibniz Laboratory for Radiometric Dating and Stable Isotope Research at Kiel University. This publication is contribution no. 77 of the Sonderforschungsbereich 574 "Volatiles and Fluids in Subduction Zones" at Kiel University. Additional support received from Volkswagen Foundation grant 1/76.

## REFERENCES CITED

Andres, R.J., and Rose, W.I., 1995, Remote sensing spectroscopy of volcanic plumes and clouds, *in* McGuire, B., Kilburn, C., and Murray, J., eds., Monitoring Active Volcanoes: Strategies, Procedures and Techniques: London, UCL Press Limited, p. 301–314.

Andres, R.J., and Kasgnoc, A.D., 1998, A time-averaged inventory of subaerial volcanic sulfur emissions: Journal of Geophysical Research, v. 103, p. 25,251–25,261, doi: 10.1029/98JD02091.

Bacon, C.R., 1982, Time-predictable bimodal volcanism in the Coso Range, California: Geology, v. 10, p. 65–69, doi: 10.1130/0091-7613(1982)10<65:TBVITC>2.0.CO;2.

Baxter, P.J., 2000, Impacts of eruptions on human health, *in* Sigurdsson, H., et al., eds., Encyclopedia of Volcanoes: New York, Academic Press, p. 1035–1044.

Beget, J.E., 2000, Volcanic tsunamis, *in* Sigurdsson, H., et al., eds., Encyclopedia of Volcanoes: New York, Academic Press, p. 1005–1014.

Belousov, A., Voight, B., Belousova, M., and Muravyev, Y., 2000, Tsunamis generated by subaquatic volcanic explosions: Unique data from 1996 eruption in Karymskoye Lake, Kamchatka, Russia: Pure and Applied Geophysics, v. 157, p. 1135–1143, doi: 10.1007/s000240050021.

Bice, D.C., 1985, Quaternary volcanic stratigraphy of Managua, Nicaragua: Correlation and source assignment for multiple overlapping plinian deposits: Geological Society of America Bulletin, v. 96, p. 553–566, doi: 10.1130/0016-7606(1985)96<553:QVSOMN>2.0.CO;2.

Blong, R.J., 1996, Volcanic hazards risk assessment, *in* Scarpa, R., and Tilling, R.I., eds., Monitoring and mitigation of volcano hazards: Berlin, Springer, p. 675–698.

Blong, R., 2000, Volcanic hazards and risk management, *in* Sigurdsson, H., et al., eds., Encyclopedia of Volcanoes: New York, Academic Press, p. 1215–1227.

Bobrowski, N., Hönninger, G., Galle, B., and Platt, U., 2003, Detection of bromine monoxide in a volcanic plume: Nature, v. 423, p. 273–276, doi: 10.1038/nature01625.

Bonadonna, C., Phillips, J.C., and Houghton, B.F., 2005, Modeling tephra sedimentation from a Ruapehu weak plume eruption: Journal of Geophysical Research, v. 110, doi:10.1029/2004JB003515.

Borgia, A., Delaney, P.T., and Denlinger, R.P., 2000, Spreading volcanoes: Annual Review of Earth and Planetary Sciences, v. 28, p. 539–570, doi: 10.1146/annurev.earth.28.1.539.

Borgia, A., and van Wyk de Vries, B., 2003, The volcano-tectonic evolution of Concepción, Nicaragua: Bulletin of Volcanology, v. 65, p. 248–266.

Carey, S.N., 1996, Modeling of tephra fallout from explosive eruptions, *in* Scarpa, R., and Tilling, R.I., eds., Monitoring and mitigation of volcano hazards: Berlin, Springer, p. 429–461.

Carey, S., Morelli, D., Sigurdsson, H., and Bronto, S., 2001, Tsunami deposits from major explosive eruptions: An example from the 1883 eruption of Krakatau: Geology, v. 29, p. 347–350, doi: 10.1130/0091-7613(2001)029<0347:TDFMEE>2.0.CO;2.

Carn, S.A., Krueger, A.J., Bluth, G.J.S., Schaefer, S.J., Krotkov, N.A., Watson, I.M., and Datta, S., 2003, Volcanic eruption detection by the Total Ozone Mapping Spectrometer (TOMS) instruments: A 22-year record of sulphur dioxide and ash emissions, *in* Oppenheimer, C., Pyle, D.M., and Barclay, J., Volcanic degassing: London, Geological Society Special Publication 213, p. 177–202.

Carr, M.J., 1984, Symmetrical and segmented variation of physical and geochemical characteristics of the Central American volcanic front: Journal of Volcanology and Geothermal Research, v. 20, p. 231–252, doi: 10.1016/0377-0273(84)90041-6.

Carr, M.J., Rose, W.I., and Stoiber, R.E., 1982, Central America, *in* Thorpe, R.S., ed., Andesites: Orogenic Andesite and Related Rocks: New York, Wiley, p. 149–166.

Cecchi, E., van Wyk de Vries, B., and Lavest, J.-M., 2005, Flank spreading and collapse of weak-cored volcanoes: Bulletin of Volcanology, v. 67, p. 72–91.

Connor, C.B., Hill, B.E., Winfrey, B., Franklin, N.M., and LaFemina, P.C., 2001, Estimation of volcanic hazards from tephra fallout: Natural Hazards Review, v. 2, p. 33–42, doi: 10.1061/(ASCE)1527-6988(2001)2:1(33).

Cronin, S.J., Neall, V.E., Lecointre, J.A., and Palmer, A.S., 1997, Changes in Whangaehu River lahar characteristics during the 1995 eruption sequence, Ruapehu volcano, New Zealand: Journal of Volcanology and Geothermal Research, v. 76, p. 47–61, doi: 10.1016/S0377-0273(96)00064-9.

Cronin, S.J., Neall, V.E., Lecointre, J.A., and Palmer, A.S., 1999, Dynamic interactions between lahars and stream flow: A case study from Ruapehu volcano, New Zealand: Geological Society of America Bulletin, v. 111, p. 28–38, doi: 10.1130/0016-7606(1999)111<0028:DIBLAS>2.3.CO;2.

Delgado, H., Navarro, M., Abimelec, I., and Alatorre, M.A., 2002a, Volcanic hazards map of Concepción volcano: Universidad Nacional Autónoma de México and the Instituto Nicaragüense de Estudios Territoriales, Managua, maps and report, in Spanish, scale 1:50,000, 78 p.

Delgado, H., Navarro, M., Abimelec, I., and Horst, M.A., 2002b, Volcanic hazards map of Masaya volcano: Universidad Nacional Autónoma de México and the Instituto Nicaragüense de Estudios Territoriales, Managua, maps and report, in Spanish, scale 1:50,000, 120 p.

Delmelle, P., Stix, J., Baxter, P.J., Garcia-Alvarez, J., and Barquero, J., 2002, Atmospheric dispersion, environmental effects and potential health hazard associated with the low-altitude gas plume of Masaya volcano, Nicaragua: Bulletin of Volcanology, v. 64, p. 423–434, doi: 10.1007/s00445-002-0221-6.

Delmelle, P., Stix, J., Bourque, C.P.A., Baxter, P.J., Garcia-Alvarez, G., and Barquero, J., 2001, Dry deposition and heavy acid loading in the vicinity of Masaya volcano, a major sulfur and chlorine source in Nicaragua: Environmental Science & Technology, v. 35, p. 1289–1293, doi: 10.5555/es000153m.

de Oviedo y Valdes, G.F., 1855, Historia General y Natural de las Indias, *in* Amador de los Rios, J., ed., Madrid, v. 3-4, p. 70–92.

Duffell, H.J., Oppenheimer, C., Pyle, D.M., Galle, B., McGonigle, A.J.S., and Burton, M.R., 2003, Changes in gas composition prior to a minor explosive eruption at Masaya volcano, Nicaragua: Journal of Volcanology and Geothermal Research, v. 126, p. 327–339, doi: 10.1016/S0377-0273(03)00156-2.

Dzurisin, D., Westphal, J.A., and Johnson, D.J., 1983, Eruption prediction aided by electronic tiltmeter data at Mount St. Helens: Science, v. 221, p. 1381–1383.

Eaton, J.P., and Murata, D.J., 1960, How volcanoes grow: Science, v. 132, p. 925–938.

Edmonds, M., Pyle, D., and Oppenheimer, C., 2001, A model for degassing at the Soufriere Hills Volcano, Montserrat, West Indies, based on geochemical data: Earth and Planetary Science Letters, v. 186, p. 159–173, doi: 10.1016/S0012-821X(01)00242-4.

Flynn, L.P., Harris, A.J.L., Rothery, D.A., and Oppenheimer, C., 2000, Landsat and hyperspectral analyses of active lava flows, *in* Mouginis-Mark, P.J., Fink, J.H., and Crisp J.A., eds., Remote sensing of active volcanism: American Geophysical Union Geophysical Monograph Series, v. 116, p. 161–177.

Flynn, L., Wright, R., Garbeil, H., Harris, A., and Pilger, E., 2002, A global thermal alert system using MODIS: Initial results from 2000–2001: Advances in Environmental Monitoring and Modelling, v. 1, p. 37–69.

Francis, P.W., and Oppenheimer, C., 2004, Volcanoes, 2nd edition: Oxford, Oxford University Press, 534 p.

Francis, P.W., Wadge, G., and Mouginis-Mark, P.J., 1996, Satellite monitoring of volcanoes, in Scarpa, R., and Tilling, R.I., eds., Monitoring and mitigation of volcano hazards: Berlin, Springer, p. 257–298.

Freundt, A., 2003, Entrance of hot pyroclastic flows into the sea: Experimental observations: Bulletin of Volcanology v. 65, p. 144–164.

Freundt, A., Wilson, C.J.N., and Carey, S.N., 2000, Ignimbrites and block-and-ash flows, in Sigurdsson, H., et al., eds., Encyclopedia of Volcanoes: New York, Academic Press, p. 581–599.

Freundt, A., Kutterolf, S., Wehrmann, H., Schmincke, H.U., and Strauch, W., 2006, Tsunami generated by a compositionally zoned plinian eruption at Chiltepe peninsula, Lake Managua (Xolotlan), Nicaragua: Journal of Volcanology and Geothermal Research, v. 149, p. 103 123, doi: 10.1016/j.jvolgeores.2005.06.001.

Gaceta Oficial de Nicaragua, 1886, Eruption of Momotombo: Managua, Gaceta Oficial de Nicaragua.

Galindo, I., and Dominguez, T., 2002, Near real-time satellite monitoring during the 1997–2000 activity of Volcan de Colima (Mexico) and its relationship with seismic monitoring: Journal of Volcanology and Geothermal Research, v. 117, p. 91–104, doi: 10.1016/S0377-0273(02)00238-X.

Galle, B., Oppenheimer, C., Geyer, A., McGonigle, A., Edmonds, M., and Horrocks, L., 2003, A miniaturised ultraviolet spectrometer for remote sensing of $SO_2$ fluxes: A new tool for volcano surveillance: Journal of Volcanology and Geothermal Research, v. 119, p. 241–254, doi: 10.1016/S0377-0273(02)00356-6.

Garofalo, K., Tassi, F., Vaselli, O., Delgado-Huertas, A., Tedesco, D., Frische, M., Hansteen, T.H., Poreda, R.J., and Strauch, W., 2006, The fumarolic gas discharges at Mombacho Volcano (Nicaragua): Presence of magmatic gas species and implications for a volcanic surveillance: Bulletin of Volcanology (in press).

Giggenbach, W.F., 1996, Chemical composition of volcanic gases, in Scarpa, M., and Tilling, R.J., eds., Monitoring and mitigation of volcanic hazards: Berlin, Springer-Verlag, p. 221–256.

Hazlett, R.W., 1987, Geology of the San Cristóbal volcanic complex, Nicaragua: Journal of Volcanology and Geothermal Research, v. 33, p. 223–230, doi: 10.1016/0377-0273(87)90064-3.

Hill, B.E., Connor, C.B., Jarzemba, M.S., and La Femina, P.C., 1998, 1995 eruptions of Cerro Negro, Nicaragua, and risk assessment for future eruptions: Geological Society of America Bulletin, v. 110, p. 1231–1241, doi: 10.1130/0016-7606(1998)110<1231:EOCNVN>2.3.CO;2.

Hodgson, K.A., and Manville, V.R., 1999, Sedimentology and flow behavior of a rain-triggered lahar, Mangatoetoenui Stream, Ruapehu volcano, New Zealand: Geological Society of America Bulletin, v. 111, p. 743–754, doi: 10.1130/0016-7606(1999)111<0743:SAFBOA>2.3.CO;2.

Hradecky, P., 2001, Informe final de consultoria para el proyecto "Actualización del mapa geológico-estructural del area de Managua": Managua, Instituto Nicaragüense de Estudios Territoriales.

Instituto Nicaragüense de Estudios Territoriales (INETER), 2005, Situación actual: www.ineter.gob.ni/geofisica/desliza/desliza.html (last accessed June 2006).

Iverson, R.M., Schilling, S.P., and Vallance, J.W., 1998, Objective delineation of lahar-inundation hazard zones: Geological Society of America Bulletin, v. 110, p. 972–984, doi: 10.1130/0016-7606(1998)110<0972:ODOLIH>2.3.CO;2.

Johnston, D., and Ronan, K., 2000, Risk education and intervention, in Sigurdsson, H., eds., et al., Encyclopedia of Volcanoes: New York, Academic Press, p. 1229–1240.

Kazahaya, K., Shinohara, H., Uto, K., Odai, M., Nakahori, Y., Mori, H., Iino, H., Miyashita, M., and Hirabayashi, J., 2004, Gigantic $SO_2$ emission from Miyakejima volcano, Japan, caused by caldera collapse: Geology, v. 32, p. 425–428, doi: 10.1130/G20399.1.

Kerle, N., 2002, Volume estimation of the 1998 flank collapse at Casita volcano, Nicaragua—a comparison of photogrammetric and conventional techniques: Earth Surface Processes and Landforms, v. 27, p. 759–772, doi: 10.1002/esp.351.

Kerle, N., van Wyk de Vries, B., and Oppenheimer, C., 2003, New insight into the factors leading to the 1998 flank collapse and lahar disaster at Casita volcano, Nicaragua: Bulletin of Volcanology, v. 65, p. 331–345.

Krueger, A.J., Schaefer, S.J., Krotkov, N., Bluth, G., and Barker, S., 2000, Ultraviolet remote sensing of volcanic emissions, in Mouginis-Mark, P.J., Crisp, J.A., and Fink, J.H., eds., Remote Sensing of Active Volcanism: American Geophysical Union, Geophysical Monograph Series, v. 116, p. 25–43.

Kutterolf, S., Schacht, U., Wehrmann, H., Freundt, A., and Mörz, T., 2006, Onshore to offshore tephrostratigraphy and marine ash layer diagenesis in Central America, in Bundschuh, J., and Alvarado, G.E., eds., Central America—Geology, Resources and Hazards: Rotterdam, Balkema (in press).

La Femina, P.C., Connor, C.B., Hill, B.E., Strauch, W., and Armando Saballos, J., 2004, Magma-tectonic interactions in Nicaragua: The 1999 seismic swarm and eruption of Cerro Negro volcano: Journal of Volcanology and Geothermal Research, v. 137, p. 187–199, doi: 10.1016/j.jvolgeores.2004.05.006.

Lavigne, F., and Thouret, J.C., 2003, Sediment transport and deposition by rain-triggered lahars at Merapi volcano, Central Java, Indonesia: Geomorphology, v. 49, p. 45–69, doi: 10.1016/S0169-555X(02)00160-5.

Lesage, Ph., Mora, M., Strauch, W., Escobar, D., Matías, O., Tenorio, V., Talavera, E., Rodríguez, A., and Alvarado, G., 2006, Volcano Seismology, in Bundschuh, J., and Alvarado, G.E., eds., Central America—Geology, Resources and Hazards: Rotterdam, Balkema (in press).

Lewicki, J.L., Connor, C., St-Amand, K., Stix, J., and Spinner, W., 2003, Self-potential, soil $CO_2$ flux, and temperature on Masaya volcano, Nicaragua: Geophysical Research Letters, v. 30, p. 1817, doi: 10.1029/2003GL017731.

McBirney, A.R., 1956, The Nicaraguan volcano Masaya and its caldera: Eos (Transactions, American Geophysical Union), v. 37, p. 83–96.

McBirney, A.R., and Williams, H., 1965, Volcanic history of Nicaragua: University of California Publications in Geological Science, v. 55, p. 1–65.

McKnight, S.B., and Williams, S.N., 1997, Old cinder cone or young composite volcano?: The nature of Cerro Negro, Nicaragua: Geology, v. 25, p. 339–342, doi: 10.1130/0091-7613(1997)025<0339:OCCOYC>2.3.CO;2.

McNutt, S.R., 1996, Seismic monitoring and eruption forecasting of volcanoes: A review of the state-of-the-art and case histories, in Scarpa, R., and Tilling, R.I., eds., Monitoring and mitigation of volcano hazards: Berlin, Springer, p. 99–146.

McNutt, S.R., 2000a, Volcanic seismicity, in Sigurdsson, H., et al., eds., Encyclopedia of Volcanoes: New York, Academic Press, p. 1015–1033.

McNutt, S.R., 2000b, Seismic monitoring, in Sigurdsson, H., et al., eds., Encyclopedia of Volcanoes: New York, Academic Press, p. 1095–1119.

Métaxian, J.P., Lesage, P., and Dorel, J., 1997, Permanent tremor of Masaya volcano, Nicaragua: Wave field analysis and source location: Journal of Geophysical Research, v. 102, p. 22,529–22,545, doi: 10.1029/97JB01141.

Miller, T.P., and Casadevall, T.J., 2000, Volcanic ash hazards to aviation, in Sigurdsson, H., et al., eds., Encyclopedia of Volcanoes: New York, Academic Press, p. 915–930.

Mothes, P.A., Hall, M.L., and Janda, R.J., 1998, The enormous Chillos Valley lahar: An ash-flow-generated debris flow from Cotopaxi volcano, Ecuador: Bulletin of Volcanology v. 59, p. 233–244, doi: 10.1007/s004450050188.

Murray, J.B., Rymer, H., and Locke, C.A., 2000, Ground deformation, gravity, and magnetics, in Sigurdsson, H., et al., eds., Encyclopedia of Volcanoes: New York, Academic Press, p. 1121–1140.

Newhall, C.G., and Self, S., 1982, The volcanic explosivity index (VEI): An estimate of explosive magnitude for historical volcanism: Journal of Geophysical Research, v. 87, p. 1231–1238.

Pérez, J., 1993, Obras Históricas Completas, 2nd edition: Managua, Fondo de Promoción Cultural, BANIC, p. 793

Pérez, W., and Freundt, A., this volume, The youngest highly explosive basaltic eruptions from Masaya Caldera (Nicaragua): Stratigraphy and hazard assessment, in Rose, W.I., Bluth, G.J.S., Carr, M.J., Ewert, J.W., Patino, L., and Vallance, J.W., Volcanic hazards in Central America: Geological Society of America Special Paper 412, doi: 10.1130/2006.2412(10).

Pierson, T.C., and Costa, J.E., 1987, A rheological classification of subaerial sediment-water flows, in Costa, J.E., and Wieczorek, G.F., eds., Debris flows/avalanches: Process, recognition and mitigation: Reviews in Engineering Geology, v. 7, p. 1–12.

Pitman, E.B., Nichita, C.C., Patra, A., Bauer, A., Sheridan, M., and Bursik, M., 2003, Computing granular avalanches and landslides: Physics of Fluids, v. 15, p. 3638–3646, doi: 10.1063/1.1614253.

Protti, M., Gundel, F., and McNally, K., 1995, Correlation between the age of the subducting Cocos plate and the geometry of the Wadati-Benioff zone under Nicaragua and Costa Rica, in Mann, P., eds., Geologic and tectonic development of the Caribbean plate boundary in southern Central America: Geological Society of America Special Paper 295, p. 309–326.

Pyle, D.M., 2000, Sizes of volcanic eruptions, in Sigurdsson, H., et al., eds., Encyclopedia of Volcanoes: New York, Academic Press, p. 263–270.

Ranero, C.R., von Huene, R., Flueh, E., Duarte, M., Baca, D., and McIntosh, K., 2000, A cross section of the convergent Pacific margin of Nicaragua: Tectonics, v. 19, p. 335–357, doi: 10.1029/1999TC900045.

Rice, A., 2000, Rollover in volcanic crater lakes: A possible cause for Lake Nyos type disasters: Journal of Volcanology and Geothermal Research, v. 97, p. 233–239, doi: 10.1016/S0377-0273(99)00179-1.

Rosi, M., 1998, Plinian eruption columns: Particle transport and fallout, in Freundt, A., and Rosi, M., eds., From magma to tephra: Modelling physical processes of explosive volcanic eruptions: Amsterdam, Elsevier, Developments in Volcanology, v. 4, p. 139–172.

Rymer, H., 1996, Microgravity monitoring, in Scarpa, R., and Tilling, R.I., eds., Monitoring and mitigation of volcano hazards: Berlin, Springer, p. 169–197.

Rymer, H., van Wyk de Vries, B., Stix, J., and Williams-Jones, G., 1998, Pit crater structure and processes governing persistent activity at Masaya volcano, Nicaragua: Bulletin of Volcanology, v. 59, p. 345–355.

Sapper, C., 1925, Los Volcanes de America Central: Halle, Germany, Max Niemeyer, 116 p.

Schilling, S.P., 1998, LAHARZ: GIS programs for automated mapping of lahar-inundation hazard zones: U.S. Geological Survey Open-File Report 98, 79 p.

Schmincke, H.U., 2004, Volcanism: Springer, Heidelberg, 324 p.

Scott, K.M., Marcias, J.L., Naranjo, J.A., Rodriguez, C., and McGeehin, J.P., 2001, Catastrophic debris flows transformed from landslides in volcanic terrains: mobility, hazard assessment, and mitigation strategies: U.S. Geological Survey Professional Paper 1630, 71 p.

Scott, K.M., Vallance, J.W., Kerle, N., Macias, J.L., Strauch, W., and Devoli, G., 2005, Catastrophic precipitation-triggered lahar at Casita volcano, Nicaragua: occurrence, bulking and transformation: Earth Surface Processes and Landforms, v. 30, p. 59–79, doi: 10.1002/esp.1127.

Scott, W., Gardner, C., Devoli, G., and Alvarez, A., this volume, The A.D. 1835 eruption of Volcán Cosigüina, Nicaragua: A guide for assessing local volcanic hazards, in Rose, W.I., Bluth, G.J.S., Carr, M.J., Ewert, J.W., Patino, L., and Vallance, J.W., Volcanic hazards in Central America: Geological Society of America Special Paper 412, doi: 10.1130/2006.2412(09).

Self, S., Rampino, M.R., and Carr, M.J., 1989, A reappraisal of the 1835 eruption of Cosigüina and its atmospheric impact: Bulletin of Volcanology v. 52, p. 57–65, doi: 10.1007/BF00641387.

Sheridan, M.F., Bonnard, C., Carreno, R., Siebe, C., Strauch, W., Navarro, M., Calero, J.C., and Truijilo, N.B., 1999, 30 October 1998 rock fall/avalanche and breakout flow of Casita volcano, Nicaragua, triggered by Hurricane Mitch: Landslide News, v. 12, p. 2–4.

Simkin, T., 1993, Terrestrial volcanism in space and time: Annual Reviews of Earth and Planetary Science, v. 21, p. 427–452, doi: 10.1146/annurev.ea.21.050193.002235.

Simkin, T., and Siebert, L., 2000, Earth's volcanoes and eruptions: An overview, in Sigurdsson, H., et al., eds., Encyclopedia of Volcanoes: New York, Academic Press, p. 249–261.

Smithsonian Institution, 2005, Smithsonian Global Volcanism Program, World-wide Holocene Volcano and Eruption Information: Washington, D.C., National Museum of Natural History, www.volcano.si.edu (last accessed June 2006).

Sparks, R.S.J., 2003, Forecasting volcanic eruptions: Earth and Planetary Science Letters, v. 210, p. 1–15, doi: 10.1016/S0012-821X(03)00124-9.

Spence, R.J.S., Kelman, I., Baxter, P.J., Zuccaro, G., and Petrazzuoli, S., 2005, Residential building and occupant vulnerability to tephra fall: Natural Hazards and Earth System Science, v. 5, p. 477–494.

Stoiber, R.E., and Carr, M.J., 1973, Quaternary volcanic and tectonic segmentation of Central America: Bulletin of Volcanology v. 37, p. 304–325.

Sussman, D., 1985, Apoyo Caldera, Nicaragua: a major Quaternary silicic eruptive center: Journal of Volcanology and Geothermal Research, v. 24, p. 249–282, doi: 10.1016/0377-0273(85)90072-1.

Swanson, D.A., Casadevall, T.J., Dzurisin, D., Holcomb, R.T., Newhall, C.G., Malone, S.D., and Weaver, C.S., 1985, Forecasts and predictions of eruptive activity at Mount St. Helens, USA: Journal of Geodynamics, v. 3, p. 397–423, doi: 10.1016/0264-3707(85)90044-4.

Tilling, R.I., and Dvorak, J.J., 1993, Anatomy of a basaltic volcano: Nature, v. 363, p. 125–133, doi: 10.1038/363125a0.

Ui, T., 1972, Recent volcanism in Masaya-Granada area, Nicaragua: Bulletin of Volcanology v. 36, p. 174–190.

U.S. Environmental Protection Agency (EPA), 2005, Air trends: More details on sulfur dioxide, based on data through 2002: http://www.epa.gov/oar/airtrends/sulfur2.html (last accessed June 2006).

Valentine, G.A., 1998, Damage to structures by pyroclastic flows and surges, inferred from nuclear weapons effects: Journal of Volcanology and Geothermal Research, v. 87, p. 117–140, doi: 10.1016/S0377-0273(98)00094-8.

Valentine, G.A., and Fisher, R.V., 2000, Pyroclastic surges and blasts, in Sigurdsson, H., et al., eds., Encyclopedia of Volcanoes: New York, Academic Press, p. 571–580.

Vallance, J.W., 2000, Lahars, in Sigurdsson, H., et al., eds., Encyclopedia of Volcanoes: New York, Academic Press, p. 601–616.

Vallance, J.W., Schilling, S.P., Devoli, G., and Howell, M.M., 2001a, Lahar hazards at Concepción Volcano, Nicaragua: U.S. Geological Survey Open-File Report 01-457, 13 p.

Vallance, J.W., Schilling, S.P., and Devoli, G., 2001b, Lahar hazards at Mombacho Volcano, Nicaragua: U.S. Geological Survey Open-File Report 01-455, 14 p.

Van der Laat, R., 1996, Ground-deformation methods and results, in Scarpa, R., and Tilling, R.I., eds., Monitoring and mitigation of volcano hazards: Berlin, Springer, p. 147–168.

van Wyk de Vries, B., Kerle, N., and Petley, D.N., 2000, Sector collapse forming at Casita volcano, Nicaragua: Geology, v. 28, p. 167–170, doi: 10.1130/0091-7613(2000)028<0167:SCFACV>2.3.CO;2.

Voight, B., and Cornelius, R.R., 1991, Prospects for eruption prediction in near-real-time: Nature, v. 350, p. 695–698, doi: 10.1038/350695a0.

Voight, B., Glicken, H., Janda, R.J., and Douglass, P.M., 1981, Catastrophic rockslide avalanche of May 18, in Lipman, P.W., and Mullineaux, D.R., eds., The 1980 eruptions of Mount St. Helens, Washington: U.S. Geological Survey Professional Paper 1250, p. 347–377.

Voight, B., Hoblitt, R.P., Clarke, A.B., Lockhart, A.B., Miller, A.D., Lynch, L., and McMahon, J., 1998, Remarkable cyclic ground deformation monitored in real-time on Montserrat, and its use in eruption forecasting: Geophysical Research Letters, v. 25, p. 3405–3408, doi: 10.1029/98GL01160.

Wadge, G., 1982, Steady state volcanism: Evidence from eruption histories of polygenetic volcanoes: Journal of Geophysical Research, v. 87, p. 4035–4049.

Walker, J.A., 1984, Volcanic rocks from the Nejapa and Granada cinder cone alignments, Nicaragua: Central America: Journal of Petrology, v. 25, p. 299–342.

Walker, J.A., Williams, S.N., Kalamarides, R.I., and Feigensohn, M.D., 1993, Shallow open-system evolution of basaltic magma beneath a subduction zone volcano: the Masaya caldera complex, Nicaragua: Journal of Volcanology and Geothermal Research, v. 56, p. 379–400, doi: 10.1016/0377-0273(93)90004-B.

Walker, J.A., Patino, L.C., Carr, M.J., and Feigenson, M.D., 2001, Slab control over HFSE depletions in central Nicaragua: Earth and Planetary Science Letters, v. 192, p. 533–543, doi: 10.1016/S0012-821X(01)00476-9.

Walther, C.H.E., Flueh, E.R., Ranero, C.R., von Huene, R., and Strauch, W., 2000, Crustal structure across the Pacific margin of Nicaragua: Evidence for ophiolitic basement and a shallow mantle sliver: Geophysical Journal International, v. 141, p. 759–777, doi: 10.1046/j.1365-246x.2000.00134.x.

Wehrmann, H., Bonadonna, C., Freundt, A., Houghton, B.F., and Kutterolf, S., this volume, Fontana Tephra: A basaltic Plinian eruption in Nicaragua, in Rose, W.I., Bluth, G.J.S., Carr, M.J., Ewert, J.W., Patino, L., and Vallance, J.W., Volcanic hazards in Central America: Geological Society of America Special Paper 412, doi: 10.1130/2006.2412(11).

Weinberg, R.F., 1992, Neotectonic development of western Nicaragua: Tectonics, v. 11, p. 1010–1017.

Weyl, R., 1980, Geology of Central America, in Bender, F., et al., eds., Beiträge zur regionalen Geologie der Erde (2nd ed.): Berlin–Stuttgart, Gebrüder Borntraeger, v. 15, p. 1–371.

Williams, S.N., 1983, Plinian airfall deposits of basaltic composition: Geology, v. 11, p. 211–214, doi: 10.1130/0091-7613(1983)11<211:PADOBC>2.0.CO;2.

Manuscript Accepted by the Society 19 March 2006

# The A.D. 1835 eruption of Volcán Cosigüina, Nicaragua: A guide for assessing local volcanic hazards

**William Scott**
**Cynthia Gardner**
Cascades Volcano Observatory, U.S. Geological Survey, 1300 SE Cardinal Court, Vancouver, Washington 98683, USA

**Graziella Devoli***
**Antonio Alvarez**
Instituto Nicaragüense de Estudios Territoriales (INETER), Código Postal 2110, Managua, Nicaragua

## ABSTRACT

The January 1835 eruption of Volcán Cosigüina in northwestern Nicaragua was one of the largest and most explosive in Central America since Spanish colonization. We report on the results of reconnaissance stratigraphic studies and laboratory work aimed at better defining the distribution and character of deposits emplaced by the eruption as a means of developing a preliminary hazards assessment for future eruptions. On the lower flanks of the volcano, a basal tephra-fall deposit comprises either ash and fine lithic lapilli or, locally, dacitic pumice. An overlying tephra-fall deposit forms an extensive blanket of brown to gray andesitic scoria that is 35–60 cm thick at 5–10 km from the summit-caldera rim, except southwest of the volcano, where it is considerably thinner. The scoria fall produced the most voluminous deposit of the eruption and underlies pyroclastic-surge and -flow deposits that chiefly comprise gray andesitic scoria. In northern and southeastern sectors of the volcano, these flowage deposits form broad fans and valley fills that locally reach the Gulf of Fonseca. An arcuate ridge 2 km west of the caldera rim and a low ridge east of the caldera deflected pyroclastic flows northward and southeastward. Pyroclastic flows did not reach the lower west and southwest flanks, which instead received thick, fine-grained, accretionary-lapilli–rich ashfall deposits that probably derived chiefly from ash clouds elutriated from pyroclastic flows. We estimate the total bulk volume of erupted deposits to be ~6 km³. Following the eruption, lahars inundated large portions of the lower flanks, and erosion of deposits and creation of new channels triggered rapid alluviation. Pre-1835 eruptions are poorly dated; however, scoria-fall, pyroclastic-flow, and lahar deposits record a penultimate eruption of smaller magnitude than that of 1835. It occurred a few centuries earlier—perhaps in the fifteenth century. An undated sequence of thick tephra-fall deposits on the west flank of the volcano records tens of eruptions, some of which were greater in magnitude than that of 1835. Weathering

---

*Present address: International Centre for Geohazards, Norge Geotekniske Institutt, Sognsveien 72, P.B. 3930, Ulleväal Stadion, N-0806 Oslo, Norway

Scott, W., Gardner, C., Devoli, G., and Alvarez, A., 2006, The A.D. 1835 eruption of Volcán Cosigüina, Nicaragua: A guide for assessing local volcanic hazards, in Rose, W.I., Bluth, G.J.S., Carr, M.J., Ewert, J.W., Patino, L.C., and Vallance, J.W., Volcanic hazards in Central America: Geological Society of America Special Paper 412, p. 167–187, doi: 10.1130/2006.2412(09). For permission to copy, contact editing@geosociety.org. ©2006 Geological Society of America. All rights reserved.

**evidence suggests this sequence is at least several thousand years old. The wide extent of pyroclastic flows and thick tephra fall during 1835, the greater magnitude of some previous Holocene eruptions, and the location of Cosigüina on a peninsula limit the options to reduce risk during future unrest and eruption.**

**Keywords:** Nicaragua, Cosigüina, historical eruption, tephra, pyroclastic flow, hazards.

## INTRODUCTION

Nicaragua's largest and most explosive eruption in historical time climaxed on 20–24 January 1835 at Volcán Cosigüina, which is located on a peninsula that forms the northwestern tip of the country (Fig. 1). Brief investigations on the north and east flanks of the volcano by Williams (1952) and Self et al. (1989) provide a convincing case that the bulk volume of pyroclastic material produced (<10 km$^3$) was much smaller than most estimates (up to 150 km$^3$) made by visitors in the years and decades following the eruption. During 2002 and 2003, we conducted reconnaissance field studies on all flanks of the volcano. Here we describe the stratigraphy and distribution of 1835 pyroclastic-fall and -flow deposits, discuss their implications for eruption character and volume, provide geochemical data for eruptive products, describe evidence of pre-1835 eruptions at the volcano, and present a preliminary hazard assessment based on the 1835 event.

Historical accounts of the eruption note the wide dispersal of ash and the audibility of explosions at great distance (see summaries in Williams, 1952; Self et al. 1989). Ashfall occurred northwestward to Chiapas, Mexico (600 km), southeastward to Costa Rica (350 km), and perhaps as far as Bogotá, Colombia (1900 km) and northeastward to Jamaica (1300 km). Booming sounds made soldiers in Guatemala and Belize think that they were under artillery attack, and noises were also heard in Oaxaca (Mexico), Jamaica, and Bogotá (Galindo, 1835a, 1835b).

Cosigüina is a broad composite cone with a 2.5-km-wide summit caldera that contains a large lake (Fig. 1). The highest point on the caldera rim is 872 m above sea level; the surface of the lake is at an altitude of ~160 m. A 200–300-m-high arcuate ridge, Filete Cresta Montosa, lies on the west flank ~2 km beyond the caldera rim. Williams (1952), Hradecký et al. (2001), and other workers interpreted this feature as a somma rim representing truncation of an older cone. Hradecký et al. (2001) described pre-Cosigüina lava flows and pyroclastic deposits on the lower flanks. A broad ridge of andesitic lava flows, which we call Loma San Juan for one of its high points, forms the lower east-southeast sector of the volcano. A range of low hills, which we call Loma El Ojochito for one of its high points, extends from Apascali to El Puente along the south flank. These features deflected 1835 pyroclastic flows onto the north and southeast flanks, which merge into broad fans of pyroclastic-flow deposits, lahars, and alluvium.

The form of the volcano and its caldera at the start of the 1835 eruption is not well constrained by observations, so we don't know what conditions were like and whether or not a lake occupied the vent. Williams (1952) implied that caldera collapse had occurred in 1835, reducing the height of the volcano by ~300 m. However, contemporary reports following the eruption (see summary in Self et al., 1989) make no mention of radical changes in the profile of the cone, three sides of which were clearly visible to mariners, so perhaps the shape of the outer cone was not greatly modified by the eruption.

Our observations build on those of Self et al. (1989) and their threefold succession of tephra deposits of the 1835 eruption studied at sites on the eastern flank of Cosigüina (Fig. 2, section A). Fall and surge processes inferred to derive from phreatic and phreatomagmatic eruptions deposited basal ash-rich beds (Set 1). An overlying coarse-grained andesitic scoria-fall deposit (Set 2) represents the major magmatic deposit of the eruption, thought to be vulcanian (on the basis of the poor vesicularity of clasts) or Plinian in nature. Subsequent pyroclastic-flow, -surge, and -fall deposits (Set 3) inferred to derive from vulcanian explosions bury the scoria-fall deposit and form fans on the lower flanks.

## STRATIGRAPHY OF 1835 DEPOSITS

We studied numerous sites on all flanks of the volcano (Figs. 1 and 2) and concluded that the deposits of the 1835 eruption fit Self et al.'s (1989) threefold succession, but with some additional complexity. A prominent scoria-fall deposit is the most conspicuous, continuous, and easily traceable of the units attributed to the 1835 eruption, so we organize the following discussion about deposits in relation to their stratigraphic position with respect to the scoria-fall deposit.

### Deposits below Scoria Fall

Self et al. (1989) designate as Set 1 a sequence of fall and surge deposits that underlies the scoria-fall deposit high on the east flank (Fig. 2; section A). In that proximal area, the lower part of Set 1 comprises thin, fine-grained ash-fall and interbedded lithic lapilli-fall deposits. The upper part contains lithic-rich surge deposits with scoriaceous lenses. Self et al. (1989) interpret these deposits as originating from initial phreatic and phreatomagmatic explosions, in which the content of juvenile components increased with time.

In more distal sites, a few centimeters or less of fine-grained fall deposit of laminated, light-gray to brownish-gray ash and fine lapilli underlie the scoria-fall deposit (Fig. 2,

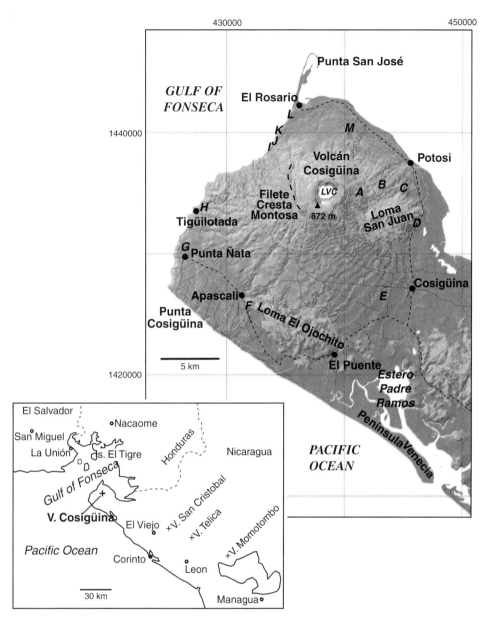

Figure 1. Shaded-relief, digital-elevation model (DEM) of Volcán Cosigüina showing location of key stratigraphic sections (letters; see Fig. 2) and major geographic features. LVC—Laguna Volcán Cosigüina in summit caldera. This DEM was made from 1972-vintage 1:50,000-scale topographic maps (Series E751; Chorrera [now Peninsula Padre Ramos], Cosigüina, Estero Real, Peninsula Venecia, and Potosi) that differ slightly from currently available maps made in the late 1980s. Roads shown in solid lines are from older maps and may not reflect current conditions. Short-dashed line shows approximate location of the current main road that connects towns on the peninsula and links to the major cities of western Nicaragua. The thick-dashed line marks a prominent arcuate ridge, Filete Cresta Montosa, that may be an old somma rim. Much of the low-relief area around Estero Padre Ramos and east of the village of Cosigüina are low-lying mangrove swamps. UTM grid lines are 10 km apart. Contour interval is 20 m. Inset map shows region around Gulf of Fonseca and locations referred to in text.

sections B–F). Its color, texture, and position with respect to the scoria-fall deposit are consistent with correlation to Set 1 of Self et al. (1989). Most of the sand-sized ash and fine lapilli comprise angular to subrounded grains of poorly to moderately vesicular, gray to black, glassy andesite. Free crystals of plagioclase, hypersthene, and augite are also present, as is rare, white to light gray pumice. Typically, a few percent of grains are variably altered hydrothermally and are likely to be accidental fragments. Our observations suggest that these deposits contain substantially more juvenile material (>75%) than Self et al. (1989) reported in the proximal deposits (<5%). Between sections F and G on the southwest flank, the basal ash thickens gradually from 2 to 50 cm. As it does so, the upper part coarsens and develops a definable subunit of laminated, gray to pinkish gray, medium to coarse ash and fine lithic lapilli. In section H, which lies on the shoulder of a broad ridge on the west flank, this unit comprises 40 cm of friable, well-bedded (thin-planar beds to low-angle cross beds), medium and coarse ash with beds and lenses of subangular to subrounded fine lithic lapilli. The deposit has some characteristics of a surge deposit, but because we don't see similar sediments elsewhere, we infer that it is a fall deposit that was partly reworked during its accumulation on a slope to form the cross beds. The distribution (Fig. 3A) and texture of this lower unit show dispersal by northeasterly to easterly winds. On the basis of average wind data for January, in which high altitude winds are chiefly southwesterly and opposite in direction to low altitude winds (Fig. 4), we infer dispersal from relatively low (<7 km) eruption columns.

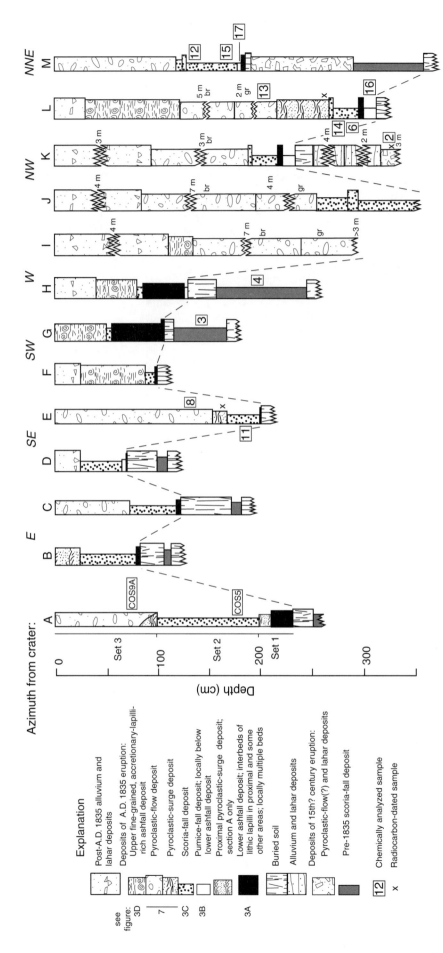

Figure 2. Key stratigraphic sections (located in Fig. 1) expose deposits from the 1835 eruption and earlier events. Section A and the designation of Sets 1–3 is from Self et al. (1989). Numbered boxes show the stratigraphic position of chemically analyzed samples (Table 1) from the sections or from units that can be physically traced to a section. X—locations of radiocarbon-dated samples (Table 3). Sections are shown to vertical scale except where thick units are broken by jagged lines and total thicknesses are given (in meters). Pyroclastic-flow deposits of 1835 are subdivided into dark gray (gr) and brownish-gray (br) units in those thick sequences. Alphanumeric symbols in left column of explanation refer to figures in which thicknesses of units are shown. Variation in width of scoria-fall and lower-ashfall deposits denotes observable change in grain size; wider units are more ash rich.

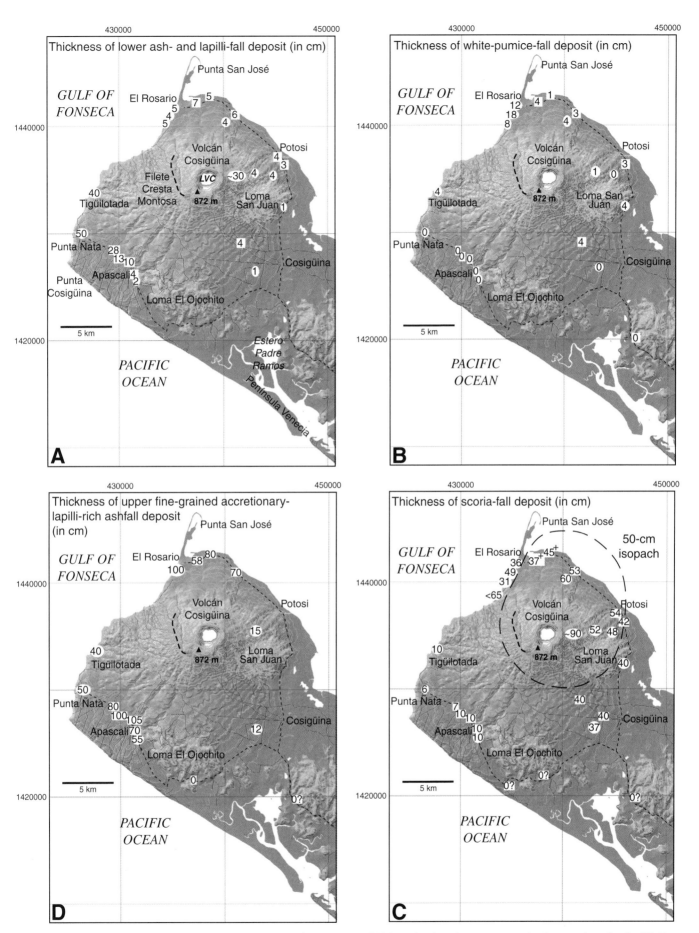

Figure 3. Maps show thickness (in centimeters) of subunits of 1835 tephra-fall deposits, from bottom to top: A—lower ash to fine lapilli; B—basal to near-basal white pumice; C—main scoria fall; D—upper fine-grained, accretionary-lapilli–rich ash. Thicknesses in white dots are from lettered stratigraphic sections (Fig. 2); those in white squares are from other sites. Approximate 50-cm isopach is shown for scoria-fall deposit. Other features are described in Figure 1.

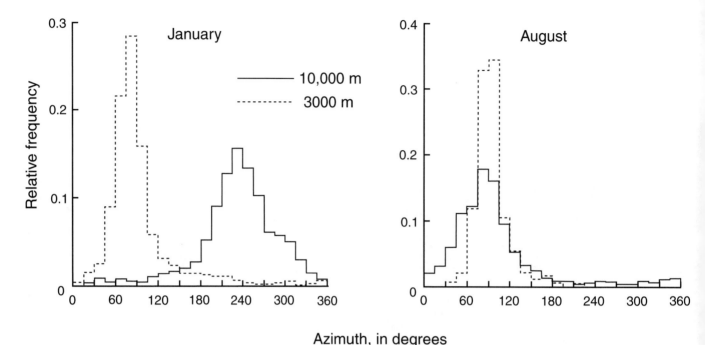

Figure 4. Relative frequency of average January (dry season) and August (wet season) wind directions for the years 1997–2005 at altitudes of ~3000 m (dashed line) and 10,000 m (solid line). Data from these altitudes are representative of low-level (below ~7000 m) and high-level (above ~7000 m) winds, respectively, in the region. Azimuth, in 15-degree bins, is the direction from which the wind blows. Data are from the Global Forecasting System, Air Resources Laboratory, National Oceanic and Atmospheric Administration (www.arl.noaa.gov/ready/metdata.html). The dry season extends from December through April, the wet season from June through October, and May and November are transitional.

A conspicuous layer of white to light gray dacitic (65 wt% $SiO_2$; Table 1, samples 16 and 17; Figs. 5 and 6) pumice, the most silicic product of the 1835 eruption, occurs in pre-scoria-fall deposits in many exposures on the lower flanks, except in southeast to southwest sectors (Figs. 2 and 3B). The location of the thickest and coarsest units of this pumice implies dispersal by southeasterly winds. Its stratigraphic position varies in relation to the basal fine ashfall deposit. In some areas, a scattering (section B) to as much as 18 cm (sections H, K, and L) of pumice lies directly on the pre-eruption surface below fall deposits of ash and fine lapilli. But elsewhere, a scattering (section C) to thin layer (sections D and M) of pumice overlies a basal gray ash and directly underlies the scoria-fall deposit. This complexity could originate from eruption of pumice both before and after eruptions that deposited the ash and fine lapilli unit, or it could result from restricted deposition and (or) differential erosion of the thin units emplaced during the early stages of the eruption. We suspect the latter, because nowhere do we see a sequence of two pumice beds separated by a gray, ash and fine lapilli–fall deposit.

**Scoria-Fall Deposit**

The andesitic scoria-fall deposit reaches a maximum observed thickness of ~1 m on the upper east flank (Set 2 of Self et al., 1989; Fig. 2, section A). At sites 5–10 km from the crater rim, thickness is typically 35–60 cm, except on the southwest flank, where it is only 5–10 cm (Fig. 3C). The restricted area within which the deposit is preserved makes construction of isopachs difficult, but the thickness data indicate dispersal chiefly to the north and east, consistent with high-altitude (>7 km) wind directions in January (Fig. 4). Limited data on clast size are also consistent with such a pattern. Diameter of the five largest scoria clasts at sites 5–10 km north and east of the caldera rim are typically 10–20 cm, whereas they are ~5 cm at a similar distance to the south and west. Maximum diameter of dense lithic clasts is ~5–10 cm at sites to north and east and ~2–5 cm to the south and west. Much of the scoria-fall deposit has been eroded from the upper flanks (Williams, 1952). Early observers noted that tropical rains created marked changes in just two years following the eruption (Roberts, 1924, p. 206–207). Furthermore, pyroclastic flows and surges that followed the scoria fall probably also scoured earlier fall deposits on the upper slopes.

In most localities, the scoria-fall deposit is a single massive bed of lapilli with a brown base and dark gray top (Fig. 5). Clasts are dominantly angular to subangular and display a wide range in vesicularity—from frothy scoria to dense, fresh fragments with few vesicles. Variably hydrothermally altered accidental clasts locally make up several percent of the unit. Reverse grading is evident locally, with an ash-rich base and the coarsest clasts about two-thirds up from the base, but grading is quite variable overall. In some localities, a few centimeters of gray scoriaceous ash cap the scoria-fall deposit (e.g., Fig. 2, section L; Fig. 5).

TABLE 1. CHEMICAL ANALYSES OF PRODUCTS OF VOLCÁN COSIGÜINA

| ID# | $SiO_2$ | $TiO_2$ | $Al_2O_3$ | FeO* | MnO | MgO | CaO | $Na_2O$ | $K_2O$ | $P_2O_5$ | Ni | Cr | Sc | V | Ba | Rb | Sr | Zr | Y | Nb | Ga | Cu | Zn | Pb | La | Ce | Th | Field sample no. | Section or locality | UTM coordinates | Description |
|---|---|---|---|---|---|---|---|---|---|---|---|---|---|---|---|---|---|---|---|---|---|---|---|---|---|---|---|---|---|---|---|
| 1 | 53.2 | 0.86 | 18.8 | 10.14 | 0.19 | 4.41 | 9.29 | 2.33 | 0.71 | 0.14 | 11 | 15 | 36 | 268 | 587 | 15 | 412 | 69 | 21 | 1.3 | 18 | 213 | 97 | 4 | 13 | 14 | 2 | 030211-3 | 150 m SSW of H | 427638E-1433539N | Old scoria fall of Tigüilotada |
| 2 | 55.3 | 0.75 | 19.1 | 7.97 | 0.18 | 3.06 | 8.47 | 3.88 | 1.05 | 0.20 | 4 | 2 | 26 | 198 | 720 | 19 | 456 | 92 | 29 | 1.5 | 19 | 110 | 87 | 5 | 12 | 24 | 2 | 030129-11 | K | 434188E-1439976N | Clast from A.D. 1400 diamict |
| 3 | 56.9 | 0.70 | 19.9 | 6.55 | 0.16 | 2.19 | 8.37 | 3.69 | 1.14 | 0.20 | 0 | 4 | 26 | 134 | 751 | 21 | 455 | 88 | 27 | 2.0 | 18 | 75 | 74 | 1 | 15 | 27 | 0 | 020205-2 | G | 426590E-1429757N | Pre-1835 scoria fall |
| 4 | 57.2 | 0.71 | 19.7 | 7.03 | 0.17 | 2.09 | 8.15 | 3.59 | 1.19 | 0.20 | 4 | 4 | 22 | 126 | 787 | 24 | 459 | 97 | 27 | 1.3 | 19 | 97 | 83 | 4 | 11 | 28 | 3 | 030211-4 | H | 427657E-1433673N | Pre-1835 scoria fall |
| 5 | 57.4 | 0.90 | 17.1 | 9.23 | 0.19 | 2.92 | 7.43 | 3.20 | 1.36 | 0.23 | 2 | 2 | 30 | 237 | 891 | 28 | 417 | 107 | 28 | 2.3 | 18 | 188 | 94 | 4 | 9 | 27 | 3 | 030211-1 | 320 m WSW of H | 427346E-1433493N | Old gray pumice of Tigüilotada |
| 6 | 57.4 | 0.74 | 19.1 | 7.09 | 0.16 | 2.28 | 8.10 | 3.67 | 1.22 | 0.20 | 3 | 4 | 23 | 147 | 778 | 24 | 447 | 96 | 28 | 1.9 | 20 | 66 | 86 | 4 | 9 | 17 | 3 | 030129-8 | Between K and L | 435022E-1440814N | Brown scoria in 1835 fall |
| 7 | 57.5 | 0.74 | 18.9 | 6.90 | 0.17 | 2.22 | 8.08 | 3.83 | 1.23 | 0.20 | 1 | 1 | 25 | 125 | 812 | 23 | 437 | 94 | 29 | 2.0 | 18 | 94 | 81 | 3 | 6 | 20 | 0 | 020206-3 | Between I and J | 433542E-1438658N | Dark gray scoria in 1835 pyroclastic flow |
| COS5 | 57.5 | 0.70 | 18.9 | 7.08 | 0.15 | 2.31 | 8.04 | 3.67 | 1.16 | 0.21 | 5 | 14 | 23 | 162 | 783 | 22 | 454 | 91 | | | | | | | | | | | Upper east flank | | Scoria in 1835 fall; Set 2 |
| 8 | 57.6 | 0.73 | 18.9 | 7.28 | 0.16 | 2.23 | 7.92 | 3.73 | 1.25 | 0.20 | 5 | 4 | 23 | 139 | 797 | 25 | 440 | 98 | 28 | 1.1 | 20 | 98 | 86 | 4 | 6 | 25 | 3 | 030212-3 | E | 443085E-1426350N | Scoria in 1835 pyroclastic flow |
| 9 | 58.0 | 0.73 | 18.8 | 7.20 | 0.17 | 2.20 | 7.71 | 3.74 | 1.29 | 0.21 | 4 | 4 | 23 | 135 | 814 | 26 | 437 | 99 | 29 | 1.6 | 20 | 97 | 83 | 7 | 7 | 26 | 1 | 030131-1 | South rim | 438489E-1433582N | Caldera rim agglutinate of 1835 |
| COS9A | 58.4 | 0.70 | 18.4 | 7.07 | 0.16 | 2.18 | 7.54 | 3.85 | 1.26 | 0.22 | 8 | 15 | 23 | 153 | 857 | 23 | 432 | 88 | | | | | | | | | | | Upper east flank | | Dense scoria in 1835 Set 3 |
| COS3 | 58.4 | 0.71 | 18.3 | 7.09 | 0.16 | 2.16 | 7.49 | 3.78 | 1.29 | 0.28 | 7 | 18 | 23 | 144 | 825 | 23 | 434 | 97 | | | | | | | | | | | East or south rim | | Caldera rim agglutinate of 1835 |
| 10 | 58.5 | 0.76 | 18.4 | 6.88 | 0.17 | 2.20 | 7.50 | 3.88 | 1.32 | 0.22 | 2 | 5 | 27 | 136 | 855 | 26 | 422 | 100 | 31 | 1.5 | 19 | 67 | 84 | 2 | 16 | 16 | 1 | 020206-1 | 700 m east of L | 436068E-1441307N | Gray scoria of 1835 pyroclastic flow |
| 11 | 58.5 | 0.75 | 18.4 | 7.06 | 0.17 | 2.19 | 7.47 | 3.87 | 1.35 | 0.22 | 3 | 4 | 23 | 132 | 843 | 27 | 430 | 106 | 30 | 1.7 | 18 | 98 | 86 | 4 | 9 | 29 | 3 | 030201-1 | 2.5 km NNW of E | 441804E-1428923N | Dense juvenile lithic in 1835 fall |
| 12 | 59.1 | 0.76 | 18.0 | 6.77 | 0.17 | 2.21 | 7.17 | 4.04 | 1.36 | 0.22 | 1 | 5 | 28 | 115 | 876 | 27 | 406 | 103 | 31 | 1.7 | 18 | 80 | 85 | 5 | 22 | 24 | 4 | 020204-6 | M | 440297E-1440139N | Black scoria in 1835 fall |
| 13 | 59.1 | 0.75 | 18.0 | 7.12 | 0.17 | 2.14 | 7.07 | 4.03 | 1.43 | 0.22 | 3 | 7 | 23 | 125 | 896 | 28 | 413 | 109 | 31 | 1.6 | 19 | 87 | 87 | 5 | 7 | 26 | 2 | 030129-1 | L | 435472E-1441663N | Scoria in 1835 pyroclastic flow |
| W | 59.4 | 0.70 | 18.0 | 7.37 | 0.23 | 2.32 | 6.86 | 3.90 | 1.19 | 0.05 | | | | | | | | | | | | | | | | | | | North flank? | | Scoria in 1835 pyroclastic flow |
| 14 | 59.7 | 0.76 | 17.7 | 7.17 | 0.17 | 2.07 | 6.73 | 4.00 | 1.49 | 0.23 | 3 | 4 | 22 | 119 | 922 | 29 | 402 | 113 | 32 | 1.3 | 17 | 78 | 94 | 6 | 12 | 23 | 2 | 030129-7 | Between K and L | 435022E-1440814N | Black scoria in 1835 fall |
| 15 | 59.8 | 0.77 | 17.6 | 6.81 | 0.17 | 2.12 | 6.71 | 4.17 | 1.46 | 0.23 | 1 | 2 | 21 | 110 | 922 | 29 | 395 | 109 | 32 | 2.5 | 19 | 74 | 85 | 0 | 18 | 19 | 2 | 020204-7 | M | 440297E-1440139N | Brown scoria in 1835 fall |
| 16 | 63.8 | 0.71 | 15.7 | 6.32 | 0.17 | 1.49 | 4.42 | 5.06 | 2.06 | 0.26 | 5 | 7 | 19 | 51 | 1192 | 42 | 330 | 148 | 41 | 2.7 | 18 | 59 | 85 | 7 | 9 | 32 | 2 | 030129-4 | L | 435346E-1441305N | White pumice in 1835 fall |
| 17 | 64.7 | 0.72 | 15.6 | 6.00 | 0.17 | 1.39 | 4.27 | 4.58 | 2.07 | 0.27 | 2 | 0 | 23 | 39 | 1256 | 43 | 314 | 148 | 40 | 3.3 | 16 | 48 | 92 | 6 | 3 | 39 | 2 | 020204-8 | M | 440297E-1440139N | White pumice in 1835 fall |
| 18 | 64.8 | 0.76 | 15.3 | 6.83 | 0.18 | 1.40 | 4.49 | 3.77 | 2.23 | 0.24 | 4 | 8 | 21 | 65 | 1344 | 46 | 347 | 155 | 37 | 2.4 | 18 | 88 | 100 | 7 | 17 | 41 | 2 | 030211-2 | 320 m WSW of H | 427346E-1433493N | Old white pumice of Tigüilotada |

*Note:* Major-element analyses are recalculated volatile-free to 100%; total Fe as FeO*. All were done by X-ray fluorescence, except for Sample W, which is from Williams (1952). Samples with COS are from Self et al. (1989). Numbered samples were processed at the WSU GeoAnalytical Laboratory, Washington State University (see Johnson et al. [1999] for information about technique, precision, and accuracy). Locations of sections are shown in Figure 1 and stratigraphic positions of samples from sections are shown in Figure 2.

Figure 5. Photograph showing the basal part of 1835 deposits near section L. A white pumice-fall deposit lies directly on the paleo-surface. The upper part of the pumice fall is stained by iron oxides and appears dark colored. The color grading of scoria-fall deposit from brown base to dark gray top is much more striking than it appears in this gray-tone image. The capping pyroclastic-flow deposit typically includes a stratified basal surge facies.

Tens of centimeters of colluvium and soil derived from reworking of 1835 deposits overlie the undisturbed scoria-fall deposit where it is not buried by deposits of pyroclastic surges or flows, ash clouds, or lahars.

Chemical analyses of five andesitic clasts from the scoria-fall deposit range over 2.4 wt% $SiO_2$ (57.4–59.8 wt% $SiO_2$; Table 1; Fig. 6), but show no systematic relationship to color, which appears to be a function of greater vesicularity in the brown clasts. Measurements from several localities show that the density of clasts from the lower brown unit average 1.00 g/cm³ ± 0.22 (1σ), whereas those from the upper, dark gray unit average 1.55 ± 0.40 g/cm³ (Table 2). A compositional gap of ~3 wt% $SiO_2$ separates the andesitic scoria from the earlier dacitic pumice. Hradecký et al. (2001) report analyses of 1835 deposits that are somewhat more mafic (~54–58 wt% $SiO_2$) than ours, but, as discussed in a later section, we disagree with them on the identification and correlation of some 1835 products.

Self et al. (1989) describe 10-m-thick flows of agglutinated, fountain-fed andesitic spatter that cover portions of the caldera rim and upper flanks. In contrast, Hradecký et al. (2001) interpret the deposit as lag-fall–breccia facies of pyroclastic flows. Samples from near the rim have composition similar to that of the scoria in fall and flow deposits (Table 1, samples 9 and COS3; Fig. 6). The agglutinate is exposed mostly on steep eroding slopes and has a discontinuous cover of lithic blocks. Nowhere have we seen below the agglutinate to judge its stratigraphic position with respect to the scoria fall or other deposits of 1835.

## Deposits above Scoria-Fall Deposit

Over much of Cosigüina's lower flanks, the scoria-fall deposit underlies a mantle of sediments related to pyroclastic flows or to rapid erosion of eruptive deposits during subsequent rainy seasons and their redeposition as lahar and alluvial sediments.

Deposits of pyroclastic flows and surges composed of scoria, dense juvenile and accidental clasts, and ash overlie the scoria-fall deposit at numerous sites on the southeast, east, north, and northwest flanks (Fig. 2, sections C, E, J–M; Williams, 1952; Self et al., 1989). They form several broad fans and narrower valley fills (Fig. 7). The pronounced arcuate ridge west of the caldera, Filete Cresta Montosa, evidently diverted most of the flows in the western sector northwestward and southward. The largest fan forms the broad, gently sloping surface west of the village of Cosigüina. Material 6–7 m thick reportedly buried nearby

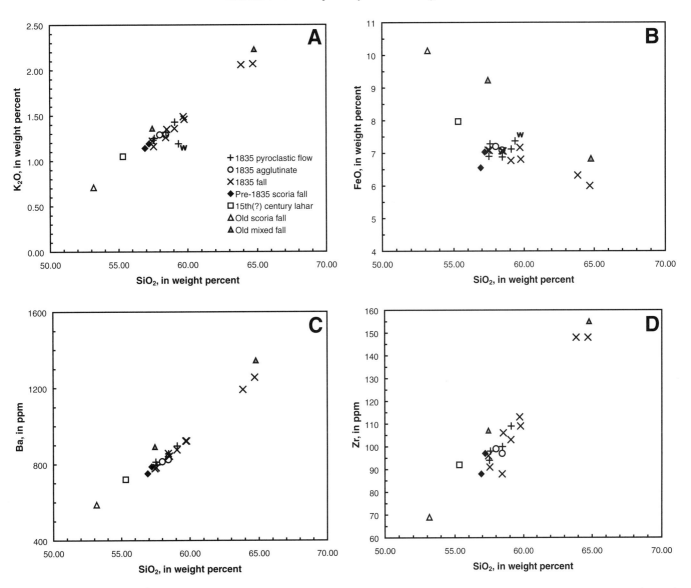

Figure 6. Plots of selected major- and trace-element data from Table 1 illustrate chemical compositions of deposits from the 1835 eruption and older events. The 1835 pyroclastic flow sample marked W is analysis from Williams (1952). Note the similar range in composition of 1835 deposits and those of the compositionally graded deposit exposed in the Tigüilotada area.

Hacienda Cosigüina in 1835 (Roberts, 1924). Outcrops within a few kilometers upslope of the hacienda site expose as much as 4 m of dark gray to pinkish- and brownish-gray pyroclastic-flow deposits with no base visible. Lesser thicknesses crop out above the scoria-fall deposit (Fig. 2, section E), and some low, broad ridges that rise a meter or two above the general level of the fan surface have only a thin (tens of centimeters) veneer of ash-cloud–surge deposits above the fall deposit.

Gray to black prismatically jointed and bread-crusted bombs (Table 1, sample 8) up to 1 m in diameter are scattered throughout the pyroclastic-flow deposits west of the village of Cosigüina and can be traced locally as high as 550 m on the upper south flank. Two small preexisting cinder cones, Cerros Chachos, whose bases

TABLE 2. DENSITY OF JUVENILE CLASTS IN DEPOSITS OF 1835 ERUPTION

| Unit of 1835 eruptive products | Number of sites/clasts | Mean clast density ± 1 standard deviation |
|---|---|---|
| Pyroclastic-flow deposits | 2/39 | 1.76 ± 0.40 |
| Dark-gray scoria-fall deposit | 3/61 | 1.55 ± 0.40 |
| Brown scoria-fall deposit | 6/108 | 1.00 ± 0.22 |
| Basal pumice-fall deposit | 4/81 | 0.61 ± 0.25 |

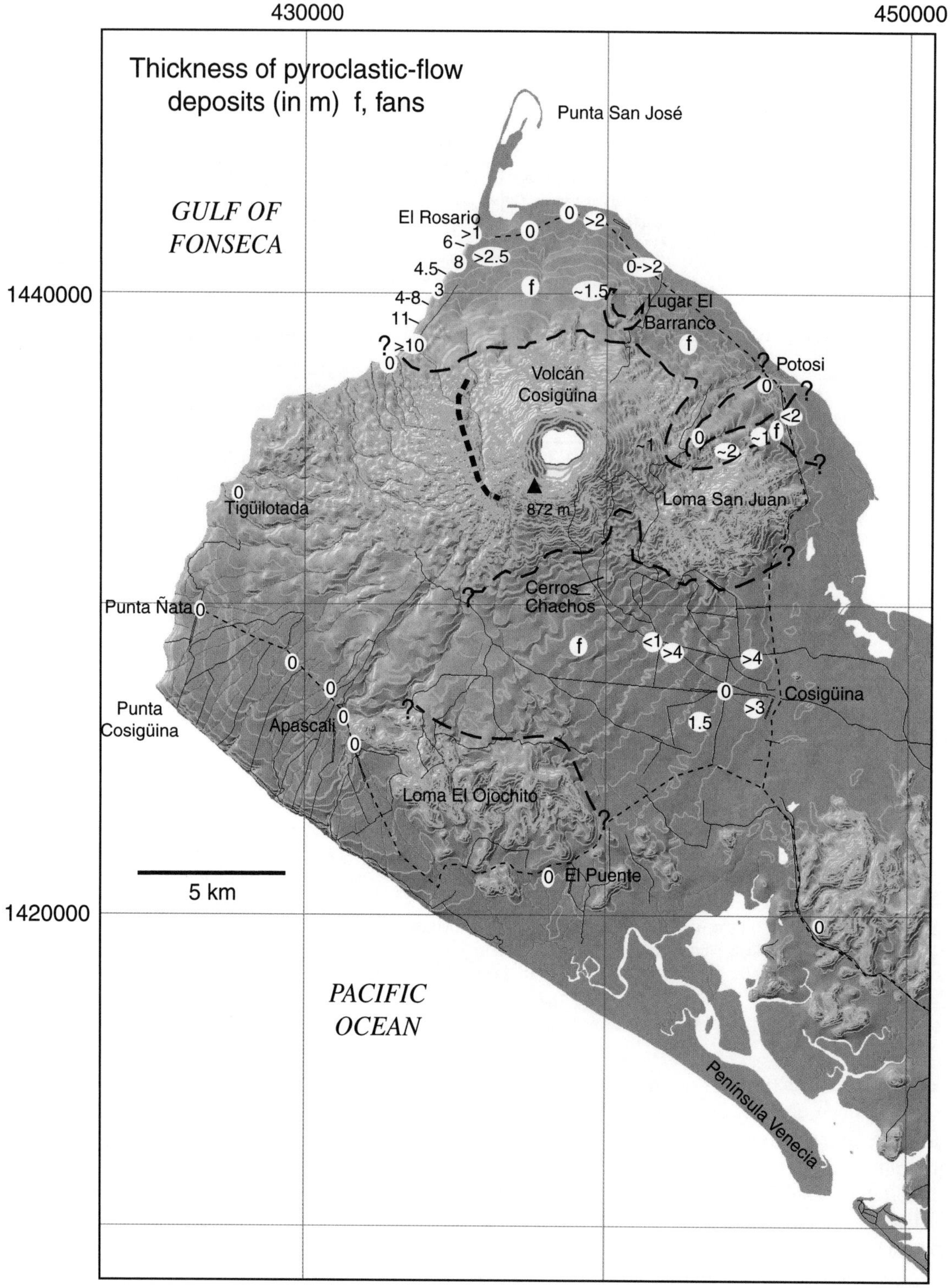

Figure 7. Approximate distribution of 1835 pyroclastic-flow and -surge deposits shown by spot-thickness measurements and approximate contacts that outline portions of thick accumulations in fans (f) on the lower flanks. The distal extent of pyroclastic-flow deposits in the southeastern sector is poorly known owing to burial by thick deposits of post-1835 lahars and alluvium and by extensive mangrove swamps.

lie at ~250 m, were scoured by the flows into slightly streamlined forms. Locally, the deposits contain abundant charcoal, ranging from small fragments to entire logs. A small, charcoalized twig in basal pyroclastic-surge deposits between the scoria-fall and pyroclastic-flow deposits in section E yielded a radiocarbon age indistinguishable from a modern age, as would be expected for a deposit dating from A.D. 1835 (Table 3).

Alluvial sand and gravel at least several meters thick derived from erosion of the pyroclastic-flow deposits underlie distal surfaces of the fan north and south of the village of Cosigüina. How far the pyroclastic flows originally extended toward the Gulf of Fonseca in these distal areas is not known. Caldcleugh's (1836) description of the eruption drawn from eyewitness accounts mentions that the shore was extended by ~250 m, but the timing of such observations is uncertain and may have followed some reworking of the deposits seaward. Therefore, the pyroclastic flows may not have traveled very far east of the village of Cosigüina, a total distance of ~10 km from the 800-m-high caldera rim.

Sea cliffs along the Gulf of Fonseca provide excellent exposures through the pyroclastic-flow fan on the northwest flank (Fig. 2, sections I–L; Fig. 8). Maximum exposed thickness of 1835 pyroclastic-flow deposits is ~15 m with no base visible within 1 m of sea level. Color and texture differentiate two massive units (Fig. 9). Their contact is planar and marked locally by thin, ash-rich surge and fall deposits. The lower unit directly overlies the scoria-fall deposit or up to 1 m of intervening scoria-rich pyroclastic-surge deposits and is dark gray to black. The upper unit, which has a distinctly browner matrix owing to a greater content of fine-grained matrix, also contains black scoriaceous clasts similar to those in the lower unit. It overlies the dark gray unit except on the higher portions of the pre-1835 surface that stood above the reach of the dark gray flows. There it overlies 1835 scoria-fall deposits mantling the older surface. At both ends of the coastal exposures shown in Figure 8, the base of the pyroclastic-flow deposits extends to some unknown depth below present sea level.

Both the dark gray and brown pyroclastic-flow deposits exposed along the cliffs must have originally extended some distance beyond the present shore. Extrapolation of the fan slope beyond the cliff suggests that the pyroclastic-flow deposits may have extended at least several hundred meters farther out into the Gulf of Fonseca, which in this region has a gently sloping bottom. Water depths 2 km offshore now range from 6 to 10 m. Reports of eyewitnesses who visited the northwest coast by boat on February 9, two weeks after the eruption, describe the destruction of the old-growth forest that mantled the volcano's slopes and the appearance of shoals and two low islands composed of pumice (Galindo, 1835b). A broad wedge of sediment topped by the spit of Punta San José comprises sand eroded from the pyroclastic-flow and younger deposits and formed as the sea cliff retreated.

Elsewhere, especially on the southwest flank (Fig. 2, sections F–H), which was protected from long-traveled pyroclastic flows by ridges, the scoria-fall deposit is overlain by light gray to light grayish-brown laminated to thinly bedded ash-fall deposits locally >1 m thick (Fig. 3D). Accretionary lapilli up to 1 cm in diameter are common throughout the ash deposit, in places as beds with little matrix (Fig. 10). The lapilli typically have cores of coarse, dark gray scoriaceous ash and rims of light-colored very fine sand- and silt-sized ash. Similar deposits, typically a few tens of centimeters thick, locally overlie pyroclastic-flow deposits along the northwest coast and northern flank. We interpret the ash as having originated

TABLE 3. RADIOCARBON AGES FROM DEPOSITS OF COSIGÜINA

| Laboratory number | Age B.P. ($^{14}$C yr ± 1σ) | Calibrated age, A.D. 1σ range (probability) | Calibrated age, A.D. 2σ range (probability) | Stratigraphic position assigned in reference | Reference |
|---|---|---|---|---|---|
| Beta-177981 | modern | – | – | 1835 surge deposits, sec L | This report |
| Beta-177982 | 30 ± 50 | 1690–1730 (17.8) 1810–1840 (12.5) 1870–1920 (31.6) 1950–1960 (6.3) | 1680–1740 (23.2) 1800–1930 (65.0) 1950–1960 (7.1) | 1835 surge deposits, sec E | " |
| Beta-177983 | 490 ± 50 | 1400–1455 (68.2) | 1300–1360 (12.6) 1380–1490 (82.8) | pre-1835 diamicts, sec K | " |
| CU-unk | 295 ± 121 | 1400–1700 (59.8) *1750–1800 (6.4) | 1400–2000 (95.4) | deposits of 1700 or 1709 eruption, UH2 | Hradecky and others (2001) |
| CU-unk | 310 ± 127 | *1400–1700 (61.8) | 1400–2000 (95.4) | " | " |
| CU-unk | 330 ± 128 | *1400–1700 (65.0) | 1400–2000 (95.4) | " | " |
| CU-unk | 374 ± 128 | 1420–1660 (68.2) | 1300–2000 (95.4) | " | " |
| CU-unk | 430 ± 128 | 1400–1640 (68.2) | *1250–1850 (94.4) | " | " |
| CU-unk | 498 ± 128 | *1290–1520 (64.5) | 1250–1700 (95.4) | deposits of 1500 eruption, UH1 | " |
| CU-unk | 505 ± 127 | *1290–1510 (65.0) | 1250–1670 (95.4) | " | " |
| CU-unk | 506 ± 127 | *1290–1500 (65.1) | 1250–1670 (95.4) | " | " |

*Note:* Calibrated by OxCal v. 3.10 (Ramsey, 2005) using data from Reimer et al. (2004). Hradecký et al. (2001) did not publish the laboratory numbers of their samples, which here are labeled CU-unk for Charles University, Prague, Czech Republic–unknown.
*Excludes extreme values that have <5% probabilities in calibration results.

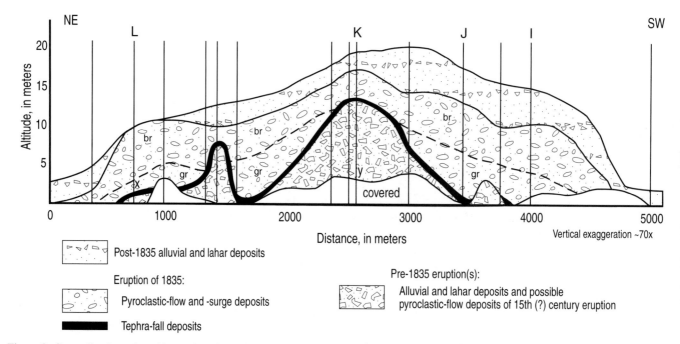

Figure 8. Generalized stratigraphic section along the northwest coast of Cosigüina shows how deposits from the 1835 eruption buried a dissected landscape formed in older deposits. Pyroclastic-flow deposits are subdivided into lower dark gray (gr) and upper brownish-gray (br) units. Pyroclastic-surge deposits are not differentiated from pyroclastic-flow deposits. The unit of 1835 tephra-fall deposits includes lower ash- and lapilli-fall deposits, a white pumice-fall deposit, and a scoria-fall deposit. Vertical lines mark sections that provide control; labeled sections are detailed in Figure 2. Note ~70× vertical exaggeration.

in at least three ways. First, its close stratigraphic relation with pyroclastic-flow deposits and the similar appearance of the ash cores and silty rims to components of those deposits suggest that most of the ash originated as coignimbrite fallout derived from ash clouds elutriated from moving pyroclastic flows (Hradecký et al., 2001). The ash was dispersed by prevailing tropospheric winds primarily southwestward, which implies relatively low (<7 km) eruption columns (Figs. 3D and 4). Second, pyroclastic flows entered the sea broadly along the northwest coast and perhaps in places along the north, northeast, and southeast coasts. Explosions driven by interaction of the hot deposits and seawater may also have produced copious ash clouds and downwind fallout (e.g., Walker, 1979). Third, historical accounts indicate several days of waning ashfall followed the climax of the eruption.

Hradecký et al. (2001) map deposits of pyroclastic flows and surges of 1835 along the entire Pacific coast on the southwest flank of the volcano. Our work shows that from sections F to G (Figs. 1 and 2), between Apascali and Punta Ñata, deposits of 1835 are limited to basal-ash and fine-lapilli fall deposits, thin scoria-fall deposits, and a meter-thick overlying sequence of fine-grained fall deposits. We find that the broad coalescing fans ending in high cliffs above the Pacific Ocean expose volcaniclastic sediments of varied ages and origins—chiefly laharic and alluvial. The arcuate ridge west of the caldera and the low range of hills, including Loma El Ojochito, to the south deflected 1835 pyroclastic flows away from the southwest flank.

The chemical compositions of gray scoriaceous bombs from pyroclastic-flow deposits are similar to those of lapilli in the scoria-fall deposit (Table 1; Fig. 6). Samples came from both the dark gray and brown deposits and from the major fans in the northwest and southeast sectors. We found none of the dacitic white to light gray pumice, common in the basal part of the 1835 eruption sequence, in the pyroclastic-flow deposits.

A mantle of diamicts and alluvial deposits locally >5 m thick overlies the pyroclastic-flow deposits along most of the northwest coast (Figs. 2, 8, 9B, and 11). We infer that they are related to erosion of the pyroclastic-flow deposits and steep upper flanks of the volcano in the years immediately following the eruption. Roberts (1924), who visited the area in the early twentieth century, reports that the descendants of the man who ran Hacienda Cosigüina in 1835 said that after two rainy seasons the land was again well vegetated and agriculturally productive and that the people who had fled had returned. As at Mount Pinatubo after the 1991 eruption, the erosion of pyroclastic-flow deposits and deposition of lahar and alluvial fans would have been most rapid during the first few years and declined substantially thereafter (Janda et al., 1996).

## Pumice Rafts?

Contemporary accounts of pumice rafts from Cosigüina encountered by ships at sea are perplexing. The two reports most often cited are secondhand. Caldcleugh (1836) relates the

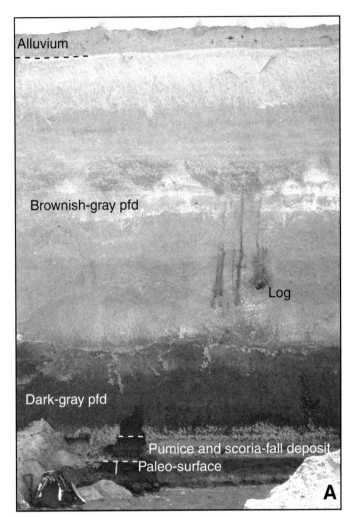

Figure 9. Photographs show typical outcrops of 1835 deposits along the northwest coast. (A) The twofold pyroclastic-flow sequence is shown overlying the pumice and scoria-fall deposits between sections K and L. Sea cliff is ~12 m high. Both pyroclastic-flow deposits (pfd) have a matrix of dark gray scoriaceous ash, but the brownish-gray unit has a conspicuous proportion of fine-grained brownish ash that makes it browner and more cohesive. Note the fluid-escape pipes emanating from a charcoalized log. (B) Dark gray and brownish-gray 1835 pyroclastic-flow deposits near section J are overlain by a wedge of alluvium and lahar deposits that represents a period of rapid erosion following the 1835 eruption.

Figure 10. Roadcut near Punta Ñata (section G) exposes 1835 tephra-fall deposits (photograph courtesy of Armin Freundt). In this sector, the scoria-fall deposit is thin and upper and lower fine-grained deposits are thick compared to thicknesses in sections at similar distance from the vent in other sectors. The upper unit contains accretionary lapilli up to 1 cm in diameter.

observations of a British captain who sailed for 65 km through floating pumice, some of which was of "considerable size," 1800 km southwest of Cosigüina. Unfortunately, the date of the encounter is not given (Caldcleugh's report was written ~7 months after the eruption). Squires (1851) was informed by a ship's captain who sailed by the volcano a few days after the eruption that the sea for ~50 leagues (250 km) was covered with a near-continuous mass of pumice. As judged from the clast-density data in Table 2, there are limited sources of floatable material, and we wonder if these reports are accurate and even if the distal raft originated from Cosigüina.

Other reports from ships passing the Pacific coast (e.g., Galindo, 1835b) describe darkness and the fall of fine-grained ash ("copious showers of dust"). Such reports are consistent with the types of deposits we find along the Pacific coast; namely, fine-grained basal ash and fine lapilli and the thicker upper unit of ash deposits related largely to pyroclastic flows.

Galindo (1835b) translated a report written by the commandant of the port of La Unión, El Salvador, which lies across the Gulf of Fonseca from the volcano, about the findings of a group who visited the shore northwest of Cosigüina just two weeks after the eruption. They describe shoals and small islands of pumice beyond the former shore, but make no mention of floating pumice rafts. If there were conspicuous pumice rafts present in the Gulf of Fonseca, which was the focus of both pumice and scoria falls and pyroclastic flows, it is remarkable that no mention is made of them, especially considering that the gulf was an important means of travel between El Salvador, Honduras, and Nicaragua.

If the eruption produced large pumice rafts, potential sources are limited. The scoria clasts in the dark gray portion of the scoria-fall deposit and in the pyroclastic-flow deposits are almost all denser than water (Table 2). The dacitic pumice-fall deposit contains a large floatable fraction, but the entire bed is typically only a few centimeters thick except in a narrow segment along the coast northwest of the volcano. The brown, basal part of the scoria-fall deposit contains a smaller fraction of floatable material than the pumice-fall deposit, but it is of greater volume. Both of these units extend chiefly northwestward to northeastward from the vent and imply that any rafts must have originated in the Gulf of Fonseca and then drifted into the Pacific. The Gulf is ~35–50 km wide, but the only reports of distal pumice fall were from Isla El Tigre, ~20 km away, where pea- to hen's-egg-sized clasts fell on 21 January (Galindo, 1835a). Could such a raft be 250 km wide off the Pacific coast or 65 km wide 1800 km away? Did the eyewitness reports exaggerate the size of the rafts?

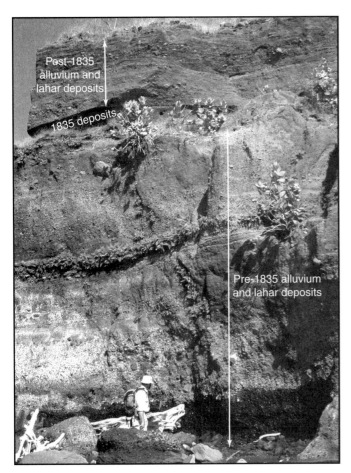

Figure 11. Sea cliff near section K exposes a sequence of pre-1835 lahar deposits and alluvium. Cliff is ~17 m high. Charred wood from the bouldery lower pre-1835 deposit yielded a calibrated radiocarbon age of A.D. 1400–1455 at 1σ uncertainty. The 1835 deposits are ~3.5 m thick, but are largely hidden by a bench in the upper part of the section. The 3–4 m of slightly indurated lahar deposits and alluvium that cap the section originated by rapid erosion of deposits on the upper flanks of the volcano following the eruption.

Certainly many other historical reports related to the eruptions were exaggerated (e.g., as discussed in Self et al., 1989). Was the 1800-km-distant raft even related to the Cosigüina eruption? At present these issues cannot be resolved.

## Volume of 1835 Eruptive Products

Estimates of the volume of tephra-fall and pyroclastic-flow deposits emplaced by the 1835 eruption have varied by several orders of magnitude. Self et al. (1989) describe how volume estimates of ≥50 km$^3$ in the decades following the eruption must represent erroneous reporting. Almost none of the nineteenth century reports, except for a few of ashfall from distal sites, are firsthand accounts. Widely repeated reports of 5 m of dust accumulating for more than 40 km from the volcano are clearly incorrect. At such a distance in eastern El Salvador and southern Honduras, several reliable descriptions note uncompacted tephra fallout of, at most, 10–20 cm. Using such constraints and limited data from the east flank of the volcano, Self et al. (1989) estimated that the maximum bulk volume of the tephra-fall deposit was 5.6 km$^3$.

We searched for Cosigüina tephra in eastern El Salvador, but we turned up no sites with recognizable deposits. We concentrated in areas near San Miguel and La Unión, which, according to eyewitness reports, received 10 cm or more of uncompacted ash (Galindo, 1835a, 1835b). On the basis of radiocarbon age, presence of hornblende, and bulk chemistry, well-preserved fine-grained white ashfall deposits up to 25 cm thick in several basin and floodplain localities are distal fallout of the A.D. 430 eruption of Ilopango caldera, near San Salvador (Escobar, 2003; Dull, 2004). In many of these sites, the original A.D. 430 fall deposit was immediately buried in alluvium or colluvium of reworked ash, thereby ensuring its preservation. None of the areas that provided good traps for the Ilopango ash had any sign of a younger deposit that could be distal tephra of the 1835 Cosigüina eruption or its reworked equivalent. A key question to us is how accurate are the tephra thicknesses reported by 1835 eyewitnesses. Perhaps exaggerated reports by terrified residents and the low bulk density of newly fallen tephra contributed to overly large estimates of the amount that fell. Coring in lakes, the Gulf of Fonseca, or coastal wetlands is probably the only way to ultimately gain confidence in accurate distal thickness estimates.

Our proximal data on the thickness of the scoria-fall deposit (Fig. 3C) can further refine and reduce the volume estimates of Self et al. (1989). For example, we find a maximum of 10 cm of scoria-fall deposit along the southern flank of the volcano (Fig. 2, sections F and G), which lies ~30–40 km inside of Self et al.'s (1989, their Figure 1) estimated circular 10-cm isopach. Thicker fine-grained fall deposits that lie below and above the climactic fall in this area originated from low-altitude eruption plumes and from ash clouds of pyroclastic flows and therefore cannot be combined with the climactic fall for purposes of extrapolating distal thickness. In addition, the data in Figure 3C allow estimation of a 50-cm isopach for the scoria-fall deposit, which is elongated in a northward direction and not circular.

A plot of thickness of the climactic scoria-fall deposit from Cosigüina versus the square root of isopach area (Fig. 12) illustrates our estimates of bulk volume compared with those of Self et al. (1989). Data for the ~8-km$^3$ 1902 Santa María (Guatemala) fall deposit (Williams and Self, 1983, their Figure 1; Fierstein and Nathenson, 1992) are also shown. A few brief observations comparing deposit thickness with distance along the dispersal axis indicate that the Cosigüina deposit is probably considerably less voluminous. The 100-, 50-, and 10-cm isopachs lie <5 km, 10 km, and <100 km, respectively, downwind of the vent at Cosigüina, whereas such isopachs lie 30 km, 55 km, and 200 km downwind at Santa María. Our revised data (line C, Fig. 12) for proximal deposits (coupled with 5- and 10-cm isopachs based on historical accounts but elongated to the north and narrowed rather than made circular) define a bulk volume of ~3 km$^3$. Great uncertainties about amounts of distal tephra confound attempts

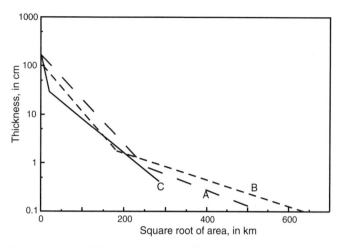

Figure 12. Plot of thickness of tephra-fall deposit versus square root of isopach area for the 1902 Santa María (Guatemala) deposit (A: bulk volume = 7.8 km$^3$ [Fierstein and Nathenson, 1992; using data of Williams and Self, 1983]), the 1835 Cosigüina deposit (B: bulk volume = 5.6 km$^3$ [Self et al., 1989; option 2]), and proximal 1835 Cosigüina (C: bulk volume = 3.0 km$^3$; this report).

to further refine estimates of the total bulk volume of tephra from the climactic scoria fall, but the 1.5 km$^3$ of distal tephra estimated by Self et al. (1989) probably provides a maximum estimate.

Reconnaissance mapping provides broad limits on the volume of pyroclastic-flow and other deposits of the 1835 eruption. Lack of exposures make thickness estimates of pyroclastic-flow deposits difficult, but a rough bulk-volume estimate is 0.5–1 km$^3$ (110 km$^2$ covered with a mean thickness of 5–10 m). Fine-grained ash-fall deposits associated with pyroclastic flows and waning stages of the eruption (Fig. 3D), whose distal extent in the Pacific Ocean is unknown, could add a few additional tenths of a cubic kilometer.

Our best estimate of the total bulk volume of eruptive products emplaced in 1835 is scoria-fall deposit from proximal data, 3 km$^3$; distal ash, maximum 1.5 km$^3$; pyroclastic-flow deposits, maximum 1 km$^3$; and fine-grained proximal ash-fall deposits, 0.2 km$^3$, for a total of roughly 6 km$^3$.

## PRE-1835 DEPOSITS

In addition to the widely reported eruption of Cosigüina in 1835, Simkin and Siebert (1994) list eruptions of uncertain magnitude and date in A.D. 1500 and 1709, an uncertain eruption in 1609, and eruptions, probably small, in 1852 and 1859. They also list an eruption in 1809, but this may be an inherited typographical error referring to the 1709 event (S. Self, 2005, personal commun.). Feldman (1993, p. 235) notes that little mention is made about Cosigüina prior to 1835 in the colonial archives of the former port, naval base, and provincial capital of El Realejo (founded in 1633 near present-day Corinto), which was located ~65 km from the volcano. If the volcano did erupt in the two centuries before 1835, the eruptions must have been nothing near the scale of the 1835 event, which was chronicled at length owing to local tephra fall, darkness, striking views of the eruption column, and loud eruption sounds. The records of other nearby cities such as El Viejo and León in Nicaragua and San Miguel in El Salvador, which were settled even earlier than El Realejo and were affected by the 1835 eruption, likewise have no mention of earlier eruptions of Cosigüina. In this section, we describe briefly the tephra and flowage deposits of older eruptions of Cosigüina that we see locally exposed below the deposits of 1835.

A recent geologic map by Hradecký et al. (2001) delineates products of Cosigüina and older rocks. They map several units of recent pyroclastic deposits as extensive as, or greater than, those of 1835 that they assign to eruptions around A.D. 1500 and 1700. Two other eruptive sequences are noted between these. As discussed below, we interpret the record differently.

In sea-cliff exposures along the northwest coast (Figs. 2, 8, and 11), pyroclastic-fall and -flow deposits of 1835 bury older diamicts and alluvium that resemble the sequence of interbedded lahar and alluvial deposits emplaced following the 1835 eruption. Complex interbedding and gradational contacts of stratified alluvial deposits with clast- to matrix-supported diamicts suggest that water played a prominent role in both. The clasts in the pre-1835 diamicts are subangular to round and composed of varied lava types rather than being dominated by a single lithology and having common prismatically jointed clasts and joint-bounded fragments as do the 1835 pyroclastic-flow deposits. The diamicts contain charred and uncharred wood fragments. One charred fragment yielded a 1σ, calibrated radiocarbon age range of A.D. 1400–1455 (Fig. 2, section K; Table 3). A rock type that appeared fresh and incipiently prismatically jointed is a basaltic andesite (Table 1, sample 2; 55 wt% SiO$_2$), more mafic than products of 1835. In the area between sections I and J, the diamicts terminate in an exhumed ancient sea cliff that was buried by 1835 deposits.

At several sections, scoria-fall and pyroclastic-flow deposits underlie the 1835 deposits, separated by a weakly developed buried soil a few tens of centimeters thick (Fig. 2; sections A–D, G, H, and M). The buried soils, which are marked by mixing of fine sediment into primary deposits and slight oxidation, are broadly similar to soils that we have examined in sixteenth- and seventeenth-century scoria-fall deposits from Volcán Momotombo and Volcán Telica, southeast of Cosigüina. Except at section M, the pre-1835 deposit is solely a uniform to slightly graded fine-lapilli scoria-fall deposit. Nothing like the 1835 fine-grained basal tephra or accretionary-lapilli–rich deposits are present. Nowhere is the older scoria-fall deposit more coarsely grained than the overlying 1835 scoria-fall deposit. These scattered observations imply that the pre-1835 eruption was of lesser magnitude than that of 1835. Broadly contemporaneous pyroclastic-surge and -flow deposits overlie the pre-1835 scoria-fall deposit in section M, but nowhere have thick pyroclastic-flow deposits on the order of several hundred years old and comparable in character to those of 1835 been found. The lack of accretionary-lapilli–rich ash-cloud deposits further suggests that there was not a great volume of pyroclastic flows produced during this penultimate eruption. Clasts from the pre-1835 scoria-fall deposit contain ~57 wt% $SiO_2$, slightly more mafic than the most mafic clasts from 1835 deposits, but more silicic than the clast from the fifteenth century diamict in section K (Table 1, samples 2–4; Fig. 6). The broad time constraint provided by the soil buried between 1835 and pre-1835 scoria-fall deposits suggests that the pre-1835 deposit could date from the fifteenth century and could perhaps be correlative with the diamicts in section K, but an age a few centuries younger or older can't be ruled out.

Tens of pyroclastic-fall and pyroclastic-flow deposits are exposed along the trail that descends steeply from section H to the coast in an area called Tigüilotada (Figs. 1 and 2). The pre-1835 scoria-fall deposit in section H is underlain by ~4–5 m of fine-grained colluvium that contains numerous strongly oxidized reddish-brown buried soils, some of which have argillic B horizons. This sequence of sediments and soils must represent at least several thousand years of accumulation and weathering. Such an interpretation is markedly different from that of Hradecký et al. (2001), who mapped the deposits in this area as all younger than A.D. 1500 (their Secuencias II to V). However, their interpretation is difficult to justify in light of the soil evidence. A scoria-fall deposit within the sequence of colluvium and soils is barely andesitic (53.2 wt% $SiO_2$; Table 1, sample 1). Below, the section is dominated by pumice-fall deposits, diamicts, and alluvial deposits. A conspicuous, compositionally graded, 1.6-m-thick pumice-fall deposit (probably pumice unit EP of Hradecký et al., 2001) about one-third of the way to the beach (UTM 0427346E/1433493N) has a light gray dacitic (64.8 wt% $SiO_2$) base and dark gray andesitic (57.5 wt% $SiO_2$) top, which closely mimics the compositional range seen in the 1835 eruptive sequence, albeit the proportion of dacite is much greater in the older deposit (Table 1; Figs. 6 and 13). In total, the section records more than 10 sizable tephra-fall eruptions, the deposits of which are all thicker and coarser than those of 1835 in section H. Perhaps one or more of these are related to events that created the somma rim of Filete Cresta Montosa. Unfortunately, there is still no radiometric-age control in the section, but the lack of prominent weathering horizons and marked disconformities in the lower part suggests that it was deposited fairly continuously prior to a hiatus of at least several thousand years, as indicated by the colluvial unit with buried soils.

We have additional differences with the map by Hradecký et al. (2001), chiefly related to their interpretation of stratigraphic relations in sea cliffs along the northwest coast and to the age and extent of pyroclastic-flow deposits in fans on the lower southeast flank of Cosigüina.

Our interpretation of the northwest coastal exposures is guided by tracing the scoria-fall deposit of 1835 around the volcano and by radiocarbon ages of charcoal within the overlying pyroclastic-flow and surge deposits (Figs. 2 and 8; Table 3). Our two radiocarbon samples that have ages indistinguishable from modern are small charcoaled twigs from surge deposits. The scoria-fall deposit mantles a dissected landscape formed chiefly in lahar and alluvial deposits. The overlying 1835 pyroclastic-flow deposits are thick in former valley bottoms and are thin or absent on former ridge tops, extend to the face of the sea cliff, and no doubt reached at least several hundred meters more seaward

Figure 13. Photograph of compositionally graded pumice-fall deposit (white arrow; 1.6 m thick) exposed near section H. Dashed line shows approximate location of gradational color change between white to light gray dacitic base and gray andesitic top. Compositional range mimics that found in 1835 deposits (Table 1, samples 5 and 18; Fig. 6).

before being eroded back. In contrast, Hradecký et al. (2001) mapped several units in this same area—pyroclastic-flow and surge deposits of A.D. 1500 (UH1), 1700 (UH2), and an intervening episode—depicted as roughly beach-parallel polygons. We do not see the stratigraphic evidence to support this interpretation. Also, they do not specify the localities of their age-dated material, but great uncertainties during the past 500 $^{14}$C yr imparted by variations in the calibration curve create poor age resolution. At a 95% confidence ($2\sigma$) level (and even at a 68% [$1\sigma$] level), the ages for units UH1 and UH2 overlap and provide only broad age constraints of several centuries (Table 3). We consider that it is not possible to assign UH1 to A.D. 1500 and UH2 to A.D. 1700 on the basis of these ages.

Hradecký et al. (2001) show pyroclastic-flow deposits of 1835 extending only ~6 km southeastward from the caldera rim; more distal deposits are assigned to eruptions ca. A.D. 1700. Section E (Fig. 2), which is located in this distal area, exposes the typical 1835 sequence of scoria-fall deposits overlain by pyroclastic-flow and -surge deposits from which we have a modern radiocarbon age. This sequence is traceable over a broad area of the southeast fan in both areas mapped as deposits from 1700 and 1835 by Hradecký et al. (2001). In addition, Hacienda Cosigüina, which is located within the pyroclastic-flow deposits ascribed to 1700, "was buried under 21 feet of drift" during the 1835 eruption (Roberts, 1924, p. 206–207). To us, this clearly means that the pyroclastic-flow deposits exposed in the southeast sector date from 1835.

## DISCUSSION

Self et al. (1989) inferred correlations between their threefold sequence of deposits on the upper east flank of the volcano and reports of eyewitnesses. We mostly agree with their interpretations, but add a few points:

**Phase 1** (following the nomenclature of Self et al., 1989): The growth of an eruption column between 0630 h and 0800 h (all local times) on 20 January was followed a few hours later by ash fall at La Unión, El Salvador, ~50 km north-northwest of Cosigüina, and at Nacaome, Honduras, ~60 km north. Basal ash and dacitic pumice-fall deposits in sections K, L, and M (Fig. 2) on the northwest to north flanks of the volcano lie directly on soil and older deposits and must correlate to this initial explosive eruption that lasted throughout the day. The "phosphoric sand" that fell in La Unión from 1600 h to 2000 h (Galindo, 1835b) may be a distal equivalent of the dacitic pumice fall, which is the coarsest-grained and lightest-colored of the basal deposits. The transport of tephra northward is consistent with high-altitude (>7 km) winds during January (Fig. 4). Thin, fine-grained basal ash present in south, southeast, and east sectors (sections B–F, Fig. 2) and thicker and coarser basal ash and fine lapilli in the southwest and west (sections G and H) are consistent with southwestward transport by low-altitude tropospheric winds.

**Phase 2:** A lull during the night of 20–21 January was followed by vigorous eruptive activity beginning on the morning of 21 January. Lapilli the size of "peas to hen's eggs" (perhaps 0.5–5 cm) fell on Isla El Tigre, 35 km north-northwest of Cosigüina (Galindo, 1835b). This represents the coarsest distal fall material reported during the eruption and most reasonably correlates with the main scoria fall that is prominent on all flanks of the volcano. During both phases 1 and 2, reports from the southeast, including El Realejo and nearby San Antonio (both near present-day Corinto), describe dramatic views of the eruption cloud, but no local ash fall. High-altitude winds continued their dominant northward and northeastward transport of tephra.

**Phase 3:** From 22 to 24 January, and perhaps through the end of the month, ash fall spread southeastward to cover most of western and southern Nicaragua. Intermittent ash fall was also reported in La Unión and other parts of eastern El Salvador. Self et al. (1989) postulated a shift in direction of high-altitude winds in at least some altitude bands during this time period to promote the southeasterly dispersal of ash. Concurrently, ships off the Pacific coast were also enveloped in ash fall. Therefore, explosions spawning pyroclastic flows and related ash clouds and also maintaining high-altitude eruption columns of chiefly fine ash characterized this time period. Thick, accretionary-lapilli–rich ash-fall deposits on the southwest flank in part represent ash clouds of pyroclastic flows that were transported by the prevailing tropospheric winds to the southwest.

Deposits of the 1835 eruption record a variety of volcanic processes. Initial events, accompanied by loud explosions, chiefly produced tephra-fall deposits. These included ash and fine-lapilli falls possibly accompanied by areally restricted surges described by Self et al. (1989) on the upper east flank. Such falls probably derived from relatively low (<7 km) eruption columns. A modest dacitic pumice fall of sub-Plinian character and the subsequent extensive andesitic scoria falls derived from higher (>7 km) eruption columns. The dacitic pumice, which is distinctly more silicic (by 3–6 wt%; Table 1, Fig. 6) than the andesite, represents a small volume of more differentiated magma that perhaps resided at a shallower level and erupted early. The andesite was the most voluminous component and, on the basis of the relatively simple character of the scoria-fall deposit, was evidently emplaced at a fairly constant rate. Early, more inflated brown scoria was rather abruptly replaced by dark gray scoria that was denser and included some nearly vesicle-free juvenile lithics. The lack of interbedding of scoria-fall and -flow deposits suggests another abrupt transition, this time to collapsing columns or fountains of scoria that fed pyroclastic flows down all flanks of the volcano. Elutriation of ash from the flows, as well as explosive interaction of the flows with seawater, produced thick deposits of accretionary-lapilli–rich ash, especially to the southwest where low-altitude winds would concentrate such deposits. The possible presence during all or part of the eruption of a caldera lake, similar to present-day conditions, suggests that surface or shallow groundwater may have played a role in quenching some of the magma to produce higher densities and some of the profound explosions that were heard far away.

## PRELIMINARY HAZARDS ASSESSMENT

During historical time (the past ~500 yr), Volcán Cosigüina has erupted much less frequently than many of the other volcanic centers in northwestern Nicaragua, such as Momotombo, Telica, or San Cristóbal (Fig. 1). But the 1835 eruption of Cosigüina was by far the most significant in terms of explosivity, extent of pyroclastic flows, and volume of tephra fall. As described above, the penultimate eruption of consequence may date from the fifteenth century, prior to European settlement. The poorly dated prehistoric record of the volcano, as viewed in the Tigüilotada section, suggests that tens of large, explosive eruptions of andesitic to dacitic magma occurred during the recent geologic past, but prior to several thousand years ago. This record of infrequent but sizable eruptions indicates that, if Cosigüina becomes restless, it should be regarded as capable of producing an eruption on the order of or larger than that of 1835.

We use the mapped extent of pyroclastic-flow and -surge deposits of 1835 (Fig. 7) as a guide for establishing a proximal hazard zone for future such events (Fig. 14). This map agrees in most aspects with the volcanic-hazards components of the geologic hazard map of Hradecký et al. (2001). In future eruptions, the arcuate ridge on the upper west flank of the volcano, Filete Cresta Montosa, will probably divert pyroclastic flows as it did in 1835. Its relatively modest height (chiefly ≤200 m) suggests that energetic flows and surges could surmount the ridge and concentrate in the small valleys that trend west and northwest from it. The major fans and valley fills of 1835 pyroclastic-flow deposits, on which are built the main villages of Cosigüina and Potosi and the small settlements, farms, and ranches along the

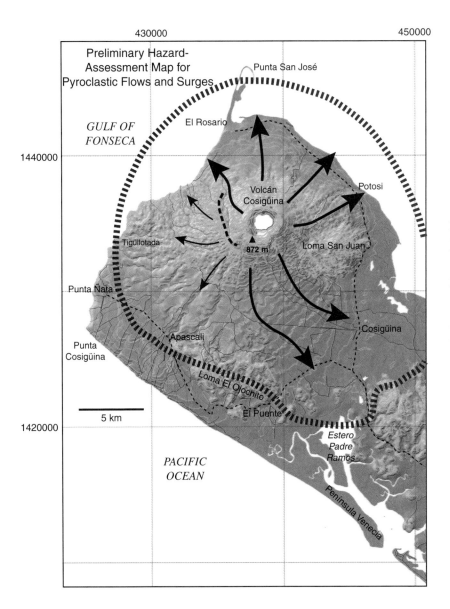

Figure 14. Preliminary hazard-assessment map for future pyroclastic flows and surges at Cosigüina based partly on the mapped extent of deposits from the 1835 eruption (Fig. 7). Small arrows depict the potential for small flows and surges surmounting the ridge west of the caldera (bold dashed line) and affecting terrain to the west and southwest, especially in valleys. Large arrows depict main flow paths that feed broad pyroclastic fans. Such areas are also expected to be affected by lahars and rapid alluviation following a significant eruption. Outer dashed line roughly outlines the area that could be affected by energetic pyroclastic surges.

road to El Rosario, are the most likely paths for future flows. The east end of Loma San Juan was not subjected to pyroclastic flows in 1835, but it was possibly swept by ash-cloud surges and is close enough to the vent (8–9 km) so as to not offer refuge. Pyroclastic flows did not reach the southwestern coast in 1835, but this area could be vulnerable to pyroclastic flows somewhat larger than those of 1835 that are able to surmount Loma El Ojochito. Nevertheless, the region from Apascali to Punta Ñata received thick falls (>1 m thick prior to settling and compaction) from ash clouds of pyroclastic flows and vent explosions in 1835. Movement of vehicles would probably be impossible under such conditions. A very unpleasant environment during and after an eruption would result as wind resuspended ash or as rain saturated ash to form deep mud. Roofs of buildings on most of the Cosigüina peninsula would be vulnerable to collapse from tephra loads of the thicknesses that accumulated in 1835, especially if the eruption were accompanied or followed by rain.

Cosigüina's location on a peninsula, with only one major road (graded gravel) allowing relatively rapid evacuation to a distance >15 km from the vent, reduces the options for risk mitigation for the ~5000 residents of the peninsula and their herds of livestock. The potentially severe impacts of an 1835-scale eruption, chiefly related to pyroclastic flows and thick tephra fall, probably make evacuation of the peninsula prior to onset of an eruption the best measure to protect humans and livestock.

From eyewitness accounts in the aftermath of 1835 and experience at other eruptions, it can be inferred that a period of rapid sediment production and channel instability as drainage networks reestablish on the new landscape would follow a significant eruption. Low areas downstream from pyroclastic-flow deposits would be most affected, but widespread tephra fall could also trigger high sediment loads in watersheds unaffected by pyroclastic flows. The broad fans on the southeast and north would likely grow and extend farther into the sea or fringing mangroves.

During future eruptions, the character of the eruption (volume, column height, etc.) and wind direction and strength will control the distribution of tephra fall. The January 1835 eruption illustrates tephra fallout under typical dry season (December–April) wind conditions (Fig. 4). Low-altitude (<7 km) winds are chiefly from the east to northeast and would carry the ash clouds of pyroclastic flows and small explosions toward Apascali, Punta Ñata, and the Pacific coast. In contrast, high-altitude (>7 km) winds blow chiefly from the west to southwest, but during the 1835 eruption must have been more southerly and carried most of the tephra to the northern sector of the volcano and to eastern El Salvador and Honduras. A shift in wind direction at higher altitudes occurred later in the eruption, accounting for tephra fall hundreds of kilometers to the southeast. Future dry-season eruptions that send eruption clouds above 7 km could cause similar widespread tephra fall depending on upper level wind patterns. During the wet season (June–October), winds below 25 km are chiefly from the east to northeast (Fig. 4), and most tephra would fall on the west to southwest flank and the Pacific Ocean. May and November are transitional months during which either wind pattern is possible.

Several additional studies could greatly help to better assess the potential hazards from future eruptions of Cosigüina. Cores from lakes and wetlands in Nicaragua, Honduras, and El Salvador are needed to obtain better measurements of thickness of distal tephra-fall deposits and thereby a better-controlled volume estimate for the 1835 eruption. The sequence of older tephras exposed on the west flank of the volcano near Tigüilotada deserves further attention—particularly to determine the age and frequency of eruptions that likely exceeded the magnitude of the 1835 event. Textural and petrologic studies of 1835 and older products would improve the understanding of eruptive processes and the magmatic system.

## CONCLUSIONS

The great Cosigüina eruption of 1835 was the largest in Nicaragua since the Spanish colonization in terms of volume, explosivity, and widespread effect. It followed a few centuries of relative quiescence; probably at least several millennia had elapsed since a period of frequent eruptions of similar or greater magnitude than that of 1835. The somma rim visible most prominently on the west flank of the volcano and deposits in the Tigüilotada area may record caldera collapse during one or more of these larger prehistoric events.

The 1835 eruption provides a useful model for developing an assessment of potential hazards during future events. The key findings reported here include:

- Detailed accounts by eyewitnesses of the 1835 eruption and its effect on the volcano's lower flanks and shoreline, coupled with lack of any contemporary reports describing major changes in the profile of the volcano, suggest that the eruption did not radically change the topography of the upper part of the cone. Whether or not a lake was present in a summit caldera at the time of the eruption remains unknown.
- Eruptions on the first day (20 January) deposited chiefly fine-grained ash and lithic lapilli over most of the Cosigüina peninsula, dacitic pumice on the north flank, and ash in eastern El Salvador and Honduras. On 21 January, the climactic andesitic scoria and ash fall affected a similar area, but with thicker accumulations. The timing of the onset of pyroclastic flows is poorly constrained but followed deposition of most, if not all, of the scoria fall.
- Scoriaceous pyroclastic flows swept to the Gulf of Fonseca in northwestern to northeastern and perhaps southeastern sectors. Topographic barriers limited their extent in the southwestern and southern sectors. Ashfall deposits derived from ash clouds of the pyroclastic flows thickly mantled the southwestern sector.

- For several days following 21 January, ashfall continued southwest of the volcano, driven by low-altitude winds, while shifting high-altitude winds caused ashfall to the north, northeast, and southeast, eventually covering most of western and southern Nicaragua.
- Although the bulk of the 1835 eruption products is andesitic (~57–60 wt% $SiO_2$), a minor amount of dacite (~64–65 wt% $SiO_2$) was erupted early. This same compositional range is found in deposits of some large prehistoric eruptions, although the fraction of dacite is much greater in the older deposits.
- The volume of distal tephra is poorly known, but our observations support an estimate of a total bulk-volume of eruptive products in 1835 of ~6 $km^3$, of which about three quarters is the climactic scoria-fall deposit and distal equivalents.
- The lack of descriptions of notable eruptive events rivaling those of 1835 in the sixteenth- to eighteenth-century records of cities within 100 km of Cosigüina suggests that any eruptions during that time were small compared to that of 1835.
- Using the 1835 eruption as a guide, we believe that much of the Cosigüina peninsula, especially on northwestern clockwise to southeastern sectors, could be affected by future pyroclastic flows and surges. Realizing that larger eruptions are recorded in the recent past, evacuation of the peninsula during a period of intense volcanic unrest is probably the best strategy to minimize risk to people and livestock. Depending on the season, ashfall could chiefly affect southwestern sectors (wet season) or affect a much broader area as in January 1835 (dry season).
- A significant eruption would be followed by a period of rapid alluviation and lahars on the volcano flanks as new deposits are eroded and moved seaward.

## ACKNOWLEDGMENTS

Field work for this study was funded by the Volcano Disaster Assistance Program, a joint effort of the Office of Foreign Disaster Assistance of the U.S. Agency for International Development and the U.S. Geological Survey. We thank Wilfried Strauch, Director de Geofísica of the Instituto Nicaragüense de Estudios Territoriales (INETER), for support and encouragement. Carlos Pullinger of the Servicio Nacional de Estudios Territoriales (SNET) of El Salvador helped with our futile search for Cosigüina tephra in eastern El Salvador and offered numerous insights regarding Central American volcanism. Craig Chesner provided stratigraphic and analytical information about the A.D. 430 Ilopango (Tierra Blanca Joven) tephra near Volcán San Miguel. Taryn Lopez patiently and ably measured clast densities. John Ewert obtained and synthesized the NOAA wind data. The manuscript benefited greatly from constructive reviews by Julie Donnelly-Nolan, James Vallance, Steve Self, Greg Bluth, and Petr Hradecký.

## REFERENCES CITED

Caldcleugh, A., 1836, Some account of the volcanic eruption of Cosigüina in the Bay of Fonseca, commonly called the Bay of Conchagua, on the western coast of Central America: Philosophical Transactions of the Royal Society of London, v. 126, p. 27–30.

Dull, R.A., 2004, Lessons from the mud, lessons from the Maya; paleoecological records of the Tierra Blanca Joven eruption, in Rose, W.I., Bommer, J.J., Lopez, D.L., Carr, M.J., and Major, J.J., eds., Natural Hazards in El Salvador: Geological Society of America Special Paper 375, p. 237–244.

Escobar, C.D., 2003, San Miguel and its volcanic hazards, El Salvador [M.S. report]: Houghton, Michigan Technological University, 181 p.

Feldman, L.H., 1993, Mountains of fire, lands that shake—earthquakes and volcanic eruptions in the historic past of Central America (1505–1899): Culver City, California, Labyrinthos, 295 p.

Fierstein, J., and Nathenson, M., 1992, Another look at the calculation of fallout tephra volumes: Bulletin of Volcanology, v. 54, p. 156–167, doi: 10.1007/BF00278005.

Galindo, J., 1835a, Eruption of the volcano Cosigüina: American Journal of Science, v. 28, p. 332–336.

Galindo, J., 1835b, On the eruption of the volcano Cosigüina, in Nicaragua: Journal of the Royal Geographical Society of London, v. 5, p. 387–392.

Hradecký, P., Havlicek, P., Opletal, M., Rapprich, V., and Sebesta, J., Sevcik, and Mayorga, E., 2001, Estudio geológico y reconocimiento de amenazas geológicas en el Volcán Cosigüina, Nicaragua: Servicio geológico de la República Checa (CGU) and Instituto Nicaragüense de Estudios Territoriales (INETER), 42 p.

Janda, R.J., Daag, A.S., Delos Reyes, P.J., Newhall, C.G., Pierson, T.C., Punongbayan, R.S., Rodolfo, K.S., Solidum, R.U., and Umbal, J.V., 1996, Assessment and response to lahar hazard around Mount Pinatubo, 1991 to 1993, in Newhall, C.G., and Punongbayan, R.S., eds., Fire and Mud: Eruptions and Lahars of Mount Pinatubo, Philippines: Quezon City, Philippine Institute of Volcanology and Seismology, and Seattle, University of Washington Press, p. 107–139.

Johnson, D.M., Hooper, P.R., and Conrey, R.M., 1999, XRF analysis of rocks and minerals for major and trace elements on a single low dilution Li-tetraborate fused bead: Advances in X-ray Analysis, v. 41, p. 843–867.

Ramsey, C.B., 2005, OxCal Program v. 3.10: http://www.rlaha.ox.ac.uk/oxcal/oxcal.htm.

Roberts, M., 1924, On the earthquake line: Minor adventures in Central America: London, Arrowsmith, 309 p.

Reimer, P.J., Baillie, M.G.L., Bard, E., Bayliss, A., Beck, J.W., Bertrand, C.J.H., Blackwell, P.G., Buck, C.E., Burr, G.S., Cutler, K.B., Damon, P.E., Edwards, R.L., Fairbanks, R.G., Friedrich, M., Guilderson, T.P., Hogg, A.G., Hughen, K.A., Kromer, B., McCormac, F.G., Manning, S.W., Ramsey, C.B., Reimer, R.W., Remmele, S., Southon, J.R., Stuiver, M., Talamo, S., Taylor, F.W., van der Plicht, J., and Weyhenmeyer, C.E., 2004, IntCal04 Terrestrial radiocarbon age calibration, 26–0 ka BP: Radiocarbon, v. 46, p. 1029–1058.

Self, S., Rampino, M.R., and Carr, M.J., 1989, A reappraisal of the 1835 eruption of Cosigüina and its atmospheric impact: Bulletin of Volcanology, v. 52, p. 57–65, doi: 10.1007/BF00641387.

Simkin, T., and Siebert, L., 1994, Volcanoes of the World, 2nd edition: Tucson, Geoscience Press, 349 p.

Squires, E.G., 1851, On the volcanoes of Central America, and the geographical and topographical features of Nicaragua, as connected with the proposed inter-oceanic canal: Proceedings of the American Association for the Advancement of Science, p. 101–122.

Walker, G.P.L., 1979, A volcanic ash generated by explosions where ignimbrite entered the sea: Nature, v. 281, no. 5733, p. 642–646, doi: 10.1038/281642a0.

Williams, H., 1952, The great eruption of Coseguina, Nicaragua, in 1835: University of California Publications in Geological Sciences, v. 29, p. 21–46.

Williams, S.N., and Self, S., 1983, The October 1902 Plinian eruption of Santa María volcano, Guatemala: Journal of Volcanology and Geothermal Research, v. 16, p. 33–56, doi: 10.1016/0377-0273(83)90083-5.

MANUSCRIPT ACCEPTED BY THE SOCIETY 19 MARCH 2006

# The youngest highly explosive basaltic eruptions from Masaya Caldera (Nicaragua): Stratigraphy and hazard assessment

**Wendy Pérez**
SFB 574, IfM-GEOMAR (Leibniz Institute for Marine Sciences), Wischhofstr. 1-3, D-24148 Kiel, Germany, and Escuela Centroamericana de Geología, Universidad de Costa Rica

**Armin Freundt**
SFB 574, IfM-GEOMAR (Leibniz Institute for Marine Sciences), Wischhofstr. 1-3, D-24148 Kiel, Germany

## ABSTRACT

The youngest highly explosive basaltic eruptions from Masaya Caldera in central western Nicaragua produced five main pyroclastic deposits: the San Antonio Tephra, La Concepción Tephra, the Masaya Triple Layer, and the Masaya Tuff with the Ticuantepe Lapilli. This tephra sequence was deposited over the past ~6000 yr. The distribution and physical characteristics of these deposits suggest they originated from the Masaya Caldera. They have volumes ranging from 0.2 km$^3$ for La Concepción Tephra to 3.9 km$^3$ for the Masaya Tuff and cover minimum areas of 600–1600 km$^2$. All deposits formed by violent eruptions discharging $10^{11}$ to $10^{12}$ kg of magma, thus reaching eruption magnitudes between 4.3 and 5.9 and volcanic explosivity indices of 3–4. An analysis of hazards for the main population centers around the Masaya Caldera shows that, if there were a similar eruption today, the most vulnerable communities would be Ticuantepe, Nindirí, and Masaya. In addition, La Concepción, southwest of the caldera, and the capital Managua, more than 15 km to the northwest, could be affected.

**Keywords:** Masaya Caldera, basaltic Plinian fall deposit, phreatomagmatic eruption, pyroclastic surge, volcanic hazards, vulnerable communities.

## INTRODUCTION

The Masaya Caldera, in Nicaragua, is a NW-SE elongated 6 × 11 km caldera (Fig. 1) partially filled by post-collapse lava flows (Walker et al., 1993) and hosting the Masaya cone, consisting of four pit craters (Rymer et al., 1998). The Masaya caldera is of very special volcanological interest because it has produced highly explosive large-magnitude basaltic eruptions (Plinian and phreatomagmatic) in the past (Williams, 1983b; Wehrmann et al., this volume). The geochemical characteristics of the eruptive products led Walker et al. (1993) to postulate a large, shallow magma chamber beneath Masaya that was replenished by magmas from depth and contaminated by crustal components.

It is still a matter of debate whether the caldera formed by subsidence following passive subterranean withdrawal of magma (McBirney, 1956; Williams and McBirney, 1979) or during a series of collapses, over an extended period of time, resulting from magma withdrawal during large, explosive, basaltic eruptions (Bice 1980, 1985) or just due to the eruption of widespread pyroclastic surges and basaltic ignimbrite deposits (Williams, 1983a).

The first stratigraphic work on the deposits of highly explosive basaltic eruptions related to Masaya Caldera (Bice, 1980,

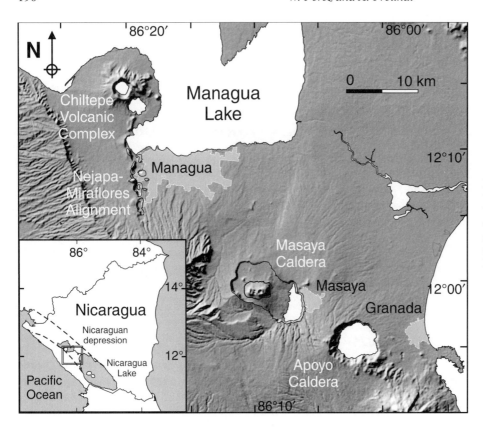

Figure 1. Location of the study area, with the digital elevation model showing important volcanotectonic structures, volcanic edifices, and the three main cities (Managua, Masaya, and Granada). Inset map shows position of study area in west-central Nicaragua and the Nicaragua Graben.

1985) was restricted to the Managua area and identified three widespread basaltic deposits. Williams (1983a) studied these pyroclastic deposits in more detail, including a geological characterization of the post-collapse products inside the caldera and of its walls. The three main basaltic deposits are (1) the older Masaya Lapilli Bed (Bice, 1985) or Fontana Lapilli (Williams, 1983b), the subject of Wehrmann et al. (this volume); (2) the younger deposits of the Masaya Triple Layer (Bice, 1985) or San Judas Formation (Williams, 1983b), all interpreted as fall deposits; and (3) the Masaya Tuff (Bice, 1985) or El Retiro Tuff (Williams, 1983b), interpreted as phreatomagmatic fall by Bice but recognized as a pyroclastic surge deposit by Williams.

The main purpose of this work is to present a detailed stratigraphy of the youngest widespread tephra sequence related to the Masaya volcanic center, which includes the Masaya Triple Layer, the Masaya Tuff, and other newly identified deposits.

The volcanic hazard aspects, which can be derived from this tephrostratigraphic study, are of great importance, because the most important cities in Nicaragua, including the capital and socioeconomic center of the country, Managua, lie only 15 km from the Masaya Caldera.

## GEOLOGICAL SETTING

The Masaya Caldera is located in central western Nicaragua inside the NW-trending, 45–50-km-wide Nicaraguan depression, which hosts Lake Nicaragua, Lake Managua, and Pleistocene-Holocene volcanic centers, the products of which have contributed to the filling of the basin (Fig. 1).

The local basement of the depression is the poorly known Las Sierras Formation, which is a group of pyroclastic and epiclastic rocks mostly of mafic composition with no clear relationship to any present volcanic structure (Bice, 1980). Radiometric ages of rocks from the upper Las Sierras Formation range in age from >100,000 yr (Kuang, 1971; Bice, 1980) to 29,200 ± 800 yr B.P. (Williams, 1983a). Because of this wide age range, Williams (1983a) considered this formation to be an assemblage of many genetically unrelated rock units.

A tephra sequence that has been called the Managua Series by Niccum (1976) overlies these older volcanic rocks. The Managua Series has been described as "a distinctly layered, thin (10–20 m total) sequence of basaltic or rhyodacitic airfall tuffs, ashes and lapilli beds" (Bice, 1985, p. 554). The sources, inferred eruption styles and compositional characteristics of this sequence range from basaltic Plinian and extensive phreatomagmatic deposits related to Masaya Caldera to rhyodacitic Plinian fall deposits derived from Apoyo, Apoyeque, and Xiloá volcanoes (Bice, 1980).

## STRATIGRAPHY

Our detailed logging of 237 outcrops around Masaya revealed a total of five depositional units of young, highly explosive eruptions originating from Masaya caldera. Lithological and bedding characteristics distinguish the San Antonio Tephra, La Concepción Tephra, the Masaya Triple Layer, the Masaya Tuff, and the Ticuantepe Lapilli.

These units crop out around the Masaya Caldera, although there are few outcrops to the east and southeast of the caldera. At the time of this writing, we had only found a few outcrops of the San Antonio Tephra; the following preliminary description may be subject to additions as our work continues.

## San Antonio Tephra

The San Antonio Tephra is the oldest unit found among the products of the youngest highly explosive Masaya Caldera eruptions. It underlies La Concepción Tephra and the Masaya Triple Layer, from which it is separated by a thick paleosol and a regional erosive unconformity. It is also separated by paleosols from underlying pumiceous fall deposits erupted at Apoyo caldera and the Chiltepe peninsula. The youngest of these is the Xiloá Tephra with a radiocarbon age of 6105 ± 30 yr B.P. (see Figure 7A in Freundt et al., this volume). Since we have dated the Masaya Triple Layer as 2120 ± 120 yr B.P. (see below), the San Antonio Tephra must have been emplaced between ~2500–6000 yr ago.

A few exposures can be found to the northwest, north, and south of the Masaya Caldera. In the most proximal outcrops at the NW sector of the caldera, there is a lowermost sequence made up by an alternation of thin (<15 cm), well-sorted, coarse-ash to fine-lapilli scoria fall beds and laminated tuffs rich in fine ash (layers A1–A5, Fig. 2). These are overlain by a well-sorted layer of vesicular fluidal-shaped black scoria lapilli (layer A6) and a very distinctive lapilli layer (A7) with a high content of yellowish and whitish hydrothermally altered lithic fragments (Fig. 2). Upward, the sequence becomes more complex, being dominated by tuff beds with strong lateral changes of thickness, dunes, and cross-bedding, as well as accretionary lapilli–rich tuffs (A8–A12, Fig. 2). At the top, there is a well-sorted deposit of mixed vesicular and dense, reddish lapilli (A13), which is found in several exposures at the south of the caldera. The lowermost sequence of the San Antonio Tephra is interpreted as the fall deposits of magmatic eruptions, interrupted by phreatomagmatic pulses, and culminating in layer A6. From layer A7 on, the dominantly phreatomagmatic eruption produced pyroclastic surges and fall.

## La Concepción Tephra

This unit is a succession of well-sorted scoria lapilli and ash layers interbedded with tuffs that crops out only to the south of the Masaya Caldera. It overlies the San Antonio Tephra above a thick paleosol and a regional unconformity and is overlain by the Masaya Tuff above an erosional unconformity. Williams (1983a, 1983b) interpreted some exposures of this unit as representing part of the proximal facies of the Masaya Triple Layer.

La Concepción Tephra is composed of eight well-sorted, black, scoria lapilli to coarse-ash layers (B1, B3, B5, B7, B9, B11, B14, B16) intercalated with tuff beds (B2, B4, B6, B8, B10, B12, B13, B15) (Figs. 3 and 4). The lower part of the sequence is dominated by lapilli layers while the upper part is mainly composed of tuff beds (Figs. 3 and 4).

The well-sorted layers B1, B3, and B5 are composed of highly vesicular black lapilli to coarse-ash, with small amounts of basaltic lava country rock fragments (~1%–3%). The scoriaceous lapilli contain 50–90 vol% vesicles, sparse plagioclase phenocrysts with abundant melt inclusions, and olivine in a sideromelane and tachylite groundmass with plagioclase and olivine microlites. Layer B5 is the thickest lapilli bed, vaguely stratified by vertically alternating grain size, composed of fluidal-textured reticulitic lapilli and ash particles, and hence is a useful marker bed in all outcrops (Figs. 4B and 4C). Above B5, the well-sorted layers consist of highly vesicular and dense to poorly vesicular scoria lapilli, with the fraction of denser clasts increasing upward through the succession, and hydrothermally-altered lithic fragments.

The tuff beds are mostly grayish and cemented (Fig. 4D). Layer B4 is a relatively well-sorted, fine grained, parallel laminated tuff. Other tuffs (B2, B6, B8, B10) are fine-grained massive beds; some of them contain accretionary lapilli, armored lapilli, dispersed glassy scoria fragments, and plant molds (Fig. 3). The uppermost tuffs (B12, B15) are thicker and poorly-sorted, with low angle cross-bedding and dune-structures and laterally changing thickness. The componentry is dominated by glassy ash, with ~15% of scoriaceous lapilli and lithic fragments mainly of pre-existent lavas with different textures. They also contain abundant molds of branches that are not systematically aligned. Layer B13 is a bed of fine ash rich in accretionary lapilli.

The well-sorted black lapilli and ash layers up to layer B5 are interpreted as magmatic fall deposits. Well-sorted layers above B5, containing poorly vesicular clasts, and the massive to laminated tuff layers are interpreted as phreatomagmatic fall deposits, with the exception of the tuffs B12 and B15, which are surge deposits.

## Masaya Triple Layer

Bice (1985, p. 560) observed this deposit in the surroundings of Managua city, where "it typically gives the appearance of two thin (1–2 cm) indurated gray tuffs sandwiched among three thicker (5–10 cm), loose layers of coarse black ash." He also mentioned that the unit thickens toward Masaya Caldera and becomes more complex, adding extra layers. Williams (1983a, 1983b) described it as well-sorted, nongraded, and composed of five to twelve scoria fall layers, with a thickness between 0.5 and 1.5 m. According to his isopach map and site descriptions, he included the deposits of La Concepción Tephra south of the Masaya Caldera, which is a similar deposit but with a different internal bedding sequence. Our radiocarbon dating of fossil plant material in the layer C3 (Figs. 5 and 6) gives an age of 2120 ± 120 yr B.P., much younger than the age of 7000 to 9000 yr estimated by Bice (1985).

The Masaya Triple Layer is mainly a fall deposit composed of seven major beds of black, well-sorted scoriaceous lapilli to coarse-ash (C1, C2, C3b, C4, C6, C8 and C10 in Figs. 5 and 6),

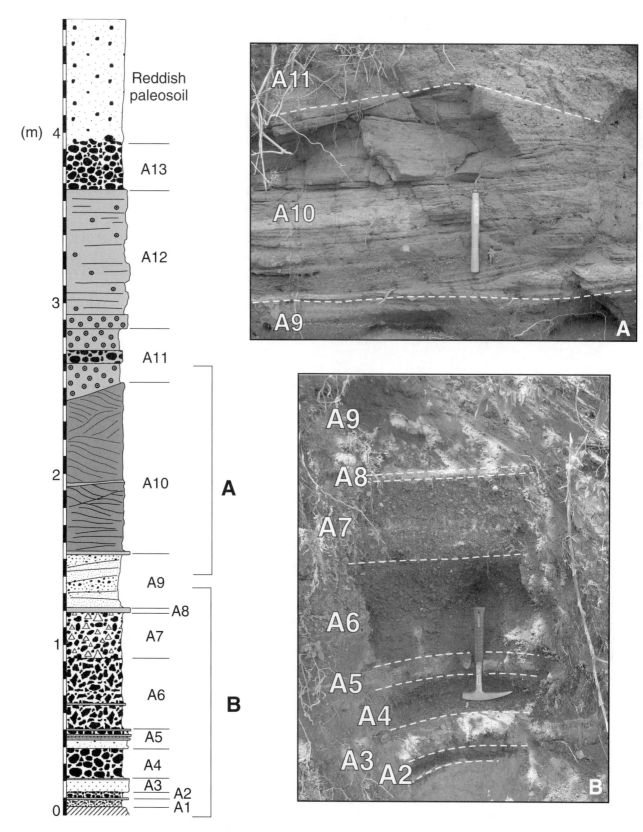

Figure 2. Stratigraphic column of the most complete exposure of the San Antonio Tephra (Loc. W67, UTM 582886W/1333379N), 7 km NW of the caldera rim. The two photographs show some of the layers. (A) Dune structures in layer A10 (scraper = 30 cm). (B) Lower part of the sequence, showing the main fall layer A6 and the overlying lithic-rich layer A7 (rock hammer = 30 cm).

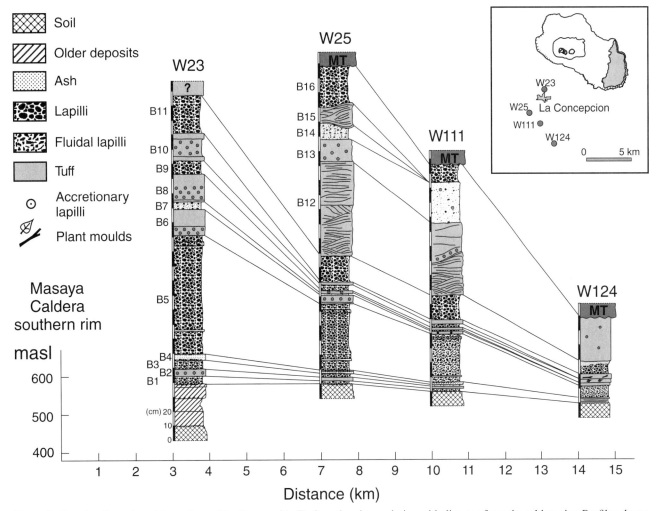

Figure 3. Correlated stratigraphic sections of La Concepción Tephra, showing variation with distance from the caldera rim. Profiles shown are from exposures 3, 7, 10, and 14 km, respectively, to the south of the caldera as shown in the inset map. The tuff layers B6, B8, and B10 thin rapidly but can be traced for more than 8 km. Layers B12–B15 merge into one single massive deposit in the distal facies.

separated by four major (C3a, C5, C7, C9) and several minor tuff layers. The lowermost lapilli layers (C1, C2) contain highly vesicular and fluidal scoria particles; the other layers (C3b, C4, C6, C8, C10) are characterized by relatively round scoria lapilli and by higher contents of lava lithic fragments.

The massive tuff layer C7 contains accretionary lapilli and is thicker than the other tuff beds, reaching a maximum thickness of 40 cm in the most proximal exposures. The tuff beds C5 and C9, as well as some of the minor thin tuff beds, have desiccation cracks on their upper surfaces, an indication of wet emplacement and of repose time between the different depositional events. Most of these tuff beds appear as thin, massive indurated tuffs in the distal facies. In the medial facies, they exhibit parallel lamination, and in the proximal facies each corresponds to a sequence of thin lapilli to ash layers.

The scoriaceous lapilli of the Masaya Triple Layer have <5 vol% of olivine, plagioclase, and clinopyroxene phenocrysts in a highly vesiculated groundmass of brown sideromelane. Some contain lithic fragments of basalt and gabbro. The abundance of clinopyroxene phenocrysts increases stratigraphically upward in the Masaya Triple Layer.

The lapilli and ash layers are interpreted as fall deposits from magmatic eruptions, while the intercalated grayish tuff layers, with evidence of wet emplacement, are the products of phreatomagmatic eruptions. Some of these tuff layers do not extend to great distances, so the stratigraphic succession gets simpler away from the vent. Overall, we suggest that the Masaya Triple Layer is the product of a spasmodic eruption, alternating subplinian to Plinian events with phreatomagmatic explosions.

**Masaya Tuff**

The first report on this deposit was by Krusi and Schultz (1979). It was later described by Bice (1980, 1985) as a thick widespread light tan poorly to moderately indurated, faintly layered to massive fine airfall basaltic tuff, with an average

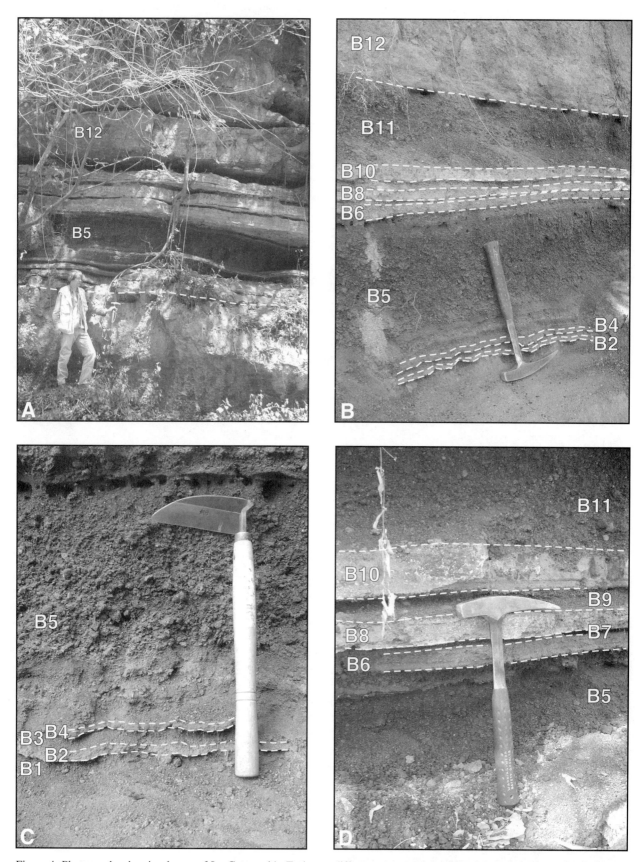

Figure 4. Photographs showing layers of La Concepción Tephra at different exposures (rock hammer and scraper used for scale: 30 cm). (A) Proximal facies at 3 km to the south from the caldera rim (W23, UTM 588222W/1320796N). Dashed white line marks base of deposit. (B) 7.5 km to the south of the caldera (W111, UTM 587770W/1316699N). (C) Lower part of La Concepción Tephra sequence at the outcrop W111, showing the most distinctive layer B5 with its vertical grain size variations. (D) Phreatomagmatic tuffs (B6, B8, B10) at 5.5 km from the caldera rim (UTM W24, 587909W/1318501N).

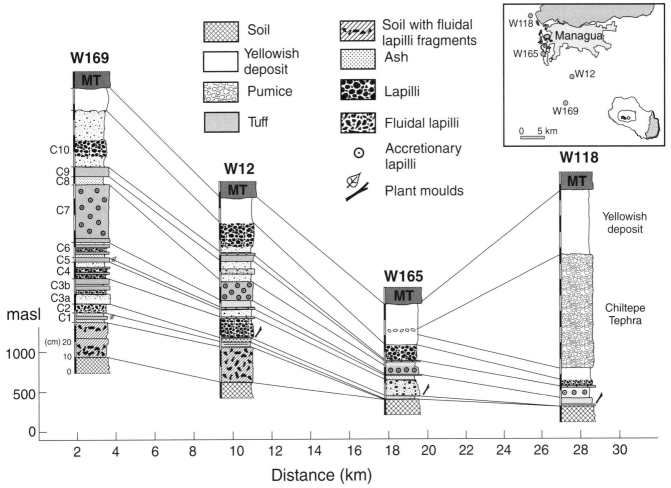

Figure 5. Correlated stratigraphic sections of the Masaya Triple Layer (MTL) of outcrops located 2, 8.5, 18, and 26.5 km NW of the caldera rim (inset map). The proximal exposures show a more complicated internal bedding, with several intercalated tuff beds. The stratigraphic succession gets simpler away from the vent. Near Chiltepe peninsula, the Masaya Triple Layer is overlain by the Chiltepe Tephra pumice fall which is followed by the Masaya Tuff (MT).

thickness between 20 and 50 cm and which constitutes the youngest large explosive deposit in the Managua area. Based on the chemical composition, petrographic similarities, and field evidence, Bice (1980) stated that the Masaya Tuff came from the site of the present Masaya Caldera and was probably genetically related to the collapse and formation of the caldera as it presently exists. He thought this fine-grained indurated tuff in Managua city was the fine fall deposit equivalent of the base-surge facies at the Masaya caldera walls. Williams (1983a) described this unit as pyroclastic surge deposits of unprecedented areal extent and volume; with a median grain size much coarser than any other measured before, as well as dune-forms and flow velocity of tremendous scale.

The Masaya Tuff is a grayish, partly indurated deposit that extends to a distance of at least 35 km from Masaya Caldera. It conformably overlies the Masaya Triple Layer, separated by a yellowish, weathered, massive fine-grained tuffaceous deposit containing scattered scoria or pumice fragments (Fig. 5), but locally it rests on an erosive unconformity cutting into the La Concepción Tephra. In outcrops west of Managua city and near Chiltepe Peninsula, the Chiltepe Tephra separates the Masaya Triple Layer and the Masaya Tuff (Fig. 5). This indicates that deposition of the Masaya Tuff occurred a significant time after the eruption of the Masaya Triple Layer, probably <2000 yr ago, and hence much more recently than the 3000–6000 yr estimated by Bice (1985).

The Masaya Tuff is relatively poorly-sorted ($\sigma > 3$), composed mainly of vitric fine ash and 5–20 vol% of lapilli to block-sized fragments of preexistent basaltic lavas and dense aphyric to scoriaceous lapilli with rare plagioclase, olivine, and pyroxene phenocrysts.

A complete section of the Masaya Tuff is exposed 3 km to the northeast from the caldera rim, where we divided the succession into four units (D1 to D4) based on intercalated accretionary-lapilli tuffs and structural changes (Fig. 7). There is an overall fining upward in grain size culminating in the fine-ash planar-bedded

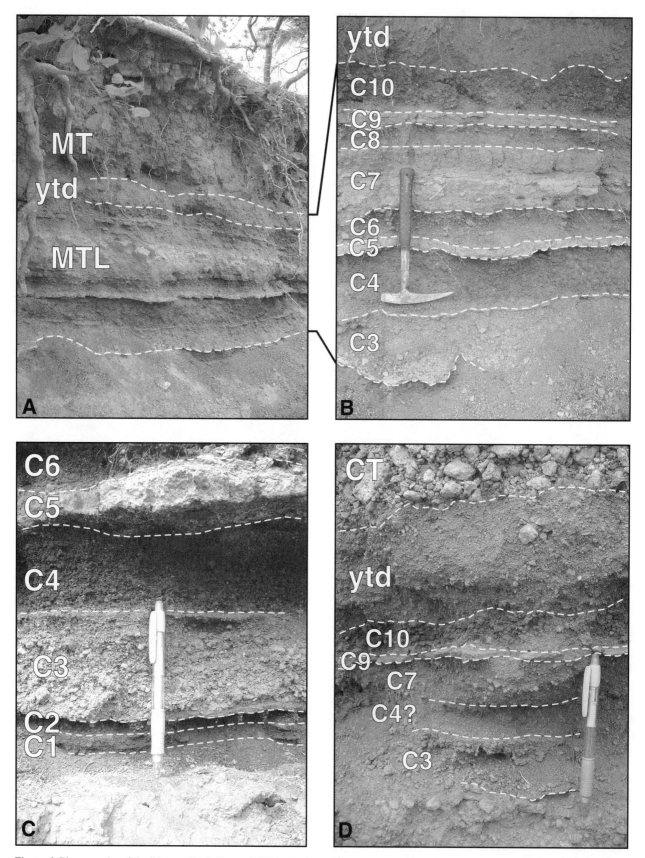

Figure 6. Photographs of the Masaya Triple Layer (MTL). (A) Complete sequence of Masaya Triple Layer at 7 km to the NW of the caldera rim (W67, UTM 582886W/1333379N). Wavy upper contact to the yellowish tuffaceous deposit (ytd) and the Masaya Tuff (MT). (B) Close-up of the Masaya Triple Layer sequence in a showing layers C3 through C10 (rock hammer = 30 cm). (C) Detail of the lower part of the Masaya Triple Layer at the outcrop W17 (9.5 km from the caldera, UTM 583230W/1336730N), showing layer C3 containing carbon dated as 2120 ± 120 yr B.P. (pencil as scale = 15 cm). (D) Distal facies of the Masaya Triple Layer at 25 km from the caldera rim, showing a very simplified sequence in which some of the layers can still be correlated (W119, UTM 571835W/1344537N). The tuffaceous yellowish deposit (ytd) is at the top, overlain by the Chiltepe Tephra (CT).

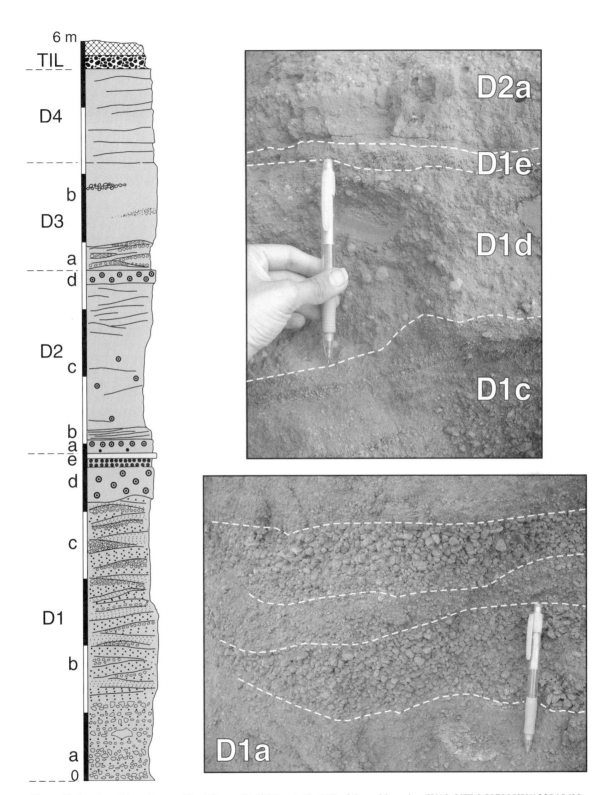

Figure 7. Stratigraphic column of the Masaya Tuff 3 km to the NE of the caldera rim (W10, UTM 597898W/1329104N). Photographs illustrate the boundary between units D1 and D2 marked by the accretionary-lapilli–rich layer D1d (top) and well-sorted lapilli-rich lenses in otherwise poorly sorted basal layer D1a (bottom). Pen (15 cm) for scale. TIL—Ticuantepe Lapilli.

tuff D4. The unit D3 only occurs in the northeastern sector of the caldera. Unit D1 constitutes the major part of the Masaya Tuff and, together with D2 and D4, occurs all around the caldera.

The proximal facies of the Masaya Tuff consists of a dense-clast breccia at the base, occurring only within ~1 km from the caldera rim. This is followed by several layers with cross-bedding and dune structures (subunits D1b and D1c) that range in grain size from lapilli to ash and contain lenses of well-sorted, rounded and abraded scoriaceous lapilli (Figs. 8A and 8B). The cumulative thickness reaches 15 m, and the top of the deposit is formed by a planar-bedded fine-ash layer containing abundant accretionary lapilli (D1d in Figs. 7 and 8D). In the medial facies at ~5–15 km from the caldera, the basal unit has a higher concentration of coarse fragments (mainly lapilli, but also minor blocks), with faint cross-bedding. The overlying tuff is finer grained (fine lapilli to medium ash) with well developed cross-bedding and progressive and regressive dune structures (Figs. 8B and 8C). This is capped by a sequence of several indurated planar bedded tuffs with accretionary lapilli (Fig. 8D). The distal facies (>15 km) to the NW is an indurated, 10–50-cm-thick planar stratified fine ash deposit.

The lithological characteristics of the Masaya Tuff, such as poor sorting, high ash content, abundance of dunes and cross-bedding, as well as the radial distribution around the caldera, support the interpretation as a pyroclastic surge deposit given by Williams (1983a). The low vesicularity of most of the juvenile clasts, the presence of armored lapilli and abundant accretionary lapilli, rims of hardened ash around branch molds, and the cementation of the deposit in most of the exposures suggest the surges were wet and formed from phreatomagmatic eruptions that probably involved lake water in the preexisting caldera.

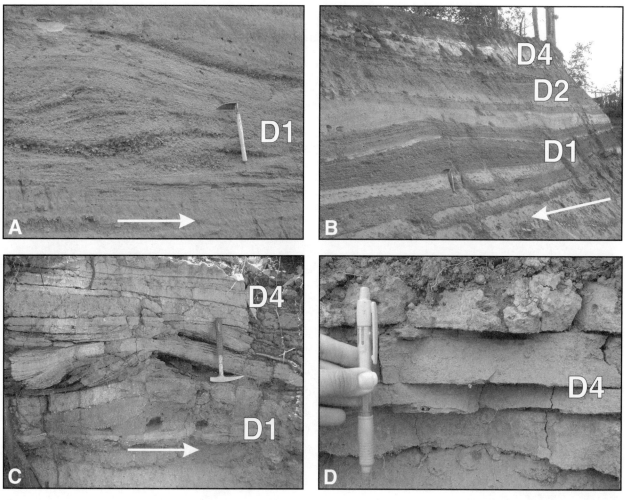

Figure 8. Photographs of the Masaya Tuff showing lateral facies variations away from the source. Arrows indicate flow direction. Scales are 30 cm for the rock hammer and scraper; 15 cm for the pencil. (A) Antidunes in an outcrop along the Pan-American Highway (W30, UTM 586397W/1333233N), 4.5 km NW of the caldera. (B) Well-developed dune structures in the lower part of the deposit and planar bedding at the top at 7.5 km to the east (W174, 604576W/1324075N). (C) The medial facies 7 km to the northwest comprises dune structures in the lower part of the deposit and planar bedding toward the top (W64, UTM 581925W/1331662N). (D) Planar-stratified fine-ash beds with accretionary lapilli 10 km to the northeast (W183, UTM 605963W/1329466N).

The end of the Masaya Tuff eruption is marked by a fall deposit, which we call the Ticuantepe Lapilli.

## Ticuantepe Lapilli

This unit could be the "upper black lapilli" of Collins et al. (1976), which was sampled by Bice (1980) and correlated with the Masaya Caldera deposits. Williams (1983a) stated that the surges (Masaya Tuff) grade up to scoria and ash fall deposits at the top of the section.

The Ticuantepe Lapilli is a well-sorted, black scoria lapilli deposit that occurs in the west of Masaya Caldera. It directly overlies the Masaya Tuff without any sign of repose time in between but with a sharp contact, and it is overlain by a grayish massive indurated tuff, an orange deposit, and younger Masaya deposits near the volcano.

The Ticuantepe Lapilli unit consists of four well-sorted lapilli layers (E1, E3, E5, E7), intercalated with three hardened ash-rich layers (E2, E4, E6 in Figs. 9 and 10). Most of the lapilli are black to dark-gray (brownish when altered) vesicular scoria, but there is also an abundance of dense, rounded juvenile clasts. The lithic content increases from bottom to top of the lowermost lapilli layer (E2), but no systematic variation in lithic content could be detected in the upper layers.

The scoria fragments consist of ~5% plagioclase, olivine, and rare clinopyroxene phenocrysts in a tachylitic groundmass with plagioclase and clinopyroxene microlites.

The Ticuantepe Lapilli is interpreted as a fallout deposit from the final stage of the Masaya Tuff eruption, which became able to generate high eruption columns probably when the availability of water for magma-water interaction decreased.

## MAGNITUDES OF THE ERUPTIONS

### Thickness and Source of the Deposits

Isopach maps were constructed for each unit with the thickness data collected from 237 outcrops. A total deposit isopach map for the San Antonio Tephra could not be constructed, due to the lack of good exposures, but a preliminary isopach map for layer A6 alone is shown in Figure 11A.

The isopach map of La Concepción Tephra shows the thickest part of the deposit at the western-southwestern side of the Masaya Caldera and a dispersal axis to the south (Fig. 11B).

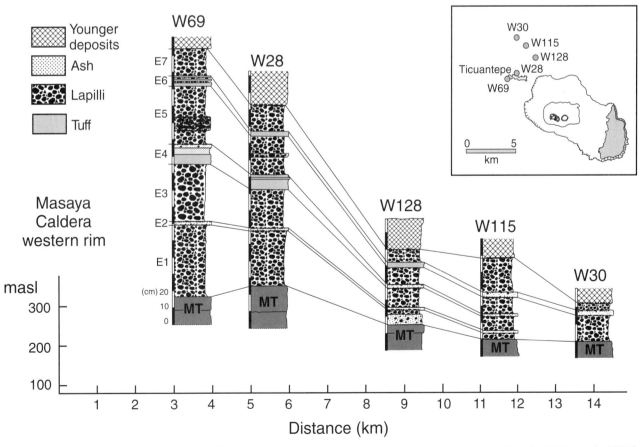

Figure 9. Correlated stratigraphic sections of the Ticuantepe Lapilli, showing the lateral variations at 3, 5, 8.5, 11 and 13.5 km to the NW from the caldera rim. The main well-sorted lapilli beds thin away from the vent and the tuff layers disappear.

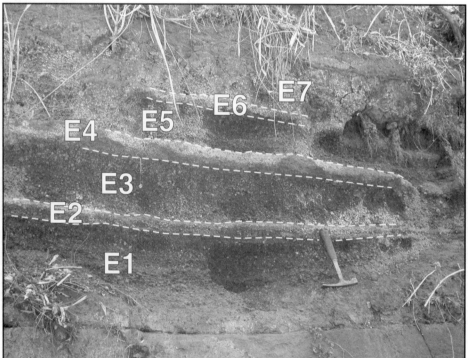

Figure 10. Photographs of the Ticuantepe Lapilli (TIL). (A) Mantle-bedding across the topography left by the Masaya Tuff (MT) in a quarry in Ticuantepe. (B) Detail of the well-sorted lapilli layers (E1, E3, E5) and the intercalated tuffs (E2, E4, E6) at a roadcut in the center of Ticuantepe.

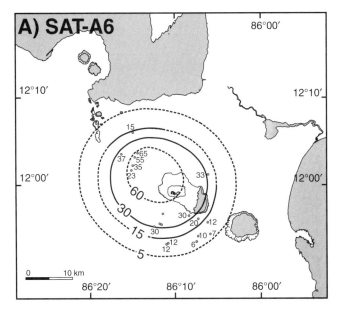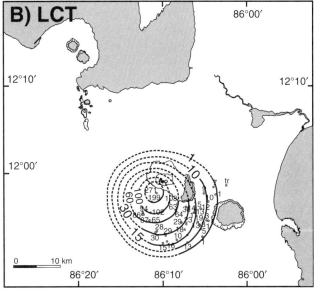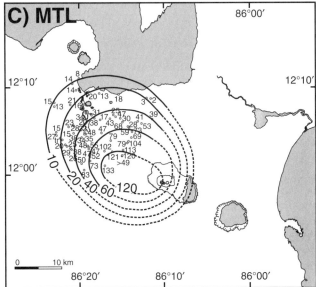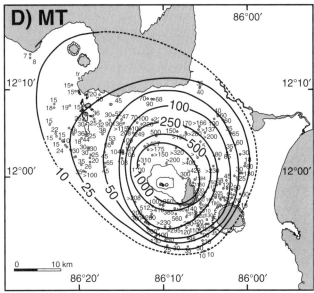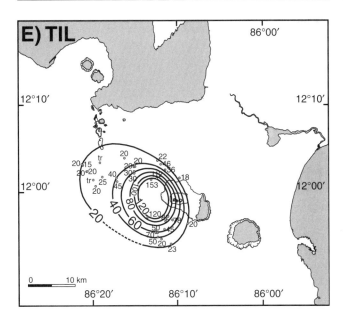

Figure 11. Isopachs maps (thickness in centimeters). Deposit traces—tr. (A) Unit A6 of San Antonio Tephra (SAT); (B) La Concepción Tephra (LCT); (C) Masaya Triple Layer (MTL); (D) Masaya Tuff (MT); and (E) Ticuantepe Lapilli (TIL). For locations, see Figure 1.

The almost round isopachs suggest low wind strength during the eruption. The total thickness ranges from 2 m at 3 km to 15 cm at a distance of almost 11.5 km along the dispersal axis, away from the caldera rim.

The 63 thickness measurements for the Masaya Triple Layer isopach map indicate a source at the NW rim of the Masaya Caldera that cannot be well constrained due to the lack of proximal exposures (Fig. 11C). The dispersal axis is directed toward the NW, and along-axis thickness ranges from 1.3 m at 10 km from the caldera rim to 15 cm at 25 km.

The Masaya Tuff isopachs show a radial pattern around the caldera and have near-circular shapes out to the 50 cm isopach (Fig. 11D). Only the more distal isopachs are considerably elongated to the NW, indicating increasing influence of the wind on distal deposition from the surge clouds. The thickness decreases from 15 m at the caldera rim to 1 m at a distance of 10 km, but a thickness of ~7 cm is still observed near Mateare, 35 km from Masaya Caldera (Fig. 11D). Bice (1980) proposed the northeast corner of the Masaya caldera as the source of the Masaya Tuff. Our new isopach maps do not constrain a specific vent site inside the caldera; this is partly because most of the proximal Masaya Tuff deposits are eroded at the top and give a minimum thickness only. Isopachs of the Ticuantepe Lapilli, interpreted to represent the terminal phase of the Masaya Tuff eruption, suggest a source in the western portion of the caldera (Fig. 11E). The Ticuantepe fall is less widely distributed than the Masaya Tuff surge deposits but was also dispersed to the NW.

## Volume and Area

A diagram of the natural logarithm of isopach thickness versus the square root of the area enclosed by each isopach was constructed for each unit (Fig. 12). Best-fit straight lines through the data were then used to calculate the bulk deposit volume after Pyle (1989) (Table 1). The dense rock equivalent (DRE) erupted magma volume was estimated by subtracting the estimated volumes of lithic fragments and the porosity of the bulk deposit. The area covered by the deposits was estimated by extrapolating the linear regressions in Figure 12 to the 1 cm isopach (Table 1).

The prominent layer A6 of the San Antonio Tephra covers an area of 1200 km$^2$ and has a bulk volume of 0.7 km$^3$ (0.4 km$^3$

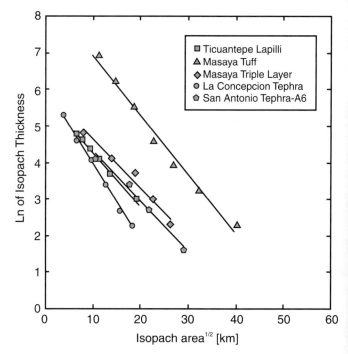

Figure 12. Diagram of natural logarithm of thickness (in cm) versus square root of the area (in km$^2$) based on isopach maps in Figure 11.

DRE). La Concepción Tephra covers an area of 600 km$^2$ and has a bulk volume of 0.2 km$^3$, corresponding to a volume of erupted magma of 0.1 km$^3$.

Williams (1983b) calculated a covered area of 200 km$^2$ and a total bulk volume of ~1.2 km$^3$ (0.35 km$^3$ DRE) for the Masaya Triple Layer, but some of his data actually correspond to the La Concepción Tephra. Bice (1980, 1985) estimated the Masaya Triple Layer volume as 1.7 km$^3$ dense rock equivalent (DRE). The total covered area from extrapolating our data to the 1 cm thickness is 1400 km$^2$. The calculated volumes of 0.7 km$^3$ bulk tephra and 0.3 km$^3$ DRE are smaller than those given by Bice (1980, 1985) and Williams (1983a).

For the Masaya Tuff, Bice (1980) calculated a volume of 4.3 km$^3$ DRE, and Williams (1983a), using Bice's data beyond the 1 m isopach, calculated a minimum bulk volume of 5.5 km$^3$. In this work, we calculate a total covered area for

TABLE 1. PHYSICAL CHARACTERISTICS OF THE UNITS

| Unit | Area (km$^2$) | Bulk volume (km$^3$) | DRE volume (km$^3$) | Erupted mass (kg) | Magnitude | VEI | Destructive potential |
|---|---|---|---|---|---|---|---|
| SAT (A6) | 1200 | 0.7 | 0.3 | 9.4 × 10$^{11}$ | 5.0 | 3 | 2.8 |
| LCT | 600 | 0.2 | 0.1 | 2.2 × 10$^{11}$ | 4.3 | 3 | 3.1 |
| MTL | 1400 | 0.7 | 0.3 | 8.1 × 10$^{11}$ | 4.9 | 3 | 3.0 |
| MT | 2900 | 3.9 | 2.7 | 7.3 × 10$^{12}$ | 5.9 | 4 | 3.2 |
| TIL | 1600 | 0.3 | 0.1 | 2.7 × 10$^{11}$ | 4.4 | 3 | 2.7 |

Note: DRE—dense rock equivalent; VEI—volcanic explosivity index; SAT—San Antonio Tephra; LCT—La Concepción Tephra; MTL—Masaya Triple Layer; MT—Masaya Tuff; TIL—Ticuantepe Lapilli.

the Masaya Tuff of 2900 km² and a total volume of 3.9 km³ (2.7 km³ DRE), the highest volume of erupted material of the studied units.

The Ticuantepe Lapilli is volumetrically the smallest deposit, with a covered area of 1600 km² and a bulk volume of 0.3 km³ (0.1 km³ DRE).

**Erupted Mass, Magnitude, and Volcanic Explosivity Index**

The erupted mass is obtained by multiplying the DRE volume of erupted magma with a basaltic magma density of 2700 kgm⁻³. This yields erupted masses between $10^{11}$ kg and $10^{12}$ kg (Table 1). Since the Ticuantepe Lapilli is considered the final product of the Masaya Tuff eruption, the combined erupted magma mass of this eruption was $7.6 \times 10^{12}$ kg.

The magnitude, M, of the eruptions was calculated after Pyle (2000) as $M = \log_{10}(m) - 7$, where $m$ is erupted mass in kg. The largest magnitude, $M = 5.9$, of the Masaya Tuff is almost 10 times the magnitudes of the fall-dominated eruptions of the other units studied (Table 1).

The volcanic explosivity index (VEI) defined by Newhall and Self (1982) is a commonly used tool to classify volcanic eruptions, based on the volume of tephra that is produced. The eruptions studied here had VEI values between 3 and 4 and were medium-large to large eruptions. The fall-producing eruptions were thus of a subplinian to Plinian nature, while the surges were produced by very large "Surtseyan" eruptions.

**Destructive Potential**

The destructive potential is defined by Pyle (2000, p. 268) to be the "area within which human structures would expect to be completely destroyed as a result of an eruption." He defined the destructiveness index as $D = \log_{10}(A)$ with $A$ being the area over which accumulations of volcanic deposits exceed a load of 1 kPa (sufficient to cause roof collapse). For the basaltic tephra studied here, the limiting load of 1 kPa corresponds to the 10 cm isopach whose area thus determines the destructiveness index.

The destructiveness indices for the Masaya eruptions range between 2.5 (La Concepción Tephra) and 3.2 (Masaya Tuff); they are thus considered moderately to highly destructive events. These values are comparable to those for well-known large explosive eruptions of the last century like the Mount St. Helens (USA) eruption of 1980, which had a destructiveness index of 2.9, or the Mount Pinatubo (Philippines) 1991 eruption with an index of 3.4 (Pyle, 2000).

**VOLCANIC HAZARDS ASPECTS**

The central western region of Nicaragua, with the cities of Managua, Masaya, and Granada, is the most densely populated area of the country. It also holds a concentration of important socioeconomic infrastructures, such as the International Airport, the Pan-American Highway, government buildings and embassies, factories, and the telephone, water, and energy supply companies. This area is surrounded by three volcanic centers that produced highly explosive eruptions in the past: the Chiltepe Peninsula volcanic centers (Apoyeque, Xiloá), Masaya Caldera, and Apoyo Caldera.

Schmincke (2004) cited examples of cities and megacities with large population densities and high growth rates within the reach of active volcanoes as places of major volcanic risk. Among these cities—which include Tokyo, Naples, and México City—he also mentioned (p. 233) Managua as being at risk because it is " built on many young volcanoes, nearby active Masaya volcano having erupted violently by magma-water interaction in the late Holocene," referring in the last sentence to the Masaya Tuff eruption.

In view of the highly explosive eruptions over the past few thousand years, we have documented here that Masaya caldera (together with the Chiltepe Volcanic Complex) has the highest probability of a similarly large eruption in the future (cf. Freundt et al., this volume). What would be the hazards if an eruption like those of the recent past took place at Masaya today?

**Hazards from Pyroclastic Fall**

The main hazard of pyroclastic fall is the collapse of buildings under the tephra load accumulating on rooftops. In this case there are two important aspects to consider, one being the thickness of tephra that accumulates in a defined area, and the other is the quality of the roofs and structures. According to Pyle (2000), it only takes an accumulation of dry tephra that exceeds 1 kPa load stress, equivalent to a dry noncompacted thickness of 10 cm of basaltic tephra, to trigger the collapse at least of low-quality roof structures.

The quality of the houses and roofs is thus important to consider in the case of Masaya. According to the 3rd Housing Census in 1995 (INEC, 1997a), 51% to 72% of the houses in the studied area have zinc-sheet roofs, 7% to 21% have fiber reinforced cement, 4% to 36% clay or cement tiles, and 2% to 7% of houses have thatch or waste material roofs. This last category of houses would not even resist a minimal 10 cm of tephra coverage. Although relevant quantitative data on the resistance of roof structures to ash loading are lacking, experience from Pinatubo (Spence et al., 1996) shows that all the structures listed above would have a high risk of partial to total damage when loaded with fallout ash.

Ballistically ejected large fragments can cause severe damage in the areas near the vent. Some ballistic blocks have been reported to reach as far as 5.5 km from the vent at Arenal Volcano in Costa Rica (Melson and Sáenz, 1973) and 8 km at Fuego Volcano in Guatemala (Rose et al., 1973, 1978). Ballistic blocks were not ejected or did not reach as far as the most proximal outcrops of both fall and surge deposits during the Masaya eruptions studied here. Nevertheless, ballistic block ejection may occur during a future eruption.

## Hazards from Pyroclastic Surges

Pyroclastic surges are turbulent, often highly energetic currents that can travel at speeds exceeding 100 km/h. They are not restricted to low topographic areas as are lahars and can surmount topographic highs. This is clearly shown by the radial distribution of the Masaya Tuff across the preexisting topography.

The main hazards from surges include destruction by the force of the current, impact by rock fragments, burial by deposits, incineration, noxious gases, and asphyxiation (Valentine, 1998; Baxter et al., 2005). At Unzen volcano, the ash cloud surge accompanying the 3 June 1991 pyroclastic density current ($5 \times 10^5$ m$^3$) killed a group of journalists and volcanologists, shifting the carbonized bodies and nearby cars ~80 m from their original position (Nakada, 2000). This surge was small compared to those that deposited the Masaya Tuff.

Pyroclastic surges are considered deadly dangers over proximal areas of high-energy lateral spreading, but turn into relatively harmless ash clouds depositing fall ash at greater distance (Wohletz, 1998). In the case of the Masaya Tuff, however, although distal deposition became increasingly affected by wind, it was still controlled by rapid lateral flow as far away as Managua city (~15 km from the caldera), as indicated by the cross-bedding and pinch and swell structures in the lowermost part of the deposit.

The phreatomagmatic eruptions forming the Masaya Tuff produced highly energetic, so-called base surges that radiated directly from the vent to great runout distance and were able to surmount topographic barriers several hundred meters high. The geologic record shows no precursor activity that may have announced their appearance, but the eruption itself started with a blast forming a thin basal layer observed in proximal to medial outcrops, which is rich in lithic fragments and quenched vitric fine-lapilli. The sudden occurrence and rapid spreading of such a surge leaves little chance for countermeasures once the eruption is in progress. Even the subplinian to Plinian, mainly fallout-producing eruptions of Masaya were repeatedly interrupted by collapse of the eruption column and production of pyroclastic surges, although these were of much lower energy and lateral runout compared to the Masaya Tuff.

## Hazards for Areas around Masaya Caldera

Nicaragua is politically divided into 17 departments and each department is subdivided into municipal units. The studied volcanic deposits from the Masaya Caldera cover parts of four Departments—Managua, Masaya, Carazo and Granada—and thereby ~26 municipal units (Fig. 13).

Some important aspects of the risk assessment of each municipal unit in the studied area are summarized in Table 2. They include number of inhabitants, population density, and number of houses. The most populated areas are Managua, Granada, and Masaya. Smaller cities located less than 2 km from Masaya

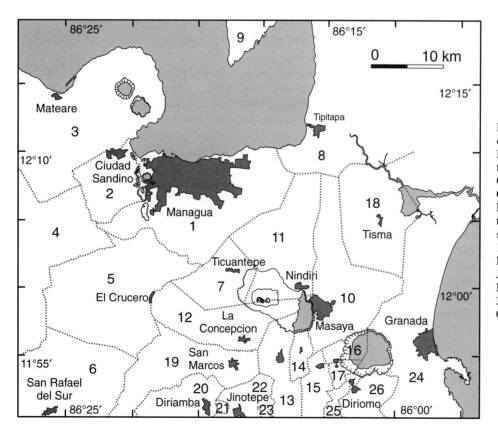

Figure 13. Political division and location of main cities at the surroundings of the Masaya Caldera. The numbers indicate the municipal units: 1—Managua; 2—Ciudad Sandino; 3—Mateare; 4—Villa Carlos Fonseca; 5—El Crucero; 6—San Rafael del Sur; 7—Ticuantepe; 8—Tipitapa; 9—San Francisco Libre; 10—Masaya; 11—Nindirí; 12—La Concepción; 13—Masatepe; 14—Nandasmo; 15—Niquinohomo; 16—Catarina; 17—San Juan de Oriente; 18—Tisma; 19—San Marcos; 20—Diriamba; 21—Dolores; 22—Jinotepe; 23—El Rosario; 24—Granada; 25—Diriá; 26—Diriomo.

Caldera's rim also have high population densities, and include Ticuantepe, Nindirí, La Concepción, Masatepe, and San Marcos (Fig. 13, Table 2).

Based on the isopach maps of the studied units and the political division, it is possible to estimate which municipal units were most affected by the most recent highly explosive eruptions at Masaya. In Table 3, we present the areal percentage of each municipal unit covered by the smallest isopach drawn for each deposit, except the San Antonio Tephra, and the range of thickness reached by the different units in the entire area of the municipal unit.

The isopach maps are, to variable degrees, limited by the lack of more distal exposures or the lack of preservation of thin distal deposits. This means that some of the political units that show no coverage by these deposits could actually have been affected by distal tephra fallout of up to 20 cm thickness, still more than enough to cause roof collapse.

Based on criteria including the cumulative thickness of the pyroclastic units studied, population density, and distance of the main cities from the caldera rim, we constructed a risk map for each municipal unit (Fig. 14). The municipal units with higher risk are those located very near the Masaya Caldera, especially in the west to northwest, the direction of the prevailing winds when most of these explosive eruptions occurred. These communities happen to be the most populated. The municipal units with the highest cumulative thickness of tephra, and thus most likely to sustain damage from similar (or even smaller) future eruptions, are Ticuantepe and Nindirí, each with more than 20 m of cumulative tephra. They are followed by Masaya (also located at the caldera rim), for which the main hazard is posed by pyroclastic surges, since it is located in the opposite direction of the prevailing winds. Smaller cities and municipal units in the south, like San Marcos, Masatepe, and Nandasmo, have average cumulative thicknesses from 3 to 6 m of tephra, so they are considered areas of medium risk. The least affected cities would be those near the Pacific coast (Villa Carlos Fonseca, San Rafael del Sur) or at distances >10 km to the east or southeast of the caldera, which might experience distal ash showers.

## CONCLUSIONS

The geological record shows that in the past 6000 yr, four widely dispersed pyroclastic fall deposits and a phreatomagmatic pyroclastic surge deposit have been produced from the Masaya Caldera. The fallout-forming eruptions (San Antonio Tephra, La Concepción Tephra, and Masaya Triple Layer) show similar characteristics: they started with a relatively weak eruption forming a basal ash bed, then quickly built up to a high eruption column emplacing widespread basaltic lapilli fall. Episodic access of water to the conduit interrupted this activity, and the eruptions became fully phreatomagmatic during their later stages, producing pyroclastic fall and surges. The last eruption (Masaya Tuff–Ticuantepe Lapilli) was an exception, because it started with the dominant phreatomagmatic pyroclastic surge phase with no observable precursory events but ended with an extended lapilli fall, probably when the availability of water for magma-water interaction decreased.

This eruptive activity is a major threat to western central Nicaragua, where four of the most important departments are located: Managua, Masaya, Granada, and Carazo, comprising 26 municipal units, some of which also have their main cities located on or very near the caldera rim. Our map of estimated volcanic risk for the municipal units and cities around the caldera indicates high risk for Ticuantepe, Nindirí, Masaya, and La Concepción, all located near the caldera rim, and also for Managua, Nicaragua's most densely populated city.

Although the more recent activity at Masaya consisted of minor pyroclastic and mainly effusive eruptions inside the caldera (Walker et al., 1993), there is no geologic evidence that large eruptions should not occur again. The types of large eruptions that one can expect from Masaya are Plinian, producing extended basaltic fall that accumulates to form heavy loads on building structures and pollutes the air with dust, and violent pyroclastic surges of great extent, capable of surmounting topographic highs

TABLE 2. GEOGRAPHICAL ASPECTS OF EVERY MUNICIPAL UNIT INSIDE THE STUDIED AREA

| Municipal unit | Area (km$^2$) | Inhabitants (projection for 2005)* | Population density inhab/km$^2$ | Number of houses[†] |
|---|---|---|---|---|
| Managua | 289 | 1015067 | 3799.3 | 162297 |
| Ciudad Sandino | 45 | 71975 | 1408.2 | 10557 |
| Mateare | 297.4 | 27142 | 91.3 | 3792 |
| Villa Carlos Fonseca | 562 | 36027 | 64.1 | 4894 |
| El Crucero | 210 | 21259 | 94.2 | 5314 |
| San Rafael del Sur | 357.3 | 43797 | 122.6 | 7827 |
| Tipitapa | 975.3 | 128840 | 132.1 | 15537 |
| San Francisco Libre | 562. | 10103 | 18.0 | 1626 |
| Ticuantepe | 60.8 | 26129 | 422.0 | 3808 |
| Masaya | 146.6 | 162868 | 1110.8 | 21719 |
| Nindirí | 142.9 | 35358 | 247.4 | 5778 |
| La Concepción | 65.7 | 30453 | 463.7 | 5036 |
| Masatepe | 59.4 | 35590 | 599.2 | 4920 |
| Nandasmo | 17.6 | 10688 | 606.2 | 1397 |
| Niquinohomo | 31.7 | 17988 | 567.6 | 2669 |
| San Juan de Oriente | 9.2 | 4335 | 471.2 | 625 |
| Catarina | 11.5 | 7106 | 618.5 | 1334 |
| Tisma | 126.2 | 13113 | 103.9 | 1896 |
| San Marcos | 118.1 | 35014 | 296.5 | 5062 |
| Diriamba | 348.9 | 56244 | 161.2 | 10209 |
| Dolores | 2.6 | 7718 | 2945.8 | 1049 |
| Jinotepe | 280.5 | 42188 | 150.4 | 7281 |
| El Rosario | 14.1 | 5121 | 363.7 | 812 |
| Granada | 592.1 | 115645 | 195.3 | 18526 |
| Diriá | 25.5 | 6647 | 260.5 | 1168 |
| Diriomo | 50.1 | 26133 | 521.8 | 3596 |

*Estimations made by the Dirección de Estadística Sociodemográfica (July 2004 revision) based on the data from the population census of 1971 and 1995 (INEC, 1997b). Source: http://www.inec.gob.ni.
[†]Data from the III Housing Census of 1995 (INEC, 1997a) and from http://www.inifom.gob.ni.

TABLE 3. PERCENTAGE OF AREA OF EVERY MUNICIPAL UNIT AFFECTED AND THE RANGE OF THICKNESSES EXHIBITED BY EVERY DEPOSIT

| Municipal Unit | Percentage of area of the Municipal Unit covered by each deposit and range of thickness | | | | | | | |
|---|---|---|---|---|---|---|---|---|
| | LCT | | MTL | | MT | | TIL | |
| | % | Th (cm) | % | Th (cm) | % | Th (cm) | % | Th (cm) |
| Managua | 0 | distal ashes | 90 | <20–>120 | 100 | 20–500 | 21 | <20–80 |
| Ciudad Sandino | 0 | distal ashes | 90 | 10–40 | 100 | <30 | 0 | – |
| Mateare | 0 | distal ashes | <1 | <10 | 40 | <20 | 0 | – |
| Villa Carlos Fonseca | 0 | distal ashes | 2 | <20 | 4 | <20 | 0 | – |
| El Crucero | 0 | distal ashes | 42 | <10–>120 | 43 | <10–100 | 37 | <20–80 |
| Tipitapa | 0 | – | 0 | – | 6 | <10–80 | 0 | – |
| Ticuantepe | 84 | <1–200 | 100 | 80–>120 | 100 | 50–>1000 | 100 | 50–>120 |
| Masaya | 26 | <20 | 0 | – | 100 | 10–>1000 | 0 | – |
| Nindirí | 39 | <15 | 44 | <10–>60 | 100 | 50–>1000 | 13 | <20–40 |
| La Concepción | 80 | <1–>200 | 83 | >10–120 | 100 | 25–>1000 | 94 | 30–>120 |
| Masatepe | 81 | <1–>100 | 0 | – | 83 | >10–1000 | 19 | <40 |
| Nandasmo | 100 | 15–40 | 0 | – | 100 | 100–500 | 0 | – |
| Niquinohomo | 56 | <15 | 0 | – | 76 | >10–250 | 0 | – |
| San Juan de Oriente | 24 | <10 | 0 | – | 100 | 25–100 | 0 | – |
| Catarina | 59 | <13 | 0 | – | 100 | 25–200 | 0 | – |
| Tisma | 0 | – | 0 | – | 81 | <10–200 | 0 | – |
| San Marcos | 39 | <100 | 4 | <10 | 48 | >10–400 | 37 | <60 |
| Diriamba | 2 | <15 | 0 | – | 1 | <20 | 0 | – |
| Dolores | 41 | <5 | 0 | – | 0 | – | 0 | – |
| Jinotepe | 9 | <45 | 0 | – | 6 | <10–80 | 0 | – |
| El Rosario | 15 | <13 | 0 | – | 3 | <10 | 0 | – |
| Granada | 0 | – | 0 | – | 5 | <50 | 0 | – |
| Diriá | 0 | – | 0 | – | 27 | <10 | 0 | – |
| Diriomo | 0 | – | 0 | – | 1 | <10 | 0 | – |

*Note*: LCT—La Concepción Tephra; MTL—Masaya Triple Layer; MT—Masaya Tuff; TIL—Ticuantepe Lapilli; Th—thickness.

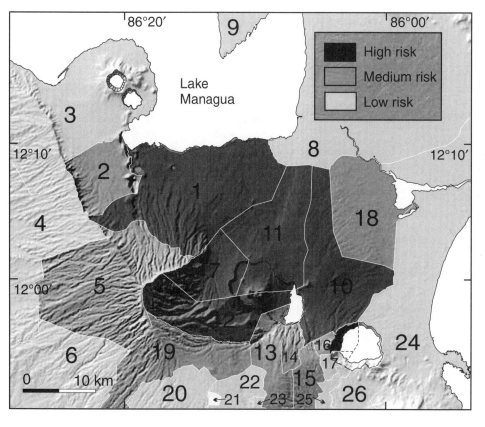

Figure 14. Risk map of the municipal units in the study area. The criteria used to evaluate risk were number of inhabitants, population density, number of houses, cumulative thickness of the studied deposits, and distance from the caldera rim. Numbers refer to municipal units in Figure 13.

and causing severe damage along their path. When an event like the one that produced the Masaya Tuff happens in the future, the surge would reach Managua city in minutes and present a deadly danger to most of the two million people living in the area.

Because of the grave hazard to large populations, it is important to continue detailed studies of the eruptive past of Nicaragua's volcanic centers and to maintain continuous monitoring. As Schmincke (2004, p. 230) stated, "Whether a volcanic eruption leads to major loss in human lives or not depends on the population density in the proximity of a volcano and its state of preparedness and organization." What is important is to recognize the potential dangers and to have in place effective and timely contingency plans; this can reduce the loss of lives.

## ACKNOWLEDGMENTS

This publication is contribution no. 67 of the Sonderforschungsbereich SFB 574 "Volatiles and fluids in subduction zones" at Kiel University. $^{14}$C dates were determined by the Leibniz Laboratory for Radiometric Dating and Stable Isotope Research at the University of Kiel, Germany. We wish to thank the people at INETER (Instituto Nicaragüense de Estudios Territoriales), Managua, for their great support during our field work. H. Wehrmann and S. Kutterolf collaborated in the fieldwork and also made a revision of the manuscript. J. White, P. Del Carlo, and L. Patiño helped to improve the paper with their opportune remarks as reviewers.

## REFERENCES CITED

Baxter, P.J., Boyle, R., Cole, P., Neri, A., Spence, R., and Zuccaro, G., 2005, The impacts of pyroclastic surges on buildings at the eruption of the Soufrière Hills volcano, Monserrat: Bulletin of Volcanology, v. 67, p. 292–313, doi: 10.1007/s00445-004-0365-7.

Bice, D.C., 1980, Tephra stratigraphy and physical aspects of recent volcanism near Managua, Nicaragua [Ph.D. thesis]: Berkeley, University of California, 422 p.

Bice, D.C., 1985, Quaternary volcanic stratigraphy of Managua, Nicaragua: Correlation and source assignment for multiple overlapping plinian deposits: Geological Society of America Bulletin, v. 96, p. 553–566, doi: 10.1130/0016-7606(1985)96<553:QVSOMN>2.0.CO;2.

Collins, D.E., Niccum, M.R., and Bice, D.C., 1976, Preliminary summary of late Pleistocene and Holocene volcanic and sedimentary stratigraphy of the Managua area, Nicaragua: Publicaciones Geológicas del Instituto Centroamericano de Investigación y Tecnología Industrial, no. V, p. 105–112.

Freundt, A., Kutterolf, S., Schmincke, H.-U., Hansteen, T., Wehrmann, H., Pérez, W., Strauch, W., and Navarro, M., this volume, Volcanic hazards in Nicaragua: Past, present, and future, in Rose, W.I., Bluth, G.J.S., Carr, M.J., Ewert, J.W., Patino, L., and Vallance, J.W., Volcanic hazards in Central America: Geological Society of America Special Paper 412, doi: 10.1130/2006.2412(08).

INEC (Instituto Nacional de Estadísticas y Censos), 1997a, VII Censo Nacional de Población y III de Vivienda 1995, Vivienda, Municipios, v. II.

INEC (Instituto Nacional de Estadísticas y Censos), 1997b, VII Censo Nacional de Población y III de Vivienda 1995, Población, Municipios, v. IV.

Krusi, A., and Schultz, J., 1979, Base surge deposits of the Nicaragua volcano Masaya: Geological Society of America Abstract with Programs, v. 11, no. 3, p. 87–88.

Kuang, S.J., 1971, Estudio geológico del Pacífico de Nicaragua: Nicaragua, Catastro e Inventario de Recursos Naturales, Informe Geológico, no. 3, 101 p.

McBirney, A.R., 1956, The Nicaraguan volcano Masaya and its caldera: American Geophysical Union Transactions, v. 37, no. 1, p. 83–96.

Melson, W.G., and Sáenz, R., 1973, Volume, energy and cyclicity of eruptions at Arenal Volcano, Costa Rica: Bulletin of Volcanology, v. 37, p. 416–437.

Nakada, S., 2000, Hazards from pyroclastic flows and surges, in Sigurdsson, H., Houghton, B., McNutt, S., Rymer, H., and Stix, J., eds., Encyclopedia of Volcanoes: San Diego, California, Academic Press, p. 945–955.

Newhall, C.G., and Self, S., 1982, The volcanic explosivity index (VEI): An estimate of explosivity magnitude for historic volcanism: Journal of Geophysical Research, v. 87, p. 1231–1238.

Niccum, M., 1976, Regional tectonic, geologic and seismic data with hazard classifications: Appendix A, unpublished report for use with Woodward-Clyde Consultants, Inc. reports in Managua, Nicaragua: San Francisco, California, Woodward-Clyde Consultants Inc.

Pyle, D.M., 1989, The thickness, volume and grain size of tephra fall deposits: Bulletin of Volcanology, v. 51, p. 1–15, doi: 10.1007/BF01086757.

Pyle, D.M., 2000, Sizes of volcanic eruptions, in Sigurdsson, H., Houghton, B., McNutt, S., Rymer, H., and Stix, J., eds., Encyclopedia of Volcanoes: San Diego, California, Academic Press, p. 263–269.

Rose, W.I., Bonis, S., Stoiber, R.E., Keller, M., and Bickford, T., 1973, Studies of volcanic ash from two recent Central American eruptions: Bulletin of Volcanology, v. 37, p. 338–364.

Rose, W.I., Anderson, A.T., Jr., Woodruff, L.G., and Bonis, S.B., 1978, The October 1974 basaltic tephra from Fuego Volcano: Description and history of the magma body: Journal of Volcanology and Geothermal Research, v. 4, p. 3–53, doi: 10.1016/0377-0273(78)90027-6.

Rymer, H., Van Wyk de Vries, B., Stix, J., and Williams-Jones, G., 1998, Pit crater structure and processes governing persistent activity at Masaya volcano, Nicaragua: Bulletin of Volcanology, v. 59, p. 345–355, doi: 10.1007/s004450050196.

Schmincke, H.U., 2004, Volcanism: Berlin and Heidelberg, Springer Verlag, 334 p.

Spence, R.J.S., Pomonis, A., Baxter, P.J., Coburn, A.W., White, M., and Dayrit, M., 1996, Building damage caused by the Mount Pinatubo eruption of June 15, 1991, in Newhall, C.G., and Punongbayan, R.S., eds., Fire and Mud: Eruptions and Lahars of Mount Pinatubo, Philippines: Seattle, University of Washington Press, p. 1055–1062.

Valentine, G.A., 1998, Damage to structures by pyroclastic flows and surges, inferred from nuclear weapons effects: Journal of Volcanology and Geothermal Research, v. 87, p. 117–140, doi: 10.1016/S0377-0273(98)00094-8.

Walker, J.A., Williams, S.N., Kalamarides, R.I., and Feigenson, M.D., 1993, Shallow open-system evolution of basaltic magma beneath a subduction zone volcano: the Masaya Caldera Complex, Nicaragua: Journal of Volcanology and Geothermal Research, v. 56, p. 379–400, doi: 10.1016/0377-0273(93)90004-B.

Wehrmann, H., Bonadonna, C., Freundt, A., Houghton, B.F., and Kutterolf, S., this volume, Fontana Tephra: A basaltic Plinian eruption in Nicaragua, in Rose, W.I., Bluth, G.J.S., Carr, M.J., Ewert, J.W., Patino, L., and Vallance, J.W., Volcanic hazards in Central America: Geological Society of America Special Paper 412, doi: 10.1130/2006.2412(11).

Williams, H., and McBirney, A.R., 1979, Volcanology: San Francisco, Freeman, Cooper and Co., 397 p.

Williams, S.N., 1983a, Geology and eruptive mechanisms of Masaya Caldera Complex, Nicaragua [Ph.D. thesis]: Dartmouth College, Hanover New Hampshire, 169 p.

Williams, S.N., 1983b, Plinian airfall deposits of basaltic composition: Geology, v. 11, p. 211–214, doi: 10.1130/0091-7613(1983)11<211:PADOBC>2.0.CO;2.

Wohletz, K.H., 1998, Pyroclastic surges and compressible two-phase flow, in Freundt, A., and Rosi, M., eds., From magma to tephra: Modelling physical processes of explosive volcanic eruptions: Developments in Volcanology 4: Amsterdam, Elsevier, p. 247–312.

MANUSCRIPT ACCEPTED BY THE SOCIETY 19 MARCH 2006

# *Fontana Tephra:* A basaltic Plinian eruption in Nicaragua

**Heidi Wehrmann**
*SFB 574 at Kiel University, Wischhofstr. 1-3, 24148 Kiel, Germany*

**Costanza Bonadonna**
*Department of Geology, University of South Florida, 4202 East Fowler Ave, Tampa, Florida, 33620, USA, and Department of Geology and Geophysics, School of Ocean and Earth Science and Technology, University of Hawaii at Manoa, 1680 East-West Road, POST Building 617a, Honolulu, Hawaii 96822, USA*

**Armin Freundt**
*IfM-GEOMAR (Leibniz Institute of Marine Sciences), RD4, Wischhofstr. 1-3, 24148 Kiel, Germany, and SFB 574 at Kiel University, Wischhofstr. 1-3, 24148 Kiel, Germany*

**Bruce F. Houghton**
*Department of Geology and Geophysics, School of Ocean and Earth Science and Technology, University of Hawaii at Manoa, 1680 East-West Road, POST Building 617a, Honolulu, Hawaii 96822, USA*

**Steffen Kutterolf**
*SFB 574 at Kiel University, Wischhofstr. 1-3, 24148 Kiel, Germany*

## ABSTRACT

Fontana Tephra was erupted from the Masaya area in west-central Nicaragua in the late Pleistocene. This basaltic-andesitic Plinian eruption evolved through (1) an initial sequence of short, highly explosive pulses emplacing thinly stratified fallout lapilli, (2) emplacement of a surge to the southwest while fallout took place in the northwesterly dispersal sectors, (3) a series of quasi-steady Plinian episodes depositing massive fallout beds of highly vesicular scoria lapilli, and (4) a terminal phase of the eruption comprising numerous subplinian eruption pulses in which varying amounts of external water were involved, forming a well-stratified sequence of lapilli beds. The Plinian episodes were repeatedly interrupted by phreatomagmatically affected pulses, evidenced by layers of higher lithic contents and scoria clasts with quenched rims, as well as by proximal cross-bedded fine to medium lapilli pyroclastic surge deposits, which left pale ash partings at distal locations.

Erupted tephra volumes, column heights, and wind velocities have been estimated for three different vent scenarios because no firm source location could be identified. The minimum total erupted tephra volume is between 1.4 and 1.8 km$^3$, much lower than previous estimates for this eruption. Eruption column heights ranging from 24 to 30 km for the Plinian eruptive phases were obtained by comparing lithic and scoria distribution data with modeling results. Consistent results from different approaches suggest that these models, which were developed for dacitic to rhyolitic Plinian eruptions, also provide good approximations for basaltic Plinian eruptions considering all sources of uncertainty.

**Keywords:** basaltic Plinian eruption, tephra volume, eruption parameters, Fontana Tephra, Nicaragua.

# INTRODUCTION

## Mafic Plinian Volcanism

Two classic papers (Williams, 1983b; Walker et al., 1984) suggested for the first time that basaltic magma can be erupted with the explosive intensity necessary to produce Plinian fall deposits. Few significant developments have taken place in the 20 years that have elapsed except for the detailed documentation of another example, the 122 B.C. eruption of Etna described by Coltelli et al. (1998). In this study, we reexamine the reportedly largest and most powerful of these documented violent basaltic eruptions, which produced Fontana Tephra in Nicaragua. Explosive basaltic events more commonly take the form of Hawaiian lava fountains, Strombolian eruptions, or phreatomagmatic Surtseyan to vulcanian events (Simkin and Siebert, 1994). Such eruptions are characterized by relatively low magma discharge rates, which limit both eruption column height and lateral dispersal of tephra. Plinian basaltic volcanism is potentially the most dangerous, yet the least well-studied type of basaltic activity. Such intense explosive eruptions have far-reaching impacts due to their wide dispersal areas, high emissions of volcanic gases, and potential for pyroclastic density currents. The risk to communities is exacerbated further because the rapid ascent rate of basaltic magma means that the warning time for a basaltic Plinian eruption (time between onset of unrest and onset of eruption) may be as short as a few hours. In addition, because such eruptions are atypical of most volcanism at basaltic centers, their precursors may be difficult to interpret and misunderstood until too late.

The factors that facilitate the eruption of basaltic magma in a Plinian fashion remain largely unknown. In most cases, open-system magma ascent and degassing leads to variably efficient separation of low-viscosity melt and gas to produce lava effusion or Hawaiian and Strombolian eruptions (Head and Wilson, 1987; Jaupart, 2000; Jaupart and Vergniolle, 1989; Mangan and Cashman, 1996; Mangan et al., 1993; Parfitt, 2004; Parfitt and Wilson, 1995; Seyfried and Freundt, 2000; Vergniolle and Jaupart, 1986; Wilson and Head, 1981). The majority of basaltic eruptions fit this model, but Plinian and subplinian eruptions are less compatible with the model, due to their sustained and intense character. Preliminary results from ongoing studies show that basaltic magma interacted with rhyolitic country rock during the 1886 Tarawera eruption (Houghton et al., 2004), whereas the 122 B.C. Etna eruption appears to have been controlled mostly by inherent magmatic processes (Coltelli et al., 1998). Thus, no common factor explaining the high intensity of basaltic Plinian eruptions has yet been identified.

Our present reinvestigation of Fontana Tephra—the Fontana Lapilli of Williams (1983b) and Masaya Lapilli Bed of Bice (1985)—addresses another question, which is especially relevant to hazard assessment at basaltic volcanoes: How do basaltic Plinian eruptions evolve? Part of this problem is to decide if commonly employed models for tephra dispersal that were designed for silicic Plinian eruptions can be applied successfully to basaltic Plinian fallout and to case studies where areal coverage of exposures is incomplete. We have divided Fontana Tephra into seven units (A–G) and document the lithology, areal thickness, and maximum grain-size distribution of each to infer the position of the now-buried vent, changes in eruptive style, and to determine eruption parameters such as ejected volume, column height, and discharge rate.

## Geologic Setting and Previous Work on Fontana Tephra

The volcanic front of western Nicaragua is part of the Central American Volcanic Arc. West-central Nicaragua, including the capital, Managua, and the cities of Masaya and Granada, is densely populated and exposed to hazards from nearby volcanic centers, particularly from Masaya volcano. The Masaya volcanic system comprises the younger Masaya composite cone with its active Santiago crater (McBirney and Williams, 1965; Walker et al., 1993; Rymer et al., 1998) and the 6 × 11 km Masaya Caldera, from which several highly explosive eruptions within the past ~6000 yr produced widespread basaltic deposits (Pérez and Freundt, this volume; Bice, 1985; Williams, 1983b). These are slightly less silicic (~50% $SiO_2$) than the basaltic-andesitic Fontana Tephra (53% $SiO_2$). Tuffaceous sediments and a well-developed paleosol separate Fontana Tephra from the concordantly overlying Lower Apoyo Tephra. The Lower and Upper Apoyo tephras are separated by an incipient paleosol but yield overlapping $^{14}$C-ages of 23,890 ± 240 and 24,650 ± 120 yr B.P., respectively (S. Kutterolf et al., 2004, personal commun.). Deposits below the Fontana Tephra, collectively named Las Sierras Formation, have not been studied in detail to date; the two available age determinations range from 100,000 yr B.P. (Bice, 1980) to 29,200 ± 800 yr B.P. (Williams, 1983a). We estimate the age of the Fontana eruption as ca. 30 ka, largely in agreement with Bice's (1985) estimate of 35–25 ka B.P., as well as with a preliminary estimate by S. Kutterolf et al. (2004, personal commun.) in the same range inferred from marine tephrostratigraphy offshore of Nicaragua.

Bice (1980, 1985) described Fontana Tephra as a uniform, nonbedded, nongraded, and well-sorted black basaltic scoria deposit with a distinctive thin, black basal ash. He also noted parallel whitish bands cutting through the deposit, which he interpreted as plant ash. Based on the direction of increasing deposit thickness, Bice (1980, 1985) tentatively identified Masaya Caldera as the source of Fontana Tephra, where it is covered by younger volcanic deposits. He supported his assumption with the chemical similarity between Fontana Tephra and the younger Masaya Triple Layer and Masaya Tuff, which both stem from Masaya Caldera. Williams (1983a, 1983b) studied these three products of highly explosive eruptions from Masaya Caldera in detail. His data on Fontana Tephra provided a useful starting point for our investigation. Williams attributed the sheet-like geometry of the deposit to a Plinian eruption with a very high eruption column estimated at 50 km and a wind velocity of 30 m/s based on the observations of the Mount St. Helens 1980 eruption. Williams (1983b) calculated a tephra volume of 12 km³ from his isopach map, within the 1 mm isopach, and estimated the discharge rate as $2 \times 10^5$ m³/s, indicating that the eruption would have lasted for two hours. He suggested that the high temperature of the basaltic magma resulted in a higher eruption column than from a silicic eruption of comparable mass flux. Williams also observed a 73-m-thick, nonwelded fall deposit at the western Masaya Caldera wall, close to his proposed vent inside the caldera. This he interpreted as a proximal cone-building phase of fallout from sufficiently great height to facilitate cooling of the clasts and prevent welding of the deposit. In contrast to Bice (1980, 1985), Williams (1983a, 1983b) interpreted the distinctive light-colored bands in the deposit to have formed by percolating groundwater.

Williams' (1983a, 1983b) quantitative estimates involve significant extrapolation of his isopach and isopleth data. Moreover, the abundance of similar-looking scoria lapilli deposits in the Masaya-Managua region requires careful identification of lithologic components and correlation of internal beds between outcrops to substantiate correlations. Our reexamination of Fontana Tephra shows that the eruption evolved through several phases of distinct eruption styles and intensities. Such details are crucial to gain a better understanding of the mechanisms of mafic Plinian volcanism.

## FONTANA TEPHRA

We have remapped Fontana Tephra over an area of 200 km², to the north and northwest of Masaya Caldera (Fig. 1). The pattern of exposure of Fontana Tephra is rather unusual. The deposit is exceptionally well-exposed over only some 150 km² to the north and west of possible source-vent locations. No exposures exist in any of the remaining sectors and northerly exposures end at the shoreline of Lake Managua. The deposit can be subdivided readily into seven units, each of which consists of several beds with interspersed ash partings. These units differ in grain size, clast populations, and bedding characteristics. Absolute and relative thicknesses of these units vary between outcrops due to different areal dispersal patterns. Isopach and isopleth maps were constructed for the prominent units of the succession. Figures 2 and 3a illustrate the succession of units as found at proximal to medial distances.

### Unit A

The base of Fontana Tephra is made up of two beds of well-sorted, black, highly vesicular, fluidally textured Hawaiian-type lapilli with very thin glass walls separating the bubbles (Fig. 3e). We found no lithic fragments in this layer, which reaches 12 cm thickness at the most proximal exposures (outcrop 15 in Fig. 1) and thins rapidly away from source.

### Unit B

Unit B is composed of dark-gray fine to medium scoria lapilli that are highly vesicular, but less so than the particles of

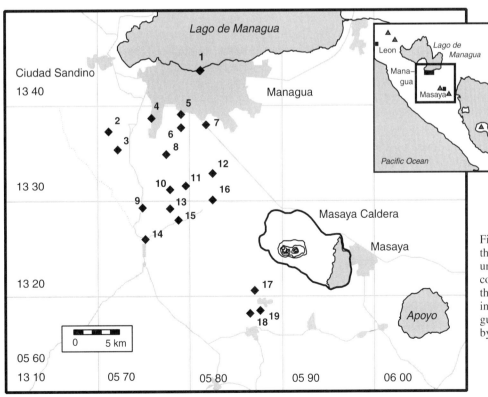

Figure 1. Fontana Tephra localities in the Managua-Masaya area. Map uses universal transverse Mercator (UTM) coordinates; field area position within the Nicaraguan volcanic front is shown in the inset map of west-central Nicaragua. Major volcanic centers are marked by triangles.

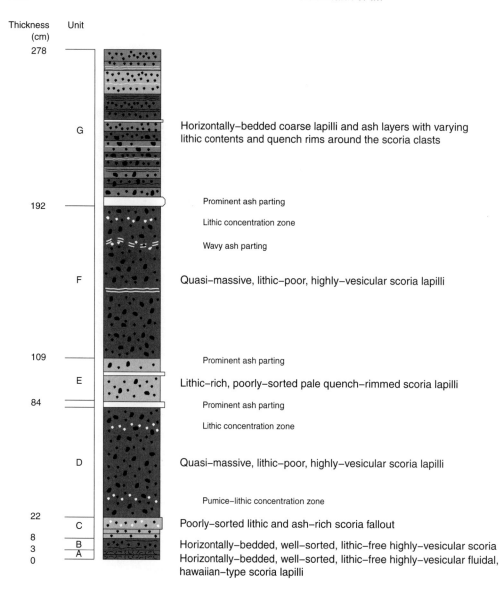

Figure 2. Stratigraphic section of Fontana Tephra at outcrop 8 illustrating variations in lithology and unit division. Units A–G are described in the text.

unit A. At outcrop 15, unit B reaches a maximum thickness of 86 cm and is crudely bedded. At more distal locations toward the north-northwest, up to eight planar beds of alternating fine lapilli and coarse ash can be distinguished (Fig. 3d). Lithics make up less than 1% of unit B (visually estimated).

## Unit C

To the south of the probable vent area, the stratigraphic position between units B and D is occupied by a cross-bedded, poorly sorted deposit that is 400 cm thick at outcrop 17, the lowermost 300 cm being distinctly coarser grained than the top 100 cm, where cross-bedding is better developed. The deposit thins rapidly over 2.5 km to 30–60 cm at outcrop 19. Lapilli-size (typically 3–10 mm in diameter) scoria and lithic particles are set in an ash matrix. The juvenile particles are ash-coated, medium-gray scoria with spherical to elongated bubbles. Lithic particles reach 130 mm in diameter, account for ~10% of the lapilli fraction, and are dominated by mafic dense lava clasts, ~70% of which are strongly hydrothermally altered. A minor amount of rhyolitic pumice clasts is present. The topmost 10 cm consist of a gray medium-coarse ash.

No such thick cross-bedded deposit is found at other proximal localities. Instead, outcrops across the medial range to the northwest show a series of thin, poorly sorted, horizontally bedded lithic-bearing layers of fine-ash–coated medium-lapilli scoria in this stratigraphic position (Figs. 3d and 3h). Scoria clasts are subangular to subrounded, medium gray, crystal-poor, and are characterized by spherical and elongate vesicles. These layers are also matrix-supported and contain lithic clasts of mafic lava and hydrothermally altered mafic lava, as well as rhyolitic pumice. The pumices are silky white-pink with tubular bubbles

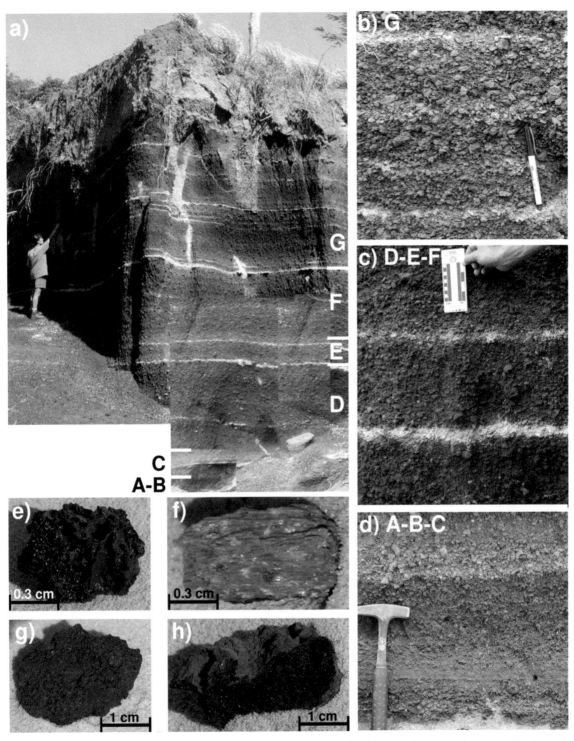

Figure 3. (a) Composite photo of Fontana Tephra at outcrop 12 (Fig. 1); (b) thin beds of unit G; (c) top of unit D through E to the base of F; (d) layered base (units A–C). Note the higher lithic contents, poorer sorting, and quench-rimmed scoria in units C, E, and G, reflecting increased interaction of external water with the magma. (e) Very highly vesicular, fluidal clast of unit A. (f) Rhyolitic pumice lithic with tubular bubbles of units C and D. (g) Highly vesicular scoria clast of the Plinian units D and F. (h) Quench-rimmed scoria of unit E.

(Fig. 3f). Despite extensive stratigraphic work in west-central Nicaragua, no source deposit for this pumice has been found (S. Kutterolf et al., 2004, personal commun.). However, this pumice type only occurs in Fontana Tephra and is thus a valuable tool to distinguish it from other scoria lapilli deposits.

At more distal outcrops, thin, poorly sorted ash beds occur. Those are composed of pale gray glass shards and a relatively large percentage (~10%–15%) of hydrothermally altered lithics.

## Unit D

Unit D is a thick, massive to vaguely stratified bed of highly vesicular dark-gray scoria lapilli, which is ubiquitously present in the entire area of dispersal. The top of this unit is marked by a prominent white ash layer (Fig. 3c). Most scoria clasts are highly vesicular with spherical bubbles (Fig. 3g); however, denser clasts with irregular, coalesced bubbles are also present. The unit reaches a thickness of 137 cm at outcrop 16.

Although the lithic content of this unit amounts to <2%, there are two horizontal zones of higher lithic concentrations (5%–7%) in which particular lithic types occur. A zone in the lower quarter of the deposit contains rhyolitic pumice lithics of the same type as in unit C (Fig. 3f). A second zone ~5–10 cm below the top of unit D contains conspicuous orange, hydrothermally altered fragments of older lavas and tuffs (Fig. 3c).

## Unit E

Unit E consists of two layers of dark gray, medium scoria lapilli with medium-gray fine-ash coatings, separated by a central prominent ash parting (Figs. 3c and 3h). In contrast to the well-sorted units D and F, unit E is only moderately well sorted due to the presence of ash in interstices in the lapilli-supported deposit. Scoria clasts display a range in vesicle shapes from spherical to irregular. The unit also contains conspicuous lithics (~5%) that are predominantly hydrothermally altered mafic lava. The unit crops out throughout the western to northern part of the area of dispersal and reaches a maximum measured thickness of 38 cm at outcrop 16.

## Unit F

Unit F is a thick layer of dark gray to black scoria lapilli. It shows weak bedding defined by intervals that are characterized by diffuse concentrations of stained or ash-coated lapilli. The layer includes several (three to ten) weak to strong, partly wavy but mostly planar, pale and brown ash partings. Highly vesicular scoria clasts with spherical to elongated vesicles coexist with moderately vesicular ones containing irregular bubbles. About 5% of dark red scoria clasts have, on average, a larger size and larger bubble diameters compared to the dark gray clasts. A few (<1%) Hawaiian-type scoria lapilli are present. In the two lithic-bearing zones, mostly hydrothermally altered mafic lavas and tuffs make up ~5%. Like Unit E, Unit F is distributed toward the west and the north of the vent. Its greatest thickness is 140 cm.

## Unit G

The upper part of the deposit is distinctly better stratified than the underlying units (Fig. 3b). It is made up of a series of thin (cm-scale), fine to coarse lapilli and ash layers, some of which are well-sorted lithic-poor beds of angular scoria, while others are only moderately sorted and characterized by higher lithic contents (up to 3%–5%). At least seven ash partings are present. Although unit G reaches a thickness of >3 m at proximal sites, it is commonly strongly eroded so that a consistent correlation of its component layers across the mapping area was not possible.

## Ash Partings

The entire deposit exhibits a spectacular set of ash partings (Figs. 3a–3c) at fairly regular decimeter-thick intervals that easily correlate throughout most outcrops. Across most of these ash partings, the grain sizes of the scoria lapilli do not change, but the pore spaces between the clasts are filled with much finer-grained white to pale gray-brown particles down to fine ash size. At one outcrop location on the dispersal axis (outcrop 13; Fig. 1), a good exposure of Unit G shows the lateral transition of such an ash parting via gradual thickening and coarsening to a cross-bedded, fine-lapilli deposit. Our tentative interpretation of this bed is emplacement by a pyroclastic surge because (a) the outcrop is located on the lee-side flank of a ridge where reworking by wind was unlikely, and (b) syneruptive downslope slumping did not affect other, thicker Fontana beds.

## DISTRIBUTION OF DEPOSITIONAL UNITS

The pattern of preservation of Fontana Tephra places some limitations on our ability to constrain the dispersal of the eruption products. Units D, E, and F are the best-preserved units in the whole dispersal area of the Fontana deposit; the isopach and isopleth maps for these units are shown in Figures 4 and 5. The contours are drawn by interpolation between observed data but smoothed to give a volcanologically reasonable pattern. Maximum scoria and lithic clast isopleths are based on the average dimension of three axes determined for the five largest clasts, respectively, as suggested by Sparks (1986).

Data for units D, E, and F have some features in common. Although dispersal axes generally point toward the northwest, elongation of contours along the axes in any reconstruction remains very limited. Fontana Tephra is exposed south of the vent at outcrops 17, 18, and 19 (Fig. 1), evidenced by stratigraphic position, petrographic characteristics, and the presence of the conspicuous pumice lithics. However, at these sites, the outcrop conditions did not allow us to correlate the deposit's units, and therefore these outcrops were not included in the isopach and isopleth maps. As no outcrop data are available to close the contour lines in the southern through eastern sectors, and proximal exposures are absent, a straight-forward identification of the Fontana vent site was inhibited.

If Fontana Tephra originated from Masaya Caldera as previously suggested (Bice, 1985; Williams, 1983b), our isopach and isopleth data would point to a source in the northwestern part of the present caldera (vent 1). However this assumption was predicated strongly on the existence of a thick proximal exposure on the Masaya Caldera wall. We found no evidence for a "proximal cone building phase" as proposed by Williams (1983a), which was critical for his reconstruction of the deposit geometry. In fact, we did not find any deposit belonging to the Fontana eruptive sequence in the northwestern wall of Masaya Caldera. Instead it is built up from scoria lapilli fallout that has been tentatively associated with the ca. 2 ka B.P. Masaya Triple Layer events (Pérez and Freundt, this volume) and underlain by several beds of lava, breccia, phreatomagmatic deposits, and other fallout tephras. The absence of the easily recognized Apoyo tephras in the caldera wall (which are exposed widely outside the caldera) suggests the key stratigraphic interval may be buried on the caldera rim. Without such exposures, there is no geologic evidence to prove Masaya Caldera is the source of Fontana Tephra. The chemical similarity to the ~30 k.y. younger Masaya Triple Layer, invoked by Bice (1985), is not, in itself, conclusive.

We have therefore considered three contrasting geometries for the Fontana isopachs and isopleths and investigated the implications for eruptive intensity and tephra volume. The first case, which follows the concept of Bice (1980, 1985) and Williams (1983a, 1983b) by extrapolating isopach contours around a vent site within Masaya Caldera, produces a rather unusual contour pattern of strong along-axis elongation close to the vent yet little elongation across the medial range (vent 1 in Figs. 4 and 5). This is atypical of Plinian dispersal patterns and would require complicated, and not very probable, explanations such as an inclined lower eruption column.

In the following sections, we therefore discuss two other possible vent sites (Figs. 4 and 5). In addition to vent 1, we have investigated a vent position to the northwest, and outside of, Masaya Caldera (vent 2 in Figs. 4 and 5). Such a vent position is better compatible with our field data and coincides with the vent area suggested for pyroclastic deposits of the Las Sierras Formation (van Wyk de Vries, 1993).

Finally, we have considered near-circular isopach and isopleth contours around the location labeled as vent 3 as an extreme end member representing conditions of no or little wind during eruption.

## EVOLUTION OF THE FONTANA ERUPTION

The choice of vent position, not surprisingly, affects the estimates of eruption parameters. To give a reasonable range of values for volume, column height, and maximum wind velocity at time of eruption, we apply the models of Pyle (1989), Carey and

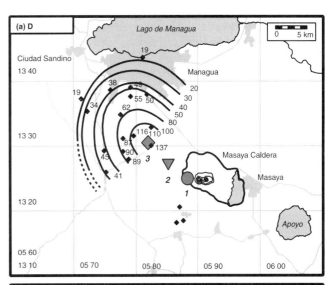
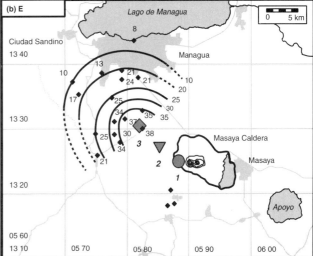
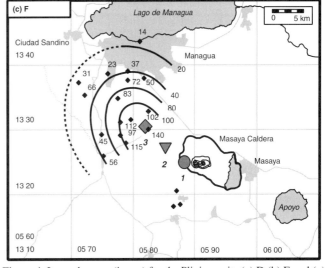

Figure 4. Isopach maps (in cm) for the Plinian units (a) D (b) E and (c) F. Dashed lines are extrapolated contours. Large symbols mark possible vent sites discussed in the text: vent 1 (circle), vent 2 (triangle), vent 3 (diamond).

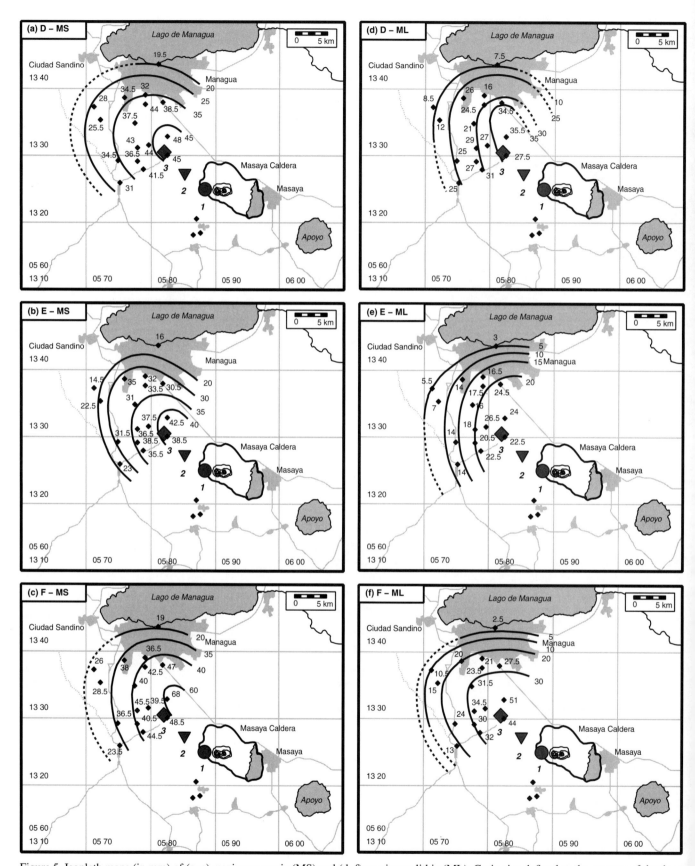

Figure 5. Isopleth maps (in mm) of (a–c) maximum scoria (MS) and (d–f) maximum lithic (ML). Grain size defined as the average of the three main axes of the five largest clasts for the Plinian units D, E, and F. Large symbols mark vent sites as in Figure 4.

Sparks (1986), and Sparks et al. (1992) to our three vent scenarios. These parameters are crucial for understanding the dynamics of each eruption phase and to assess the risk from future eruptions of this type. Model application must, however, be considered within the qualitative constraints on eruptive behavior deduced from the deposit characteristics.

## Variations in Eruptive Style

The Fontana eruption was not continuous, but proceeded in numerous successive eruptive pulses that were mainly Plinian in style. The eruption began with several short, explosive events ejecting ragged and fluidal pyroclasts (Hawaiian-type). These pulses were followed by formation of a pyroclastic surge toward the southwest, while fallout dominated the other dispersal sectors.

Fairly continuous magma discharge characterizes each of the following Plinian phases (units D to F). The wide dispersal area of these units (Figs. 4 and 5) suggests high eruption columns. Interbedded layers with poorer sorting, higher contents of various lithic populations, and quench-rims around the scoria clasts indicate repeated but limited interaction of the magma with external water. Transient access of water caused some fluctuations in the eruption, but very similar dispersal characteristics of units D to F suggest that after each incursion of water Plinian eruption columns recovered to steady conditions.

The stratified alternation between lithic-poor, well-sorted dark gray scoria layers with more lithic-rich, moderately sorted pale ash-coated scoria beds, combined with strongly varying grain-size distributions in the terminal phase (unit G) indicates rapidly changing eruptive conditions, probably due to periodic partial collapse of conduit walls, generating short-lived eruption pulses.

## Erupted Volumes

Tephra volumes were obtained for units D, E, and F from the isopach maps in Figure 4, where closure of the isopachs toward the southeast was constructed with respect to previously observed, typical depositional fan geometries of Plinian eruptions. The calculation of erupted volumes used the technique of Pyle (1989), assuming that the deposit thins exponentially away from source. This allows extrapolating from the well-constrained medial part of the depositional fan toward the source as well as to distal locations. Volumes of fall deposits thus determined are known to represent only a minimum erupted volume as the deposit thinning is not well constrained where distal thickness data is lacking (e.g., Pyle, 1995; Rose, 1993; Bonadonna and Houghton, 2005). However, this is the most reliable method available to determine the volume of poorly constrained deposits. Plots of natural logarithm of the thickness versus the square root of each isopach area display subparallel straight lines for units D, E, and F in the three vent scenarios (Fig. 6). Table 1 shows the resulting volume estimates for these units based on each vent scenario. Volumes obtained by summing the products of isopach area times thick-

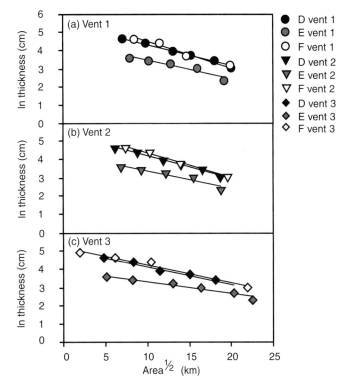

Figure 6. Natural logarithm of thickness (cm) versus square root of area (km) of the isopachs plots of units D, E, and F in the three scenarios considered in the text.

TABLE 1. ERUPTED VOLUMES OF THE FALLOUT UNITS ($km^3$)

|  | Volume estimates* | | | Volume estimates (cumulative) | | |
| --- | --- | --- | --- | --- | --- | --- |
|  | Vent 1 | Vent 2 | Vent 3 | Vent 1 | Vent 2 | Vent 3 |
| D | 0.34 | 0.29 | 0.34 | 0.34 | 0.30 | 0.35 |
| E | 0.16 | 0.16 | 0.24 | 0.16 | 0.16 | 0.25 |
| F | 0.34 | 0.31 | 0.37 | 0.36 | 0.32 | 0.40 |
| Total (incl. units A,B,C,G) | 1.68 | 1.36 | 1.78 | | | |

*After Pyle (1989).

ness are also shown for comparison. Volumes range between 0.29 and 0.34 $km^3$ for unit D, 0.16 and 0.24 $km^3$ for unit E, and 0.31 and 0.37 $km^3$ for unit F.

Dispersal parameters provide quantitative evidence of the Plinian nature of the eruptive phases that produced units D, E, and F. The Fontana units produce the same low thickness decay (shallow slope) in the ln (thickness) versus (area)$^{1/2}$ diagram that is typical of other Plinian deposits, in particular also of the basaltic Plinian tephras at Tarawera and Etna (Fig. 7). Deposits of Hawaiian, Strombolian, and subplinian eruptions yield distinctly steeper slopes. Moreover, thickness half-distances ($b_t$ = 6–11 km) and maximum clast half-distances ($b_c$ = 9–14 km) at

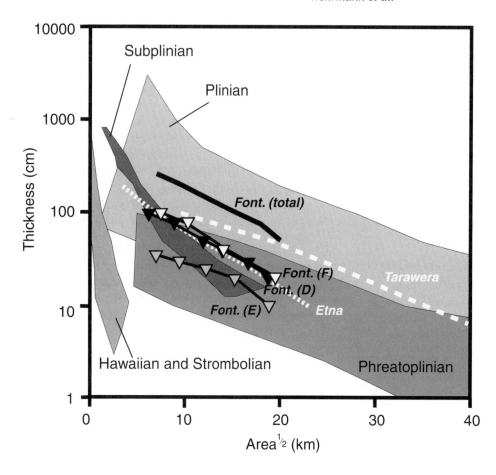

Figure 7. Semi-logarithmic plot of the thicknesses of units D, E, F for vent 2, and of the total thickness of Fontana Tephra versus the square root of the respective isopach areas, in comparison to dispersal patterns of other Plinian, phreatoplinian, Strombolian, and Hawaiian eruptions (after Houghton et al. 2000), including the basaltic Plinian 1886 A.D. Tarawera (dashed white line) and 122 B.C. Etna (dotted white line) deposits (Houghton et al. 2004).

medial distances plot in the Plinian field of the $b_c/b_t$ versus $b_t$ diagram after Pyle (1989).

## Eruption Column Heights and Wind Speeds

Eruption column heights and wind conditions at the time of deposition can be estimated by comparing the downwind and crosswind clast-size dispersal with theoretical modeling results (Carey and Sparks, 1986). Such models were initially based on the physics of particle transport of clasts with various densities resulting from height, spread, and lateral drift of the eruption cloud in response to exit velocity and advection by wind. Subsequently, the validity of the numerical approach was verified through experimental studies and direct observations of recent explosive eruptions and the corresponding depositional fans (Rosi, 1998; and references therein). Established models of Carey and Sparks (1986), and Sparks et al. (1992) turned out to be the most suitable for our data set. Specifically, the approach of Carey and Sparks (1986) yields peak eruption column heights for each phase based on the maximum clast size of both juvenile and lithic clasts (albeit averages from five clasts). As with the volume calculations, we have determined results for the three different vent locations.

For the wind-affected tephra dispersal (vents 1 and 2), eruption column heights have been determined from downwind-range and half-crosswind range data for different isopleths (Fig. 8). Column heights in the no-wind scenario (vent 3) are inferred through plots of clast diameter against the area enclosed by the respective isopleths (Fig. 9).

During the first Plinian phase D, the eruption column reached an estimated height of 26–30 km. It dropped to 25–27 km for E, then rose again to 24–29 km in the Plinian phase F. In contrast to column heights, corresponding wind velocities strongly depend on the vent scenario chosen. For example, using unit D, wind velocity is 12–30 m/s for vent 1, 7–18 m/s for vent 2, and zero for vent 3. Similarly large differences result for the other units. For each of the vent 1 or vent 2 scenarios, however, wind strength and direction did not apparently change during eruption from unit D through unit F.

Data from different isopleths of a single unit do not parallel the model curves in Figures 8 and 9 but follow a shallower trend such that larger grain sizes indicate lower column height and higher wind speed. Such deviation has also been noted by Rosi (1998) for other deposits and can be attributed to different trajectories of clasts deposited from the vertically rising versus the laterally spreading (umbrella) regions of an eruption column. For example, assuming a column height of 24 km, clasts within ~6 km of the vent would have fallen from the vertically rising eruption column. Rosi (1998) concluded that the modeling

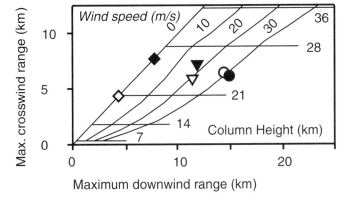

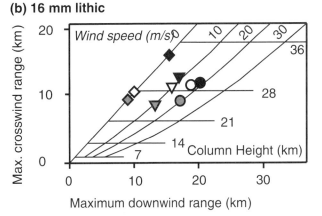

Figure 8. Diagrams of crosswind range versus downwind range for lithics (2500 kg/m³) of (a) 8 mm, (b) 16 mm, and (c) 32 mm diameter for units D, E, and F in the three scenarios considered. Horizontal grid lines indicate eruption column heights (in km) and diagonal grid lines wind velocity (in m/s) as theoretically calculated by Carey and Sparks (1986).

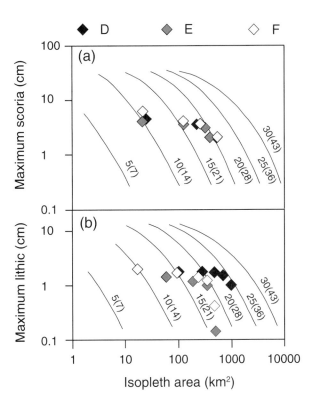

Figure 9. Diagrams of isopleth area versus clast size (a) of scoria (density of ~1000 kg/m³) and (b) of lithics (~2500 kg/m³) for units D, E, and F in a no-wind scenario (vent 3). Values along the lines correspond to column height at neutral buoyancy level and total column height (in parentheses) as predicted by the analytical model of Carey and Sparks (1986).

captures well the fallout beyond the column edge but is less reliable for more proximal fallout.

Another approach to estimate eruption column heights uses the plume-sedimentation model of Sparks et al. (1992). In contrast to the model above, in which data from maximum clasts indicate maximum column heights, the sedimentation model uses a whole grain–size distribution and thus indicates the average column height. The model of Sparks et al. (1992) does not account for wind advection, variation of particle Reynolds number, variation of particle density and aggregation processes, which have proved crucial to our understanding of tephra dispersal (Bonadonna and Phillips, 2003). However, we have applied the model of Sparks et al. (1992) to support our determination of eruption parameters. We are planning to carry out a more detailed investigation for dispersal of tephra from basaltic Plinian eruptions to account for all these factors.

The Sparks et al. (1992) model relates the rate of decrease of sedimentation rate away from the vent to the column height. As the sedimentation rate is proportional to the thickness of the deposit by a factor of time, we used thickness relationships to determine the respective column heights. The natural logarithms of the thicknesses normalized to the extrapolated maximum thicknesses versus radial distances from vent for the three different scenarios are presented in Figure 10.

During the first Plinian eruptive phase D, the average eruption column ranged in height between 22 and 26 km. For phase E, heights of 25–28 km are inferred, whereas phase F is fairly similar to D with the eruption column between 21 and 26 km. The lowest values of column height are obtained in the no-wind scenario 3, which reflects the conditions the model is designed for.

For units D and F, these average column heights are a few kilometers lower than the maximum column heights estimated above, as they should be. The gradients of the data in Figure 10 agree well with model expectations. However, for unit E the average column height equals the maximum column height estimated above. Moreover, while the maximum clast approach suggests a lower eruption column during phase E compared to phases D and F, the sedimentation model implies the opposite behavior. These inconsistencies shed some doubt on the applicability of the models to basaltic Plinian tephra.

## Mass Discharge Rate and Eruption Duration

The eruption column height, $H$, is controlled by the thermal mass flux, $Q = C \cdot T \cdot M_0$, such that $H$ is proportional to $Q^{0.25}$, where $C$ is an average specific heat, $T$ the temperature difference to ambient, and $M_0$ the mass discharge rate (Sparks et al., 1997).

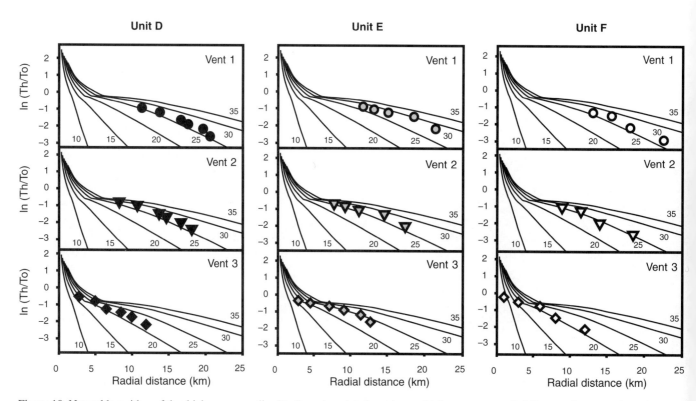

Figure 10. Natural logarithm of the thickness normalized to the extrapolated maximum thickness versus radial distance from vent for units D, E, and F in the three different scenarios, projected onto sedimentation rate decrease curves after Sparks et al. (1992). The curves reflect the expected average column heights.

As an approximation, ignoring details such as atmospheric stratification, Wilson and Walker (1987) determined an average column height

$$H = 0.236 \, [M_0]^{0.25} \; H \, [\text{km}], \; M_0 \, [\text{kg s}^{-1}] \quad (1)$$

from their modeling using an eruption temperature of 750 °C and 1–5 wt% released water. The higher eruption temperature of a mafic magma (say, 1100 °C) may be accounted for by simply introducing a factor $f = (1100/750)^{0.25} = 1.1$ into Equation (1). This suggests that Equation (1) would underestimate column heights from mafic magma eruptions by ~10%. Keeping this uncertainty in mind, Equation (1) is used to estimate mass discharge rates from eruption column heights determined by other methods above. The resulting mass discharge rates vary between 2.3 and $2.6 \times 10^8$ kg s$^{-1}$ for unit D, between 1.3 and $1.7 \times 10^8$ kg s$^{-1}$ for unit E, and ~$2.0 \times 10^8$ kg s$^{-1}$ for unit F of Fontana Tephra. These mass fluxes are in the same range as those of the other known basaltic Plinian eruptions; the Tarawera 1886 A.D. eruption reached ~$2 \times 10^8$ kg s$^{-1}$ at a column height of 28–34 km (Walker et al., 1984; Carey and Sparks, 1986), and the Etna 122 B.C. eruption yielded ~$7 \times 10^7$ kg s$^{-1}$ at 24–26 km (Coltelli et al., 1998).

Preliminary investigations on clast density and vesicularity for units D, E, and F suggest that the average deposit density for these units is ~800 kg m$^{-3}$. Based on the volume determined for vent 2 in Table 1, the corresponding minimum duration of the sustained phase for units D, E, and F is 15–18 min, 16–21 min, and 20 min, respectively.

## DISCUSSION AND CONCLUSIONS

### Applicability of Eruption Models

The models of Pyle (1989), Carey and Sparks (1986), and Sparks et al. (1992) are typically used for the determination of volume, column height, and wind velocity at time of eruption for rhyolitic to andesitic tephra fall deposits. Therefore, the application of these models to Fontana Tephra provides interesting insights into their applicability to mafic Plinian volcanism.

The volume results are very consistent for the three units (D, E, and F) and for the three vent positions considered, in line with the suggestion that plotting the square root of the isopach area eliminates the effects of wind on the resulting contours for the volume calculation (Pyle, 1989). Applying the model of Pyle (1989) to older deposits of limited exposure can underestimate the tephra volume for the reasons discussed by Bonadonna et al. (1998) and Rose (1993), but our data show that it can still give consistent results even in those situations in which the position of the vent is uncertain. Our data suggest that the model can readily be applied to Plinian fallout deposits of mafic composition. The resulting total volume of Fontana Tephra of 1.4–1.8 km$^3$ is much smaller than the estimate of 12 km$^3$ by Williams (1983b).

The results for eruption column heights using the method of Carey and Sparks (1986) are more complex. Estimated column heights are internally consistent and are little influenced by the assumed vent position, which does, however, strongly affect estimated wind velocities. Deviations from the model curves such as those of the Fontana data trends in Figure 9 had also been noted by Carey and Sparks (1986) in their data, with the most prominent deviation in the data of the basaltic Plinian Tarawera 1886 A.D. deposit. They explained such deviation by the breakage of the larger clasts upon impact on the ground. Rosi (1998) argued that the model does not describe proximal fallout from within the column edge as successfully as more distal fallout from the umbrella region; this deficiency would apply to rhyolitic and basaltic eruptions alike.

Carey and Sparks (1986) model the clast distribution assuming an eruption column reached a certain height. As far as clast dispersal from that height is concerned, the model is insensitive as to whether rhyolitic or basaltic magma was erupted. The difference in composition does, however, affect the modeled column height for any given discharge rate because basaltic magma, ~300 °C hotter than rhyolitic magma, provides greater buoyancy to the column. According to model results of Woods (1988), the basaltic column would be <5 km higher than a rhyolitic one at otherwise identical conditions. The higher magma temperature affects the vertical velocity distribution in the column such that the buoyancy-driven acceleration zone extends (Woods and Bursik, 1991), thereby controlling the clast size distribution in the deposit, particularly at proximal locations. These factors are not fully considered when applying the models to basaltic eruptions. The errors due to a basaltic composition involved in the above column height estimates for the Fontana eruption is, however, not larger than other, composition-independent uncertainties such as degree of ash-gas thermal disequilibrium or atmospheric properties (Sparks et al., 1997).

The plume-sedimentation model of Sparks et al. (1992) assumes an initial grain-size distribution that we cannot verify for Fontana Tephra. The underlying eruption column dynamics also would be affected to some extent by the higher temperature of basaltic magma. Most of all, however, cross-wind effects are not considered in the model but would shift the data in Figure 10 parallel to the distance axis. We have therefore only compared the gradients in the data with the gradients of the model curves, and reading column heights from Figure 10 is thus less precise than from Figure 8. The thickness decay patterns shown by the Fontana data are, however, quite compatible with the modeled results.

In summary, there are some uncertainties in comparing the Fontana field data with deposit properties derived from eruption column modeling. The observed deviations and uncertainties are, however, mainly due to deficiencies in the input parameters (e.g., ignorance of crosswind) or in the field data (incomplete exposure) and less so to magma composition. Within the uncertainties that must be accepted with deposits of any composition, the models can be applied to basaltic Plinian tephra to obtain approximate values of eruption parameters.

## Conclusions

The dispersal characteristics of Fontana Tephra prove its deposition during a basaltic Plinian eruption. This eruption evolved through seven phases that successively emplaced the depositional units A to G. Unsteady eruption phases at the beginning (units A to C) and end (unit G) of the eruption bracketed the major Plinian activity producing units D to F. These are generally lithic-poor fallout of highly vesicular scoria lapilli but contain horizons in which distinct lithic lithologies, quench-rimmed scoria lapilli, and higher ash content attest to limited water access to an unstable conduit. Such transient disturbances could not, however, prevent the eruption from restoring stable Plinian conditions. Due to a lack of proximal outcrop sites, the position of the eruptive vent for Fontana Tephra remains uncertain but is possibly not within the Masaya caldera, as previously favored. We prefer vent 2 in Figures 4 and 5 as the site most compatible with the observed tephra distribution. Based on this scenario, the eruption of the major units D, E, and F took place at wind velocities of 8–20 m/s, and the total erupted volume of all units amounts to at least 1.4 km$^3$. The eruption column peaked at ~30 km height during phase D, slightly waned to ~26 km for phase E, then reached ~28 km in phase F (applying the method of Carey and Sparks, 1986).

The Fontana Plinian eruption was a multi-phase event that also produced pyroclastic surges extending to at least 7 km around the vent. Due to their extremely rapid lateral movement largely independent of wind velocity or direction, they potentially endanger areas in all directions from the vent (Cas and Wright, 1988). Younger basaltic Plinian eruptions of the Masaya area (Pérez and Freundt, this volume) produced even larger surges and show that basaltic Plinian activity is not that unusual in this volcanic setting.

The densely populated greater Managua area, located downwind from the Fontana source area, is home to roughly two million people, of whom more than one million live in the capital. The entire area is subject to intense population growth. The Fontana eruption led to devastation of the greater Managua region by emplacement of a several meter–thick tephra sheet; more than one meter was emplaced in the capital city itself. This is likely to happen again during similar future eruptions. Moreover, the repeated temporary waning or cessation of eruptive activity that mafic Plinian eruptions show might mislead the authorities to premature lowering of alert levels, and subsequent reawakening or intensifying of activity might come unexpectedly.

Because such eruptions are considered atypical and not much is known about possible unrest prior to eruption, their precursors may be misinterpreted as the forerunners to relatively weak Strombolian activity. The risk to communities is exacerbated because the low-viscosity mafic magmas ascend considerably faster than their silicic counterparts (Houghton et al., 2004). The rapid ascent rate means that the warning time; i.e., the time elapsing between potential precursory signals and the onset of the basaltic Plinian eruption, may be extremely short; Houghton et al. (2004) estimated it at a few hours for Tarawera 1886. Taken together, Fontana Tephra, younger deposits from Masaya Caldera (Pérez and Freundt, this volume), and the abundance of apparently widespread mafic lapilli beds observed during reconnaissance of the Las Sierras Formation, indicate that mafic Plinian-style eruptions may be a rather common exception to the rule in the Masaya area. This amplifies the need for a well-established volcano monitoring with particular regard to early warning systems.

## ACKNOWLEDGMENTS

Strong thanks are due to the staff of the Instituto Nicaragüense de Estudios Territoriales (INETER) in Managua for their enthusiastic support during our extensive field work. Bonadonna and Houghton were funded by U.S. National Science Foundation grants EAR-0310096 and EAR-0125719. Radiocarbon dates were generated at Leibniz Laboratory for Radiometric Dating and Stable Isotope Research, University of Kiel, Germany. We are grateful to Lina Patino and Steve Carey for the helpful reviews of this manuscript. This publication is contribution No. 71 to the Sonderforschungsbereich 574 "Volatiles and Fluids in Subduction Zones" at University of Kiel.

## REFERENCES CITED

Bice, D.C., 1980, Tephra stratigraphy and physical aspects of recent volcanism near Managua, Nicaragua [Ph.D. Thesis]: Berkeley, University of California, 422 p.

Bice, D.C., 1985, Quaternary volcanic stratigraphy of Managua, Nicaragua: Correlation and source assignment for multiple overlapping plinian deposits: Geological Society of America Bulletin, v. 96, p. 553–566, doi: 10.1130/0016-7606(1985)96<553:QVSOMN>2.0.CO;2.

Bonadonna, C., and Phillips, J.C., 2003, Sedimentation from strong volcanic plumes: Journal of Geophysical Research, v. 108, no. B7, 2340, 5-1.

Bonadonna, C., and Houghton, B.F., 2005, Total grainsize distribution and volume of tephra-fall deposits: Bulletin of Volcanology, v. 67, p. 441–456, doi: 10.1007/s00445-004-0386-2.

Bonadonna, C., Ernst, G.G.J., and Sparks, R.S.J., 1998, Thickness variations and volume estimates of tephra fall deposits: the importance of particle Reynolds number: Journal of Volcanology and Geothermal Research, v. 81, no. 3-4, p. 173–187, doi: 10.1016/S0377-0273(98)00007-9.

Cas, R.A.F., and Wright, J.V., 1988, Volcanic Successions—Modern and Ancient: London, Chapman and Hall, 528 p.

Carey, S., and Sparks, R.S.J., 1986, Quantitative models of the fallout and dispersal of tephra from volcanic eruption columns: Bulletin of Volcanology, v. 48, p. 109–125, doi: 10.1007/BF01046546.

Coltelli, M., Del Carlo, P., and Vezzoli, L., 1998, Discovery of a Plinian basaltic eruption of Roman age at Etna volcano, Italy: Geology, v. 26, p. 1095–1098, doi: 10.1130/0091-7613(1998)026<1095:DOAPBE>2.3.CO;2.

Head, J.W., and Wilson, L., 1987, Lava fountain heights at Pu'u O'o, Kilauea, Hawaii: indicators of amount and variations of exsolved magma volatiles: Journal of Geophysical Research, v. 92, p. 13,715–13,719.

Houghton, B.F., Wilson, C.J.N., and Pyle, D.M., 2000, Pyroclastic fall deposits, in Sigurdsson, H., et al., eds., Encyclopedia of Volcanoes: Academic Press, San Diego, p. 555–570.

Houghton, B.F., Wilson, C.J.N., Del Carlo, P., Coltelli, M., Sable, J.E., and Carey, R., 2004, The influence of conduit processes on changes in style of basaltic Plinian eruptions: Tarawera 1886 and Etna 122 BC: Journal of Volcanology and Geothermal Research, v. 137, p. 1–14, doi: 10.1016/j.jvolgeores.2004.05.009.

Jaupart, C., 2000, Magma ascent at shallow levels, in Sigurdsson, H., et al., eds., Encyclopedia of Volcanoes: Academic Press, San Diego, p. 237–245.

Jaupart, C., and Vergniolle, S., 1989, The generation and collapse of a foam layer at the roof of a basaltic magma chamber: Journal of Fluid Mechanics, v. 203, p. 347–380, doi: 10.1017/S0022112089001497.

Mangan, M.T., and Cashman, K.V., 1996, The structure of basaltic scoria and reticulite and inferences for vesiculation, foam formation, and fragmentation in lava fountains: Journal of Volcanology and Geothermal Research, v. 73, no. 1-2, p. 1–18, doi: 10.1016/0377-0273(96)00018-2.

Mangan, M.T., Cashman, K.V., and Newman, S., 1993, Vesiculation of basaltic magma during eruption: Geology, v. 21, p. 157–160, doi: 10.1130/0091-7613(1993)021<0157:VOBMDE>2.3.CO;2.

McBirney, A., and Williams, H., 1965, Volcanic history of Nicaragua: University of California Publications in Geological Sciences, v. 55, p. 1–65.

Parfitt, E., 2004, A discussion of the mechanisms of explosive basaltic eruptions: Journal of Volcanology and Geothermal Research, v. 134, p. 77–107, doi: 10.1016/j.jvolgeores.2004.01.002.

Parfitt, E.A., and Wilson, L., 1995, Explosive volcanic eruptions—IX: The transition between Hawaiian-style lava fountaining and Strombolian explosive activity: Geophysical Journal International, v. 121, p. 226–232.

Pérez, W., and Freundt, A., this volume, The youngest highly explosive basaltic eruptions from Masaya Caldera (Nicaragua): Stratigraphy and hazard assessment, in Rose, W.I., Bluth, G.J.S., Carr, M.J., Ewert, J.W., Patino, L., and Vallance, J.W., Volcanic hazards in Central America: Geological Society of America Special Paper 412, doi: 10.1130/2006.2412(10).

Pyle, D.M., 1989, The thickness, volume and grainsize of tephra fall deposits: Bulletin of Volcanology, v. 51, p. 1–15, doi: 10.1007/BF01086757.

Pyle, D.M., 1995, Assessment of the minimum volume of tephra fall deposits: Journal of Volcanology and Geothermal Research, v. 69, p. 379–382, doi: 10.1016/0377-0273(95)00038-0.

Rymer, H., van Wyk de Vries, B., Stix, J., and Williams-Jones, G., 1998, Pit crater structure and processes governing persistent activity at Masaya Volcano, Nicaragua: Bulletin of Volcanology, v. 59, p. 345–355.

Rose, W.I., 1993, Comment on another look at the calculation of fallout tephra volumes: Bulletin of Volcanology, v. 55, p. 372–374, doi: 10.1007/BF00301148.

Rosi, M., 1998, Plinian eruption columns: particle transport and fallout, in Freundt, A. and Rosi, M., eds., From Magma to Tephra: Elsevier, Amsterdam, p. 139–176.

Seyfried, R., and Freundt, A., 2000, Analog experiments on conduit flow, eruption behaviour, and tremor of basaltic volcanic eruptions: Journal of Geophysical Research, v. 105, p. 23,727–23,740, doi: 10.1029/2000JB900096.

Simkin, T., and Siebert, L., 1994, Volcanoes of the World: Smithsonian Institution and Geoscience Press, 349 p.

Sparks, R.S.J., 1986, The dimensions and dynamics of volcanic eruption columns: Bulletin of Volcanology, v. 48, p. 3–15, doi: 10.1007/BF01073509.

Sparks, R.S.J., Bursik, M.I., Ablay, G.J., Thomas, R.M.E., and Carey, S.N., 1992, Sedimentation of tephra by volcanic plumes. Part 2: controls on thickness and grain-size variations of tephra fall deposits: Bulletin of Volcanology, v. 54, p. 685–695, doi: 10.1007/BF00430779.

Sparks, R.S.J., Bursik, M.I., Carey, S.N., Gilbert, J.S., Glaze, L.S., Sigurdsson, H., and Woods, A.W., 1997, Volcanic Plumes: New York, John Wiley, 574 p.

van Wyk de Vries, B., 1993, Tectonics and magma evolution of Nicaraguan volcanic systems [Ph.D. Thesis]: Open University, Milton Keynes, UK, 328 p.

Vergniolle, S., and Jaupart, C., 1986, Separated two-phase flow and basaltic eruptions: Journal of Geophysical Research, v. 91/B, 12,842–12860 p.

Walker, G.P.L., Self, S., and Wilson, L., 1984, Tarawera, 1886, New Zealand—A basaltic Plinian fissure eruption: Journal of Volcanology and Geothermal Research, v. 21, p. 61–78, doi: 10.1016/0377-0273(84)90016-7.

Walker, J.A., Williams, S.N., Kalamarides, R.I., and Feigeson, M.D., 1993, Shallow open-system evolution of basaltic magma beneath a subduction zone volcano: the Masaya Caldera Complex, Nicaragua: Journal of Volcanology and Geothermal Research, v. 56, p. 379–400, doi: 10.1016/0377-0273(93)90004-B.

Williams, S.N., 1983a, Geology and Eruptive Mechanisms of Masaya Caldera Complex, Nicaragua [Ph.D. Thesis]: Hanover, New Hampshire, Dartmouth College, 169 p.

Williams, S.N., 1983b, Plinian airfall deposits of basaltic composition: Geology, v. 11, p. 211–214, doi: 10.1130/0091-7613(1983)11<211:PADOBC>2.0.CO;2.

Wilson, L., and Head, J.W., 1981, Ascent and eruption of basaltic magma on the earth and moon: Journal of Geophysical Research, v. 86, p. 2971–3001.

Wilson, L., and Walker, G.P.L., 1987, Explosive volcanic eruptions, VI, Ejecta dispersal in plinian eruptions—The control of eruption conditions and atmospheric properties: Geophysical Journal of the Royal Astronomical Society, v. 89, p. 657–679.

Woods, A.W., 1988, The fluid dynamics and thermodynamics of eruption columns: Bulletin of Volcanology, v. 50, p. 169–193, doi: 10.1007/BF01079681.

Woods, A.W., and Bursik, M.I., 1991, Particle fallout, thermal disequilibrium and volcanic plumes: Bulletin of Volcanology, v. 53, p. 559–570, doi: 10.1007/BF00298156.

Manuscript Accepted by the Society 19 March 2006

# Tephra deposits for the past 2600 years from Irazú volcano, Costa Rica

**Scott K. Clark**
*Department of Geology, University of Illinois, Urbana, Illinois 61801, USA*

**Mark K. Reagan**
*Department of Geoscience, University of Iowa, Iowa City, Iowa 52242, USA*

**Deborah A. Trimble**
*St. Charles School, 850 Tamarack Ave, San Carlos, California 94070, USA*

## ABSTRACT

We report the first detailed study of recent tephra deposits at Irazú volcano, Costa Rica. These ash-fall deposits consist of unconsolidated, moderately to well-sorted, mostly juvenile ash of porphyritic basalt to basaltic andesite. Ash accumulations are thickest SW of the crater, an area that includes the headwaters of the Reventado River, which flows through the city of Cartago. With increasing eruption intensities, deposition shifts more westerly—toward the capital city, San José. Of seventeen historic eruptions, only two have left distinct ash deposits. At least eight other ash-fall deposits from the past 2600 yr are preserved on the SW flank of Irazú. Carbon-14 based correlations of deposits indicate that the ash accumulation rate has been relatively consistent during this period (e.g., ≈18 cm/century, 5 km SW of the crater). This consistency combined with the historic preservation ratio and correlated prehistoric deposits implies that Irazú may have erupted >85 times during the past 2600 yr. Most of these would have been small, volcanic explosivity index (VEI) ≤2 eruptions, with only ten or so VEI = 3 eruptions likely occurring every 200–400 yr. The largest historic eruption occurred in 1963–1965, and we estimate a minimum tephra volume of $3 \times 10^7$ m$^3$ for that eruption. The 1963–1965 eruption was not quite as energetic as some eruptions of the past 2600 yr, but it is of the same order of magnitude, and, based on its thickness, it approximates the size and duration of the larger eruptions of the past 2600 yr.

**Keywords:** Irazú volcano, Costa Rica, tephra, ash-fall deposits.

## INTRODUCTION

Irazú volcano, Costa Rica, is situated 24 km east of the capital city, San José, and 15 km northeast of Costa Rica's second largest city, Cartago (Fig. 1). With an elevation of 3432 m and an approximate volume of 227 km$^3$ (Carr et al., 2003), Irazú is one of the largest volcanoes in Central America. Mapping by Krushensky (1972) showed the volcanic edifice primarily to consist of Pleistocene lavas, mudflows, landslides, and ash-flow tuffs of the Irazú Group (Krushensky, 1972). During the late Pleistocene, lava erupted from the current summit area and craters located on the southern flank, 4–5 km from the summit, with the most recent lava flows emplaced some 14,500 yr ago (Sáenz et al., 1982). Due to the thin nature of the recent ash fall deposits, Krushensky's 1972 map does not include the recent ash fall deposits.

Irazú has erupted at least seventeen times since 1561 A.D. (Alvarado-Induni, 1989; Siebert and Simkin, 2002; Simkin and Siebert, 1994). Although Siebert and Simkin (2002) list eruptions

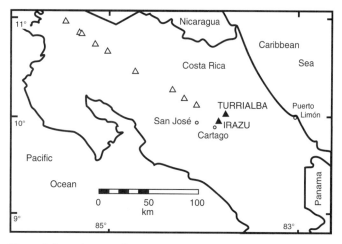

Figure 1. Location map for Irazú volcano. Note its location ~25 km E of San José, and 15 km NE of Cartago, the two largest cities in Costa Rica. Solid triangles represent volcanoes specifically mentioned in text. Open triangles represent other Costa Rican volcanoes.

in 1977 and 1994, we do not include these two events in our calculations because records at the Observatorio Vulcanológico y Sismológico de Costa Rica show only fumarolic activity in 1977, and the volcanic event in 1994 was a small, volcanic explosivity index (VEI) <<1 (see Newhall and Self, 1982, for details on the VEI), phreatic explosion with no magma or ash deposition near the summit (E. Malavassi, 2005, personal commun.). Historic eruptions have been strombolian and vulcanian, with VEI's of ≤3. Mudflows have accompanied some eruptions (Alvarado-Induni, 1989; Murata et al., 1966) and can occur independently of eruptions in association with landslides and debris avalanches. Eruptions in 1561, 1723, 1917–1921, and 1963–1965 deposited ash as far as San José (Alvarado-Induni, 1989; Sáenz et al., 1982). The 1963–1965 eruption caused ten fatalities (Siebert and Simkin, 2002), and ash accumulation to the west of Irazú in the Central Valley impacted the Costa Rican economy by lowering yields of coffee and other commodities (Murata et al., 1966).

Irazú is one of Central America's most problematic volcanoes because of its frequent explosive eruptions, its proximity to Cartago, and its position upwind of Costa Rica's other major population centers and international airport. Irazú, therefore, poses a considerable hazard to the population and economy of Costa Rica. To better understand the nature of these hazards, we have investigated the recent eruptive history of Irazú as recorded in the character and distribution of volcanic deposits at 71 locations on the slopes of Irazú and in the eastern Central Valley. Most of these measured sections were at road cuts within a 25-km-long by 10-km-wide band to the west-southwest of Irazú (Fig. 2), so chosen because this area coincides with newspaper reports of ash-fall zones for many historical eruptions (Fig. 3; Barquero-Hernandez, 1975). Correlation of these deposits allows us to address two important questions: (1) To what extent can the relatively well-documented 1963–1965 eruption be used as a proxy for typical, large, prehistoric, ash-fall eruptions of Irazú?

(2) How frequently have VEI > 2 eruptions occurred at Irazú during the past 2600 yr?

## NATURE OF DEPOSITS

Typical measured sections consist of alternating layers of gray ash ± lapilli and soils ± dispersed ash and lapilli. Ash deposits are commonly well-sorted, fine-grained, and crystal-rich (average 50% crystals), dark gray to black basalt and basaltic andesites. The relative proportions of minerals vary, but a typical deposit consists of plagioclase (60%), clinopyroxene (18%), olivine (14%), orthopyroxene (7%), and opaques (1%).

Most deposits lack internal features or distinct boundaries and grade into underlying and overlying soils. Where present, millimeter- to centimeter-thick layering ranges from discontinuous and slightly wavy to continuous and horizontal. Significant bedding discontinuities were found in only three units at scattered locations on the flanks of Irazú. These discontinuities include trough-shaped cross-beds and beds with variable thickness, both of which crosscut horizontal ash layers of the same unit. Based on a comparison of these deposits with descriptions of bedding associated with reworking of tephras erupted in 1963–1965 (Murata et al., 1966), these units likely were reworked by erosion and redeposition, perhaps during periods of heavy rain. Fine-grained ash deposits quickly degrade into clays and form soils in tropical climates, and soils formed on the flanks of Irazú typically contain remnant clasts from ash-fall tephras. Bioturbation has also disturbed layering and blurred bedding contacts.

Many of the fine-ash layers are vesiculated. Vesicles within the ash layers are irregularly shaped and vary from less than one millimeter up to one centimeter across. Vesiculated ash deposits indicate that the ash was saturated with water during deposition (Cas and Wright, 1988). This vesiculation can result from steam during hot deposition of phreatomagmatic eruptions or cool compaction of ash by rainfall or coalescence of accretionary lapilli (Rosi, 1992; Walker, 1981). Syneruption rainfalls were documented during the 1963–1965 eruption (Murata et al., 1966), and it is reasonable to assume that rainfall accompanied many other eruptions at Irazú.

## TEPHRA DISPERSION

The nature of individual ash deposits varies in grain size and thickness from section to section, depending on the relationship of the outcrop to the dispersion axes of the eruptions and the extent of post-depositional reworking. In general, tephra deposits have a limited aerial extent of <20 km downwind. Tephra deposits are best preserved within 7 km of the crater. Deposits found 7–12 km away from the crater are generally thin (<5 cm), and indistinct. Beyond ~12 km, only two thin layers of ash were observed. Deposits are thickest southwest of the summit, where they are incised by the headwaters of the Reventado River. Beyond 2 km from the crater, all deposits measure <1 m thick, and beyond 5 km, <50 cm thick (Fig. 2). Most lapilli were found within

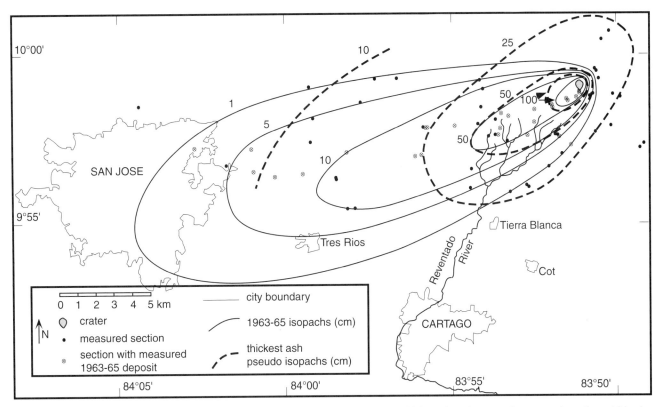

Figure 2. Isopachs for the 1963–1965 eruption and pseudo-isopachs based on the thickest deposits at each measured section within the past 2600 yr. These pseudo-isopachs are not correlated to any single eruption.

10 km of the crater and typical maximum clast size decreases to fine ash beyond 15 km (Fig. 4). During the 1963–1965 eruption, bombs >30 cm diameter fell within 4 km of the crater (Paniagua and Soto, 1988), and small pumice fell 18 km west of the summit (Murata et al., 1966).

Ash-fall dispersion axes based on isopachs measured in this study (Fig. 2) and ash-fall maps for historic eruptions (Fig. 3) demonstrate the major dispersion axis of ash-fall has been consistently oriented west-southwest. Dispersion axes for preserved 1963–1965 deposits show a SW trend close to the crater but rotate more westerly for farther dispersed tephras (Fig. 2). This rotation appears to be due to a long-lived and consistent split in the predominant wind direction a few hundred meters above the summit of Irazú, with low altitude winds dominantly blowing southwesterly and higher altitude winds blowing more westerly. Murata et al. (1966) noted this discontinuity in wind direction during the 1963–1965 eruptions, and the eruptions mapped by Barquero-Hernandez (1975) show similar trends for prior eruptions (Fig. 3).

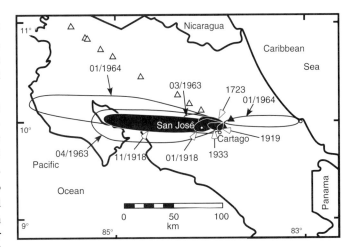

Figure 3. Areal extent of ash falls for the largest recorded historical eruptions. Delineated areas are based on historical accounts, not preserved deposits (after Barquero-Hernandez, 1975).

## STRATIGRAPHIC CORRELATIONS

Stratigraphic correlations of measured sections (Figs. 5–8) are based on layer thickness, maximum grain size, petrology, layering patterns, seven $^{14}$C dates of charcoal (Table 1), an easily recognizable "salt-and-pepper" tephra deposit from nearby Turrialba volcano dated at 1970 cal yr B.P. (Reagan et al., this volume), and one pottery fragment on the NE flank of Irazú from the Curridabat phase (F. Corrales, 1992, personal commun.), dated at 1200–1500 yr B.P. (Aguilar, 1976). This pottery age is supported by a $^{14}$C age date of 1156 cal yr B.P. on charcoal found

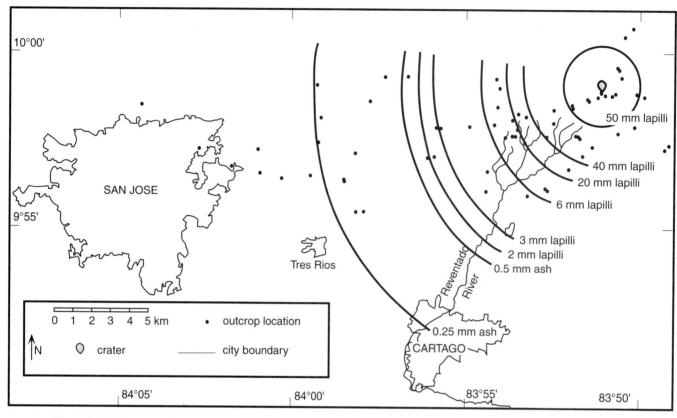

Figure 4. Maximum areal extent of pyroclasts of differing sizes, compiled from all deposits erupted within the past 2600 yr.

with the pottery. Factors inhibiting ready identification of units include similarities in bedding characteristics, grain size, and dispersion patterns for tephra units, as well as a lack of distinct and well-dispersed marker layers, bioturbation, and the rapid breakdown of the ash into clays and soils. Erosion of the units, including the 1963–1965 deposits, commonly causes deposits to have uneven distributions. For example, sections 91-53, 91-54 (Fig. 6) and some other measured sections SW of the crater (not shown) would be expected to have ≥50 cm of 1963–1965 ash based on isopachs in Figure 2, but only indistinct ash-rich soils were present near the surface. Nevertheless, the $^{14}C$ ages and several marker horizons have provided enough stratigraphic control to allow confidence in our correlations. Even though the 1963–1965 deposit has been eroded in some locations, in many other areas, the 1963 through June 1964 tephra sequence, as described by Murata et al. (1966) at an exposure 800 m SW of the crater, is recognizable several kilometers from the summit and is a good marker horizon. This sequence is characterized by lower and upper thirds that are finer grained and have better stratification than the middle, coarser third, and, in this study, was recognized in the stratigraphy at sites up to 9 km from the crater. Another distinct ash-fall deposit that overlays charcoal dated at 389 cal yr B.P. correlates to the 1561 eruption recorded by early Spanish settlers near Cartago, Costa Rica (Alvarado-Induni, 1989). Within 4 km SW of the summit of Irazú, this deposit consists of a ≥5 cm layer of lapilli and ash overlain by a ≥9 cm ash layer. Beyond ~6 km, the lapilli grades into coarse and then fine ash, and the overlying ash grades into an ash-rich soil (Fig. 6).

Roughly ten eruptions from the past 2600 yr are preserved as distinct deposits on the western and southern flanks of Irazú. Eight of these tephras were correlated between numerous sections (Figs. 5 and 6; Table 2). As discussed above, beyond ~12 km, only two thin layers of ash were observed. The upper ash is related to the 1963–1965 eruption, and the lower ash is related to the 1561 A.D. eruption. This means that the outer, fine-ash deposits of older, large eruptions have been weathered and reworked to the extent that they generally are no longer recognizable. Deposits on the northeast flank were locally correlated (Figs. 5 and 7). However, due to the small volume and lobate nature of Irazú's deposits, these deposits could not be correlated to deposits on the southwest flank.

## ERUPTION INTENSITIES, VOLUMES, AND FREQUENCIES

Historic eruptions deposited ash as far as San José in 1561, 1723, 1917–1921, and 1963–1965 (Alvarado-Induni, 1989; Sáenz et al., 1982). We have calculated the tephra volume for the 1963–1966 eruption to be ≈$3 \times 10^7$ m$^3$ based on an exponential linear regression of the log thickness of isopachs from Figure 2

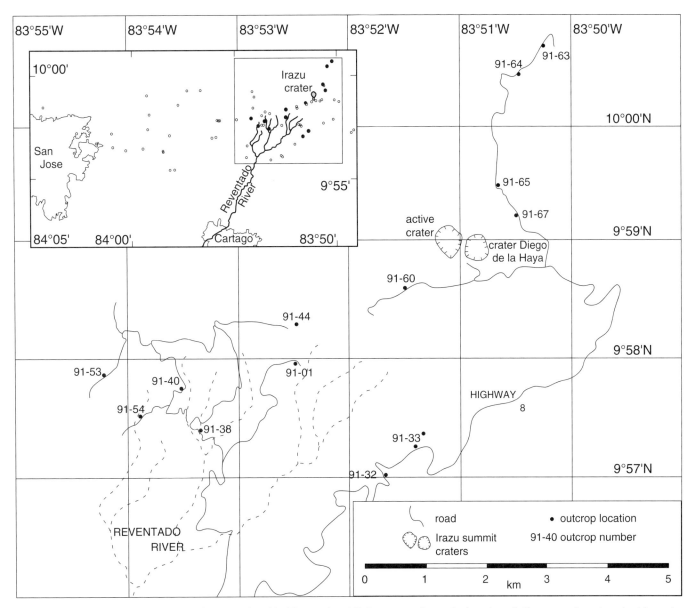

Figure 5. Map showing the tephra locations correlated in Figures 6 and 7. Inset map shows the location of all measured sections in this study. Closed dots indicate which sections are noted in the larger map. Latitude and longitude of the correlated deposits are given in Table 2.

plotted against the square root of the area enclosed by the isopachs (plot not shown) and using the slopes of two straight-line segments of the plotted data to determine the tephra volume (Fierstein and Nathenson, 1992; Pyle, 1989). Even though ash fall from the 1963–1965 eruption reached both the Pacific and Caribbean coasts of Costa Rica (Barquero-Hernandez, 1975; Murata et al., 1966), these distal deposits were too thin to leave a discernable record and were not used in the calculation. Moreover, we mapped the deposits 26 yr after the 1963–1965 eruption and are unsure how much ash was eroded in the intervening years. Bearing in mind this missing data and the limitations of estimating volumes based on straightforward integration of tephra isopach data (Pyle, 1995), this calculated volume must be considered a minimum. Nevertheless, this volume agrees with Siebert and Simkin (2002) and is consistent with a VEI = 3 (Newhall and Self, 1982). Thus, it probably is a reasonable estimation of the actual volume of deposited tephra.

The 1561 A.D. deposits are as much as 75% thinner than the 1963–1965 deposits but contain larger clasts farther from the crater (e.g., Fig. 6), implying the 1561 eruption was a VEI = 3 eruption of less duration but with a higher eruptive column (Carey and Sparks, 1986) than the 1963–1965 eruption. Siebert and Simkin (2002) report that eruptions in 1723 and 1917–1921 were also VEI = 3, but unlike the 1963–1965 and 1561 eruptions, we did not observe distinct deposits from either the 1723 or 1917–1921 eruptions, and these eruptions may have had VEI = 2.

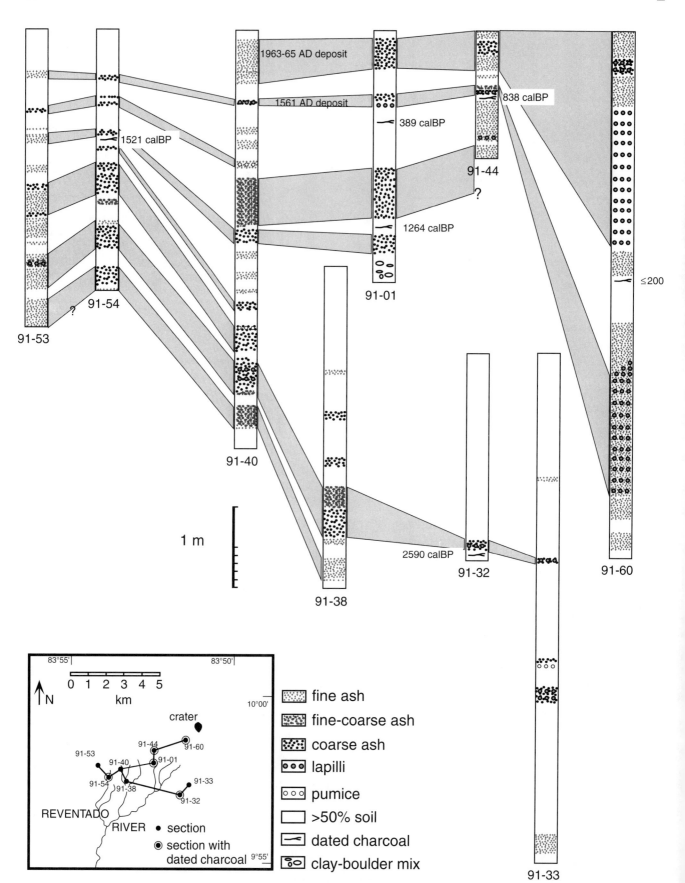

Figure 6. Correlation of deposits SW of the crater. Ages are cal yr B.P. from ¹⁴C dates of charcoal samples. Two-sigma uncertainty ranges of ¹⁴C ages are listed in Table 1.

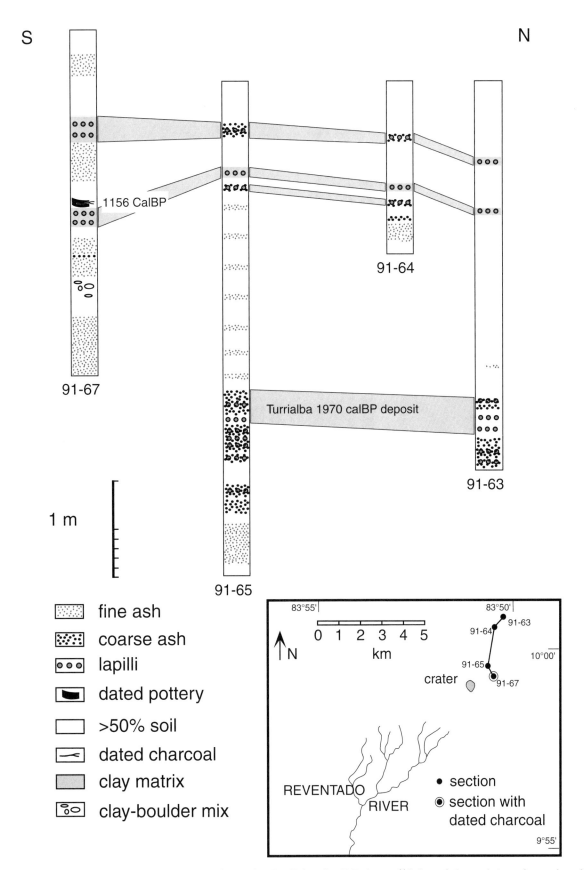

Figure 7. Correlation of deposits NE of the crater. Age for section 91-67 is cal yr B.P. from a $^{14}$C date of charcoal. Ages for sections 91-63 and 91-65 are from an easily recognizable deposit from Turrialba (Reagan et al., this volume). Two-sigma uncertainty range of the $^{14}$C age is listed in Table 1.

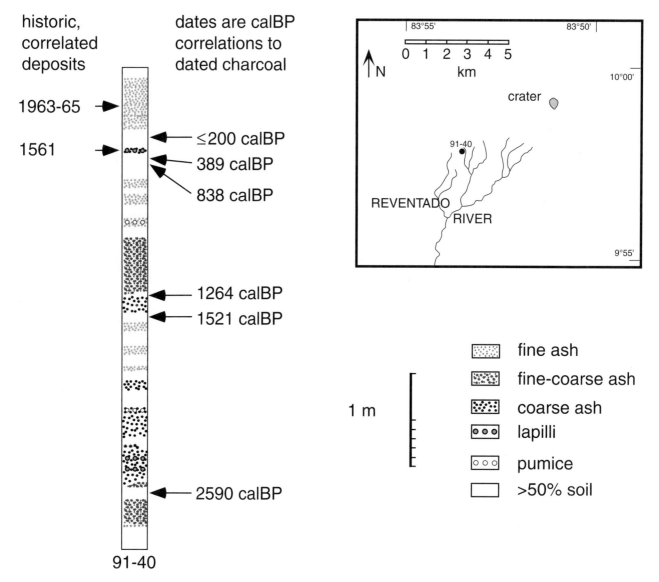

Figure 8. Section 91-40, located 5 km SW of the crater. Ages are cal yr B.P. and are based on correlations to sections with $^{14}$C dates of interbedded charcoal samples (ref. Fig. 6). Two-sigma uncertainty ranges of $^{14}$C ages are listed in Table 1.

To address whether or not the 1963–1965 eruption is representative of the larger eruptions of the past 2600 yr, we compared isopachs for the 1963–1965 eruption against pseudo-isopachs that we generated by mapping the thickest discernable deposits exposed at the measured sections, regardless of stratigraphic position (Fig. 2). Although the 1963–1965 deposits are not as thick or as extensive as the thickest recorded deposits, the 1963–1965 deposits are comparable overall to the scale of the pseudo-isopachs. For example, section 91-40, which is in-line with the general direction of the main dispersion axis and 5 km SW of the crater, shows all eight of the post-2600 B.P. correlated deposits (Fig. 8). A coarse-ash directly above a 1264 cal yr B.P. dated charcoal is the only deposit as thick as the 1963–1965 deposit, which consists of fine ash at this site. Based on relative clast sizes (Carey and Sparks, 1986), this indicates that although the 1963–1965 was not the most energetic eruption within the past 2600 yr, its explosivity was within the same order of magnitude as the more energetic eruptions, and based on thickness, it was one of the most voluminous.

Age correlations and ash thicknesses at section 91-40 indicate that Irazú has produced ash at a rather consistent rate during the past 2600 yr (Fig. 9). At this location, the accumulation rate has been linear and has averaged ≈18 cm/century. Because this value is based on the preserved ash thicknesses, it represents a minimum ash deposition rate. It also assumes that all of the ash is from Irazú. However, Turrialba volcano is located only 11 km NE of Irazú, and during this period, four eruptions from Turrialba were large enough that they could have contributed ash fall at this location (Reagan et al., this volume). In this study, a 9-cm-thick deposit from the distinctive, VEI = 4 (Siebert and Simkin, 2002), 1970 cal yr B.P. Turrialba eruption was seen ≈9 km WSW of Irazú's crater. Based on this measurement and isopachs in Reagan et al. (this volume), a few cm of ash from this eruption must have fallen at section 91-40. However, the tephra was not observed and must have been eroded. Based on thicknesses of the three youngest significant ash-fall deposits for Turrialba (Reagan et al., this volume), and assuming ash-fall dispersion axes were predominantly due west of Turrialba, it is unlikely that enough ash from these eruptions fell at section 91-40 to significantly affect the estimated accumulation rate for this site.

Of Irazú's seventeen documented historical eruptions since 1561 (see Alvarado-Induni, 1989; Siebert and Simkin, 2002), only the 1561 and 1963–1965 are correlated to preserved ash layers. This means ~12% of historic eruptions have left discernible deposits. The stratigraphic correlations in Figures 6 and 7 indicate that during the past 2600 yr, roughly ten eruptions have left preserved deposits on the west-southwest flank of Irazú. Assuming a consistent preservation rate for historic and prehistoric eruptions, Irazú may have erupted >85 times during the past 2600 yr, with 90% of those eruptions being smaller than the 1963–1965 eruption. Historic eruptive phases have, on average, occurred every 26 yr (Siebert and Simkin, 2002) and continued for periods of hours to >3 yr. Quiescent periods have lasted from less than a year up to 162 yr between the 1561 and 1723 eruptions. We note that it is quite possible that small, VEI = 1 eruptions may have occurred between 1561 and 1723, but were not noted in historical accounts. For example, some of the eruptions recorded during the 1920s through 1940s were so small that people living in the highlands of Coronado, 5 km west of the crater, only knew of an ash-fall event by observing ash on white clothes left hanging out to dry during the night (E. Malavassi, 2005, personal commun.). Approximately ten VEI = 3 eruptions during the past 2600 yr and a consistent ash accumulation rate imply an average of one, large, VEI = 3 eruption every 200–400 yr.

The nearly linear ash accumulation rate of basalt and basaltic andesites from numerous small eruptions indicates a consistency of Irazú's activity and magma composition over the past 2600 yr, which is a rather remarkable feature of this volcano and suggests that its magma input, storage, differentiation, and eruption rates are dynamically balanced to produce a nearly steady-state output. Using trace element ratios and variations in $^{226}$Ra-$^{230}$Th-$^{238}$U disequilibrium, Clark et al. (1998) showed that the 1963–1965 eruption tapped basaltic andesites from two separate magma chambers, and it seems likely that the eruptions of the past 2600 yr are repeatedly tapping multiple small magma chambers under Irazú.

## FUTURE ACTIVITY

Based on eruptions dating back to 1723, Stine and Banks (1991) predicted that the most likely future eruptions of Irazú will be strombolian to vulcanian, producing ash-fall tephras with associated mudflows. The stratigraphic record for the past 2600 yr

TABLE 1. $^{14}$C AGES OF CHARCOAL SAMPLES

| Section and sample no. | Calibrated age cal yr B.P. | +/−* | Conventional age $^{14}$C yr B.P. | ± | $\delta^{13}$C |
|---|---|---|---|---|---|
| 91-01.05 | 1264 | 38/42 | 1325 | 35 | −24.4 |
| 91-01.10 | 389 | 67/40 | 315 | 20 | −23.4 |
| 91-32.01 | 2590 | 408/300 | 2530 | 170 | −25.0 |
| 91-44.09 | 838 | 97/114 | 920 | 60 | −25.0 |
| 91-54.02 | 1521 | 406/349 | 1600 | 180 | −25.0 |
| 91-60.06 | MODERN† | n/a | MODERN | n/a | −25.0 |
| 91-67.05 | 1156 | 132/106 | 1230 | 70 | −25.0 |

Note: Calibrated ages determined using Calib 5.0 (Stuiver et al., 2005) with the INTCAL04 dataset (Reimer et al., 2004).
*2σ standard deviation.
†Age of sample is ≤200 yr old.

TABLE 2. LOCATIONS OF DESCRIBED DEPOSITS

| Section no. | Latitude | Longitude | Site Details |
|---|---|---|---|
| 91-01 | 9°57′58″N | 83°52′28″W | S side of road |
| 91-32 | 9°57′00″N | 83°51′40″W | E side of road |
| 91-33-a | 9°57′15″N | 83°51′24″W | S and N side of road |
| 91-33-b | 9°57′21″N | 83°51′20″W | ~150 m NE of 91-33-a |
| 91-38 | 9°57′22″N | 83°53′20″W | E side of road |
| 91-40 | 9°57′44″N | 83°53′30″W | W side of road |
| 91-44 | 9°58′18″N | 83°52′28″W | hand-dug trench on top of Cerro Retes |
| 91-53 | 9°57′52″N | 83°54′24″W | W side of road |
| 91-54 | 9°57′29″N | 83°53′51″W | W side of road |
| 91-60 | 9°58′37″N | 83°51′30″W | N side of road, along a gully |
| 91-63 | 10°00′42″N | 83°50′15″W | E side of road |
| 91-64 | 10°00′27″N | 83°50′29″W | E side of road |
| 91-65 | 9°59′29″N | 83°50′40″W | E side of road |
| 91-67 | 9°59′14″N | 83°50′30″W | W side of road |

Note: 91-33 is a composite section. Latitude and longitude were determined from two topographic maps: Istaru, hoja (sheet) 3445 IV, Edicion 2-IGNCR, 1981; and Carrillo, hoja (sheet) 3446 III, Edicion 1-IGNCR, 1967.

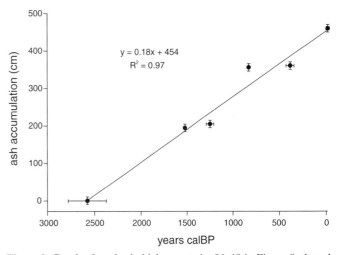

Figure 9. Graph of total ash thickness at site 91-40 in Figure 8 plotted against ages of deposits in calibrated $^{14}$C years. Error bars are 1σ for age and 10 cm for thickness. Note the linearity of the correlation, suggesting that Irazú has had a consistent eruption rate for the past 2600 yr of ~0.18 cm/yr at this site.

provides a basis to refine estimates of potential dispersion patterns, eruption intensities, and frequencies of activity. Preserved tephra deposits at Irazú record more energetic eruptions than that of 1963–1965. However, based on thickness and distribution patterns, we estimate that no eruption within the past 2600 yr has lasted significantly longer or been more than an order of magnitude larger than the 1963–1965 eruption (i.e., VEI = 3). Lapilli accumulation is unlikely to extend beyond 10 km from the crater. Heavy ash fall accumulation can be expected for nearby river drainages, with the Reventado River to the southwest and the Tiribi River and Virilla-Durazno River basins to the west expected to have the highest potential for heavy ash accumulation and resulting mudflows. The Sucio River to the north and the Toro Amarillo River to the northeast are upwind of the prevailing trade winds, making them less vulnerable to heavy accumulations of ash.

Even though no pyroclastic flows were observed in the stratigraphic record on the flanks of Irazú, base surges during the 1963–1965 eruption were deposited in an inactive crater adjacent to the erupting crater (Hudnut, 1983), and small pyroclastic flows have been reported for the 1723, 1917, and 1963–1965 eruptions (Murata et al., 1966; Siebert and Simkin, 2002). Lava last erupted at Irazú some 14,500 yr ago (Sáenz et al., 1982), but magma rose to within 100 m of the summit in December 1963 (Murata et al., 1966), and eruptions at nearby Turrialba volcano have produced voluminous andesite lava flows during the past 9300 yr (Reagan et al., this volume). Thus, pyroclastic and lava flows are both possible during eruptions of Irazú, but ash fall, mudflows, and mass wasting are the most probable hazards during future activity.

During the past 2600 yr, volcanic activity at Irazú has been frequent and rather consistent, with a nearly linear ash accumulation rate from ≥25 and possibly >85 strombolian to vulcanian eruptions, all with VEI ≤ 3. If future activity follows this pattern, small, VEI ≤ 2, eruptions are likely to occur every 20–25 yr, with larger, VEI = 3 eruptions possible every 200–400 yr.

## ACKNOWLEDGMENTS

Clark thanks Jorge Brennes for assistance in the field and the Observatorio Vulcanológico y Sismológico de Costa Rica for logistical support during field work in Costa Rica. Fieldwork was supported by a grant to Reagan by UNOCAL. Comments by Eduardo Malavassi, David Pyle, Lina Patino, Luis Gonzalez, Jim Faulds, and an anonymous reviewer greatly improved this manuscript.

## REFERENCES CITED

Aguilar, C.H., 1976, Relaciones de las culturas Precolombinas en el intermontano central de Costa Rica: Vinculos, v. 2, p. 75–86.

Alvarado-Induni, G.E., 1989, Los volcanes de Costa Rica: San José, Costa Rica, Editorial Universidad Estatal a Distancia (EUNED), 175 p.

Barquero-Hernandez, J., 1975, Cronología de la actividad volcanica del Irazú [Tesis de Licenciatura en Geografía Física]: Heredia, Costa Rica, Universidad Nacional, 36 p.

Carey, S., and Sparks, R.S.J., 1986, Quantitative models of the fallout and dispersal of tephra from volcanic eruption columns: Bulletin of Volcanology, v. 48, no. 2–3, p. 109–125, doi: 10.1007/BF01046546.

Carr, M.J., Feigenson, M.D., Patino, L.C., and Walker, J.A., 2003, Volcanism and Geochemistry in Central America: Progress and Problems: Geophysical Monograph, v. 138, p. 153–179.

Cas, R.A.F., and Wright, J.V., 1988, Volcanic successions, modern and ancient; a geological approach to processes, products, and successions: London, Allen & Unwin, 528 p.

Clark, S.K., Reagan, M.K., and Plank, T., 1998, Trace element and U-series systematics for 1963–1965 tephras from Irazú Volcano, Costa Rica: Implications for magma generation processes and transit times: Geochimica et Cosmochimica Acta, v. 62, no. 15, p. 2689–2699, doi: 10.1016/S0016-7037(98)00179-3.

Fierstein, J., and Nathenson, M., 1992, Another look at the calculation of fallout tephra volumes: Bulletin of Volcanology, v. 54, no. 2, p. 156–167, doi: 10.1007/BF00278005.

Hudnut, K.W., 1983, Geophysical survey of Irazú Volcano [senior honors thesis]: Hanover, New Hampshire, Dartmouth College, 93 p.

IGNCR (Instituto Geográfico Nacional de Costa Rica), 1967, Carrillo, hoja (sheet) 3446 III, Edicion 1-IGNCR: Instituto Geográfico Nacional, Costa Rica, scale 1:50:000, 1 sheet.

IGNCR (Instituto Geográfico Nacional de Costa Rica), 1981, Istaru, hoja (sheet) 3445 IV, Edicion 2-IGNCR: Instituto Geográfico Nacional, Costa Rica, scale 1:50:000, 1 sheet.

Krushensky, R.D., 1972, Geology of the Istarú Quadrangle, Costa Rica: U.S. Geological Survey Bulletin 1358, 46 p.

Murata, K.J., Dondoli, C., and Sáenz, R., 1966, The 1963–65 eruption of Irazú Volcano, Costa Rica: Bulletin Volcanologique, v. 29, p. 765–796.

Newhall, C.G., and Self, S., 1982, The volcanic explosivity index (VEI): an estimate of explosive magnitude for historical volcanism: Journal of Geophysical Research, Oceans & Atmospheres, v. 87, p. 1231–1238.

Paniagua, S., and Soto, G., 1988, Peligros volcanicos en el Valle Central de Costa Rica: Ciencia y Tecnología, v. 12, p. 145–156.

Pyle, D., 1989, The thickness, volume and grainsize of tephra fall deposits: Bulletin of Volcanology, v. 51, p. 1–15, doi: 10.1007/BF01086757.

Pyle, D., 1995, Assessment of the minimum volume of tephra fall deposits: Journal of Volcanology and Geothermal Research, v. 69, p. 379–382, doi: 10.1016/0377-0273(95)00038-0.

Reagan, M., Duarte, E., Soto, G.J., and Fernández, E., this volume, The eruptive history of Turrialba volcano, Costa Rica, and potential hazards from future eruptions, in Rose, W.I., Bluth, G.J.S., Carr, M.J., Ewert, J.W., Patino, L.C., and Vallance, J.W., Volcanic hazards in Central America: Geological Society of America Special Paper 412, doi: 10.1130/2006.2412(13).

Reimer, P.J., Baillie, M.G.L., Bard, E., Bayliss, A., Beck, J.W., Bertrand, C.J.H., Blackwell, P.G., Buck, C.E., Burr, G.S., Cutler, K.B., Damon, P.E., Edwards, R.L., Fairbanks, R.G., Friedrich, M., Guilderson, T.P., Hogg, A.G., Hughen, K.A., Kromer, B., McCormac, F.G., Manning, S.W., Ramsey, C.B., Reimer, R.W., Remmele, S., Southon, J.R., Stuiver, M., Talamo, S., Taylor, F.W., van der Plicht, J., and Weyhenmeyer, C.E., 2004, IntCal04 Terrestrial radiocarbon age calibration, 26–0 ka BP: Radiocarbon, v. 46, p. 1029–1058.

Rosi, M., 1992, A model for the formation of vesiculated tuff by the coalescence of accretionary lapilli: Bulletin of Volcanology, v. 54, p. 429–434, doi: 10.1007/BF00312323.

Sáenz, R., Barquero, J., and Malavassi, E., 1982, Excursion al volcan Irazú, USA-CR joint seminar in volcanology: Boletín de Vulcanología, v. 14, p. 101–116.

Siebert, L., and Simkin, T., 2002, Volcanoes of the world: An illustrated catalog of Holocene volcanoes and their eruptions: Smithsonian Institution, Global Volcanism Program Digital Information Series, GVP-3, http://www.volcano.si.edu/gvp/world/.

Simkin, T., and Siebert, L., 1994, Volcanoes of the world; a regional directory, gazetteer, and chronology of volcanism during the last 10,000 years, 2nd edition: Tucson, Geoscience Press, 349 p.

Stine, C.M., and Banks, N.G., 1991, Costa Rica volcano profile: U.S. Geological Survey Open-File Report 91-591, 67 p.

Stuiver, M., Reimer, P.J., and Reimer, R.W., 2005, CALIB 5.0: http://radiocarbon.pa.qub.ac.uk/calib/

Walker, G.P.L., 1981, Characteristics of two phreatoplinian ashes, and their water-flushed origin: Journal of Volcanology and Geothermal Research, v. 9, p. 395–407, doi: 10.1016/0377-0273(81)90046-9.

MANUSCRIPT ACCEPTED BY THE SOCIETY 19 MARCH 2006

Geological Society of America
Special Paper 412
2006

# The eruptive history of Turrialba volcano, Costa Rica, and potential hazards from future eruptions

**Mark Reagan**
*Department of Geoscience, University of Iowa, Iowa City, Iowa 52242, USA*

**Eliecer Duarte**
*Observatorio Vulcanológico y Sismológico de Costa Rica (OVSICORI), Universidad Nacional, Apdo. 2346-3000, Heredia, Costa Rica*

**Gerardo J. Soto**
*Apdo. 360-2350, San Francisco 2 Ríos, Costa Rica*

**Erick Fernández**
*Observatorio Vulcanológico y Sismológico de Costa Rica (OVSICORI), Universidad Nacional, Apdo. 2346-3000, Heredia, Costa Rica*

## ABSTRACT

Turrialba volcano's high summit elevation and steep slopes, its position upwind of the Central Valley, and its record of explosive eruptions all suggest that it poses a significant threat to Costa Rican population and economy. To better understand the nature and significance of this threat, the geology, stratigraphy, and recent eruptive history of Turrialba were investigated. Outcrops of lava and pyroclastic units from at least 20 eruptions of basalt to dacite are recorded in Turrialba's summit area. The majority of these eruptions preceded a major erosional period that may have involved glaciation and that produced a prominent northeast-facing valley at Turrialba. This period also was apparently marked by a dearth of volcanism. The post-erosional period began with eruptions of massive andesite to dacite lava flows ca. 9300 yr B.P. Five of the six most recent eruptions, including the eruption of 1864–1866 A.D., were small volume (<0.03 km$^3$) phreatic and phreatomagmatic explosive eruptions involving basalt and basaltic andesite. The exception was a Plinian eruption of silicic andesite at ca. 1970 yr B.P. with a volume of ~0.2 km$^3$. Turrialba's next eruption will likely be similar to the recent eruptions of basaltic to basaltic andesitic composition, although a larger volume and more destructive eruption of silicic andesite to dacite also is possible.

**Keywords:** Costa Rica, volcanism, tephrostratigraphy, volcanic risk, glacial erosion, Turrialba.

# INTRODUCTION

Turrialba, the southeasternmost young volcano in Costa Rica (Fig. 1), is Central America's second tallest volcano after nearby Irazú. These volcanoes are ~10 km apart and their combined volumes comprise the largest stratovolcano complex in Central America (Carr et al., 1990). Turrialba is upwind of the Central Valley of Costa Rica, which includes San José and most of Costa Rica's other large population centers, as well as its main international airport. Therefore, an eruption of Turrialba could pose a severe hazard to the economy of Costa Rica. Moreover, the frequent rainfall, dense radial hydrographic network, relatively sparse vegetation coverage on its upper flanks, and limited official measures for local emergency response suggest that an eruption of Turrialba could pose a particularly serious threat to its local population (Duarte, 1990).

Turrialba last erupted between 1864 and 1866, producing ash fallout deposits that extended into the Central Valley and a few pyroclastic surges that affected the summit of Turrialba. Lahars associated with this eruption flowed down river drainages, extending into towns surrounding the volcano. More recently, periodic seismic swarms and enhanced fumarolic activity have been observed at Turrialba since 1998 (Barboza et al., 2000; Fernández et al., 2002; Barboza et al., 2003; Mora et al., 2004). With this record of historical eruptions and recent activity, there is a strong likelihood that Turrialba will erupt explosively again.

This paper discusses the geology and tephra stratigraphy of the Turrialba summit region. It covers all of the geological units exposed in and near Turrialba's summit crater complex, but emphasizes units erupted in the past ~3400 yr because of their better representation in the rock record. The reconstructed history is based on geological and stratigraphic observations, geochemical correlations between lavas and tephras (see Appendix Table A1), and radiocarbon ages (Table 1). The paper concludes with an assessment of the hazards posed by Turrialba volcano to its surrounding area.

# TURRIALBA'S VOLCANIC HISTORY

The Turrialba massif overlies the complexly deformed Tertiary sediments of the Limón Basin, which are capped by andesite lavas dated at 2.15 ± 0.30 Ma (Tournon 1984; Soto, 1988). The edifice that was built during the current phase of volcanism at Turrialba overlies deeply eroded older volcanoes that in combination represent ~1 m.y. of arc-related volcanic activity (Bellon and Tournon, 1978; Gans et al., 2003). The currently active edifice has an elliptical crater that faces to the northeast and contains three interior craters labeled w (west), c (central), and e (east) based on their geographic position, as well as three exterior peaks: Cerro San Carlos to the north, Cerro San Enrique to the east, and Cerro San Juan to the southwest (Fig. 2). Normal faults cut through the summit region from near Cerro San Enrique to the southwest for several kilometers. Two pyroclastic cones named Tiendilla and El Armado, which are generally aligned with these northeast structural trends, lie 2–3 km southwest of Cerro San Juan and Cerro San Enrique (Soto, 1988).

At least 20 eruptions are recorded in the crater region of Turrialba, some of which are associated with major lava flows exposed along Turrialba's northern, western, and southern flanks (see Soto, 1988). The volcanic units associated with these eruptions are separated into two groups based on their stratigraphic relationship to a significant erosional period that may have been associated with glaciation (see below). The minimum age for the oldest volcanism recorded in the summit region is 9300 yr B.P., based on the age of the oldest post-erosional lavas. The maximum age is poorly constrained, but may be ca. 170,000 yr B.P. based on $^{40}$Ar-$^{39}$Ar dating of Central Valley volcanics (Gans et al., 2003). The oldest pre-erosional stratigraphic units cropping out in the crater region are medium- to high-K calcalkaline basalts, andesites, and dacites (see Table A1) that were vented by at least nine eruptions (units 15 and 14; Figs. 3 and 4). Lava flows and breccias of basaltic andesite (unit 13, 55 wt% $SiO_2$) and dacite (unit 12, 64 wt% $SiO_2$) that crop out along the northwest wall of crater c (Fig. 3) and the overlying andesitic 55–57 wt% $SiO_2$ pyroclastic breccias of unit 11 progressively overlie the lowermost units. These breccias, which also crop out in crater w and the southeast wall of crater c, likely represent several eruptions. Lake bed deposits consisting of finely bedded white to light gray clay and altered talus are exposed in the wall separating craters w and c, postdate unit 14 (Fig. 5), and predate the excavation and collapse that resulted in the modern crater complex. Medium-K calcalkaline basaltic andesite lavas (54–55 wt% $SiO_2$; see Table A1) that dip inward toward the center of crater w are designated as unit 10 (Fig. 6).

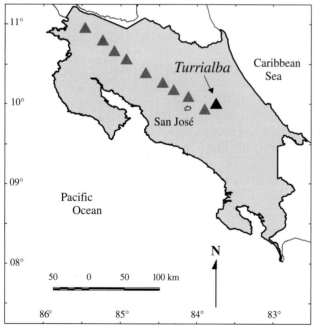

Figure 1. Location map for Turrialba volcano in Costa Rica.

TABLE 1. RADIOMETRIC AGES FOR SAMPLES FROM TURRIALBA VOLCANO

| Sample | Unit | Material | Radiocarbon age (yr B.P.) | Calibrated calendar year ages[§] |
|---|---|---|---|---|
| 89-3* | 3 | 10 cm diameter peat log in phreatic explosion debris | 1415 ± 75 | 644 A.D. (597–668 A.D.) |
| T-20-m-4 | 4 | <1 cm charcoal fragments in pyroclastic flow (pf) | 1860 ± 100 | 131 A.D. (31–318 A.D.) |
| T-100[†] | 4 | 3 cm diameter charcoal branch in fallout deposit | 1975 ± 45 | 27–49 A.D. (38 B.C.-76 A.D.) |
| T-109-7[†] | 4 | <1cm charcoal fragments in overlying pf from Irazú | 2010 ± 60 | 36 B.C.–1 A.D. (87 B.C.-63 A.D.) |
| T-27-b-10 | 4 | Outer 4 cm of a 15–20-cm-thick peat log in fallout deposit | 2330 ± 90 | 397 B.C. (479–260 B.C.) |
| 5-2-89* | 5 | Small charcoal fragments in base of overlying soil | 2495 ± 135 | 760–560 B.C. (802–401 B.C.) |
| T-26-b-10 | 5 | Small charcoal fragments in base of overlying soil | 2590 ± 180 | 796 B.C. (905–409 B.C.) |
| 1-6-89* | 5 | Small charcoal fragments in base of overlying soil | 2705 ± 85 | 831 B.C. (966–800 B.C.) |
| 11-9-89* | 6 | Small charcoal fragments in underlying soil | 2995 ± 215 | 1258–1218 B.C. (1493–917 B.C.) |
| 4-10-89* | 6 | Small charcoal fragments in underlying soil | 3115 ± 140 | 1406 B.C. (1520–1133 B.C.) |
| T-80 | 8b | 1–3 cm charcoalized wood fragments in pf | 8250 ± 300 | 7314–7203 B.C. (7514–6826 B.C.) |

*Analyzed at Krueger Enterprises.
[†]Analyzed at USGS Radioanalytical Laboratory; all others at Teledyne Isotopes.
[§]Best fit calendar ages using the correction curves from Stuiver et al. (1998). Ages in parentheses represent the maximum and minimum ages defined by 1σ total error.

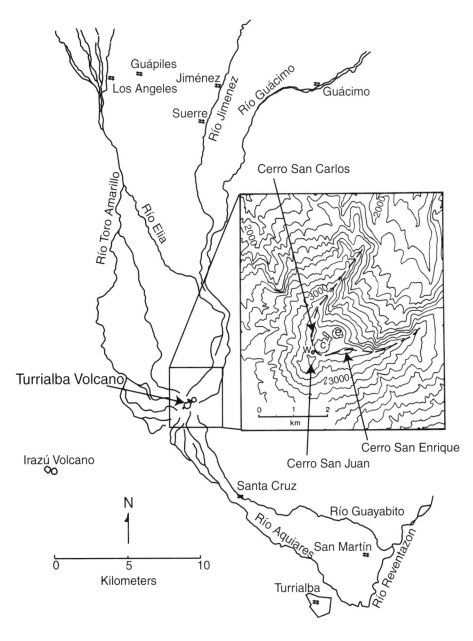

Figure 2. Regional sketch map of geographic features near Turrialba volcano, including cities (indicated by number signs) and major drainages. The inset is a topographic map of Turrialba's summit region. Contours are in 100 m intervals. The rim of the erosional valley is outlined with a dashed line. The three craters of Turrialba's summit are marked (w) for the west crater, (c) for the central crater, and (e) for the eastern crater east craters. Also labeled are the three summit peaks of Turrialba.

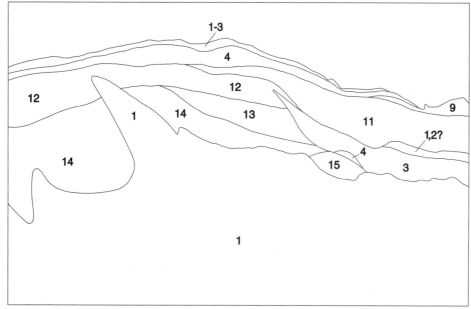

Figure 3. Photograph and sketch of the northwest wall of crater c. Perspective is from the southern rim of crater c. Unit numbers are labeled on the sketch maps.

Medium-K silicic andesite (57–59 wt% $SiO_2$) lavas of unit 9 make up the majority of the crater walls beneath Cerro San Enrique, as well as the ridge-crest to the northeast of Cerro San Juan (Fig. 4). Another outcrop is found ~300 m south-southwest of crater w. The presence of outcrops on three sides of Turrialba's erosional valley indicates that this eruption was one of Turrialba's most voluminous and that it built a tall central summit cone. The minimum age for this unit is the 9300 yr B.P. age assigned to unit 8. The maximum age for this lava is ca. 50,000 yr B.P. based on a K-Ar age that is indistinguishable from zero (J.B. Gill, 1986, personal commun.).

Much of the northeast side of Turrialba (~1 km³) was removed by erosion between the eruptions of units 9 and 8 (i.e., between 50,000 and 9300 yr B.P.). The embayment formed by this erosion resembles a debris avalanche crater (Fig. 2). However, a major debris avalanche deposit has not been identified on the laharic plane to the north of Turrialba volcano, suggesting that the embayment was produced by a different mechanism and/or in piecemeal fashion by a number of mass wasting events.

The time constraints on the erosional period coincide with the Chirripó and Talamanca glacial episodes of Costa Rica's Cordillera de Talamanca, 50–80 km south to southeast (Orvis and Horn, 2000; Lachniet and Seltzer, 2002). Extrapolation of Turrialba's present-day slopes toward a central peak indicates that its former elevation could have been >3500 m before erosion. This height may have been sufficient to allow ice accumulation

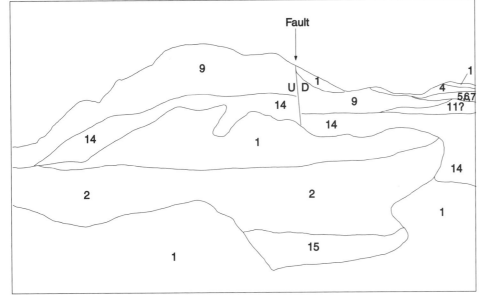

Figure 4. Photograph and sketch of the southeastern wall of crater c. Perspective is from the west rim of crater c. Unit numbers are labeled on the sketch maps. U and D indicate the relative movement on the crater wall fault.

during the maximum Talamanca (Lachniet and Seltzer, 2002) or Chirripó IV (Orvis and Horn, 2000) glacial events, which had equilibrium line altitudes as low as 3300 m in the Cordillera de Talamanca. Therefore, the horseshoe-shaped morphology of the crater rim may have begun as a cirque that was carved by ice into the northeast-facing slope of a peak centered near the location of crater c before 12,000 yr B.P. (see Orvis and Horn, 2000). Following this glacial erosion, subsequent slope failures, including one significant enough to leave a hummocky deposit within the Río Elia–Río Guácimo valley ~3 km northeast of crater e (Soto, 1988), and stream erosion widened and deepened this valley.

The first eruptions after the erosional period produced three large lava flows of high-K silicic andesite and dacite. These flows have been grouped into unit 8 (Fig. 7). The oldest of these lavas is an olivine-augite-orthopyroxene silicic andesite (subunit 8a) that is exposed along the Río Guácimo and a dacite (subunit 8b) that flowed through the mouth of the Río Elia–Río Guácimo valley and forms a 200-m-ridge 12 km to the north of Turrialba's crater complex. A dacitic pyroclastic flow overlying the outcrop of unit 9 southwest of crater w may be the stratigraphic equivalent of subunit 8b. If so, then subunit 8b erupted ca. 9300 yr B.P. based on a radiocarbon age of 8250 ± 300 yr (Table 1) corrected to calendar years using the method of Stuiver et al. (1998). Massive lava flows of 57–58 wt% $SiO_2$ hornblende-orthopyroxene-augite-olivine andesite overlie these two lavas and make up the northeast and east walls of crater e (subunit 8c, Fig. 5). These

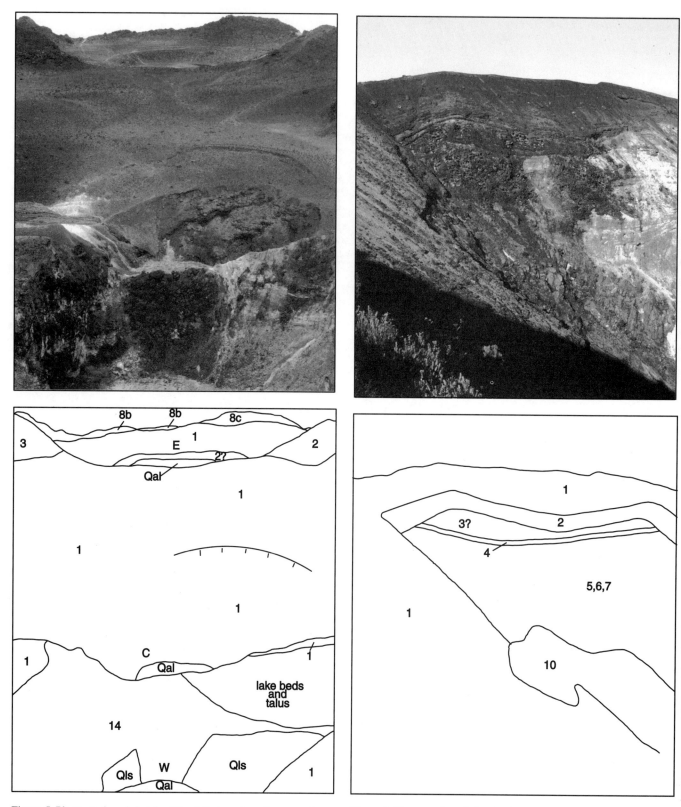

Figure 5. Photograph and sketch of Turrialba's craters. C—central crater; E—east crater; W—west crater. Perspective is from the southwest rim of crater W. Unit numbers are labeled on sketch maps. Qal—Quaternary alluvium; Qls—Quaternary landslide.

Figure 6. Photograph and sketch of the south side of crater w. Perspective from its east rim. Unit numbers are labeled on the sketch maps.

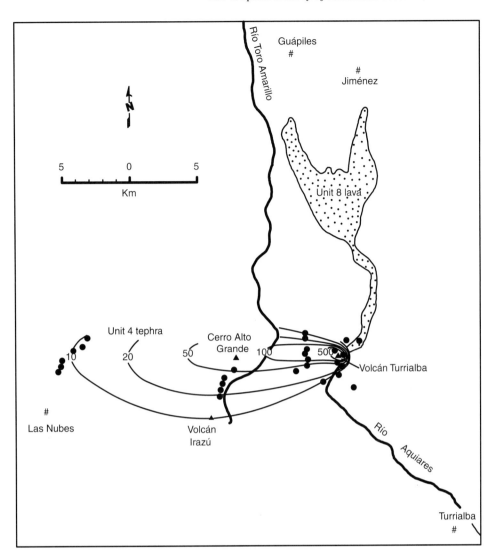

Figure 7. Distribution of unit 4 fallout tephras (thickness in cm) and unit 7 lavas (stippled). Small filled circles show tephra sites.

flows erupted from a vent that was located in what is now crater e and flowed down the northeast-facing valley and eventually traveled up to 20 km from the vent (Fig. 7), where flow-fronts are 60–80 m tall. The total volume of unit 8 is at least 4.5 km$^3$, based on the assumption that its thickness averages a minimum of 150 m. This is one of the largest lava flow complexes recorded in Central America, and may have been the result of magma ponding and differentiation during the erosional era followed by rapid extrusion of the stored magma once the glacier and summit area was removed.

A fundamental change in Turrialba's eruptive style occurred between the eruptions of units 8 and 7. Whereas the eruptions of unit 8 produced relatively large-volume andesite to dacite lava flows, succeeding eruptions were more predominantly explosive and, with the exception of unit 4, were generally less silicic. Although tephra deposits from these eruptions are relatively widespread, the associated lavas are confined to the interior of Turrialba's crater complex.

Unit 7 consists of basaltic andesitic (~56 wt% $SiO_2$, see Table A1) scoria in pyroclastic fall and surge beds that underlie unit 6 on the south side of crater c (Fig. 4). This scoria consists of red to dark gray augite-olivine basalt and contains <1 cm blebs of porphyritic dacite pumice. This dacite may have been the last remnant of the silicic unit 8 magmas flushed from the system by eruption of the basaltic andesite. Charcoal from an overlying soil horizon exposed in a road cut below the radio tower on Turrialba's southeast side was dated at 2995 ± 215 radiocarbon years (Table 1), which corrects to a calendar age of ca. 3200 yr B.P. Thus, the minimum age of unit 7 is 3200 yr B.P. Ash found within a peat deposit 35 km northeast from Turrialba volcano at El Silencio has a radiocarbon age of 3370 yr B.P. (Obando and Soto, 1993) and may be the equivalent of unit 7.

On the southeast rim of crater c, unit 6 consists of interbedded well to moderately sorted and laminated dark gray juvenile ash and poorly sorted ash, lapilli, and block deposits with dense juvenile clasts up to 1 m in length. On Turrialba's flanks (Fig. 8), the base of the deposit consists of yellow clay that originally was phreatic explosion debris. Overlying this are weakly to highly weathered gray to dark gray fine to coarse juvenile ash deposits. Where undisturbed, these deposits are usually finely bedded and rarely cross-bedded. Juvenile clasts are olivine-augite basalt or basaltic andesite, with minor augite-orthopyroxene silicic andesite to dacite and hybrids of these two compositions. Well-developed orange-brown soil horizons underlie and overlie this deposit. The maximum age of unit 6 is 3400 yr B.P. based on the $3115 \pm 140$ yr B.P. radiocarbon date of charcoal fragments from the underlying soil horizon (Table 1). The minimum age of unit 6 can be estimated by the age of the soil atop unit 5.

Unit 5 comprises thin, poorly-sorted ash and lapilli deposits found on Turrialba's upper flanks (Fig. 9). This deposit is highly weathered, consisting of dense angular andesitic lapilli in a light gray to tan clay matrix, and probably was deposited by pyroclastic flows or surges. No well-sorted deposits clearly related to ash fallout were observed, although such deposits may have been obscured by weathering. Charcoalized wood fragments <2 cm in length from the base of a well-developed soil that overlies unit 5 were dated between $2705 \pm 85$ and $2495 \pm 135$ radiocarbon years (Table 1). The charcoal's location at the base of the soil suggests that it was produced in the eruption or from a fire relatively shortly after eruption. Thus, the best estimate for the calendar year age of unit 5 is a calibrated age based on the oldest of these radiocarbon ages, ca. 2800 yr B.P., and the maximum repose time between unit 5 and 6 was ~600 yr.

Unit 4 consists of air fall, surge, and pyroclastic flow deposits of high-K, 58–59 wt% $SiO_2$ andesite that erupted from a vent in the vicinity of crater c. This andesite can be distinguished from those of other units by its abundant phenocrysts (30%) of plagioclase, dark green augite, orthopyroxene, and rare olivine in frothy pumice, and its relatively high $SiO_2$ concentrations. It is useful as a marker horizon for central Costa Rica (see Clark et al., this volume). The deposits are up to 40 m thick on Cerro San Carlos, 1.3 m thick 2.5 km from the summit, and 0.1 m thick ~20 km west of Turrialba (Fig. 7). Pyroclastic flow and surge deposits from this same eruption overlie the fallout deposit on the south and west flanks of Turrialba. Radiocarbon ages of the fallout and pyroclastic-flow deposits range from $2330 \pm 90$ yr B.P. to $1860 \pm 100$ yr B.P. The most precise date ($1975 \pm 45$ yr B.P.) is from a 3-cm-thick charcoalized branch found in fallout pumice near the summit of Turrialba. The most likely calendar age of this branch and the fallout deposit is ca. 1970 yr B.P. (see Table 1). Melson et al. (1985) report a similar radiocarbon age ($1970 \pm 90$) for a phreatic explosion layer that is overlain by a pumiceous fall deposit and underlain by a paleosol bearing 2000 yr B.P. ceramic fragments. Therefore, the repose time between the eruptions of units 4 and 5 was ~800 yr.

On the east rim of crater c, unit 4 comprises interlayered buff, gray, and dark-gray beds consisting almost entirely of angular clasts of juvenile siliceous andesite up to 40 cm in length (Fig. 10). The buff layers are poorly sorted to well-sorted, consisting of lapilli and blocks of juvenile pumice, with lesser amounts of dense, dark gray juvenile clasts (5%–15%) and altered volcanic lithics (<5%). Large pumice blocks grade outward from a dark-brown core through a 1–5 cm pink zone to a 1–3 cm thick buff rim (Plate 1). This pumice is interbedded with gray, poorly sorted layers of dense to coarsely vesicular juvenile clasts and minor altered lithic fragments.

Downwind to the west, along the crest of the fallout dispersal pattern, and within 1–10 km of the summit, unit 4 is normally graded overall and consists of 4 layers, labeled A–D from bottom to top (Fig. 10; Plate 2). Layer A is well-sorted, normally graded, and is mostly buff pumice lapilli with subordinate amounts of dark gray, dense juvenile clasts and accidental ejecta. Within 2 km of the summit, thinly bedded surge deposits of gray ash sometimes underlie layer A. The base of layer B usually consists of <1 cm of fine pink ash. Most of layer B is moderately sorted, weakly bedded, medium gray, and consists of buff pumice lapilli and gray ash. Layer C is like layer A but is finer-grained. Layer D is a moderately sorted gray ash that contains sparse pumice lapilli. Farther downwind, near the town of Las Nubes, unit 4 consists of one well-sorted bed consisting of lapilli and coarse ash with a distinctive salt-and-pepper appearance. On Turrialba's east, west, and south flanks, and within 3 km of the vent, the fallout tephras of unit 4 are usually overlain by interbedded pyroclastic flow and surge deposits from the same eruption (Fig. 11).

The minimum volume of the fallout deposit was calculated to be 0.2 km$^3$ using the method of Pyle (1989) and the contours shown in Figure 7. Adding the over-thickened proximal fallout deposits and the relatively minor volumes of pyroclastic flows and surges associated with this eruption would bring the total volume to no more than about double this amount.

The pyroclastic deposits of unit 3 overlie a well-developed soil horizon atop unit 4 (Fig. 12). This soil contains charcoal dated at $1630 \pm 160$ radiocarbon years by William Melson (1990, personal commun.). On the west rim of crater c, unit 3 has a buff to tan basal portion, which is massive and poorly sorted at its base and moderately sorted and well-bedded on top. Overlying this is a dark gray upper portion that consists of interbedded near-vent, surge, and fallout deposits, with the proportion of fallout deposits increasing down-section. Greater than 20 m of gray near-vent pyroclastic deposits make up unit 3 in northern crater c, at the head of the Río Elia (Fig. 12), suggesting that pyroclastic surges of unit 3 may have flowed down the Río Elia Valley. On the western rim of crater c, the lowermost juvenile clasts consist of white bands of nearly pure dacite intermingled with black bands of a 60% basaltic andesite–40% silicic andesite mixture. Juvenile clasts in overlying deposits are mostly basaltic andesite.

On Turrialba's flanks, unit 3 has a yellow to tan base entirely consisting of highly altered volcanic clasts (Plate 3). The base generally makes up one-third to two-thirds of the total deposit and usually is massive and poorly sorted, although moderately well-sorted and laminated strata with rare cross-beds are occasionally present. The overlying beds are dark gray and massive to planar with rare cross-beds and consist of moderately to well-sorted mostly juvenile basaltic andesite ash. An orange- to

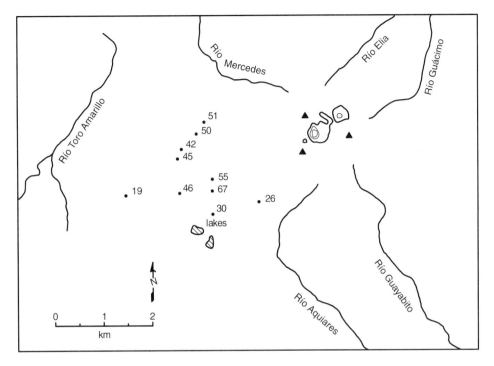

Figure 8. Thicknesses of unit 6 pyroclastic deposits in centimeters.

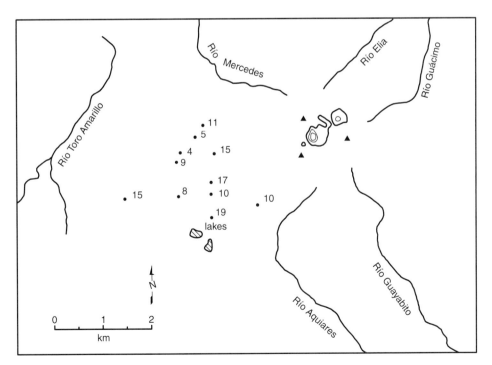

Figure 9. Thicknesses of unit 5 pyroclastic deposits in centimeters.

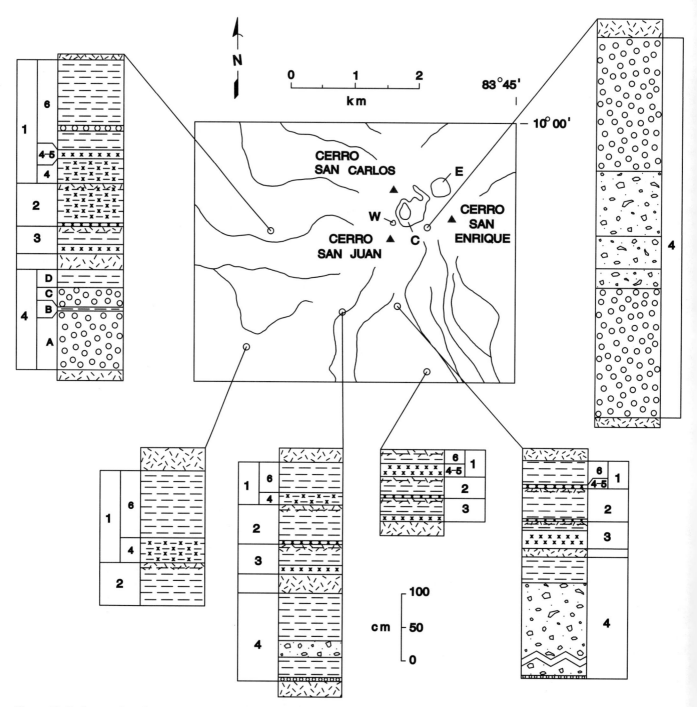

Figure 10. Tephra stratigraphy at representative sites. Unit numbers are listed next to stratigraphic columns. Fallout tephras with juvenile pyroclasts are shown by fields of open circles. Well-bedded pyroclastic surge and fallout deposits with juvenile clasts are shown by fields of horizontal dashed lines. Similar deposits, but with mostly nonjuvenile clasts are shown by fields of dashed lines and x's; x's alone represent poorly sorted deposits of nonjuvenile debris. Fields with randomly oriented angular forms are pyroclastic flow deposits. Randomly oriented dashes represent soil horizons. Subdivisions within unit 1 represent the estimated dates of eruption: 4 = 1864; 4–5 = 1864–1865; 6 = 1866. Subdivisions A–D for unit 4 are described in text. C—central crater; E—east crater; W—west crater.

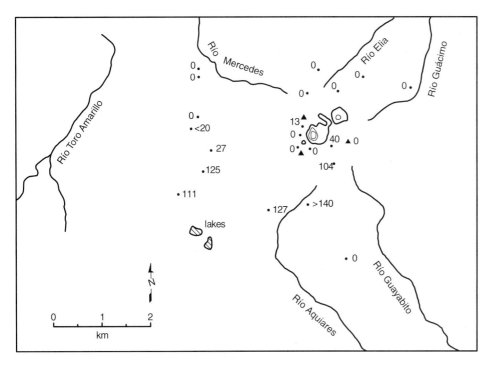

Figure 11. Thicknesses of unit 4 pyroclastic flow and surge deposits in centimeters.

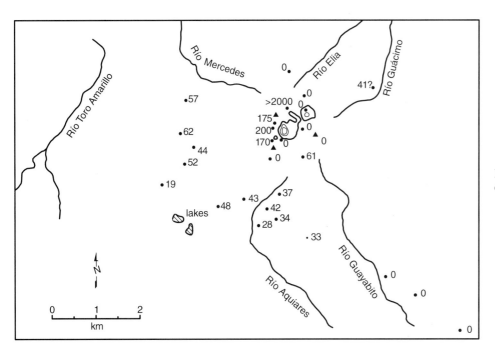

Figure 12. Thicknesses of unit 3 pyroclastic deposits in centimeters.

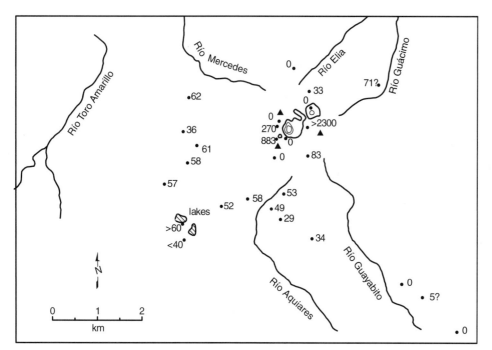

Figure 13. Thicknesses of unit 2 pyroclastic deposits in centimeters.

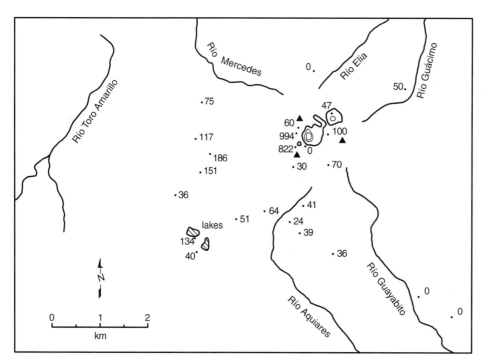

Figure 14. Thicknesses of unit 1 pyroclastic deposits in centimeters.

red-brown soil that is typically thinner than the underlying soil caps unit 3.

A 10 cm thick peat log found within the basal nonjuvenile surge beds of unit 3 yielded a radiocarbon age of 1415 ± 75 (Table 1), which corrects to a calendar age of ca. 1360 yr B.P. (Stuiver et al., 1998). Thus, the repose time between the eruptions of units 3 and 4 was ~600 yr.

Within Turrialba craters, pyroclastic deposits of unit 2 overlie unit 3 in the southeast wall of crater e and the south wall of crater w and unit 15 in the east inner wall of crater c (Figs. 5, 6, and 13). The deposits are generally well-bedded, well-sorted to poorly sorted, and buff to gray. Nonjuvenile, variably altered volcanic clasts dominate most beds. Nevertheless, juvenile basaltic andesite clasts are found in exposures around craters c and e. Where bedding is well-preserved, unit 2 is moderately to well sorted and well bedded, and has rare cross beds. The upper 5–20 cm usually has a pervasive pale orange tint and some more deeply orange horizons and veins (Plate 3). This likely represents a weakly developed soil and suggests that the repose time between the eruptions of units 1 and 2 is significantly less than the intervals between the eruptions of units 2–6.

Unit 1, which erupted in 1864–1866 A.D., mantles most of Turrialba's summit region (Fig. 14). Juvenile lavas and pyroclasts of unit 1 are 52.5–53 wt% $SiO_2$ basalt (see Table A1). On the floor of crater c, unit 1 comprises agglutinates and rootless lava flows overlain by surficial fallout and near-vent flowage deposits. On the crater rims, it consists of alternating beds of well-sorted to poorly sorted pyroclastic deposits (Plate 4). Its thickness is highly irregular, reflecting syn- and post-eruption erosion, as well as variable depositional thickness. Its base often consists of poorly sorted deposits of ash- to block-sized clasts of buff to yellow altered volcanic rock. Overlying deposits are well bedded and poorly to well sorted, with the proportion of juvenile clasts increasing up-section. Some of these beds are indurated and vesicular, and some have low angle cross beds. About 10%–30% of the sections along Turrialba's crater rims consist of open-network well-sorted ash or lapilli deposits. Within 200–500 m of crater w, the uppermost layer is a fallout deposit of highly altered volcanic debris that erupted from crater w.

Outside of the crater, to the east of the Turrialba crater complex, the base of unit 1 consists of buff to yellow ash and lapilli of altered volcanic debris. Most of this deposit is poorly sorted, but some well-sorted ash beds are present. To the west and south, the base of unit 1 is a varicolored horizon consisting of interlayered yellowish and light to dark gray beds made up of more and less altered volcanic clasts. The top of this bed often has some discontinuous orange-stained horizons. This is overlain by a thin, poorly sorted yellowish deposit rich in nonjuvenile clasts. The upper of two-thirds to nine-tenths of unit 1 consists of plane parallel layers of moderately to well-sorted ash with rare cross beds (Plate 5). Some of these beds are indurated and vesicular. Cross bedding is particularly conspicuous in deposits associated with topographic lows (Plate 6). The ash in these beds is mostly juvenile basalt with 10%–40% nonjuvenile volcanic debris. One well-sorted fall bed composed of scoria lapilli is present in sections to the west and southwest of Turrialba's summit. Poorly sorted deposits <2 m thick that likely represent lahars from the 1864–1866 eruption are found in patches along the banks of the Río Aquiares near the town of Aquiares and along the Río Guacimo and Río Roca valleys 7 km north of Turrialba's summit.

Although the thicknesses of units 1, 2, 3, 5, and 6 are highly irregular, they appear to be similar to or less than the thicknesses measured at similar distances for basaltic tephras erupted in 1963–1965 from Irazú volcano. Thus, all of these units probably have volumes ≤0.03 km$^3$, which is the estimated volume for the 1963–1965 eruption (Clark et al., this volume).

## THE 1864–1866 ERUPTION

The following discussion is our interpretation of the 1864–1866 eruption sequence based mostly on the eyewitness accounts reported in González-Viquez (1910). Following a period of enhanced fumarolic activity beginning on or before 1723 A.D., Turrialba began erupting significantly on 17 August 1864 and continued erupting until at least February 1866.

On 26 February 1864, an expedition to Turrialba found a summit region with an active crater >100 m deep with more than 100 fumaroles in its walls and bottom (see Figs. 2 and 10). This may have been a precursor to today's crater w. The interior four walls of the crater were steep and black to yellow and white in color. The west crater wall and crater floor were intensely altered, and the west crater wall was hot to the touch. No vegetation was present within the crater area, and much of the vegetation to the west and northwest was dead because of acid rain. Areas near the fumaroles were covered with sulfur. The vapors from all of the fumaroles combined and rose ~200 m above the summit. The accounts suggest that the summit had two "dead" craters, one to the east and one to the northeast. These probably correlate to the area occupied by craters c and e today.

The explosive eruption of Turrialba began on 17 August 1864. Ash fell on Costa Rica's Central Valley throughout 16–21 September, and ash was detected as far west as Atenas and Grecia. Enough ash fell in San José that it caused alarm and could be scooped up for a chemical analysis, but apparently not enough to cause major hardship. The thickness of ash in San José, therefore, was probably less than ~2 mm.

A second expedition trekked past a lake with more than a third of a meter of ash nearby and reached Turrialba's summit on 30 September 1864. Although the location of the lake is not well described, the only lake close enough to Turrialba's summit to have received this much ash is a transient lake 3 km southwest of crater c. A stratigraphic column of unit 1 from the lake area (Fig. 10) shows a 36-cm-thick varicolored basal unit bearing cross-beds that may correspond to the deposit seen by the expedition.

The summit region and the fumarolic activity had changed substantially since the first expedition. Where there were 100 fumaroles before, there was now one, which was a belching green

Plate 1. Clasts in unit 4 near radio tower at Turrialba's summit.

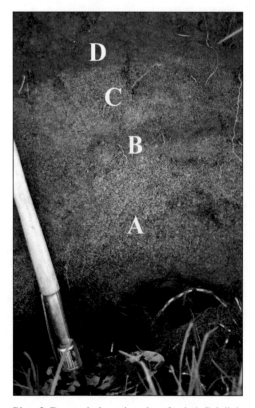

Plate 2. Downwind stratigraphy of unit 4. Subdivisions A–D for unit 4 are described in text.

Plate 3. Units 2 and 3 in roadcut 2 km south of Turrialba's crater complex.

Plate 4. Unit 1 near rim of crater w.

Plate 5. Unit 1 tephra in roadcut 2 km west of Turrialba's crater complex.

Plate 6. Cross-bedding in unit 1.

to black plume about twice as tall as the one seen during the first trip. Blue flames, possibly from the burning of sulfur, were associated with this plume. The roar of this vent was very loud, and the mountain was continually shaking. Part of the summit was missing, and it was suggested that it fell into the new crater and was thrown back out as ash. A lot of this ash apparently went to the north and northwest. Large blocks lay on the surface in the entire summit region. Vegetation was destroyed between Turrialba volcano and near where the Río Tortuguero (now named the Río Toro Amarillo) starts, and for many kilometers to the northwest and west of Turrialba volcano.

It appears that after the 16–20 September eruption, there was still only one active crater (crater w), but it had been significantly widened. This excavated material was deposited to the north, south, and west of the volcano by ash fallout and pyroclastic surges. It is likely that little magma was involved in this part of the eruption and that the lowermost varicolored beds of unit 1 found west and southwest of Turrialba are associated with this eruption.

A third expedition visited Turrialba on 9 March 1865 after more ash had fallen on San José during late 1864 to 8 March 1865 (von Seebach, 1865). This expedition found a widened westernmost crater with a funnel shape like today's crater w. The additional material excavated from crater w was probably deposited as the yellow to tan beds overlying the varicolored beds in unit 1. Although this expedition observed some bombs near the craters, it is likely that juvenile pyroclasts still did not make up a significant proportion of the ejecta.

Eruptions in January and February 1866 were apparently the largest of the 1864–1866 eruptive period. Ash fell in the Central Valley for four days in January and three days in February, and fell as far as Puntarenas (125 km away) ca. 1 February. The juvenile-clast–rich upper portions of unit 1 probably erupted during this period, including the thick scoriaceous deposits on the south side of crater w and the rootless lava flows and agglutinates on the floor of crater c. Most of the material probably erupted from the interior crater of crater c. Two concentric tephra rings that run along the northeastern floor of crater c show that the vent crater was larger and located farther east during early stages of the eruption. The distribution of tephra deposits suggests that much of the area within 5 km of the volcano was buried in thick ash by the eruption. The eruption concluded with a small phreatic explosion from crater w.

In summary, the most recent eruption of Turrialba volcano began on 17 August 1864 and lasted into February 1866. The eruption was entirely explosive. Phreatic explosions from crater w characterized the first year of the eruption. During the final two months of eruption, juvenile basalt was expelled in phreatomagmatic eruptions from crater c, followed by a last phreatic explosion from crater w. The phreatic and phreatomagmatic explosions produced eruption columns that rained ash and lapilli in the Turrialba summit region and downwind to the west. Pyroclastic surges may have occurred in the summit region during the eruption.

## TURRIALBA'S PYROCLASTIC DEPOSITS AND ASSOCIATED ERUPTIONS

Internal structures and distributions of many of the deposits from Turrialba volcano indicate that they were deposited by fallout of ash and lapilli. For example, the salt and pepper lapilli beds of unit 4 and the well-sorted scoria beds of unit 1 are good examples of fallout beds. In contrast, the block, lapilli, and ash deposits of unit 4 clearly were deposited by pyroclastic flows and surges. However, the deposits associated with many of the explosive basaltic and andesitic tephras of Turrialba (e.g., the cross-bedded and well-sorted deposits in unit 1) have characteristics that are intermediate between those expected for pyroclastic falls and surges.

Along Turrialba's crater rims, tephra deposits are well bedded with typical thicknesses of 10–50 cm. The beds range from continuous to discontinuous, and sorting is variable. Well-sorted beds composed completely of ash or lapilli are common, as are poorly sorted block, lapilli, and ash beds. Dropstones are commonly present. At distances >1 km from its crater complex, Turrialba's basaltic and basaltic andesitic deposits are bedded on a millimeter to 10 cm scale. They are ubiquitously fine-grained, consisting of moderately sorted to well sorted ash, which suggests that they were ejected by phreatomagmatic explosions (e.g., Fischer and Schmincke, 1984). The bedding is commonly laterally continuous on an outcrop scale, although variations in bedding thickness and low-angle dune- to steep ripple-scale cross bedding is sporadically present (Plate 6). Some well-sorted ash beds are friable, but many beds are indurated, and some of these consist of coarse ash fragments coated with fine ash. These beds are commonly vesicular, indicating the presence of condensed water during deposition. In contrast to typical fallout sequences, which typically are lobate and thicken toward the source and in the downwind (west to northwest) direction, these deposits have more uneven depositional patterns (Figs. 8–14).

The 1963–1965 eruption of basaltic andesite from Irazú volcano (Murata et al., 1966; Krushensky and Escalante, 1967; Alvarado, 1993; Clark et al., this volume) produced pyroclastic deposits that are similar to those of Turrialba and may be good analogues for the recent eruptions of Turrialba. The 1963–1965 eruption was characterized by periodic Strombolian to vulcanian explosions, whose intensities ranged up to a VEI (volcanic explosivity index; Newhall and Self, 1982) of 3 with intervening quiescent periods that varied over day-long to month-long stages (Sáenz et al., 1982; Alvarado, 1989; Simkin and Siebert, 2002; Clark et al., this volume). The eruption began with phreatic explosions that tossed dense blocks more than a kilometer and covered the western flanks of Irazú with ash. Over the succeeding two years, explosions produced fine-grained ash beds similar to those of the mafic eruptions from Turrialba. The frequent heavy rainfalls also generated destructive lahars and floods of ash-choked water. The descriptions of Murata et al. (1966) and Krushensky and Escalante (1967) indicate that most of the laminated ashes erupted in 1963–1965 were fallout deposits. They attributed the ripples they observed to aeolian processes, although pyroclastic surges produced similar deposits on north side of the volcano (see Hudnut, 1983). The high degree of tephra fragmentation reflects recycling of conduit wall, as well as a high degree of magma-water interaction during the eruption (Alvarado, 1993).

These observations suggest that the majority of the laminated deposits from explosive basaltic and basaltic andesitic eruptions at Turrialba were fallout tephras from vulcanian to Strombolian explosions with a VEI of 3 or less. Their somewhat uneven distributions can be attributed to syneruption erosion. Some of the cross bedding observed in units 1–3 may have resulted from reworking of unconsolidated ashes by wind, sheet-wash, and stream-flow. Nevertheless, the dropstone-bearing cross-bedded deposits found in Turrialba's crater complex were probably associated with pyroclastic surges. Thus, pyroclastic surges did occur during Turrialba's recent eruptions of basalt and basaltic andesite and constitute a risk during future eruptions.

## IMPLICATIONS FOR THE POTENTIAL HAZARDS OF TURRIALBA

Beginning in May 1996, seismic activity recorded by a single station installed at the summit of Turrialba in 1990 escalated from 0 to 91 earthquakes per month to a maximum of 2000 earthquakes per month in October 2000. This seismicity included long-period events. From 2000 to 2004, eight additional seismic pulses were documented by a four station local network that was installed in 2000 (Barboza et al., 2000, 2003). This period of enhanced seismicity was accompanied by fumarolic activity that increased in area, flux, and concentration of magmatic constituents in craters c and w. Burned vegetation, opening of ground cracks, small landslides, rock alteration, and soil heating around both craters also has been reported (Fernández et al., 2002). This recent activity and the relatively short repose period indicated by the weakly developed soil between units 1 and 2 suggest that Turrialba may erupt again within the next several years to several decades.

Five of the six most recent significant eruptions, including the last three, ejected ≤0.03 km³ of altered volcanic debris and juvenile basalt or andesite tephra in VEI ≤3 explosions from the summit crater complex. The most probable significant future eruption of juvenile magma, therefore, would probably occur from the summit craters and would be of similar volume and type. The presence of relatively young cinder cones on Turrialba's southwest flank shows that eruptions along the crater lineament, but away from the summit, are also possibility. Like past eruptions of Turrialba, the initial phase of renewed activity would likely involve phreatic explosions. These explosions would produce fallout deposits that could significantly blanket areas within 2–3 km of Turrialba's summit with ash and generate pyroclastic surges affecting the summit region. A few millimeters of ash could fall on the Central Valley during these types of eruptions.

Although it is possible that the eruption would end during the phreatic phase, it is likely that a phreatomagmatic vulcanian to Strombolian phase would follow. Area similar to that affected by the 1864–1866 eruption would be affected by such an eruption. Ash fall would be heavy in the summit area. Downwind to the west, in the area between Irazú and Cerro Alto Grande, <10 cm of ash would be expected. In the Central Valley, near San José and Heredia, the ash fall would probably be <1 cm. Pyroclastic surges would likely accompany an eruption and could affect areas within a few kilometers of the summit. Low-topography areas, such as river valleys, would have the highest risk of pyroclastic surge damage. Mudflows and floods could occur in many of the drainages leading off of the summit, including Río Toro Amarillo, Río Mercedes, Río Elia, Río Roca, Río Guácimo, Río Guayabo, Río Guayabito, and Río Aquiares. It is possible, though unlikely, that the eruption would conclude with lava effusion.

The young tephra deposits of the historically more active Irazú volcano are similar to those of Turrialba. At Irazú, only 10% of historic eruptions have left discernable deposits (Clark et al., this volume), suggesting that many small eruptions with VEI values of 2 or less may have occurred at Turrialba and were not recorded in its stratigraphy. Thus, minor phreatic and phreatomagmatic eruptions associated with shallow magma intrusions that would only affect the summit region of Turrialba are also possible.

Relatively large volumes of silicic andesite tephra (unit 4) and massive silicic andesite and dacite lava flows (unit 8) have erupted since 9300 yr B.P. Thus, destructive eruptions of silicic andesite and dacite are possible at Turrialba. However, the likelihood of such eruptions at Turrialba is lower than for an eruption of more mafic lava, as Turrialba's last two eruptions ejected juvenile basalts and basaltic andesites and the third oldest eruption involved only a minor amount of silicic andesite mixed with basaltic andesite.

A Plinian eruption of a silicic andesite, like that of unit 4, would be significantly more destructive than an eruption of basalt or basaltic andesite. Ash fall in the Central Valley could reach several centimeters. Pyroclastic flows or surges could extend >10 km from the summit. Potentially destructive lahars would likely be generated by such an eruption. Large volume lava flows, such as those of unit 8, are also a remote possibility with this type of eruption.

There is a risk of debris avalanching during future eruptions of Turrialba. Features that promote failure of a volcanic edifice and debris avalanching are (1) a steep slope, (2) massive lavas overlying weak pyroclastic substratum, (3) widespread hydrothermal alteration of the core of a volcano (e.g., Reid et al., 2001), (4) the migration of vents in a direction parallel to the axis of the avalanche caldera, and (5) water saturation of the edifice (Siebert, 1984; Siebert et al., 1987; McGuire, 1996). Turrialba has all of these features. Turrialba's northeast side has the greatest danger of failure (Fig. 2) because it has its steepest slope, it is aligned with Turrialba's vents and faults, and has young, dense lavas overlying an altered volcanic core. If such an avalanche occurred, it would flow down the Río Elia–Río Guácimo valley onto the northern plain. Debris avalanches usually travel ~2–20 times as far as the vertical drop (Ui, 1983; Siebert, 1984). The vertical drop at Turrialba would be ~2.5 km. Thus, the region containing Jiménez, Suerre, and Guápiles could be inundated by such an avalanche. Avalanching of the relatively steep southern, western, and northwestern portions of the edifice is also possible. Indeed, Alvarado et al. (2004) documented a 0.75–1.4 km³ debris avalanche down the Turrialba river valley ~17,000 yr ago that resulted from the collapse of the highland between Turrialba and Irazú volcanoes.

Volcanic blasts are often, though not necessarily, associated with debris avalanches (e.g., the 18 May 1980 eruption of Mount St. Helens; Mullineaux and Crandell, 1981), and such an event could occur at Turrialba. If it did, it would probably be in the same direction as the debris avalanche initially, but would not follow topography as closely and could spread out over ridges onto the northern plane for several kilometers.

## CONCLUSIONS

At least 20 eruptions of basaltic to dacitic lavas and tephras are recorded in the stratigraphy of Turrialba volcano's summit region. A period of enhanced erosion and apparently diminished volcanism divides this stratigraphy into pre-erosional and post-erosional units. This erosion-dominated period approximately coincided with the Chirripó and Talamanca glacial episodes of Costa Rica's Cordillera de Talamanca. This, the likelihood that Turrialba had a central peak that exceeded 3500 m in elevation before the erosional period and the morphology of its prominent northeast-facing valley suggest that Turrialba may have hosted a glacier during the last glacial maximum.

The first significant eruption of the post-erosional period was marked by the eruption of massive andesite to dacite lava flows at ca. 9300 yr B.P. These lavas flowed down the northeast-facing valley and out onto the laharic plane as much as 20 km from Turrialba's summit. Suppression of volcanic activity during glaciation followed by enhanced volcanism during interglacials has been noted elsewhere and probably results from a change in the stress distribution in the crust due to ice loading (e.g., Hall, 1982; Nakada and Yokose, 1992; McGuire et al., 1997; Glazner et al., 1999). At Turrialba, the removal of a cirque glacier and the upper volcanic edifice could have decreased the vertical component of the stress field between 1 and 5 MPa depending on the size, shape, and positioning of the central summit cone produced by the eruption of unit 9 and glacier. Rebound above the active vent system may have been sufficient to trigger eruption of magma that had ponded and differentiated beneath the tall edifice built by the last pre-glacial eruption.

Six explosive eruptions are recorded in the tephra stratigraphy at Turrialba over the past 3400 yr. Most were small volume (≤0.03 km³) phreatic and phreatomagmatic explosive eruptions involving mostly basalt and basaltic andesite. The exception was a Plinian eruption of ~0.2 km³ of silicic andesite at ca. 1970 yr B.P. Repose periods between significant eruptions over the past 3400 yr have principally been in the range of 800–600 yr. However, the weak soil horizon between the second oldest tephra and the tephra erupted in 1864–1866 A.D., as well as the recent seismicity, fumarolic, and deformational activity observed beneath Turrialba (Barboza, et al., 2000, 2003; Fernández et al., 2002; Mora et al., 2004) suggest that Turrialba may have entered a more active phase.

Then next eruption probably will be similar to the 1864–1866 and other recent basaltic to basaltic andesitic eruptions. Nevertheless, a larger volume and more destructive eruption of silicic andesite also is possible. Turrialba's steep slopes, particularly the northeast slope of the cone filling the erosional valley, could be susceptible to debris avalanching, and Turrialba should be monitored carefully for deformation during any future eruption.

## ACKNOWLEDGMENTS

Eduardo Malavassi is thanked for logistical assistance and for translation of González-Viquez (1910). Field assistance by Joe Hill helped delineate eruptive units 5 and 6. Debbie Trimble is thanked for the radiocarbon analyses from the U.S. Geological Survey. Bill Melson is thanked for sharing his unpublished radiocarbon ages. Funding from UNOCAL and the International Program at the University of Iowa for Reagan is gratefully acknowledged. Soto thanks the Universidad de Costa Rica for funding his Turrialba research. Jim Gill is thanked for comments on an early version of this study. This paper benefited greatly from reviews by John Ewert and C. Dan Miller

## REFERENCES CITED

Alvarado, G.E., 1989, Los volcanes de Costa Rica: San José, Costa Rica, Editorial Universidad Estatal a Distancia (EUNED), 175 p.
Alvarado, G.E., 1993, Volcanology and Petrology of Irazú Volcano, Costa Rica [PhD thesis]: Germany, University if Kiel, XXXV + 261 p.
Alvarado, G.E., Vega, E., Chaves, J., and Vásquez, M., 2004, Los grandes deslizamientos (volcánicos y no volcánicos) de tipo debris avalanche en Costa Rica, in Soto, G.J., and Alvarado, G.E., eds., La Vulcanología y su entorno geoambiental: Special volume, Revista Geológica de América Central, v. 30, p. 83–99.
Barboza, V., Fernández, E., Martínez, M., Duarte, E., and Van der Laat, R., E., Marino, T., Hernández, E., Valdés, J., Sáenz, R., and Malavassi, E., 2000, Volcán Turrialba: Sismicidad, Geoquímica, Deformación y nuevas fumarolas indican incrementos en la actividad (abstract), in Los retos y propuestas de la investigación en el III milenio (CONINVES): Memoria, Editorial UNED, p. 78.
Barboza, V., Fernández, E., Duarte, E., Sáenz, W., Martínez, M., Moreno, N., Marino, T., van der Latt, R., Hernández, E., Malavassi, E., and Valdés, J., 2003, Changes in the activity of Turrialba Volcano; seismicity, geochemistry, and deformation: Seismological Research Letters, v. 74, p. 215.
Bellon, H., and Tournon, J., 1978, Contribution de la géochronométrie K-Ar à l'étude du magmatisme de Costa Rica, Amérique Centrale: Bulletin de la Société Géologique de France, v. 7, p. 955–959.
Carr, M.J., Feigenson, M.D., and Bennett, E.A., 1990, Incompatible element and isotopic evidence for tectonic control of source mixing and melt extraction along the Central American arc: Contributions to Mineralogy and Petrology, v. 105, p. 369–380, doi: 10.1007/BF00286825.
Clark, S.K., Reagan, M.K., and Trimble, D.A., this volume, Tephra deposits for the past 2600 years from Irazú volcano, Costa Rica, in Rose, W.I., Bluth, G.J.S., Carr, M.J., Ewert, J.W., Patino, L.C., and Vallance, J.W., Volcanic hazards in Central America: Geological Society of America Special Paper 412, doi: 10.1130/2006.2412(12).
Duarte, E., 1990, Algunos aspectos del riesgo volcanico en el Volcan Turrialba [Licenciado en Geografia thesis]: Costa Rica, Universidad Nacional, 123 p.

Fernández, E., Duarte, E., Sáenz, W., Malavassi, E., Martínez, M., Valdés, J., and Barboza, V., 2002, Cambios en la Geoquímica de las Fumarolas del Volcán Turrialba, Costa Rica: 1998–2001 (abstract): Colima Volcano: Eighth Internacional Meeting, p. 84.

Fischer, R.V., and Schmincke, H.-U., 1984, Pyroclastic rocks: Berlin, Springer-Verlag, 472 p.

Gans, P.B., Alvarado, G., Pérez, W., MacMillan, I., and Calvert, A., 2003, Neogene evolution of the Costa Rican Arc and development of the Cordillera Central: Geological Society of America Abstracts with Programs, v. 35, no. 4, p. 74.

Glazner, A.F., Manley, C.R., Marron, J.S., and Rojstaczer, S., 1999, Fire or ice: Anticorrelation of volcanism and glaciation in California over the past 800,000 years: Geophysical Research Letters, v. 26, p. 1759–1762, doi: 10.1029/1999GL900333.

González-Víquez, C., 1910, Temblores, terremotos, inundaciones, y erupciones volcánicas en Costa Rica 1608–1910: Tipografia de Avelino Alsina, San José de Costa Rica, p. 35–41.

Hall, K., 1982, Rapid deglaciation as an initiator of volcanic activity: An hypothesis: Earth Surface Processes and Landforms, v. 7, p. 45–51.

Hudnut, K.W., 1983, Geophysical survey of Irazú Volcano [Bachelor's thesis]: Dartmouth College, 82 p.

Krushensky, R.D., and Escalante, G., 1967, Activity of Irazu and Poas Volcanoes, Costa Rica, November 1964–July 1965: Bulletin of Volcanology, v. 31, p. 75–84.

Lachniet, M.S., and Seltzer, G.O., 2002, Late Quaternary glaciation of Costa Rica: Geological Society of America Bulletin, v. 114, p. 547–558, doi: 10.1130/0016-7606(2002)114<0547:LQGOCR>2.0.CO;2.

McGuire, W.J., 1996, Volcano instability; a review of contemporary themes, in McGuire, W.J., Jones, A.P., and Neuberg, J., eds., Volcano instability on the Earth and other planets: London, Geological Society Special Publication 110, p. 1–23.

McGuire, W.J., Howarth, R.J., Firth, C.R., Solow, A.R., Pullen, A.D., Saunders, S.J., Stewart, I.S., and Vita-Finzi, C., 1997, Correlation between rate of sea-level change and frequency of explosive volcanism in the Mediterranean: Nature, v. 389, p. 473–476, doi: 10.1038/38998.

Melson, W.G., Barquero, J., Sáenz, R., and Fernández, E., 1985, Erupciones explosivas de importancia en volcanes de Costa Rica: Costa Rica: Boletín de Vulcanología, v. 16, p. 15–22.

Mora, R., Ramírez, C., and Fernández, M., 2004, La actividad de los volcanes de la Cordillera Central, Costa Rica, entre 1998–2002, in Soto, G.J. and Alvarado, G.E., eds., La Vulcanología y su entorno geoambiental: Revista Geológica de América Central, v. 30, p. 189–197.

Mullineaux, D.R., and Crandell, D.R., 1981, The eruptive history of Mount St. Helens, in Lipman, P.W., and Mullineaux, D.R., eds., The 1980 eruptions of Mount St. Helens, Washington: U.S. Geological Survey Professional Paper 1250, p. 3–16.

Murata, K.J., Dóndoli, C., and Sáenz, R., 1966, The 1963–65 eruption of Irazu Volcano, Costa Rica (the period of March 1963 to October 1963): Bulletin of Volcanology, v. 29, p. 765–796.

Nakada, M., and Yokose, H., 1992, Ice age as a trigger of active Quaternary volcanism and tectonism: Tectonophysics, v. 212, p. 321–329, doi: 10.1016/0040-1951(92)90298-K.

Newhall, C.G., and Self, S., 1982, The volcanic explosivity index (VEI): An estimate of explosive magnitude for historical volcanism: Journal of Geophysical Research, Oceans and Atmospheres, v. 87, p. 1231–1238.

Obando, L.G., and Soto, G.J., 1993, La turbera del Río Silencio (El Cairo, Siquirres, Costa Rica): Paleoambientes lagunares influenciados por las cenizas del Volcán Turrialba: Revista Geológica de América Central, v. 15, p. 41–48.

Orvis, K.H., and Horn, S.P., 2000, Quaternary glaciers and climate on Cerro Chirripo, Costa Rica: Quaternary Research, v. 54, p. 24–37, doi: 10.1006/qres.2000.2142.

Pyle, D.M., 1989, The thickness, volume and grain size of tephra fall deposits: Bulletin of Volcanology, v. 51, p. 1–15, doi: 10.1007/BF01086757.

Reagan, M.K., 1987, Turrialba Volcano, Costa Rica: Magmatism at the southeast terminus of the Central American arc [Ph.D. dissertation]: Santa Cruz, University of California, 216 p.

Reagan, M.K., and Gill, J.B., 1989, Coexisting calcalkaline and high-niobium basalts from Turrialba volcano, Costa Rica: Implications for residual titanates in arc magma sources: Journal of Geophysical Research, v. B94, p. 4619–4633.

Reagan, M.K., Gill, J.B., Malavassi, E., and Garcia, M.O., 1987, Changes in lava composition at Arenal Volcano, Costa Rica, 1968–85: Realtime monitoring of open-system differentiation: Bulletin of Volcanology, v. 49, p. 415–434.

Reid, M.E., Sisson, T.W., and Brien, D.L., 2001, Volcano collapse promoted by hydrothermal alteration and edifice shape, Mount Rainier, Washington: Geology, v. 29, p. 779–782, doi: 10.1130/0091-7613(2001)029<0779:VCPBHA>2.0.CO;2.

Sáenz, R., Barquero, J., and Malavassi, E., 1982, Excursión al volcán Irazú, USA-CR joint seminar in volcanology: Boletín de Vulcanología, v. 14, p. 101–116.

Siebert, L., 1984, Large volcanic debris avalanches: Characteristics of source areas, deposits, and associated eruptions: Journal of Volcanology and Geothermal Research, v. 22, p. 163–197, doi: 10.1016/0377-0273(84)90002-7.

Siebert, L., Glicken, H., and Ui, T., 1987, Volcanic hazards from Bezymianny and Bandai-type eruptions: Bulletin of Volcanology, v. 49, p. 435–459, doi: 10.1007/BF01046635.

Simkin, T., and Siebert, L., 2002, Global Volcanism FAQs: Smithsonian Institution, Global Volcanism Program Digital Information Series, GVP-5, http://volcano.si.edu/gvp.faq.

Soto, G., 1988, Estructuras volcano-tectónicas del Volcán Turrialba, Costa Rica, América Central, in Actas Quinto Congreso Geológico Chileno, Santiago, 8–12 de agosto de 1988: Tomo III, p. I-163–I-175.

Stuiver, M., Reimer, P.J., Bard, E., Beck, J.W., Burr, G.S., Hughen, K.A., Kromer, B., McCormac, G., van der Plicht, J., and Spurk, M., 1998, INTCAL98 radiocarbon age calibration, 24,000–0 cal BP: Radiocarbon, v. 40, p. 1041–1083.

Tournon, J., 1984, Magmatismes du Mesozoique a l'actuel en Amerique Centrale: L'exemple de Costa Rica, des ophiolites aux andesites [PhD Dissertation]: Paris, Université Pierre and Marie Curie, IIX + 335 p.

Ui, T., 1983, Volcanic dry avalanche deposits-identification and comparison with non-volcanic debris stream deposits, in Aramaki, S., and Kushiro, I., eds., Arc Volcanism: Journal of Volcanology and Geothermal Research, v. 18, p. 135–150.

von Seebach, K., 1865, Besteigung des Vulkans Turrialba in Costa Rica: Petermann's Mitteil, v. 9, p. 321–324.

MANUSCRIPT ACCEPTED BY THE SOCIETY 19 MARCH 2006

TABLE A1. MAJOR AND TRACE ELEMENT DATA FOR TURRIALBA'S LAVAS AND TEPHRAS

| Sample | T-1 | T-24-2 | T-37-1 | T-37-5 | T-51-3 | T-51-5 | T-131 | T-85 | T-86 | T-62 | T-51-1 | T-55-i | T-102 | T-38-5 | T-38-5b | T-59 |
|---|---|---|---|---|---|---|---|---|---|---|---|---|---|---|---|---|
| Unit | 1 | 1 | 1 | 1 | 1 | 1 | 1 | 1 | 1 | 2 | 2 | 2 | 2 | 3 | 3 | 3 |
| Type | cinder | lava | cinder | cinder | cinder | cinder | cinder | dike | dike | lava | cinder | cinder | lava | pumice | pumice | cinder |
| $SiO_2$ | 52.51 | 52.55 | 52.64 | 52.31 | 52.39 | 52.13 | 52.37 | 51.73 | 51.96 | 56.55 | 54.87 | 55.21 | 57.58 | 61.71 | 57.34 | 54.71 |
| $TiO_2$ | 1.04 | 1.04 | 1.04 | 1.39 | 1.05 | 1.05 | 1.05 | 1.91 | 1.06 | 0.91 | 0.91 | 0.92 | 0.77 | 0.74 | 1.13 | 1.44 |
| $Al_2O_3$ | 17.18 | 17.19 | 17.02 | 16.82 | 17.62 | 17.9 | 17.14 | 15.91 | 17.61 | 16.66 | 18.1 | 17.08 | 17.58 | 17.16 | 15.98 | 15.63 |
| $Fe_2O_3$ | 8.52 | 8.33 | 8.48 | 8.78 | 8.5 | 8.43 | 8.62 | 9.69 | 8.53 | 7.25 | 7.78 | 7.46 | 6.87 | 5.24 | 6.84 | 7.85 |
| MnO | 0.14 | 0.14 | 0.14 | 0.14 | 0.14 | 0.14 | 0.14 | 0.14 | 0.14 | 0.12 | 0.12 | 0.12 | 0.12 | 0.1 | 0.12 | 0.12 |
| MgO | 6.24 | 6.24 | 6.41 | 5.96 | 6.04 | 6.26 | 6.16 | 6.48 | 6.1 | 5.32 | 5 | 5.55 | 4.02 | 2.84 | 5.05 | 6.19 |
| CaO | 9.23 | 9.19 | 9.27 | 9.44 | 9.15 | 9.18 | 9.48 | 8.93 | 9.57 | 7.5 | 8.16 | 8.28 | 7.42 | 5.06 | 6.63 | 7.67 |
| $Na_2O$ | 3.37 | 3.47 | 3.19 | 3.2 | 3.26 | 3.14 | 3.29 | 3.6 | 3.3 | 3.59 | 3.22 | 3.48 | 3.31 | 3.92 | 3.83 | 3.39 |
| $K_2O$ | 1.42 | 1.47 | 1.45 | 1.46 | 1.46 | 1.4 | 1.37 | 1.09 | 1.37 | 1.78 | 1.52 | 1.59 | 2.08 | 2.94 | 2.56 | 2.37 |
| $P_2O_5$ | 0.37 | 0.4 | 0.38 | 0.52 | 0.4 | 0.38 | 0.39 | 0.53 | 0.38 | 0.32 | 0.32 | 0.31 | 0.26 | 0.3 | 0.52 | 0.63 |
| LOI | -0.02 | 0.12 | -0.01 | 0.67 | -0.05 | -0.05 | -0.24 | -0.17 | 0.12 | 0.77 | 1 | -0.01 | 0.2 | 0.98 | 0.2 | 1.76 |
| total* | 100.78 | 100.67 | 100.33 | 100.35 | 100.1 | 99.45 | 99.74 | 100.34 | 99.35 | 99.95 | 100.7 | 100.85 | 99.26 | 100.22 | 99.31 | 100.1 |
| Ni | 54.2 | 58 | 59.2 | 58.3 | 49.9 | 50.3 | 51.5 | 101.5 | 53 | 73.7 | 44.6 | 54.4 | 35.7 | 33 | 84.9 | 111.4 |
| Sc | 26.1 | 26.4 | 27.1 | 26.5 | 27 | 27.5 | 27.6 | 21.6 | 27.7 | 19.7 | 23.2 | 22.5 | 18.1 | 10.9 | 16.1 | 18.5 |
| Cr | 126 | 142 | 144 | 127 | 125 | 124 | 130 | 186 | 135 | 144 | 86 | 116 | 40 | 54 | 151 | 222 |
| Ce | 67 | 67 | 69 | 87 | 68 | 65 | 65 | 82 | 64 | 69 | 56 | 62 | 62 | 87 | 105 | 123 |
| Nd | 33.8 | 30.3 | 34.6 | 39.9 | 34.1 | 32.7 | 32.7 | 38.1 | 31.8 | 30.9 | 27.1 | 29.1 | 28.3 | 34 | 43.8 | 53 |
| V | 232 | 228 | 241 | 233 | 239 | 244 | 243 | 218 | 240 | 187 | 221 | 200 | 157 | 96 | 146 | 184 |
| Ba | 674 | 670 | 691 | 680 | 679 | 657 | 662 | 524 | 661 | 798 | 706 | 716 | 845 | 1059 | 964 | 858 |
| La | 35.9 | 33.3 | 37.8 | 46.9 | 37.9 | 36.4 | 35.1 | 42 | 37.3 | 38.8 | 30.6 | 31.6 | 32.3 | 46.7 | 57.9 | 68.4 |
| Nb | 14 | 13.1 | 14 | 25.4 | 13.9 | 13.2 | 14.2 | 35.8 | 13.9 | 19.5 | 12 | 8.9 | 11.1 | 18.5 | 26.5 | 33.1 |
| Zr | 142 | 142 | 148 | 175 | 143 | 139 | 144 | 194 | 145 | 169 | 132 | 146 | 157 | 245 | 249 | 248 |
| Y | 22 | 21.8 | 21.6 | 23.3 | 22.2 | 20.3 | 22 | 25.6 | 21.6 | 18.1 | 17.7 | 18.3 | 16.2 | 22.4 | 23.1 | 25.1 |
| Sr | 784 | 802 | 799 | 976 | 773 | 825 | 784 | 887 | 792 | 776 | 832 | 854 | 1010 | 744 | 883 | 999 |
| Rb | 29.2 | 28.7 | 29.3 | 29.9 | 29.2 | 30.8 | 29.6 | 22.2 | 29.3 | 38.4 | 31.5 | 33.9 | 44.5 | 74.4 | 62.5 | 53 |
| Th | — | 5.4 | — | 5.9 | — | — | — | 5.5 | — | — | — | — | — | 13.7 | 13 | 12.4 |
| U | — | 1.8 | — | 2 | — | — | — | 1.6 | — | — | — | — | — | 4.7 | 4.1 | 3.7 |

*Continued*

TABLE A1. MAJOR AND TRACE ELEMENT DATA FOR TURRIALBA'S LAVAS AND TEPHRAS (continued)

| Sample | T-101-1 | T-101-2 | T-3 | T-4 | T-35-1 | T-35-2 | T-35-4 | T-35-5 | T-103-1-3 | T-103-6 | T-103-7a | T-63 | T-129 | T-124 | T-126 |
|---|---|---|---|---|---|---|---|---|---|---|---|---|---|---|---|
| Unit | 3 | 3 | 4 | 4 | 4 | 4 | 4 | 4 | 4 | 5 | 7 | 8b | 8b | 8c | 8c |
| Type | cinder | cinder | pumice | clast | pumice | cinder | clast | pumice | pumice | cinder | cinder | lava | lava | lava | lava |
| $SiO_2$ | 59.61 | 58.21 | 58.26 | 57.59 | 58.49 | 58.16 | 58.21 | 58.95 | 58.41 | 55.97 | 56.12 | 62.58 | 64.09 | 57.23 | 57.21 |
| $TiO_2$ | 1.24 | 1.1 | 0.89 | 0.92 | 0.9 | 0.92 | 0.91 | 0.86 | 0.92 | 0.85 | 0.87 | 0.67 | 0.63 | 0.86 | 0.83 |
| $Al_2O_3$ | 19.5 | 16.98 | 17.3 | 16.88 | 16.93 | 16.89 | 16.87 | 16.86 | 16.78 | 17.18 | 17.07 | 16.76 | 16.53 | 17.16 | 17.42 |
| $Fe_2O_3$ | 4.8 | 6.51 | 6.45 | 6.62 | 6.57 | 6.66 | 6.62 | 6.29 | 6.64 | 7.79 | 8.04 | 5.00 | 4.62 | 6.97 | 6.97 |
| MnO | 0.08 | 0.11 | 0.12 | 0.12 | 0.14 | 0.12 | 0.12 | 0.12 | 0.12 | 0.13 | 0.13 | – | 0.09 | 0.12 | 0.12 |
| MgO | 3.63 | 4.4 | 3.91 | 4.08 | 3.96 | 4.09 | 3.94 | 3.82 | 3.86 | 5.03 | 5.02 | 3.02 | 2.22 | 4.29 | 4.06 |
| CaO | 6.1 | 7.51 | 6.66 | 7.02 | 6.56 | 6.69 | 6.92 | 6.5 | 6.68 | 8.39 | 8.17 | 5.05 | 4.64 | 7.26 | 7.49 |
| $Na_2O$ | 2.66 | 2.66 | 3.75 | 4.15 | 3.75 | 3.8 | 3.7 | 3.85 | 3.9 | 2.66 | 2.63 | 3.92 | 3.88 | 3.74 | 3.53 |
| $K_2O$ | 1.96 | 2.08 | 2.31 | 2.29 | 2.36 | 2.34 | 2.37 | 2.43 | 2.37 | 1.71 | 1.67 | 2.77 | 3.09 | 2.06 | 2.08 |
| $P_2O_5$ | 0.43 | 0.45 | 0.35 | 0.34 | 0.34 | 0.34 | 0.35 | 0.33 | 0.33 | 0.3 | 0.29 | 0.24 | 0.23 | 0.31 | 0.3 |
| LOI | 7.29 | 2.74 | 1 | 0.26 | 0.33 | 0.26 | 0.11 | 0.2 | 0.12 | 2.09 | 3.18 | 1.37 | 0.26 | −0.05 | 0.27 |
| total* | 100.8 | 100.94 | 100.9 | 99.83 | 100.33 | 100.33 | 99.83 | 100.07 | 100.01 | 100.9 | 100.84 | 100.24 | 100.5 | 99.74 | 99.71 |
| Ni | 47.8 | 67.2 | 27.8 | 29.4 | 27.4 | 28.1 | 29 | 28.4 | 26.3 | 53.9 | 65.8 | – | 16 | 32.1 | 32.3 |
| Sc | 22.3 | 19.5 | 16.8 | 17.9 | 16 | 17.8 | 17.5 | 16.9 | 16.9 | 25.2 | 23.4 | – | 10.6 | 17.5 | 19.5 |
| Cr | 240 | 179 | 55 | 58 | 60 | 61 | 61 | 58 | 59 | 205 | 191 | – | 21 | 58 | 50 |
| Ce | 98 | 97 | 84 | 83 | 80 | 84 | 85 | 81 | 83 | 73 | 78 | – | 85 | 77 | 71 |
| Nd | 44.7 | 45 | 36.8 | 36.4 | 33.5 | 36.6 | 36.9 | 35.1 | 36.2 | 37.9 | 41.8 | – | 33.6 | 34.7 | 32.3 |
| V | 328 | 217 | 155 | 162 | 158 | 161 | 158 | 156 | 158 | 209 | 201 | – | 77 | 156 | 159 |
| Ba | 774 | 810 | 947 | 936 | 925 | 949 | 955 | 955 | 941 | 740 | 746 | – | 1117 | 913 | 881 |
| La | 50.4 | 49.8 | 46.8 | 46.1 | 41.9 | 46.5 | 44.6 | 44.1 | 44.9 | 39.8 | 41.9 | – | 43.8 | 38.7 | 38.6 |
| Nb | 23.8 | 21.9 | 22.3 | 21.3 | 21.2 | 21.9 | 21.3 | 21.1 | 21.5 | 14.2 | 14.1 | – | 19.7 | 11.5 | 13.9 |
| Zr | 202 | 197 | 215 | 204 | 208 | 215 | 213 | 217 | 212 | 167 | 167 | – | 250 | 179 | 171 |
| Y | 28.2 | 32.8 | 22.2 | 22.4 | 21.8 | 22.8 | 23 | 22.7 | 22.6 | 30 | 32 | – | 20.3 | 20.4 | 18.9 |
| Sr | 895 | 856 | 759 | 798 | 752 | 781 | 774 | 730 | 748 | 645 | 645 | – | 710 | 892 | 946 |
| Rb | 48.6 | 48.8 | 55.4 | 54.8 | 55.3 | 58.2 | 57.8 | 58 | 57.8 | 40.6 | 38.5 | – | 78.4 | 47 | 47.2 |
| Th | – | – | – | – | 10.3 | – | – | – | – | – | – | – | – | – | – |
| U | – | – | – | – | 3.5 | – | – | – | – | – | – | – | – | – | – |

*Continued*

TABLE A1. MAJOR AND TRACE ELEMENT DATA FOR TURRIALBA'S LAVAS AND TEPHRAS (continued)

| Sample | T-65 | T-64 | T-5 | T-22-9 | T-69-d | T-70-6 | T-70-7 | T-70-12 | T-70-13 | T-76 | T-89 | T-90 | T-7 | T-132 | C-3 |
|---|---|---|---|---|---|---|---|---|---|---|---|---|---|---|---|
| Unit | 8c | 8c | 9 | 9 | 9 | 9 | 9 | 9 | 9 | 9 | 10 | 10 | 11 | 11 | 11 |
| Type | lava | lava | lava | lava | lava | lava | lava | lava | lava | lava | lava | lava | lava | lava | cinder |
| $SiO_2$ | 57.19 | 57.49 | 58.03 | 58.27 | 57.81 | 56.95 | 57.1 | 57.92 | 57.4 | 57.52 | 54.97 | 54.31 | 56.94 | 57.74 | 56.43 |
| $TiO_2$ | 0.79 | 0.79 | 0.71 | 0.71 | 0.73 | 0.77 | 0.76 | 0.7 | 0.74 | 0.71 | 0.91 | 0.9 | 0.86 | 0.72 | 1.08 |
| $Al_2O_3$ | 17.75 | 17.85 | 17.98 | 18.11 | 18.15 | 18.07 | 18.22 | 18.03 | 18.02 | 18.11 | 18.46 | 18.51 | 17.61 | 18.07 | 18.73 |
| $Fe_2O_3$ | 6.87 | 6.92 | 6.64 | 6.71 | 6.65 | 6.9 | 6.95 | 6.82 | 6.74 | 6.76 | 7.81 | 8.18 | 6.86 | 6.57 | 7.82 |
| MnO | 0.12 | 0.12 | 0.11 | 0.11 | 0.11 | 0.11 | 0.12 | 0.12 | 0.11 | 0.11 | 0.13 | 0.12 | 0.11 | 0.1 | 0.11 |
| MgO | 4.17 | 3.96 | 3.77 | 3.79 | 3.68 | 4.04 | 4.23 | 3.8 | 4.13 | 3.85 | 4.61 | 4.44 | 4.55 | 4.13 | 4.62 |
| CaO | 7.4 | 7.05 | 7.12 | 7.01 | 7.36 | 7.58 | 7.16 | 7.13 | 7.2 | 7.36 | 8.37 | 8.57 | 7.35 | 7.43 | 6.8 |
| $Na_2O$ | 3.3 | 3.39 | 3.76 | 3.52 | 3.55 | 3.74 | 3.63 | 3.66 | 3.65 | 3.71 | 3.3 | 3.38 | 4 | 3.85 | 2.44 |
| $K_2O$ | 2.13 | 2.15 | 1.63 | 1.51 | 1.68 | 1.55 | 1.53 | 1.57 | 1.75 | 1.6 | 1.11 | 1.26 | 1.41 | 1.11 | 1.65 |
| $P_2O_5$ | 0.28 | 0.29 | 0.26 | 0.27 | 0.29 | 0.3 | 0.3 | 0.26 | 0.27 | 0.28 | 0.33 | 0.34 | 0.31 | 0.28 | 0.32 |
| LOI | 1.06 | 0.69 | 0.34 | 0.35 | 0.55 | 0.35 | 0.41 | 0.59 | 0.48 | 0.52 | 1.32 | 1.76 | 0.7 | 0.81 | 3.92 |
| total* | 99.63 | 100.65 | 101.21 | 99.86 | 99.72 | 100.32 | 100.3 | 100.79 | 100.74 | 99.51 | 99.46 | 99.6 | 99.92 | 100.19 | 101 |
| Ni | 37.9 | 34.4 | 24.5 | 23.5 | 24.7 | 28.6 | 28.6 | 23.9 | 25.8 | 24.5 | 30.1 | 29.2 | 49.8 | 44.1 | 22.7 |
| Sc | 18.5 | 17.5 | 17.4 | 15.9 | 16.5 | 18.4 | 17.3 | 16.4 | 16.9 | 16.4 | 25.4 | 23.1 | 18.6 | 16.7 | 23.9 |
| Cr | 40 | 44 | 16 | 11 | 16 | 29 | 21 | 10 | 18 | 10 | 51 | 46 | 56 | 60 | 79 |
| Ce | 54 | 64 | 35 | 35 | 35 | 40 | 41 | 35 | 35 | 36 | 58 | 52 | 52 | 43 | 58 |
| Nd | 26.5 | 28.4 | 17.8 | 17.7 | 17.9 | 21.9 | 22.6 | 18.4 | 18.7 | 20 | 27.9 | 25.9 | 24.1 | 20.7 | 28.1 |
| V | 160 | 149 | 152 | 138 | 148 | 167 | 174 | 142 | 155 | 148 | 210 | 216 | 179 | 156 | 247 |
| Ba | 819 | 859 | 713 | 699 | 716 | 700 | 702 | 722 | 711 | 718 | 689 | 693 | 794 | 694 | 684 |
| La | 30.3 | 35.4 | 17.5 | 15 | 15.7 | 18.4 | 17.6 | 18 | 14.7 | 13.9 | 25.6 | 28.6 | 26.1 | 17.6 | 31.5 |
| Nb | 10.3 | 11.9 | 6.2 | 7.2 | 6.4 | 6.4 | 6.6 | 6.1 | 6.3 | 6.4 | 9.5 | 9.3 | 9.5 | 7 | 11.7 |
| Zr | 156 | 168 | 98 | 99 | 96 | 91 | 93 | 96 | 95 | 94 | 109 | 111 | 117 | 93 | 142 |
| Y | 16.9 | 17.1 | 13.1 | 13.7 | 14.5 | 16.1 | 16 | 13.5 | 13.7 | 12.8 | 17.3 | 18.2 | 15.4 | 10.9 | 22.7 |
| Sr | 987 | 978 | 1069 | 1088 | 1070 | 1087 | 1090 | 1095 | 1062 | 1083 | 886 | 883 | 933 | 999 | 761 |
| Rb | 43.9 | 46.8 | 29.8 | 27.4 | 29.9 | 27 | 27.2 | 28.9 | 28.8 | 28.4 | 21.5 | 24.1 | 27.4 | 20.4 | 36.6 |
| Th | – | 9.3 | – | – | – | – | – | 3 | – | – | – | – | 3.3 | – | – |
| U | – | 2.8 | – | – | – | – | – | 1.4 | – | – | – | – | 1.3 | – | – |

*Continued*

TABLE A1. MAJOR AND TRACE ELEMENT DATA FOR TURRIALBA'S LAVAS AND TEPHRAS (continued)

| Sample | T-68 | T-67 | T-2 | T-66 | T-69-g | T-70-2 | C-8 | T-10 | T-57 | T-11 | T-47 | T-81 | T-91 | T-108-1 | T-108-2 |
|---|---|---|---|---|---|---|---|---|---|---|---|---|---|---|---|
| Unit | 12 | 13 | 14 | 14 | 14 | 14 | 15 | 15 | 15 | 15 | 15 | 15 | 15 | 15 | 15 |
| Type | lava | lava | lava | lava | lava | lava | lava | clast | lava | lava | lava | lava | lava | lava | lava |
| $SiO_2$ | 64.11 | 55.18 | 59.47 | 59.46 | 59.53 | 58.75 | 63.28 | 51.95 | 59.12 | 56.24 | 64.87 | 53.37 | 51.18 | 51.54 | 52.73 |
| $TiO_2$ | 0.6 | 0.87 | 0.74 | 0.74 | 0.74 | 0.76 | 0.57 | 1.13 | 0.76 | 0.98 | 0.67 | 1 | 1.15 | 1.14 | 0.95 |
| $Al_2O_3$ | 16.65 | 16.93 | 17.19 | 17.77 | 17.54 | 17.46 | 16.04 | 18.19 | 17.04 | 17.53 | 15.76 | 17.61 | 17.63 | 17.73 | 18.38 |
| $Fe_2O_3$ | 4.81 | 7.99 | 6.29 | 6.26 | 6.34 | 6.6 | 4.78 | 9.12 | 6.94 | 7.34 | 4.42 | 7.8 | 8.51 | 8.54 | 8.06 |
| MnO | 0.09 | 0.13 | 0.11 | 0.1 | 0.1 | 0.09 | 0.09 | 0.15 | 0.11 | 0.13 | 0.08 | 0.13 | 0.15 | 0.14 | 0.13 |
| MgO | 1.93 | 6.07 | 3.99 | 3.52 | 3.67 | 3.99 | 3.47 | 5.53 | 3.91 | 4.37 | 2.45 | 5.5 | 6.32 | 6.03 | 5.21 |
| CaO | 4.51 | 8.32 | 6.27 | 6.32 | 5.94 | 6.62 | 4.7 | 9.23 | 5.94 | 6.73 | 4.06 | 8.97 | 9.5 | 9.71 | 9.48 |
| $Na_2O$ | 3.88 | 2.76 | 3.53 | 3.33 | 3.62 | 3.49 | 3.78 | 3.06 | 3.39 | 4.09 | 4.05 | 3.76 | 3.79 | 3.57 | 3.55 |
| $K_2O$ | 3.11 | 1.42 | 2.13 | 2.19 | 2.21 | 1.93 | 3.09 | 1.33 | 2.48 | 2.17 | 3.41 | 1.48 | 1.34 | 1.21 | 1.19 |
| $P_2O_5$ | 0.33 | 0.34 | 0.28 | 0.31 | 0.31 | 0.31 | 0.2 | 0.32 | 0.31 | 0.42 | 0.25 | 0.39 | 0.43 | 0.41 | 0.33 |
| LOI | 1.06 | 1.62 | 2.09 | 1.48 | 1.28 | 2.56 | 0.37 | 0.2 | 1.31 | 1.05 | 0.3 | 0.87 | 0.12 | 0.34 | 0.49 |
| total* | 100.44 | 100.01 | 100.78 | 99.81 | 99.85 | 99.81 | 100.36 | 99.63 | 99.95 | 99.36 | 99.67 | 99.92 | 101.38 | 100.17 | 100.35 |
| Ni | 12.4 | 78.8 | 47.6 | 48.9 | 46.9 | 61.9 | 51.8 | 39.2 | 63.4 | 44.2 | 23.7 | 51.9 | 36.9 | 40.8 | 28.7 |
| Sc | 9.6 | 24.1 | 17 | 16.7 | 15.8 | 18.4 | 10.8 | 26.9 | 18.5 | 17.1 | 9.6 | 23.4 | 26.5 | 27.5 | 25.6 |
| Cr | 12 | 206 | 77 | 55 | 77 | 103 | 137 | 51 | 129 | 79 | 42 | 93 | 96 | 100 | 63 |
| Ce | 98 | 65 | 68 | 66 | 71 | 66 | 71 | 40 | 89 | 95 | 99 | 70 | 72 | 70 | 49 |
| Nd | 39.4 | 32.6 | 30.4 | 28.7 | 30.1 | 30.6 | 30.5 | 23.6 | 42.8 | 42.6 | 40.1 | 35.2 | 32.1 | 36.4 | 25 |
| V | 84 | 213 | 161 | 163 | 151 | 158 | 92 | 269 | 164 | 176 | 87 | 251 | 251 | 265 | 232 |
| Ba | 1191 | 685 | 977 | 923 | 918 | 875 | 1027 | 542 | 928 | 1014 | 1218 | 741 | 643 | 668 | 665 |
| La | 56.8 | 36.6 | 36.9 | 37.5 | 36.4 | 37.8 | 38.9 | 23 | 49.3 | 57.4 | 55.4 | 40.7 | 37.7 | 41.4 | 26.6 |
| Nb | 23.2 | 13.2 | 14.9 | 14.5 | 15.4 | 13.5 | 18.8 | 9.5 | 17.5 | 20 | 25.6 | 12.4 | 12.4 | 12.5 | 9.2 |
| Zr | 280 | 157 | 166 | 170 | 182 | 165 | 243 | 115 | 212 | 199 | 293 | 132 | 131 | 128 | 110 |
| Y | 24.6 | 30.8 | 22.4 | 21.3 | 20.4 | 18.8 | 22.1 | 23.2 | 35.5 | 22.7 | 27 | 20.8 | 22.9 | 22.4 | 19.1 |
| Sr | 695 | 633 | 771 | 794 | 745 | 752 | 575 | 944 | 634 | 880 | 565 | 963 | 912 | 898 | 921 |
| Rb | 82.5 | 25.9 | 46.6 | 50.8 | 50 | 43.4 | 80.9 | 22.8 | 60 | 48.8 | 90 | 24.9 | 23.3 | 22 | 20.8 |
| Th | — | — | — | — | — | — | — | — | — | — | — | — | — | — | — |
| U | — | — | — | — | — | — | — | — | — | — | — | — | — | — | — |

Note: All major and trace element data except U and Th were collected by XRF techniques descried in Reagan et al. (1987). U and Th were analyzed by alpha spectrometry techniques described in Reagan et al. (1989). Sample locations are in Reagan (1987) and can be obtained from the first author.
*All major elements are normalized to 100% the listed total is before normalization.

*Recent volcanic history of Irazú volcano, Costa Rica: Alternation and mixing of two magma batches, and pervasive mixing*

Guillermo E. Alvarado*
*Área de Amenazas y Auscultación Sísmica y Volcánica, Instituto Costarricense de Electricidad, Apdo. 10032-1000, Costa Rica*

Michael J. Carr*
Brent D. Turrin
Carl C. Swisher III
*Department of Geological Sciences, Rutgers University, New Brunswick, New Jersey 08903, USA*

Hans-Ulrich Schmincke
*Leibniz Institute for Marine Science, IfM-GEOMAR (Leibniz Institute for Marine Sciences), Wischhofstr. 1-3, D-24148 Kiel, Germany*

Kenneth W. Hudnut
*U.S. Geological Survey, 525 South Wilson Ave., Pasadena, California 91106-3212, USA*

## ABSTRACT

$^{40}Ar/^{39}Ar$ dates, field observations, and geochemical data are reported for Irazú volcano, Costa Rica. Volcanism dates back to at least 854 ka, but has been episodic with lava shield construction peaks at ca. 570 ka and 136–0 ka. The recent volcanic record on Irazú volcano comprises lava flows and a variety of Strombolian and phreatomagmatic deposits, with a long-term trend toward more hydrovolcanic deposits. Banded scorias and hybridized rocks reflect ubiquitous magma mixing and commingling. Two distinct magma batches have been identified. One magma type or batch, Haya, includes basalt with higher high field strength (HFS) and rare-earth element contents, suggesting a lower degree melt of a subduction modified mantle source. The second batch, Sapper, has greater enrichment of large ion lithophile elements (LILE) relative to HFS elements and rare-earth elements, suggesting a higher subduction signature. The recent volcanic history at Irazú records two and one half sequences of the following pattern: eruptions of the Haya batch; eruptions of the Sapper batch; and finally, an unusually clear unconformity, indicating a pause in eruptions. In the last two sequences, strongly hybridized magma erupted after the eruption of the Haya batch. The continuing presence of two distinct magma batches requires two active magma chambers. The common occurrence of hybrids is evidence for a small, nearer to the surface chamber for mixing the two batches. Estimated pre-eruptive temperatures based on two-pyroxene geothermometry range from ~1000–1176 °C in basalts

---

*E-mails: Alvarado: galvaradoi@ice.go.cr; Carr: carr@rutgers.edu; Turrin: bturrin@rci.rutgers.edu; Swisher: cswish@rci.rutgers.edu; Schmincke: h-u.schmincke@t-online.de; Hudnut: hudnut@usgs.gov

Alvarado, G.E., Carr, M.J., Turrin, Brent D., Swisher, C.C., Schmincke, H.-U., and Hudnut, K.W., 2006, Recent volcanic history of Irazú volcano, Costa Rica: Alternation and mixing of two magma batches, and pervasive mixing, *in* Rose, W.I., Bluth, G.J.S., Carr, M.J., Ewert, J.W., Patino, L.C., and Vallance, J.W., Volcanic hazards in Central America: Geological Society of America Special Paper 412, p. 259–276, doi: 10.1130/2006.2412(14). For permission to copy, contact editing@geosociety.org. ©2006 Geological Society of America. All rights reserved.

to 922 °C in hornblende andesites. Crystallization occurred mainly between 4.6 and 3 kb as measured by different geobarometers. Hybridized rocks show intermediate pressures and temperatures. High silica magma occurs in very small volumes as banded scorias but not as lava flows. Although eruptions at Irazú are not often very explosive, the pervasiveness of magma mixing presents the danger of larger, more explosive hybrid eruptions.

**Keywords:** Costa Rica, Irazú volcano, magma mixing and commingling, magma batches, recent stratigraphy, $^{40}Ar/^{39}Ar$ dating, hazards.

## INTRODUCTION

Irazú is a complex volcano located close to a highly populated area. Understanding its volcanic and petrological evolution is important for economic development and, more importantly, is a prerequisite for appreciating its potential hazards and facilitating effective hazard mitigation.

Irazú, the highest (3432 m) volcano in Costa Rica and one of the largest (~600 km³) and most active volcanoes in Central America, is also famous for being the only place in Central America where one can see both the Caribbean Sea and Pacific Ocean. Eight new $^{40}Ar/^{39}Ar$ age determinations, presented here, together with previous dates (U/Th, $^{14}C$), provide new insights into how and when the Irazú volcano evolved since the late early Pleistocene to present. Numerous recent eruptions, a well-defined stratigraphy in the summit region, and lateral vents allow a detailed reconstruction of its magmatic history. During the past 136 k.y., magma compositions varied widely from basalts–basaltic andesites (lavas and tephras) to andesites and rare dacites. At least two chemically distinct magma batches alternated with each other. Mixing between these two batches, called Haya and Sapper, produced a variety of hybrids, ranging from commingled to nearly homogenized magmas. By selecting the least mixed samples, we attempt to trace the evolution of two end-member magma batches via fractional crystallization. The coexistence of mafic with silicic magmas at shallow crustal levels, the periodic injection of mafic magmas, and the alternation of "dry" and "wet" events during the recent eruptive history are critical factors at Irazú because they cause explosive eruptions, a major hazard to nearby towns.

The recognition of magma mixing and mingling at Irazú is the key to understanding the volcano's history. These processes are well known from the volcanic front (VF) in Central America, for example at the Atitlán composite volcano (Halsor and Rose, 1991), the Masaya caldera complex (Walker et al., 1993), the Nejapa and Granada cinder cones (Walker, 1984), and in Panama at El Valle volcano (Defant et al., 1991). Magma mixing events have also been reported from Arenal (Reagan et al., 1987) and Poás (Prosser and Carr, 1987) volcanoes in Costa Rica.

## RECENT VOLCANIC ACTIVITY

Irazú, a prehispanic name possibly meaning "the mountain of the earthquakes and the thunder," is a volcano that poses a substantial risk to major cities in Costa Rica. The earliest reported eruption of Irazú in 1723 produced violent Strombolian fountains followed by phreatomagmatic explosions. All subsequent verifiable eruptive phases took place during the last century: 1917–1921, 1924, 1928, 1930, 1933, 1939–1940, and 1963–1965. Most reports of eruptions during the eighteenth and nineteenth centuries are doubtful and no plausible deposits have been found so far (Alvarado, 1993). Despite their low volcanic explosivity index (VEI 3 or less), Irazú's eruptions caused fatalities, serious agricultural losses, and damage to infrastructure, mostly by lahars and persistent ash fall on the Valle Central, the most populated area of Costa Rica (Alvarado and Schmincke, 1994). According to historic data, the total estimated damage through eruptions of Irazú and volcanic mass flows into Cartago and other towns was on the order of 150 ± 50 million U.S. dollars. The most damaging volcano-tectonic earthquakes associated with Irazú occurred during the 1723 eruption (Alvarado, 1993). The largest historic tectonic earthquake at Irazú hit the WNW flank of the volcano (Montero and Alvarado, 1995) on 30 December 1953 and had a magnitude of 5.9 (surface wave magnitude, $M_s$). Seismic swarms are common beneath the volcano, but most hypocenters are located at depths shallower than 7 km and none have depths exceeding 14 km (Barquero et al., 1995). Since May 1991, the persistence of fumaroles and low-frequency seismic events in and near the main crater indicate ongoing volcanic unrest. In fact, on the northwest flank of Irazú, a small phreatic (hydrothermal) explosion followed by a gravity slide, or small debris avalanche, and debris flows occurred on 8 December 1994 in an area of structural weakness previously identified by Alvarado (1993).

Pre-Columbian tephra deposits erupted during the past 2500 yr, including one that occurred ~1561 A.D., were studied by Clark et al. (this volume).

## GEOLOGIC SETTING

The tectonic framework of Costa Rica is complex because four plates and microplates interact: the Cocos, Caribbean, and Nazca plates and the Panama block (Fig. 1). The Costa Rican VF is associated with the northeastward subduction of the Cocos plate beneath the Caribbean plate along a well-defined Benioff zone. The dip of the subducting slab shallows dramatically in central Costa Rica from 60° northwest of Arenal volcano to 30° toward the southeast (Burbach et al., 1984). The Panama fracture zone (i.e., the Cocos-Nazca plate boundary) occurs just southwest of the Cocos Ridge (Fig. 1).

Seismological, magnetic, and geodetic data suggest the presence of two small, shallow magma chambers ~0.6–1.8 km and 3.2–4 km beneath the summit (Alvarado, 1993). Several NNW-trending active faults and N-S volcanic alignments dominate the structure of the volcano, including a line of parasitic vents on the south flank (Fig. 2). The summit of the volcano is marked by two pit craters: the recently active crater, "Cráter Activo," to the west and an older crater, "Diego de la Haya," to the east (Fig. 2). A prehistoric scoria cone and two tuff rings are located farther to the east. Another composite crater is represented by prominent cliffs immediately south of the Activo and Diego de la Haya vents. The remaining floor of this structure is called "Playa Hermosa."

There are many older volcanic vents on the N80°W-striking summit ridge (Fig. 3). A series of semicircular escarpments on this ridge was the source of several Pleistocene and Holocene debris avalanches observed on the flanks (Alvarado et al., 2004; Siebert et al., this volume).

Krushensky (1972), Hudnut (1983), Thomas (1983), Tournon (1984), and Alvarado (1993) defined the general stratigraphy of part of Irazú and the surrounding area and provided many chemical analyses of Irazú volcanics.

## PETROLOGY OF CENTRAL COSTA RICAN VOLCANOES AND PREVIOUS WORK AT IRAZÚ

Costa Rica is characterized by a strong geochemical gradient along the volcanic front (Carr et al., 1990). To the northwest, the magma source is more depleted mantle with a metasomatic overprint from the subducted slab, whereas in central Costa Rica, lavas are derived from an enriched oceanic island basalt (OIB)-like source with minor contributions from the subducting slab (Carr et al., 1990; Herrstrom et al., 1995). The presence of an enriched or OIB source mantle beneath the Cordillera Central is also supported by the occurrence of alkalic lavas with OIB-like isotope ratios behind the Costa Rican VF (Carr et al., 1990; Feigenson et al., 2004). The $^{87}Sr/^{86}Sr$ and $^{143}Nd/^{144}Nd$ ratios of these alkaline basalts overlap with samples from Irazú whose Sr and Nd isotopic ratios are ~0.7036 and 0.51296, respectively (Carr et al., 1990; Clark et al., 1998; Feigenson et al., 2004).

Previous isotopic work at Irazú was conducted without the benefit of detailed stratigraphy. Nevertheless, it provided valuable general constraints on magmatic evolution. Initial $^{230}Th/^{232}Th$ activity ratios of lavas increase regularly with time, implying that the erupted products are all differentiates of a common parent magma rather than successive partial melts of a common source (Allègre and Condomines, 1976; Gill, 1981). Irazú andesites or their more mafic parents are more likely partial melts of peridotite with Th/U = 3–4 than of subducted mid-oceanic-ridge basalt (MORB) (Gill, 1981; Tournon, 1984). The temporal increase in Th/U ratios requires fractionation of U into vapor as well as into crystal phases in order to suppress U-enrichment in the residual melt (Allègre and Condomines, 1976, 1982; Gill, 1981). The

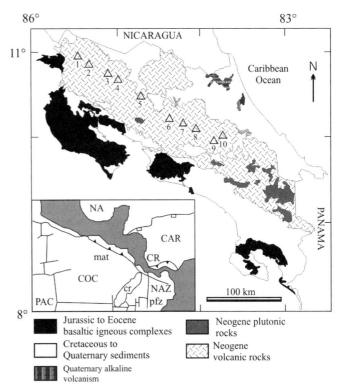

Figure 1. Simplified geological map of Costa Rica (CR). The 10 Quaternary volcanoes are as follows: 1—Orosí; 2—Rincón de la Vieja; 3—Miravalles; 4—Tenorio; 5—Arenal; 6—Platanar; 7—Poás; 8—Barva; 9—Irazú; and 10—Turrialba. Inset map illustrates the plate tectonic setting of Costa Rica; subduction of the Cocos Plate at the Middle America Trench (mat) is limited to the south by the Cocos ridge (cr) and Panama fracture zone (pfz). Tectonic plates: NA—North American Plate; CAR—Caribbean Plate; COC—Cocos Plate; NAZ—Nazca Plate; PAC—Pacific plate.

excess $^{230}Th$ in many central Costa Rican lavas probably reflects the greater incompatibility of Th during relatively dry melting of a garnet-bearing mantle source (Feigenson and Carr, 1993; Herrstrom et al., 1995; Thomas et al., 2002). The low B and $^{10}Be$ contents of selected Irazú lavas (summarized by Leeman et al., 1994) indicate that the contribution of slab material at Irazú was small. Clark et al. (1998) suggest that a water-rich carbonatitic fluid generated by dehydration of the subducting slab may have migrated into the garnet-bearing mantle wedge, where it triggered melting. Overall, the isotope and trace element geochemistry of Irazú indicates partial melting of a garnet-bearing source followed by fractional crystallization. There is no evidence for crustal assimilation based on isotopic and trace element systematics (Feigenson et al., 2004). Magma mixing was reported mainly only in 1963–1965 tephras (Carr and Walker, 1987). Estimates of magmatic water in Irazú were based on basaltic andesite tephra from the 1723 and 1963–1965 eruptions. Water measured in olivine-hosted melt inclusions by ion microprobe range from 0.81 wt% to 3.29 wt% and correlate with Cl and S (Benjamin et al., 2004).

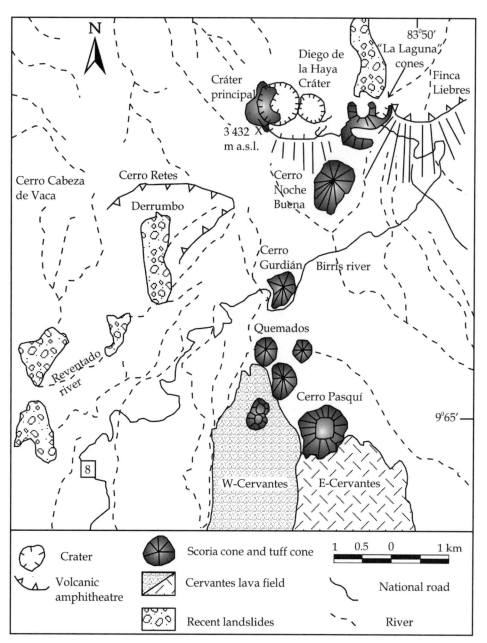

Figure 2. Principal volcanic structures at Irazú.

## RECENT STRATIGRAPHY

Irazú volcano consists mostly of basalts to andesites that overlie dacitic ignimbrites (Krushensky, 1972). There are few outcrops on the flanks of Irazú, and much of the northern sector of the volcano is not easily accessible and is covered by dense rain forest. Sparse alpine vegetation allows good exposure at the summit. The deposits of the latest eruptions are well-exposed in the summit ridge of Irazú, including the main craters and surrounding areas. The known stratigraphy for the uppermost sections of Irazú is summarized in Table 1 and Figure 4. Table 2 lists the new $^{40}Ar/^{39}Ar$ dates. There are clear local unconformities between the major lava and tephra deposits, which correspond to noneruptive intervals of varying duration (Fig. 5A). However, more field work (especially of the Reventado basin), chemical analyses, and physical dating are needed to verify the stratigraphy proposed here and to better define the main lithological units.

### San Jerónimo Unit

Some isolated outcrops are present along the Reventazón river and its tributaries. One of these is the San Jerónimo ignimbrite. It is a dark gray to black, partially welded tuff with abundant broken phenocrysts of plagioclase, common euhedral biotite, and

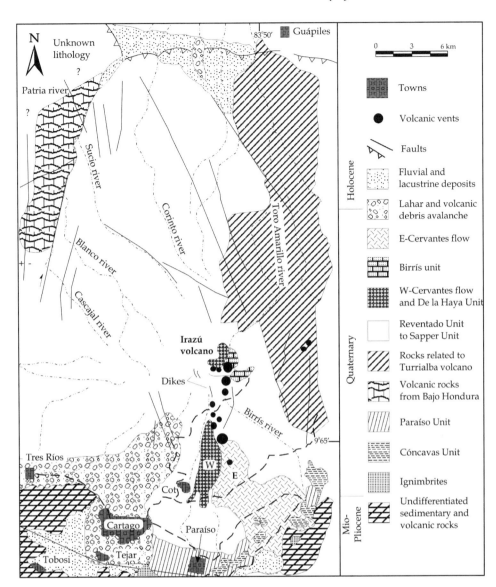

Figure 3. Simplified geologic map of Irazú volcano (after Alvarado, 1993).

rare pyroxenes in an abundant glassy matrix (Krushensky, 1972). New radiometric age determination using the $^{40}Ar/^{39}Ar$ method yields an average value of 854 ka (see Table 2 for details).

## Paraíso and Pico de Piedra Units

The Paraíso and Pico de Piedra Units are the oldest exposed Irazú sequence (594 ± 16 ka and 569 ± 6 ka, respectively; Table 2). The Paraíso Unit forms an older platform morphology. It includes coarse porphyritic andesites and basaltic andesites with plagioclase megacrysts, corresponding to the former flanks of an andesitic shield volcano. It is best exposed in several escarpments and waterfalls near the main rivers. The Pico de Piedra Unit, although similar in age, constitutes the proximal facies of an erosional edifice, deeply affected by edifice-failure events and hydrothermal (fumarolic) alteration.

## Reventado Unit

The Reventado Unit includes many lava flows, principally basaltic andesites to hornblende-bearing andesites, interbedded with epiclastic and tephra deposits. A sample from the upper part of the Reventado unit is ca. 110 ka according to U/Th radiometric data (Allègre and Condomines, 1976). One new $^{40}Ar/^{39}Ar$ date of 136 ± 5 ka for an andesite with rare biotite phenocrysts (Table 2) is in agreement with the U/Th date.

## Diego de la Haya Unit

The Diego de la Haya Unit consists of many 1–2-m-thick basalt lava flows (CHLF10 to CHLF5) interbedded with agglutinated and densely welded spatter and scoria deposits (Figs. 4 and 5A), covering 50 km² of the summit area. The Diego de La

TABLE 1. STRATIGRAPHIC SUMMARY (SEE ALSO FIG. 4) AT IRAZÚ VOLCANO

| Unit | Lithology | Thickness (m) | Estimated age | $SiO_2$ |
| --- | --- | --- | --- | --- |
| Historic tephras | Fallout and deposits | 10–40 | 1723–1965 A.D | 52.5%–55.7% (58.6%–60%) |
| Alfaro | Phreatic breccia | 4 | pre-1723 | nonjuvenile component |
| Tristán | Agglutinate | 11 | Holocene | 56.4%–58% (63%) |
| Dóndoli | Predominantly phreatomagmatic tephras | 28 | Holocene | 53%–56% |
| González | Phreatic and phreatomagmatic deposits | 2.5–4 | Holocene | 51%–43% |
| Birrís | Agglutinates, lava flows | ~60 | late Pleistocene? | 55.2%–56.2% (62%) |
| E-Cervantes | Composite basaltic andesite lava flows | 120 | 16,840 yr B.P. | 55.6%–56.6% |
| W-Cervantes | Composite basaltic lava flows | ~30 | ca. 57 ka | 50.7%–53% |
| Sapper | Lava flows and pyroclastic | 20–50 | ca. 68–36 ka | 53%–60% |
| Diego de la Haya | Lava flows and agglutinates | 50 | late Pleistocene | 49.7%–51.1% |
| Reventado | Lava flows and pyroclastic rocks | 600 | ca. 136–110 ka | 58.5%–63% |
| Paraíso-Pico de Piedra | Lava flows | ~170 | 594–569 ka | 56% |
| San Jerónimo | Dacite ignimbrite | ~15 | ca. 854 ka | 63.3% |

*Note:* Parentheses note silicic inclusions—(pumice enclaves).

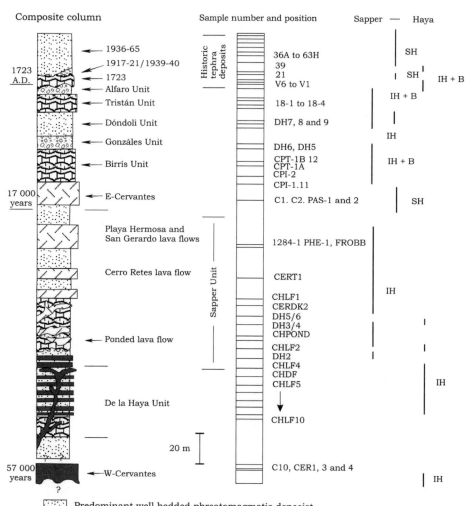

Figure 4. Generalized stratigraphic section of the most recent part of Irazú volcano.

TABLE 2. NEW $^{40}Ar/^{39}Ar$ DATES OF IRAZÚ

| Sample and unit no. | Material | Plateau (ka) | Total fusion (ka) | Isochron (ka) | Initial ratio | MSWD | Location |
|---|---|---|---|---|---|---|---|
| CR-IZ-02-1 W-Cervantes | wr | 8 ± 7 | 26 ± 15 | **57 ± 13** | 287 ± 2 | 1.3 | Boquerón, 9°53.437–83°51.308 |
| CR-IZ-02-2 E-Cervantes | wr | **20 ± 12** | 7 ± 19 | 90 ± 30 | 289 ± 3 | 0.8 | Oratorio, 9°53.939–83°50.091 |
| CR-IZ-02-5 Paraíso | wr | **594 ± 16** | 580 ± 50 | 598 ± 20 | 293 ± 7 | 1.9 | Naranjo, 9°53.046–83°45.045 |
| CR-IZ-02-17 San Jerónimo | biotite | **855 ± 6** | 850 ± 8 | 866 ± 9 | 292 ± 3 | 1.2 | San Jerónimo, 9°51.726–83°47.718 |
| CR-IZ-02-17 San Jerónimo | biotite | **862 ± 9** | 862 ± 10 | 833 ± 16 | 304 ± 8 | 1.1 | San Jerónimo, 9°51.726–83°47.718 |
| CR-IZ-02-17 San Jerónimo | plagioclase | 824 ± 8 | 814 ± 9 | **847 ± 11** | 284 ± 3 | 0.56 | San Jerónimo, 9°51.726–83°47.718 |
| CR-IZ-02-19 Reventado | biotite | **136 ± 5** | 134 ± 10 | 146 ± 10 | 293 ± 2 | 1.1 | Laguna, Tiribí river, 9°57.116–83°55.742 |
| CR-IZ-02-20 Pico de Piedra | wr | 540 ± 3 | 513 ± 6 | **569 ± 6** | 269 ± 4 | 0.74 | Las Nubes, Hacienda Abigail, 9°59.191–83°56.561 |

*Note:* The step heating data from which these dates were obtained is available in a file called CAGeochron.zip at www.rci.rutgers.edu/~carr/. Ages in bold are considered the most reliable. See Appendix for the criteria used in selecting the most reliable dates. MSWD—mean square of weighted deviates; wr—whole rock.

Haya flows are best exposed in the north wall of the Diego de la Haya crater. They are also exposed high up in the section on the north and south flanks of Irazú. The Diego de la Haya lavas represent a useful marker unit because they are well-exposed and have distinctly high MgO contents. On the eastern wall of Diego de la Haya crater, a 46-m-wide basaltic dike intruding into densely welded agglutinate shows many bifurcations (Fig. 5B; sample CHDK).

## Sapper Unit

A >500-m-thick tephra sequence with interbedded lava flows directly but unconformably overlies the uppermost Diego de La Haya flows. U/Th radiometric data by Allègre and Condomines (1976) on samples from the Sapper Unit yield ages of 68,000 ± 26,000 yr and 36,000 ± 14,000 yr. Cerro Alto Grande is an erosional remnant of several basaltic andesite and andesite lava flows located 4 km NNE of Cráter Principal. The erosion that formed Cerro Alto Grande and the Río Toro Amarillo valley lying between Turrialba volcano and Cerro Alto Grande occurred before the upper Sapper flows came down the north flank. In the upper part of the Diego de la Haya crater, there are two basaltic andesite lava flows, which are chemically similar to the Sapper unit and are, therefore, included into the unit. Outside the crater, these basaltic andesite lava flows are laterally continuous for several kilometers. The south wall of the Diego de la Haya crater, just below the tourist overlook, shows an erosional unconformity representing the inside of a crater, against which a basaltic andesite flow ponded. The pond is exposed south of the Cráter Principal and accounts for the flat surface of the Playa Hermosa bench there. The 5-m-thick Cerro Retes basaltic andesite flow is exposed beyond the erosional cliffs at the end of the summit ridge and is also part of the Sapper unit.

The Sapper dikes are a set of three 30–70-m-high spines extending northward from the west end of the summit ridge. The dikes cross-cut the upper Reventado flows and tephra layers. The largest of the three dikes (sample CERDK2) contains many small xenoliths of wall rock, giving it a brecciated appearance. Another large dike, located ~1 km south of the highest point of Irazú, cuts the Sapper Unit.

## Cervantes Units

Cervantes is the most recent lava flow complex at Irazú and one of the largest flow fields in Central America, covering 42 km² and having a volume of 1 km³ (Figs. 2 and 3). Compound lava flows erupted from two of several N-S aligned vents on the southeastern slope of Irazú volcano (Thomas, 1983; Tournon, 1984). The Western Cervantes Unit (10.5 km², 0. 17 km³) is basaltic with 5 wt% less silica than the Eastern Cervantes Unit (31.5 km², 0.88 km³) of basaltic andesite composition. The western flow contains phenocrysts of olivine (7%), clinopyroxene (4%), and rare orthopyroxene (0.5%). The groundmass has a distinct pilotaxitic texture with microlites of olivine, pyroxene, titanomagnetite, and ilmenite. The eastern composite flow contains phenocrysts of plagioclase (23%–30%), clinopyroxene (6%–9%), orthopyroxene (1%–4%), and some olivine and titanomagnetite. As much as 70% of the groundmass consists of plagioclase microlites, with the remainder including small amounts of clinopyroxene, titanomagnetite, and ilmenite in brown, rhyolitic glass or partly devitrified matrix. Carbonized vegetation collected in the Pleistocene lake and fluvial beds immediately below the eastern lava flow provide a maximum age of 13,000 ± 300 yr (Murata et al., 1966) or 14,260 ± 160 yr (see Alvarado, 1993; Alvarado et al., 2004). A calibrated and average age is ca. 16,840 yr B.P. (Alvarado et al., 2004). A Cervantes lava sample (unknown locality) yielded

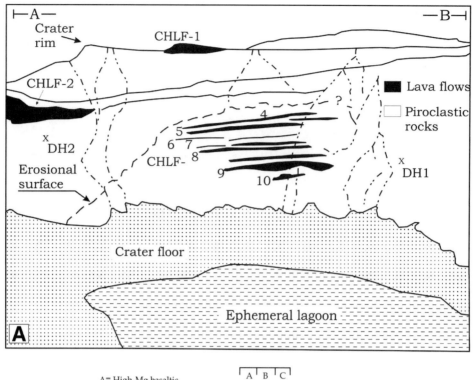
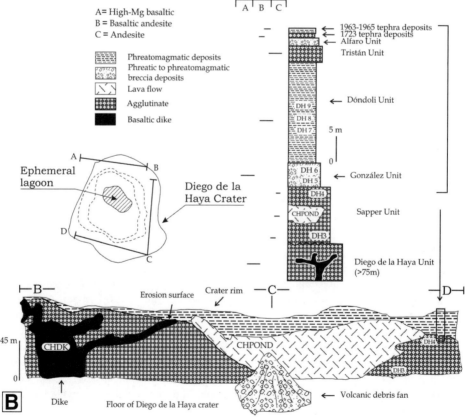

Figure 5. (A) Stratigraphic section of the north wall Diego de la Haya crater. Note the erosion surface between lavas 4 and 2. Samples CHLF10 to CHLF5, DH1 and DH2 are basalts. Samples CHLF4, CHLF2, and CHLF1 are basaltic andesites. The location of the section is shown in B. (B) East and south walls of Diego de la Haya crater showing part of the lava and pyroclastic rock succession.

a $^{238}$U/$^{230}$Th age of 23,000 yr (Allègre and Condomines, 1976). These ages are similar to our new $^{40}$Ar/$^{39}$Ar ages for the eastern flow (20 ± 12 ka). The western flow, although morphologically similar, is slightly older (57 ± 13 ka; Table 2).

## Birrís Unit

Exposed directly east of the Diego de La Haya crater in the Laguna cinder cone, the Birrís Unit consists of agglutinates, ash, lapilli, and breccia deposits of "Hawaiian," Strombolian, and phreatomagmatic eruptions. Interbedded lava flows include spatter fragments and banded scoria.

## Gonzáles Unit

This unit consists of widespread phreatic explosion breccias composed of large, angular fragments of yellow, hydrothermally altered basalt and overlies the ponded flow of the Sapper Unit. The unit consists of two breccia layers separated by dry surge deposits <1.5 m thick. The lower breccia (1 m thick) has symmetric grading (reverse to normal), while the upper breccia (3 m thick) begins with reverse grading and ends with normal grading. The composition of two samples (DH5, DH6) suggests altered Diego de La Haya basalt.

## Dóndoli Unit

Well exposed in the south wall of Diego de la Haya crater, The Dóndoli unit consists of 28 m of gray, brown, and pink, thin, unconsolidated, interbedded phreatomagmatic and rare phreatic tephra deposits. There are also some reworked tephras, isolated bombs, juvenile blocks, and layers of hydrothermally altered lithics and cinders. In the Playa Hermosa rim, there are some pyroclastic flows and reworked tephra deposits. These are in compositionally and texturally discrete stratigraphic packets interbedded with primary tephra deposits that have been strongly affected by reverse and normal faults (Fig. 5B).

## Tristán Unit

The best exposures of the Tristán Unit are in the southern part of Cráter Principal, where the thickness is 11 m. Thickness decreases toward the Diego de la Haya crater. About 95% of this crudely layered sequence consists of a massive coarse-grained fall deposit composed of angular to subangular glassy clasts, highly vesicular bombs, and lapilli and lithics, which are hydrothermally altered near the base. A layer of bombs shows banded scoria (ALGI41). Petrographically, these rocks are basaltic andesites with a matrix of brown glass.

## Alfaro Unit

The Alfaro Unit consists of bedded, variably colored (orange, red, gray, brown, white), fine-grained phreatic tephra formed by wet surges, alternating with breccias deposited by ballistic fall that contain clasts ranging from a few centimeters to 0.5 m. The clasts include rare juvenile blocks and several types of hydrothermally altered lithics. The total thickness in the southern part of the Cráter Principal is ~4 m.

## Historic Tephra Deposits

The historic tephra deposits begin with a scoria layer, up to 6.5 m thick, deposited by Strombolian eruptions that initiated the 1723 A.D. eruption. This layer consists of a non-graded fallout composed of highly vesicular, black to dark brown bombs and lapilli (hereafter referred to as "1723 scoria") with a few hydrothermally altered lapilli and rare white, andesitic, vesiculated lapilli (ALGI35 and ALGI36). This layer is overlain by a 1.2-m-thick, laminated, lapilli-bearing, gray ash showing plastic deformation and sag structures. This layer is interpreted as a phreatomagmatic deposit. Above this layer are bomb and block-bearing coarse ash deposits also ~1.2 m thick. The total thickness of the 1723 deposits is ~9 m. The next layers are thin ash deposits from eruptions in 1917–1920 and 1939–1940. These yellow to pale red laminated ashes have cross to dune lamination but no significant grading and are interpreted as wet surge deposits. The uppermost layers are the phreatomagmatic deposits of the 1963–1965 eruption (hereafter referred to as "1963–1965 scoria"). These consist of pyroclastic surge and fallout deposits with some slide blocks and interbedded, reworked tephra. Several authors misinterpreted the 1723 tephra deposits as part of the 1917–1920 or 1963–1965 explosive deposits; however, stratigraphic, petrographic, and volcanological evidences suggests a clear contrast between these deposits (Alvarado, 1993).

## METHODS

About 150 stratigraphically controlled lava and tephra samples were studied petrographically. Granulometric analyses of tephra were carried out at the Instituto Costarricense de Electricidad (ICE) and at the Dipartimento Geomineralogico of the Università degli studi di Bari, Italy, and were examined by scanning electron microscope. About 400 mineral and glass analyses were obtained with the ARLSEMQ electron microprobe of the Centro di Studio per la Stratigrafia e Petrografia delle Alpi Centrali (Milan University). These new data supplement the detailed petrographic and mineral data of Tournon (1984). The data used here include 71 rock analyses from Irazú. Major, trace, and rare earth element data were determined by direct current plasma–atomic emission spectroscopy at Rutgers University using methods described in Feigenson and Carr (1985). Data are in two files located at M.J. Carr's Web site, www-rci.rutgers.edu/~carr/index.html. CAGeochem.zip is a comprehensive list of Central American geochemical data from Rutgers and related laboratories. Irazu.zip has the same format but contains only the geochemical data from Irazú.

## Petrography and Mineral Chemistry

The Diego de la Haya and Cervantes basalts are high-K basalts and basaltic andesites and contain large (up to 6 mm) euhedral plagioclase phenocrysts, riddled with oriented inclusions. Some plagioclase crystals are reversely zoned ($An_{50}$ to $An_{70}$) or show patchy zoning. Olivine phenocrysts are euhedral with high-Mg and Ni content in the cores ($Fo_{90}$ with up to 0.4% NiO). Rims are zoned ($Fo_{80}$-$Fo_{70}$). Clinopyroxene is augite with similar core and rim compositions of $Wo_{39}En_{44}Fs_{16}$. Rare euhedral orthopyroxene is rimmed by clinopyroxene. The Fe-Ti oxides are chromian magnetite and titanomagnetite. Relatively homogeneous Cr-rich spinel or picotite ($Cr_2O_3$ = 37–39 wt%, 24–25 wt% $Al_2O_3$; 14 wt% MgO; 0.5–0.7 wt% $TiO_2$; Tournon, 1984) occurs as rare small euhedral crystals enclosed within euhedral olivine phenocrysts. The groundmass is mainly plagioclase ($An_{70}$ to $An_{60}$), olivine ($Fo_{60}$ to $Fo_{66}$), clinopyroxene ($Wo_{44}En_{44}Fs_{11}$), titanomagnetite and ilmenite, rare biotite (<0.5 modal %), and interstitial alkali feldspar.

The basaltic andesites from the 1723 eruption contain two populations of plagioclase; the dominant one is $An_{89}$-$An_{78}$, and the minor one is $An_{69}$-$An_{54}$. Some clinopyroxenes from basalts or the basaltic andesites of the 1723 scorias are twinned, complexly zoned, or have anomalous interference colors with oscillatory and sector zoning similar to titanaugites, but the $TiO_2$ contents are <1.5%, much lower than true titanaugites.

Moderate-K basaltic andesites are the most common lavas at Irazú. Plagioclase phenocrysts, commonly with sieve textures, vary greatly in composition and degree and kind of compositional zoning. Some olivine phenocrysts, up to 2 mm long, have picotite inclusions and/or are altered to iddingsite. Olivine is present even in some andesitic scorias. Some samples contain large olivines with inclusions of opaques and rims of orthopyroxene; other olivines have rims composed of orthopyroxene and tabular microphenocrysts of plagioclase, pyroxene (principally opx), and rare Fe-Ti oxides and biotite (i.e., Tristán Unit). Augite occurs as euhedral microphenocrysts and phenocrysts up to 4 mm in size. Orthopyroxene ($En_{73}$-$En_{69}$) is usually rimmed by augite. The Fe-Ti oxide is titanomagnetite. The pilotaxitic to intersertal groundmass contains abundant microlites of plagioclase, clinopyroxene, titanomagnetite and ilmenite, rare olivine, and biotite set in often partially devitrified glass.

The andesites are highly phyric with phenocrysts of plagioclase ($An_{45}$-$An_{51}$) up to 5 mm long that include oikocrysts of pyroxene, opaques, rare glass, and apatite. Augite (core $Wo_{42}En_{45}Fs_{14}$) and orthopyroxene form large phenocrysts. Rare biotite (with opaque rims), and apatite occur as microphenocrysts or as inclusions in pyroxene. Biotite phenocrysts occur only in andesitic, high-potassium rocks (e.g., Cerro Sapper, Gurdián, and Tierra Blanca lava flows). Amphiboles with ferroan pargasite compositions are restricted to rocks with $SiO_2 \geq 59.5\%$, and most have decomposition rims of oxide. The groundmass of the andesites consists of plagioclase, clinopyroxene, orthopyroxene, titanomagnetite, rare ilmenite, alkali feldspar, and cristobalite. Dacitic rocks are found as banded scoria from the Tristán and Birrís units.

Plagioclase phenocrysts in typical examples from the 1963–1965 basaltic andesite scorias comprise two types: Type A and Type AB. Type A phenocrysts with sharp boundaries include rounded or embayed grains characterized by the general absence of dusty or cellular textures and overgrowths. Type A phenocrysts are clean and uniform but some are twinned and slightly zoned ($An_{45}$-$An_{55}$) with a narrow variation within each crystal of less than $An_5$. The largest phenocrysts in any given sample are almost exclusively Type A. Type AB plagioclases are commonly irregularly shaped and have clear central cores surrounded by a dusty and cellular zone that in turn is bounded by a clear overgrowth. Optically continuous twinning traverses the entire grain. Zoning is mostly normal ($An_{78}$-$An_{50}$) and seldom reverse or complex. Most cores are $An_{49}$-$An_{57}$, and the rim (second generation) is $An_{58}$-$An_{70}$. Clinopyroxene crystals have an average composition of $Wo_{39}En_{47}Fs_{14}$. No differences between these composite crystals and isolated phenocrysts were found. Two types of olivines can be distinguished texturally and chemically in the 1963–1965 scorias: olivine ($Fo_{78}$-$Fo_{75}$) with augite reaction rims and MgO-rich olivines ($Fo_{90}$-$Fo_{87}$) without rims. A few olivines have an orthopyroxene rim with Fe-Ti oxides having lamellar structure (symplectite-texture). Inclusion patterns (plagioclase + oxides) are similar to those in clinopyroxene, and strong resorption is rare. Plagioclase, orthopyroxene, opaques, and glass inclusions in the cores are common.

## GEOCHEMISTRY

### Identification of Groups and Fractional Crystallization

Irazú lavas show several petrographic features consistent with fractional crystallization from a basaltic parent. Phenocryst assemblages and mineral proportions change systematically with increasing $SiO_2$; olivine is replaced by orthopyroxene, then amphibole; biotite occurs in the most siliceous lavas. Groundmass plagioclase is systematically more sodic than phenocrysts, and the minerals in glomeroporphyritic aggregates are identical to the phenocrysts. Changes in fractionating mineral assemblages are suggested by inflections in Harker diagrams (Fig. 6).

However, the major and trace element data for Irazú span a wide range that cannot be explained by crystal fractionation from a common parent alone. The clearest distinction among these groups occurs in Ba versus $SiO_2$ (Fig. 7). By combining stratigraphic, geochemical, and petrographic observations, we divide the samples into three subsets: *Haya batch*, high-K basalts and basaltic andesites (filled symbols, Fig. 7); *Sapper batch*, moderate to high-K basaltic andesites and andesites (open symbols, Fig. 7); *Hybrids*, mostly basaltic andesites (x and + symbols, Fig. 7).

The Haya batch includes the Diego de La Haya unit, the Western Cervantes composite lava field, and the 1723 scorias. The Sapper batch includes the Sapper, Birrís, and Tristán units.

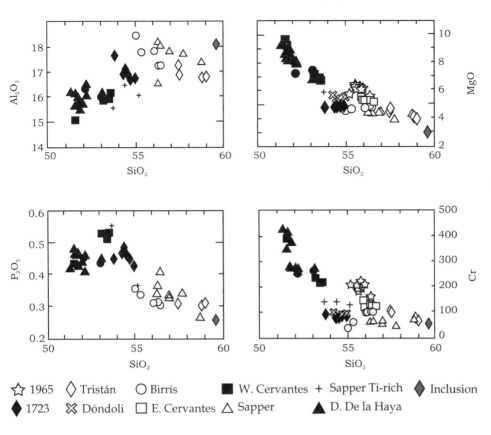

Figure 6. Harker variation diagrams. Fractionation is dominated by olivine and in relatively minor grade by clinopyroxene and Fe-Ti oxides in rocks with $SiO_2 < 53\%$, by the removal of ol + px + mt + plag + apat in rocks with $59\% > SiO_2 > 53\%$ and by px + plag + mt + hb in rocks with $SiO_2 > 59\%$.

The Eastern Cervantes lavas and the 1963–1965 scorias show abundant field and petrographic evidence for mixing, and their compositions lie between the other two groups. Solid symbols (Fig. 7) define the high-Ba, high-K trend (Haya batch). Open circles, diamonds, and triangles define a moderate-Ba, moderate-K trend (Sapper batch). Open symbols (x and +) between these groups appear to be primarily mixtures. The small pluses are Ti-rich lavas that erupted near the stratigraphic transition between the Diego de la Haya and Sapper units. Except for Ti, they are very similar to the rest of the Sapper rocks. Rocks with elevated Ti and slightly elevated Fe most likely represent an interval during which magnetite fractionation was suppressed.

The Haya and Sapper batches differ most clearly in incompatible element contents, but there are variations in major elements, indicating differences in crystal fractionation. For example, in Figure 8 we plot molar element ratios and display vectors representing the direction of magma evolution resulting from removal of phenocrysts. Sapper rocks, comprised mostly of more silicic lavas, define a line with a gentle slope, indicating primarily plagioclase and clinopyroxene fractionation. These two minerals are required, but orthopyroxene and olivine cannot be ruled out. In contrast, the Haya basalts define a clearly separate line with a steeper slope that requires substantial olivine and or orthopyroxene fractionation in addition to plagioclase and clinopyroxene. Again, there is a clear separation between the two trends.

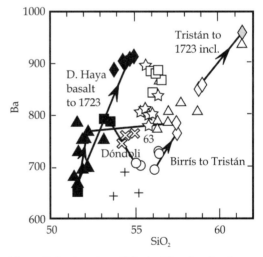

Figure 7. Ba (ppm) vs. $SiO_2$ (wt%) at Irazú volcano. The lines show a proposed mixing relationship between the two parent magmas. The arrows show the fractionation trends of Irazú. Symbols as in Figure 6.

In general terms, the inflection in the $Al_2O_3$ versus $SiO_2$ plot (Fig. 6) indicates that plagioclase removal becomes significant at 54 wt% $SiO_2$. The scatter suggests accumulation of plagioclase. Separation of relatively low density plagioclase phenocrysts from melts is less efficient than the separation of the denser, mafic phe-

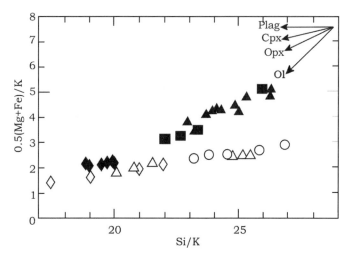

Figure 8. Molar element ratios of 0.5(Mg+Fe)/K versus Si/K. Vectors represents the direction of magma evolution resulting from removal of phenocrysts (Russell et al., 1990). Symbols as in Figure 6. Cpx—clinopyroxene; Plag—plagioclase; Ol—olivine; Opx—Orthopyroxene.

nocrysts, and this is likely one reason that plagioclase dominates the phenocryst assemblage of Irazú basalts. An inflection in $P_2O_5$ suggests that apatite begins to crystallize and to be removed at ~53–54 wt% silica. MgO and Cr form two linear segments, a steep one for the Haya batch, followed by a less steep one for the more siliceous Sapper batch. The hybridized samples fall in the corner between the two batches (Figs. 6 and 7).

Fractional crystallization and mixing processes that might explain the major oxide variations were tested using least-squares calculations (Bryan et al., 1969). A plausible solution was obtained for various parent-daughter combinations when the fit between the observed and calculated parent lava was within analytical error of the major elements. Secondarily, the Ba, Sr, and Rb contents were expected to be within 15% of the observed parent composition. Because small variations in mafic minerals can greatly change the V, Cr, and Ni contents, these elements were ignored. Mineral analyses of olivine or orthopyroxene, magnetite, plagioclase, and clinopyroxene were taken from the parent lava or from lavas with similar silica content. Successful parent-daughter relationships were difficult to find because of the presence of two magma series. We found no way to cross the gap between the Haya and Sapper batches. Successful parent-daughter combinations are shown in Figure 7 and described as follows.

The Diego de La Haya unit and the Western Cervantes flow include primitive phyric basalts with 8–9 wt% MgO (CHLF9, CHLF10 and C1O), likely parental compositions. These basalts are the most primitive lavas of the Costa Rican VF. We chose primitive Diego de la Haya basalts (either CHLF 10 or CHLF 9) as parent and derived the more evolved rocks of the Diego de La Haya unit by 8% to 17% fractional crystallization. The other important member of the Haya batch, the 1723 scoria, can be derived from the Diego de La Haya basalts by 28% fractional crystallization. The subordinate andesite inclusions (felsic pumice) in the 1723 scoria cannot be derived by fractionation from the host basaltic andesite, but by ~17% fractionation from the basaltic andesite of the Tristán Unit. The chemical composition and petrography of the Tierra Blanca andesite lava flow (cf. Tournon, 1984) are similar to the andesite pumice inclusions in the 1723 scoria.

The Tristán Unit, a thick (up to 11 m) basaltic andesite tephra sequence covering the Playa Hermosa volcanic terrace, shows an upward increase in $SiO_2$ content and a decrease in $Al_2O_3$ and MgO. The mafic end of this deposit can be derived from Birrís-type magma by 10% fractional crystallization of plagioclase, clinopyroxene, and some olivine and magnetite (Alvarado, 1993).

## Magma Mixing and Commingling

Banded scorias are typical macroscopic indicators of magma commingling at Irazú. Birrís and Tristán tephra commonly contains two types of pumice and scoria, one white or light brown andesite to dacite ($SiO_2$ = 60–64 wt%), the other black or red basaltic andesite ($SiO_2$ = 53–58 wt%). The bands range in thickness (<0.5 mm to several cm) and are intimately mingled and strongly vesiculated. Both components were molten at the time of eruption judging from the occurrence of vesicles, flowage structures, and glass.

Several phenocrysts are partly resorbed and therefore out of equilibrium with the surrounding melt. The occurrence of orthopyroxene ± olivine ± hornblende in some layers of the 1723 scoria suggests minor mixing between hornblende andesite and basaltic andesite magmas, because the dominant 1723 basaltic andesite lacks amphibole.

Olivine ($Fo_{89}$) with picotite inclusions abundant in basaltic andesites (e.g., Eastern Cervantes and 1963–1965 scoria) are compositionally similar to olivine phenocrysts ($Fo_{91}$) in the most primitive basalts found at Irazú. Ewart (1982) reports that most olivine phenocrysts in andesites lie in the range $Fo_{70}$-$Fo_{85}$ and olivines with picotite inclusions in andesites are inherited from a basalt component. The magnesian olivine at Irazú could not have been in equilibrium with a basaltic andesite liquid, much less the andesite to rhyolite liquids. The 1963 scoria contains two olivine populations: euhedral magnesian olivine Mg# (0.87–0.90) without reaction rims, whose Mg# is higher than that of coexisting pyroxenes (0.70–0.85), and olivine with augite rims that have a lower Mg#, compatible with that of pyroxenes in the same rock. The olivine phenocrysts rimmed by clinopyroxene are in approximate equilibrium with the liquid. Forsteritic olivines were in equilibrium with more mafic magma at an earlier stage, and there was insufficient time to reequilibrate prior to eruption. Additional geochemical evidence of magma mixing is provided by the peculiar composition of the 1963–1965 scoria, which has higher MgO, Cr, and Ni contents than lavas of comparable $SiO_2$ content. The high concentrations of Ni (71–101 ppm) and Cr (156–217 ppm) preclude an origin by crystal fractionation involving mainly olivine (Taylor et al., 1969a, 1969b). Such high Ni and Cr values are not found in normal basaltic andesites or andesites at Irazú but are common in the basalts.

Potential hybrid lavas were modeled with the same least squares technique that was used to model fractional crystallization. However, a larger error was allowed because we used only two components, compared to as many as five for the fractional crystallization model. More than 30 different mixing models were tested to determine a realistic combination that would reproduce the 1963–1965 magma. The best and geologically most realistic one indicates that these basaltic andesites are derived from a mix of MgO-rich basalt (32%), such as the Western Cervantes or Diego de la Haya Unit, with a basaltic andesite (68%) such as the underlying thick scoria deposit of the Tristán unit. The sum of squared residuals is 0.13 (Alvarado, 1993), so this model is compatible with the available data. Some variability in composition could reflect slight modification by low-pressure crystal fractionation. Pure mafic or basaltic andesite end members were not erupted in the 1963–1965 event. A representative sample of the hybrid Dóndoli tephra (Fig. 7) can similarly be derived from a mix of the most evolved Western Cervantes lava (44% of sample CER-3, Haya batch) and a typical Birrís tephra (56% of sample I1, Sapper batch). The sum of squared residuals is 0.11. In this case, the end members were the most recently erupted representatives of the Haya and Sapper batches, just below the Dóndoli section in the stratigraphy.

## Geothermometry and Geobarometry

Applying geothermometry and barometry to Irazú lavas is in many cases prohibited by the disequilibrium state of the phenocryst assemblage and, where attempted, requires care and restraint. Two pyroxenes (cpx + opx) are present in all samples from Irazú. The thermometer formulated by Brey and Köhler (1990) was used to obtain magmatic temperatures calculated from phenocryst rim compositions, which presumably record magmatic temperatures at the time of eruption, from the cores of pyroxenes, and from orthopyroxene jacketed by a thin clinopyroxene rim. The temperatures obtained were similar, ranging from 1000 to 1176 °C in the basalts, 1004 to 1095 °C in the basaltic andesites from 1963–1965 scoria, and 922 °C in the Tierra Blanca hornblende andesite. All estimates assumed a pressure of 5 kb. A change from 5 to 2 kb in this model has a negligible effect on temperatures. Using the temperature estimates from two-pyroxene thermometry, it is possible to estimate pressure following Herzberg (1978). The results are ~4.2–4.6 kb for the basalts and ~3 kb for a hornblende andesite.

A pseudo-Quaternary projection from plagioclase after Grove et al. (1982) provides additional qualitative pressure information (Fig. 9). Most Irazú basalts, basaltic andesites, and andesites plot within the olivine field determined from 1-atm experiments. The high-K Haya basalts are farthest from the 1-atm cotectics, either because they come from a deeper chamber or, more likely, they have accumulated olivine. The Sapper batch defines arrays roughly parallel to the cotectics, indicating fractional crystallization at moderate pressure. Only the 1723 samples approach the 1 atm cotectic. Because this eruption included several unusually large shallow volcanic earthquakes, Alvarado (1993) inferred that this eruption came from a distinctly shallower chamber.

## DISCUSSION

### Edifice Construction-Destruction

Reconstruction of the volcanic history of Irazú since the late early Pleistocene is based on geological and geochronological data ($^{40}Ar/^{39}Ar$, U/Th, $^{14}C$). The eruption of the San Jerónimo ignimbrite (ca. 854 ka) was followed by the construction of the Paleo-Irazú (ca. 594–569 ka) and, after a prolonged volcanic gap, by the formation of Neo-Irazú (136 ka–present). These ages are in complete agreement with the comprehensive regional $^{40}Ar/^{39}Ar$ investigation by Gans et al. (2003), in which the volcanoes at the Cordillera Central grew in three major constructive phases: Proto-Cordillera (980–850 ka), Pre-Cordillera (630–400 ka), and Neo-Cordillera (170–0 ka), separated by dormant (erosion) intervals (100–300 ka) and huge explosive ignimbrites (920, 570, 490, 440, and 325 ka). Thus, the last extensive eruptions at Irazú started at the end of the Middle Pleistocene (ca. 136 ka) and extended to the present, representing most of the less eroded flanks. More recent flank lavas, for example, the Western Cervantes flow (~0.17 km³) of basaltic composition (ca. 57 ka), possibly were contemporaneous with the Diego de la Haya basalts at the summit (~3 km³), suggesting a strong mafic event. In a similar way, the Eastern lava flow (~0.88 km³) of Cervantes (16,840 yr B.P.) is contemporaneous with a series of basaltic andesite cinder cones in a N-S trend, representing extensive volcanism at the end of the Pleistocene.

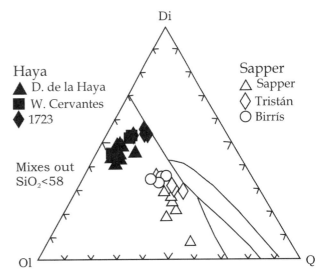

Figure 9. Pseudo-Quaternary projection from plagioclase after Grove et al. (1982).

On the other hand, there are well-documented but poorly dated large debris avalanche deposits (see Siebert et al., this volume). There are at least four avalanche deposits since the Middle Pleistocene, and nearly a dozen semicircular escarpments, suggesting that sector collapse and related debris avalanches and debris flows are common phenomena at Irazú. A recent small magnitude slide, accompanied by phreatic (hydrothermal) explosions, occurred on 8 December 1994.

## Nature of Magma Plumbing System

The major conclusions from the petrography, geochemistry, geobarometry, and geophysics are that two magma systems or batches are coexisting, evolving, and occasionally mixing (Fig. 10). The mineral chemical data indicate that the Irazú basalts equilibrated at 1000–1186 °C and at pressures of 4.2–4.6 kb; basaltic andesites at ~1000–1100 °C and at similar pressure; and a hornblende andesite at ≥920 °C and ~3 kb. Thus, pressures obtained in different rock types suggest a depth of crystallization between 10 and 24 km, most probably ~12–15 km. Mixing at this level between a primitive magma and a slightly evolved magma may be the cause of different generations of phenocrysts in basalts. Petrographic evidence for magma mixing in basaltic rocks at Irazú includes (1) mineralogical disequilibrium textures such as olivine phenocrysts with orthopyroxene rims, or resorption in olivine, and dusty zones on the margin of plagioclase phenocrysts; (2) magnesian olivine with a high Mg/(Mg+Fe) ratio (0.90–0.73) compared to coexisting pyroxenes (0.72–0.76) or to Mg/Fe ratios of the enclosed basalt; and (3) reverse zoning in plagioclase phenocrysts ($An_{72}$-$An_{53}$). Mixing of basaltic magmas can be rapid because of similar densities and viscosities.

In the high-Mg basaltic magmas at Irazú, spinel grew first, immediately followed by olivine, indicating that the chemical composition of the primary melt was close to the olivine-spinel cotectic. This magma was repeatedly mixed into a basaltic andesite magma reservoir, and phenocrysts from both magmas are preserved in some mixtures. Dispersed picotite-bearing olivine in several basaltic andesites at Irazú indicates recurrent episodes of basaltic intrusion and mixing with basaltic andesite magma bodies. Extended storage or ascent may cause nearly homogeneous hybrids (Barton et al., 1982). This appears to be the case for the Eastern Cervantes lava and the 1963–1965 scoria. Complete hybridization is most feasible where viscosity and temperature contrasts are small. Sakuyama (1984) and Kouchi and Sunagawa (1985), however, showed that the combined effect of mechanical mixing, molecular diffusion, and size of the crystals is very efficient in producing a homogeneous melt even if the initial chemical and viscosity contrast is large. The presence of phenocrysts >0.5 mm in diameter, a size common at Irazú, considerably increases the efficiency of mixing.

Two or more generations of olivine, plagioclase, and clinopyroxene phenocrysts separated by textural and compositional breaks occur in several units, including the two larger historic eruptions (1723 and 1963–1965). Mixing is the simplest mechanism that accounts for bimodal populations of phenocrysts. Euhedral olivine phenocrysts and skeletal or anhedral phenocrysts are present in the same thin section. The variation in texture suggests that the ascending magmas underwent both slow and fast cooling and/or mixing of a cooler magma with a hotter magma. Apparently both processes occurred and the cooling rate at the upper crustal level was too high to allow more extensive reequilibration.

In Holocene Strombolian deposits at Irazú, andesite-dacite inclusions (pumice enclaves) physically commingled with basaltic andesite. These banded scorias indicate that separate magmas coexisted at different levels in the plumbing system and were physically mixed shortly before and during eruption. The hornblende bearing dacite component is subordinate (<20%) in prehistoric tephra deposits or rare (<1%) in andesite inclusions (pumice enclaves) in the 1723 scorias. In 1723 basaltic andesitic scorias, hornblende xenocrysts (from assimilation of the andesitic inclusions) are unrimmed euhedral hornblendes, indicating that magma ascended rapidly during a few days or hours (Rutherford, 1993). This short interval between mixing and eruption requires the andesite magma chamber to be of small volume, of the same order or less than that of the introduced mafic magma supplied from depth (Sakuyama, 1984). The contrasts in density and temperature between the two magmas (olivine-basaltic andesite and hornblende-andesite: 53%–55% versus 59%–60% $SiO_2$) was probably large, suggesting that the intrusion of basaltic andesite magma in a small and shallow andesite magma chamber triggered the eruption. These features suggest that the volume of the silicic component in the magma chamber was always small. The absence of andesitic and dacitic lavas during the Holocene supports this conclusion. Several studies, including this study, indicate that banded tephras usually represent mechanical mixing events at a depth <8 km (i.e., Wörner and Schmincke, 1984; Wörner and Wright, 1984; Melson et al., 1990; Freundt and Schmincke, 1992; Rutherford, 1993).

The origin of the Haya and Sapper magma batches cannot be established from the limited trace element data for Irazú. Both groups display a geochemical variation common in Central America. Both are enriched in elements such as Pb, Sr, U, Ba,

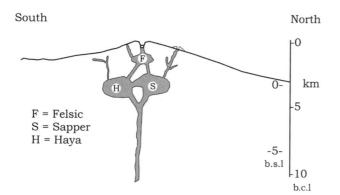

Figure 10. Simplified magma plumbing model for Irazú volcano (see text for details).

K, Rb and Th, which are likely to be carried in a hydrous fluid released from a subducted plate (Fig. 11). The Sapper basaltic andesite has slightly higher Pb, Sr, U, Ba, K, Rb and Th, but lower high field strength elements (HFSE) and rare earth elements (REE). This has been interpreted to mean higher flux from the subducted slab, leading to a higher degree of melting in the mantle (e.g., Carr et al., 1990; Leeman et al., 1994). The high flux provides high Pb, Sr, Ba, etc., and the high degree of melting leads to lower HFSE and REE because these elements come primarily from the mantle.

The 1723 scoria (misinterpreted as 1963–1965 tephra deposits by Clark et al., 1998) was generated most likely by dehydration partial melting of garnet-bearing mantle that had been enriched previously by a slab fluid (similar to carbonatite).

## Nature of Eruptive Cycle and Potential Hazards

At Irazú, each main tephra cycle starts with a relatively "dry" eruption (Hawaiian-like or Strombolian eruptions) that produce coarse, juvenile, ballistic-dominated pyroclastic deposits. Later in the eruption, "wet" (phreatic and phreatomagmatic) explosions occur that generate wet and dry surge and fallout deposits. Because no systematic chemical difference exists between the deposits from Strombolian and phreatomagmatic eruptions, this implies that the main control on the eruptive style of Irazú is the magmatic gas content and the interaction between magma and water.

The stratigraphic sequence and petrological data suggest a rough alternation between Sapper batch and Haya batch magmas with some hybrid magmas in between. Close inspection of Figure 4 suggests a crude eruption sequence that has occurred two to three times at Irazú. Haya batch magma eruptions are followed by either hybridized magma or by a prolonged eruption of Sapper batch magma. Following the last two Sapper batch eruptions, there are distinct erosion surfaces, suggesting long pauses in activity. The sequence appears to start with Haya basalt, followed by a Sapper basaltic andesite, and then a period of dormancy. If this model is valid, the next eruption will be either more hybrids, like the 1963–1965 eruption, or a switch to the Sapper batch of moderate-K basaltic andesite. Because we document only the most recent 2.5 sequences, this temporal pattern is very tenuous.

The paucity of recent lava flows suggests a long-term eruptive style characterized by dominantly explosive activity, agglutinate and bomb deposits with rare lava flows, interlayered with thick phreatomagmatic deposits. Within the well-exposed stratigraphy, which covers roughly the past 40,000 yr, the phreatomagmatic deposits appear to be increasing in volume.

Although Holocene eruptions of Irazú produced primarily basaltic andesites and had a low volcanic explosivity index (VEI ≤ 3), there is persistent evidence of hornblende-andesite/dacite magma, usually in the form of small inclusions (pumice enclaves). The presence of hornblende suggests there is an $H_2O$-rich magma. It is thus important to better understand the depth and volume of these silicic magmas and their hazard implications. Siliceous

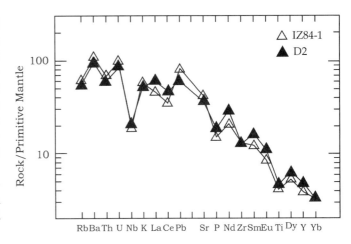

Figure 11. Chondrite normalized incompatible and trace element abundance diagram for representative Irazú lavas. Open triangles are basaltic andesites and filled triangles are basalt.

magma, especially if located in a shallow magma chamber, could lead to very explosive eruptions if suddenly unroofed by gravity sliding and associated debris avalanches or if triggered by mixing in of hot basaltic magma.

In addition, river valleys are highly susceptible to hazards from lahars, due to remobilization of pyroclastic deposits (Alvarado and Schmincke, 1994), or debris avalanches as a result of heavy syneruption or post-eruption rainfall (particularly during the rainy season from May to December), or storms, or after a strong earthquake.

Using prehistoric estimates and historic data, is possible to conclude that on average 2.6 eruptions happened every century, and one moderate eruption (VEI = 3) about every 185 yr (Alvarado et al., 2000). The economic losses in future eruptions similar to the 1723 or 1963–1965 phases will be significantly greater because of the enormous increase in population and industrial and land development of the Valle Central, as well as the large number of telecommunication, TV, and radio installations located on the summit and flanks of Irazú.

## APPENDIX

Accurate ages can be obtained on plagioclase, biotite, and whole-rock matrix via $CO_2$ laser step-heating techniques. However, the young ages and low-K contents of the basalts and andesites that make up Irazú volcano require careful evaluation of the Ar isotopic data used for age calculations; thus, a rigorous approach to selection of $^{40}Ar/^{39}Ar$ dating results is applied and described as follows.

To obtain reliable ages, $^{40}Ar/^{39}Ar$ analyses must meet the following tests. The first criteria are derived from the step-heating spectra. Individual apparent ages derived from the argon isotopes released at discrete temperature steps are plotted relative to the cumulative percentage of $^{39}Ar_K$ released in each increment relative to the total $^{39}Ar_K$ released from the completed experiment (Fig. 12). Those experiments that result in release patterns forming age plateaus are considered more reliable than those whose apparent ages vary widely relative to temperature. The plateau age is defined by gases released from a minimum of

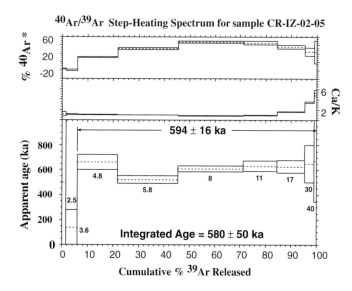

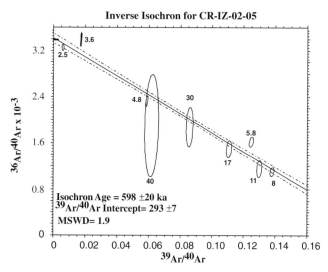

Figure 12. Step-heating spectra for sample from Irazú. Sample CR-IZ-02-05 meets all the criteria discussed in the Appendix for a reliable age. A seven-step plateau age of 594 ± 16 ka was obtained for this sample. The plateau age, defined by 94% of the total $^{39}Ar_K$ released, is concordant with the total fusion age of 580 ± 50 ka at the 95% confidence level. In addition, the isochron age of 598 ± 20 ka is concordant with the plateau and total fusion age. The isotopic data indicate an initial $^{40}Ar/^{36}Ar$ ratio of 293 ± 7, analytically indistinguishable from the accepted atmospheric ratio of 295 ± 1. On the inverse isochron, the low temperature steps plot near the y-intercept, corresponding to an accepted atmospheric ratio of 295. With each increasing temperature step, the isotopic composition moves along a mixing line toward the $^{39}Ar/^{40}Ar$ ratio, corresponding to an age of 598 ± 20 ka until the 8 W step (~890 °C). Then the isotopic composition evolves back toward the atmospheric ratio intercept along the sample mixing line. The preferred age for sample CR-IZ-02-05 is plateau age of 594 ± 16 ka. MSWD—mean square of weighted deviates.

three consecutive temperature increments whose relative ages overlap at the 95% confidence level and total 50% or more of the total $^{39}Ar_K$ released (Fleck et al., 1977). Plateaus incorporating more increments and greater total percentages of the total $^{39}Ar_K$ released during the experiment are considered of better quality and reliability.

The second group of criteria requires plotting the Ar isotopes obtained from the incremental heating experiments on inverse isochron plots, where the y-axis ($^{36}Ar/^{40}Ar$) relates to the atmospheric component of the isotopically measured gases and the x-axis ($^{39}Ar/^{40}Ar$) relates to the radiogenic component of the measured gases (Fig. 12). We consider those analyses most reliable whose isochron-derived age is analytically indistinguishable from the plateau age and the total fusion age (weighted sum of the individual relative increment ages). In addition, the initial $^{40}Ar/^{36}Ar$ ratio derived from the y-intercept should be analytically indistinguishable from the atmospheric ratio, 295 ± 1 (e.g., samples CR-IZ-02-2 wr, -5 wr, -17 bio and -19 bio in Table 2).

The quality of the incremental heating experiments can be evaluated using the goodness of fit of the $^{36}Ar/^{40}Ar$ and $^{39}Ar/^{40}Ar$ isotopic data on the isochron plot. The better defined the mixing line between the initial $^{36}Ar/^{40}Ar$ ratio and the $^{39}Ar/^{40}Ar$ ratio, the better the fit of the slope, thus the higher the quality and more reliable the isochron age. This goodness of fit is defined as the mean sum of the weighted deviates of the individual temperature ages, or MSWD (York, 1968). The MSWD is essentially the ratio of the measurement error and the observed scatter about the regression line through the points on the isochron plot. A value of 1.0 indicates that the scatter about the regression line is accounted for by the measurement errors. A value less than one suggests that the errors for the individual measurements may be overestimated. Empirical measurements of uniform isotopic material indicate that MSWD values as high as two are acceptable; however, we consider ages whose isochron plots resulted in MSWD values close to 1.0 to be the most reliable.

An additional criterion, providing further indication of the reliability of a sample's age, is the order that the isotopic data appear on the isochron plot relative to increasing or decreasing $^{36}Ar/^{40}Ar$ and $^{39}Ar/^{40}Ar$ ratios for the individual step-heating measurements (Fig. 12). These points define the "mixing line" between the initial $^{36}Ar/^{40}Ar$ ratio and the $^{39}Ar/^{40}Ar$ ratio (age) of the sample. With increasing temperature, the isotopic composition commonly moves along a mixing line from the initial $^{36}Ar/^{40}Ar$ (atmospheric) ratio toward the $^{39}Ar/^{40}Ar$ (radiogenic) ratio corresponding to the sample age until the sample reaches a unique temperature or step. In cases where increased atmospheric components are released at high temperatures, the isotopic composition should evolve back toward the initial $^{36}Ar/^{40}Ar$ intercept along the same mixing line as obtained from lower temperature increments (Fig. 12).

In the case of samples for which the isochron analysis does not indicate an atmospheric ratio (Table 2, samples CR-IZ-02-1 wr, -17 plagioclase, and -20 wr) but the MSWD is ≤2 and the mixing line is well defined, the isochron age is considered the most reliable age measurement because no assumptions are made about the isotopic composition of the initial Ar in the sample.

Three of the six Irazú samples measured indicate a nonatmospheric $^{40}Ar/^{36}Ar$ ratio and slightly disturbed step-heating spectra. The protocol described above was an important first step toward our goal of obtaining reliable ages.

## ACKNOWLEDGMENTS

Our colleagues at the Área de Amenazas y Auscultación Sísmica y Volcánica at Instituto Costarricense de Electricidad (ICE), the Universidad Nacional (UNA) in Costa Rica, the former Abteilung Vulkanologie und Petrologie (GEOMAR Forschungszentrum) at Kiel, Rutgers University, and Dartmouth

College supported our work logistically and scientifically in many ways. A. Freundt, J. Dehn, and J. Tournon critically reviewed earlier versions of this manuscript. The reviews by J. Brophy, M. Barton, J. Walker, and W.G Melson led to considerable improvement of the manuscript and are greatly appreciated. Thanks to S. Chiesa, L. Prosperi, and G. Civelli (Universitá degli Studi di Milano, Italy) for helping generously with the microprobe analyses, and L. Patiño (at that time at Rutgers Univ.) for helping with the chemical analyses. Figures were kindly drawn in different stages by W. Pérez and J. Chaves. The work of Alvarado in Kiel was supported by the German Academic Exchange Service (DAAD), that of Schmincke by the Deutsche Forschungschaft (from Dchm 250/474), and for Carr, part of the work was financed by Rutgers University. $^{40}Ar/^{39}Ar$ ages from the Rutgers noble gas laboratory was funded through National Science Foundation grant EAR-0203388. All of this is gratefully acknowledged.

## REFERENCES CITED

Allègre, C.J., and Condomines, M., 1976, Fine chronology of volcanic processes using $^{238}U-^{230}Th$ systematics: Earth and Planetary Sciences Letters, v. 28, p. 395–406, doi: 10.1016/0012-821X(76)90201-6.

Allègre, C.J., and Condomines, M., 1982, Basalt genesis and mantle structure studied through Th-isotopic geochemistry: Nature, v. 299, p. 21–24, doi: 10.1038/299021a0.

Alvarado, G.E., 1993, Volcanology and Petrology of Irazú Volcano, Costa Rica [Ph.D. thesis]: Kiel, Christian-Albrechts-Universität zu Kiel, 261 p.

Alvarado, G.E., and Schmincke, H.-U., 1994, Stratigraphic and Sedimentological aspects of the rain-trigged lahares of the 1963–1965 Irazú eruption, Costa Rica: Zentralblatt für Geologie und Paläontologie, v. I, no 1/2, p. 513–530.

Alvarado, G.E., Pérez, W., and Sigarán, C., 2000, Vigilancia y peligro volcánico, in Denyer, P., and Kussmaul, S., eds., Geología de Costa Rica: Editorial Tecnológica de Costa Rica, p. 251–272.

Alvarado, G.E., Vega, E., Chaves, J., and Vázquez, M., 2004, Los grandes deslizamientos (volcánicos y no volcánicos) de tipo *debris avalanche* en Costa Rica: Revista Geológica de América Central, no. 30, p. 83–99.

Barquero, R., Lesage, P., Metaxian, J.P., Creusot, A., and Fernández, M., 1995, La crisis sísmica en el volcán Irazú en 1991 (Costa Rica): Revista Geológica de América Central, no. 18, p. 5–18.

Barton, M., Varekamp, J.C., and van Bergen, M.J., 1982, Complex zoning of clinopyroxenes in the lavas of Vulsini, Latium, Italy: Evidence for magma mixing: Journal of Volcanology and Geothermal Research, v. 14, p. 361–388, doi: 10.1016/0377-0273(82)90070-1.

Benjamin, E.R., Plank, T., Hauri, E.R., Kelley, K.A., Wade, J.A., Alvarado, G.E., 2004, Water content of a hypothetically dry magma: the 1723 and 1963–65 eruptions of Irazu volcano, Costa Rica: Eos (Transactions, American Geophysical Union), v. 85, no. 17, International Assembly Supplement, Abstract 2607-V054B-02.

Brey, G.P., and Köhler, T.P., 1990, Geothermometry in four-phase lherzolites II. New thermobarometers and practical assessment of existing thermobarometers: Journal of Petrology, v. 31, p. 1353–1378.

Bryan, W.B., Finger, L.W., and Chayes, F., 1969, Estimating proportions in petrographic mixing equations by least-squares approximations: Science, v. 163, p. 926–927.

Burbach, G.V., Frohlich, C., Pennington, W.D., and Matumoto, T., 1984, Seismicity and Tectonics of the subducted Cocos plate: Journal of Geophysical Research, v. 89, no. 9, p. 7719–7735.

Carr, M.J., and Walker, J.A., 1987, Intra-eruption changes composition of some mafic to intermediate tephras in Central America: Journal of Volcanology and Geothermal Research, v. 33, p. 147–159, doi: 10.1016/0377-0273(87)90058-8.

Carr, M.J., Feigenson, M.D., and Bennett, E.A., 1990, Incompatible element and isotopic evidence for tectonic control of source mixing and melt extraction along the Central American arc: Contributions to Mineralogy and Petrology, v. 105, p. 369–380, doi: 10.1007/BF00286825.

Clark, S.K., Reagan, M.K., and Plank, T., 1998, Trace element and U-series systematics for 1963–1965 tephras from Irazú Volcano, Costa Rica: Implications for magma generation processes and transit times: Geochimica et Cosmochimica Acta, v. 62, no. 15, p. 2689–2699, doi: 10.1016/S0016-7037(98)00179-3.

Clark, S.K., Reagan, M.K., and Trimble, D.A., this volume, Tephra deposits for the past 2600 years from Irazú volcano, Costa Rica, in Rose, W.I., Bluth, G.J.S., Carr, M.J., Ewert, J.W., Patino, L.C., and Vallance, J.W., Volcanic hazards in Central America: Geological Society of America Special Paper 412, doi: 10.1130/2006.2412(12).

Defant, M., Clark, L.F., Stewart, R.H., Drummond, M.S., de Boer, J.Z., Maury, R.C., Bellon, H., Jackson, T.E., and Restrepo, J.F., 1991, Andesite and dacite genesis via contrasting processes: the geology and geochemistry of El Valle Volcano, Panama: Contributions to Mineralogy and Petrology, v. 106, p. 309–324, doi: 10.1007/BF00324560.

Ewart, A., 1982, The mineralogy and petrology of Tertiary-Recent orogenic volcanic rocks: With special reference to the andesitic-basaltic compositional range, in Thorpe R.S. ed., Andesites: New York, John Wiley & Sons, p. 25–95.

Feigenson, M., and Carr, M.J., 1985, Determination of major, trace and rare earth elements in rocks by DCP-AES: Chemical Geology, v. 51, p. 19–27.

Feigenson, M.D., and Carr, M.J., 1993, The source of Central American lavas: Inferences from geochemical inverse modeling: Contributions to Mineralogy and Petrology, v. 113, p. 226–235, doi: 10.1007/BF00283230.

Feigenson, M.D., Carr, M.J., Maharaj, S.V., Juliano, S., and Bolge, L.L., 2004, Lead isotope composition of Central American volcanoes: Influence of the Galapagos plume: Geochemistry, Geophysics, Geosystems, v. 5, p. Q06001, doi: 10.1029/2003GC000621.

Fleck, R.J., Sutter, J.F., and Elliot, D.H., 1977, Interpretation of discordant $^{40}Ar/^{39}Ar$ age-spectra of Mesozoic tholeiites from Antarctica: Geochimica et Cosmochimica Acta, v. 41, p. 15–32, doi: 10.1016/0016-7037(77)90184-3.

Freundt, A., and Schmincke, H.-U., 1992, Mixing of rhyolite, trachyte and basalt magma erupted from a vertically and laterally zoned reservoir, composite flow P1, Gran Canaria: Contributions to Mineralogy and Petrology, v. 112, p. 1–19, doi: 10.1007/BF00310952.

Gans, P.B., Alvarado-Induni, G., Perez, W., MacMillan, I., and Calvert, A., 2003, Neogene evolution of the Costa Rican arc and development of the Cordillera Central: Geological Society of America Abstracts with Programs, v. 35, no. 4, http://gsa.confex.com/gsa/2003CD/finalprogram/abstract_51871.htm.

Gill, J.B., 1981, Orogenic Andesites and Plate Tectonics: Berlin, Springer-Verlag, 390 p.

Grove, T.L., Gerlach, D.C., and Sando, T.W., 1982, Origin of calc-alkaline series lavas at Medicine Lake Volcano by fractionation, assimilation and mixing: Contributions to Mineralogy and Petrology, v. 80, p. 160–182.

Halsor, S.P., and Rose, W.I., 1991, Mineralogical relations and magma mixing in calc-alkaline andesites from Lake Atitlán, Guatemala: Mineralogy and Petrology, v. 45, p. 47–67, doi: 10.1007/BF01164502.

Herrstrom, E.A., Reagan, M.K., and Morris, J.D., 1995, Variations in lava composition associated with flow of asthenosphere beneath southern Central America: Geology, v. 23, no. 7, p. 617–620, doi: 10.1130/0091-7613(1995)023<0617:VILCAW>2.3.CO;2.

Herzberg, C.T., 1978, Pyroxene geothermometry and geobarometry: Experimental and thermodynamic evaluation of some subsolidus phase relations involving pyroxenes in the system $CaO-MgO-Al_2O_3-SiO_2$: Geochimica et Cosmochimica Acta, v. 42, p. 945–957, doi: 10.1016/0016-7037(78)90284-3.

Hudnut, K.W., 1983, Geophysical Survey of Irazú Volcano: [B.S. thesis]: Hanover, Dartmouth College, 82 p.

Kouchi, A., and Sunagawa, I., 1985, A model for mixing basaltic and dacitic magmas as deduced from experimental data: Contributions to Mineralogy and Petrology, v. 89, p. 17–23, doi: 10.1007/BF01177586.

Krushensky, R.D., 1972, Geology of Irazú Quadrangle: Washington, U.S. Geological Survey Bulletin 1353, 46 p.

Leeman, W.P., Carr, M.J., and Morris, J.D., 1994, Boron geochemistry of the Central American Volcanic Arc: Constraints on the genesis of subduction-related magmas: Geochimica et Cosmochimica Acta, v. 58, p. 149–168, doi: 10.1016/0016-7037(94)90453-7.

Melson, W.G., Allan, J.F., Jerez, D.R., Nelen, J., Calvache, M.L., Williams, S.W., Fournelle, J., and Perfit, M., 1990, Water contents, temperatures and diversity of the magma of the catastrophic eruption of Nevado del Ruiz, Colombia, November 13, 1985: Journal of Volcanology and Geothermal Research, v. 41, p. 97–126, doi: 10.1016/0377-0273(90)90085-T.

Montero, W., and Alvarado, G.E., 1995, El terremoto de Patillos del 30 de diciembre de 1952 (Ms = 5,9) y el contexto neotectónico de la región del volcán Irazú, Costa Rica: Revista Geológica de América Central, v. 18, p. 25–39.

Murata, K.J., Dóndoli, C., and Sáenz, R., 1966, The 1963–65 eruption of Irazú Volcano, Costa Rica (the period of March 1963 to October 1964): Bulletin Volcanologique, v. 29, p. 765–796.

Prosser, J.T., and Carr, M.J., 1987, Poás volcano, Costa Rica: Geology of the summit region and spatial and temporal variations among the most recent lavas: Journal of Volcanology and Geothermal Research, v. 33, p. 131–146, doi: 10.1016/0377-0273(87)90057-6.

Reagan, M.K., Gill, J.B., Malavassi, E., and Garcia, M.O., 1987, Changes in magma composition at Arenal volcano, Costa Rica, 1968–1985: Real-time monitoring of open system differentiation: Bulletin of Volcanology, v. 49, p. 415–434, doi: 10.1007/BF01046634.

Russell, J.K., Nicholls, J., Stanley, C.R., and Pearce, T.H., 1990, Pearce element ratios: Contributions to Mineralogy and Petrology, v. 99, p. 25–35.

Rutherford, M.J., 1993, Experimental petrology applied to volcanic processes: Eos (Transactions, American Geophysical Union), v. 74, no. 5, p. 49, 55.

Sakuyama, M., 1984, Magma mixing and magma plumbing systems in Island Arcs: Bulletin Volcanologique, v. 47, no. 4, p. 685–703, doi: 10.1007/BF01952339.

Siebert, L., Alvarado, G.E., Vallance, J.W., and van Wyk de Vries, B., this volume, Large-volume volcanic edifice failures in Central America and associated hazards, in Rose, W.I., Bluth, G.J.S., Carr, M.J., Ewert, J.W., Patino, L.C., and Vallance, J.W., Volcanic hazards in Central America: Geological Society of America Special Paper 411, doi: 10.1130/2006.2412(01).

Taylor, S.R., Capp, A.C., Graham, A.L., and Blake, D.H., 1969a, Trace element abundances in andesites II. Sapipan, Bouganville and Fiji: Contributions to Mineralogy and Petrology, v. 23, no. 1, p. 1–26, doi: 10.1007/BF00371329.

Taylor, S.R., Kaye, M., White, A.J.R., Duncan, A.R., and Ewart, A., 1969b, Genetic significance of Co, Cr, Ni, Sc and V in andesites: Geochimica et Cosmochimica Acta, v. 33, p. 275–286, doi: 10.1016/0016-7037(69)90144-6.

Thomas, K.E., 1983, An investigation of the Cervantes Formation of Irazú Volcano, Costa Rica [B.A. thesis]: Hanover, Dartmouth College, 29 p.

Thomas, R.B., Hirschmann, M.M., Cheng, H., Reagan, M.K., and Edwards, R.L., 2002, ($^{231}$Pa/$^{235}$U)-($^{230}$Th/$^{238}$U) of young mafic volcanic rocks from Nicaragua and Costa Rica and the influence of flux melting on U-series systematics of arc lavas: Geochimica et Cosmochimica Acta, v. 66, p. 4287–4309, doi: 10.1016/S0016-7037(02)00993-6.

Tournon, J., 1984, Magmatismes du mesozoique à l' actuel en Amerique Centrale: L'exemple de Costa Rica, des ophiolites aux andesites [Ph.D. thesis]: Paris, Univiversité Pierre et Marie Curie, 335 p.

Walker, J.A., 1984, Volcanic rocks from the Nejapa and Granada cinder cone alignments, Nicaragua, Central America: Journal of Petrology, v. 25, p. 299–342.

Walker, J.A., Williams, S.N., Kalamarides, R.I., and Feigenson, M.D., 1993, Shallow open-system evolution of basaltic magma beneath a subduction zone volcano: The Masaya caldera complex, Nicaragua: Journal of Volcanology and Geothermal Research, v. 56, p. 379–400, doi: 10.1016/0377-0273(93)90004-B.

Wörner, G., and Schmincke, H.-U., 1984, Petrogenesis of the zoned Laacher See tephra: Journal of Petrology, v. 25, no. 4, p. 836–851.

Wörner, G., and Wright, T.L., 1984, Evidence for magma mixing within the Laacher See magma chamber (East Eifel, Germany): Journal of Volcanology and Geothermal Research, v. 22, p. 301–327, doi: 10.1016/0377-0273(84)90007-6.

York, D., 1968, Least squares fitting of a straight line with correlated errors: Earth and Planetary Science Letters, v. 5, p. 320–324, doi: 10.1016/S0012-821X(68)80059-7.

MANUSCRIPT ACCEPTED BY THE SOCIETY 19 MARCH 2006

# Index

## A

Acajutla debris avalanche, 8–9, 17–23
Acajutla Peninsula, 8
Acatenago volcano
    debris avalanches from, 4–6, 18, 110, 112–19
    geochemistry of, 112–15, 117–19
    geochronology of, 6, 110
    geological setting of, 108, 110
    magma chamber in, 119
    maps of, 3, 106, 110
    petrographic analysis of, 110
    photographs of, 109
    satellite images of, 5
    tephra eruptions from, 118–19
    vent alignment, 119
Achiguate River, 110
actinolite, MORB phase diagram for, 31
active volcanoes, definition of, 156
Agua Caliente debris avalanche
    area of, 18, 51
    Coatán River and, 2, 51
    geochronology of, 2, 51
    H/L ratio of, 18
    length of, 18, 51
    map of, 3
    orientation of, 18
    San Rafael River and, 51
    satellite images of, 5
    thickness of, 2, 51
    Tochab River and, 51
    volume of, 2, 18, 51
Agua Caliente wedge, 42, 49
Agua Cascada, 65, 71
Agua Sabina, 64, 70
Aguas Zarcas cinder cones, 14
Agua volcano
    debris avalanches from, 109–10, 112–14, 118–19
    edifice collapse on, 17
    geochemistry of, 112–14, 116, 118, 119
    geochronology of, 118
    geological setting of, 108, 109–10
    height of, 17
    lahars from, 110
    maps of, 29, 106, 112, 123
    petrographic analysis of, 110
    phase diagram for, 31
    photograph of, 109
    photomicrographs of, 111
    $P$-$T$ profile for, 31
    thermal model of, 30
Ahogados River, 12
Alaska
    Augustine volcano, 21
    subduction zones in, 34
Alfaro Unit, 264, 267
Almolonga volcanic field
    Cerro Quemado. *See* Cerro Quemado
    debris avalanches from, 2–4, 18
    isotopic studies of, 70
    springs, studies of, 65

Amatitlán caldera, 5–7, 123
amphibole, MORB phase diagram for, 31
Angostura debris avalanche, 15–16, 19
anthigorite, phase diagram for, 31
Antilles Islands, 21, 134
Apoyeque Tephra, 149, 151
Apoyeque volcano, 146–49, 190
Apoyo caldera
    diameter of, 143
    eruption of, 149
    geological setting of, 10
    Laguna de Apoyo, 144, 153
    magnitude of eruption, 146
    Managua Series and, 190
    maps of
        digital elevation, 147, 190
        isopach, 147
        Plinian eruption of, 145
        pyroclastic activity, 148
        topographic, 142
    photographs of, 144
    Plinian eruption of, 145–50
    pyroclastic activity at, 145–48
    quiescent periods of, 156
    tsunamis from, 150
    volcanic explosivity index of, 156
Apoyo Tephra, 149, 210, 215
Aquiares River
    maps of, 237, 241, 243, 245, 246
    Turrialba volcano and, 247, 251
Arenal volcano
    ballistic blocks from, 203
    basement of, 20
    debris avalanches from, 23
    magma from, 260
    map of, 261
    structure of, 20
argon for geochronology, 273–74
Armadillo, 65, 70
Aseses Peninsula, 10–12
ash
    destructiveness index for, 146
    edifice type and, 143
    inhalation of, 147
    load stress from, 146
    Los Chocoyos, 6, 107, 108
Asososca maar
    aerial photograph of, 150
    Laguna Asososca, 144, 153
    magma from, 145
    photographs of, 144
Atitlán volcano
    crater lake, 5, 23
    cross-sectional model of, 53
    debris avalanches from, 4, 6, 18
    geochronology of, 6
    Los Chocoyos ash, 6, 107, 108
    magma from, 260
    map of, 3
    Pacaya volcano and, 6
    satellite images of, 5

    structural geometry model of, 52
    tsunami in caldera lake, 23
Augustine volcano, 21
avalanches
    block fracturing in, 2
    debris. *See* debris avalanches
    ridges and, 8
Ayarza caldera, 5, 122, 123
Azufrales debris avalanches, 13, 19
Azufrales vent, 64, 70

## B

backarc basin basalt, oxygen isotope levels in, 129
Bajo de Chiqueros, 14
Bajos debris avalanche, 16, 19
Bálsamo Formation, 8, 9
Bálsamo Range, 8
Bandai volcano, 22
Banos Aguas Amargas, 64, 70
Banos Chicovix, 65, 70
Banos Cirilo Flores Well, 65, 70
Banos De Ocos, 63, 65, 71, 76–77
barium/lanthanum ratio, crust thickness and, 136
barium/niobium ratio, magma contamination and, 31
Barú volcano
    aerial photograph of, 16
    debris avalanches from, 16, 19
    geochronology of, 17
    glaciation of, 20
    height of, 16
    maps of, 3, 13
    pyroclastic flows from, 16
    satellite images of, 17
Barva volcano
    debris avalanches from, 14–15, 19, 23
    glaciation of, 20
    maps of, 3, 13, 261
basalt
    MORB. *See* mid-oceanic-ridge basalt (MORB)
    oxygen isotope levels in, 129
    phase diagrams for, 31
    phase transformation of, 31, 33
bicarbonate-rich water, 63, 77–78
Birrís River, 262, 263
Birrís Unit, 264, 267–73
Blanco River, 263
block-and-ash flows, 54, 147
block-facies material, 2, 8
blueschist, MORB phase diagram for, 31
boron/beryllium ratio, crust thickness and, 136
boron, magma composition and, 123
bromine, measurement of, 158
brucite, phase diagram for, 31

## C

Cabeza de Vaca debris avalanche, 15, 19
Cabeza de Vaca volcano, 15, 19
caldera, definition of, 143
Caliente vent, 61
Canal Fracture Zone, 3

carbon dioxide, in crater lakes, 153
Carey and Sparks models, 218–21
Caribbean/Cocos/NA plate junction
  arc magmatism at, 135
  convergence rate of, 28, 42
  El Chichón magma and, 28
  location of, 42
  maps of, 3, 29, 41, 261
  Moho Discontinuity and, 30
  structure of, 28
  Wadati-Benioff zone, 28
Caribbean, edifice failure direction in, 21
Caribbean Plate
  Cocos Plate/NA Plate and. *See* Caribbean/Cocos/NA plate junction
  edifice failure direction and, 21
  El Chichón magma and, 28
  maps of, 3, 29
Cascadia subduction zone, 34, 75
Cascajal River, 263
Casita volcano
  construction near, 159
  edifice collapse on, 154–55
  Hurricane Mitch and, 143, 154
  lahars from, 12, 22, 143, 154–55
  landslides on, 22
  maps of, 3, 142
  photograph of, 155
Ceniza River, 109
Central American volcanic arc
  Chiapanecan volcanic arc and, 33
  Chorotega Block, 3, 17
  Chortis Block, 3, 17
  Cocos Plate beneath. *See* Cocos Plate
  edifice collapse in, 17
  evolution of, 32, 33
  gas isotope ratios in, 72, 74, 76
  geochemistry of, 122–37, 261
  geochronology of, 28, 33
  isotopic studies of, 122–24, 126, 129–38
  length of, 40
  magma from
    closed vs. open system, 130–34
    geochemistry of, 261
    plumbing systems for, 122, 136–37
    signature of, 33
    Tacaná volcanic complex, 28
  maps of, 29, 31, 41
  Middle American Trench and, 33
  Moho Discontinuity and, 30
  MORB phase diagram for, 31
  orientation of, 40
  *P-T* profile for, 33
  thermal model of, 30–33
  triple junction near. *See* Caribbean/Cocos/NA plate junction
Central Valley Lavina debris avalanche, 15
Cerro Alto Grande, 241, 265
Cerro Cabeza de Vaca, 262
Cerro Cacao, 12–13, 19
Cerro Chino, 7
Cerro Colorado, 3, 13, 17, 19
Cerro Grande, 7
Cerro Gurdián, 262
Cerro Los Achiotes, 8, 18

Cerro Mongoy, 110
Cerro Negro volcano
  classification of, 145
  gas emissions from, 158
  lava flows from, 151, 157
  maps of, 142
  photographs of, 144, 145
  pyroclastic activity at, 145
  quiescent periods of, 156
  risk scenario for, 159
  seismic studies of, 156–57
  volcanic explosivity index of, 156
Cerro Noche Buena, 262
Cerro Pasquí, 262
Cerro Pelón, 17
Cerro Quemado
  debris avalanches from, 2–4, 18
  gas emissions from, 62, 72–75, 78–79
  hydrothermal activity in, 78–79
  lateral blast from, 4, 22
  lava domes in, 20
  lava flows from, 4
  maps of
    digital elevation, 4
    geographic, 3, 61
    Walker sample sites, 63
    water types, 67
  modern eruption of, 62
  pyroclastic flows from, 4
  satellite images of, 5
  Siete Orejas volcano and, 4
  Xela caldera and, 79
Cerro Quemado debris avalanche, 2–4, 18
Cerro Retes, 262
Cerro Retes lava flow, 264, 265
Cerro San Carlos
  geological setting of, 236
  lithology of, 242
  maps of, 237, 244
  photograph of, 238
Cerro San Enrique
  geological setting of, 236
  lithology of, 238
  maps of, 237, 244
  photograph of, 239
Cerro San Juan
  geological setting of, 236
  lithology of, 238
  maps of, 237, 244
Cerros Chachos, 175–76
Cerro Totuma, 17
Cervantes lava flow
  Angostura debris avalanche and, 16
  area of, 265
  composition of, 264, 272
  Diego de la Haya Unit and, 271
  geochemistry of, 264, 265, 268–73
  geochronology of, 264–66
  geothermobarometry of, 271
  map of, 262
  Río Birrís debris avalanches and, 15
  thickness of, 264
  volume of, 265
Chamalecón couleé, 50
Chanjale caldera

  cross-sectional model of, 53
  geochemistry of, 48, 49
  geochronology of, 47, 49
  lithology of, 49
  map of, 45
  photograph of, 50
  Tacaná graben and, 52
Chespal area, 49
Chiapanecan volcanic arc
  Central American volcanic arc and, 33
  Cocos Plate beneath. *See* Cocos Plate
  crustal thickness at, 42
  evolution of, 32, 33
  geochronology of, 28, 33–34
  geological setting of, 28
  magma from, 28, 31
  magnetic studies of, 34, 35
  maps of, 29, 31, 37
  Middle American Trench and, 28, 30, 35, 36
  Moho Discontinuity and, 30, 34, 35
  MORB phase diagram for, 31
  Sierra Madre volcanic arc and, 28
  thermal model of, 30–36
Chicabal volcano, 3, 5, 8, 18
Chichicaste, 64–65, 70
Chichuj volcano
  block-and-ash flows from, 2, 50, 54
  cross-sectional model of, 43
  debris avalanches from, 2, 44, 50
  edifice collapse on, 2, 54
  geochronology of, 2, 50
  height of, 42
  hydrothermal activity in, 54
  lava flows from, 44, 50, 54
  lithology of, 50
  map of, 3
  photograph of, 43
  pyroclastic flows/surges from, 54
  Tacaná volcano and, 50
  trend of, 52
Chiltepe Tephra, 149, 195, 196
Chiltepe volcanic complex
  Apoyeque volcano, 146–49, 190
  eruptions of, 145–50
  geochronology of, 149
  Laguna Xiloá, 145, 147, 148, 151
  maps of
    digital elevation, 147, 190
    isopach, 147
    pyroclastic activity, 148
    topographic, 142
  probability tree for eruptions from, 159, 161
  pyroclastic activity at, 146–48
  quiescent periods of, 156
  volcanic explosivity index of, 156
Chinameca volcano, 10
Chiquihuat River, 9
Chiriquí Viejo River, 16, 17
chloride-rich water in springs, 63, 76–78
chlorite, phase diagram for, 31
Chocabj village, 50
Choco Block, 3
Chocosuela debris avalanche, 14, 19
Chorotega Block, 3, 17
Chortis Block, 3, 17

cinder cones
    Aguas Zarcas, 14
    vs. composite cones, 122
    formation of, 122, 136–37
    geochemistry of, 127–28, 130
    isotope levels in, 122, 129
Citlaltépetl, 20, 21
closed magmatic systems, 122, 130–34
Coastal Batholith (Chiapas), 46
Coatán River
    Agua Caliente debris avalanche and, 2, 51
    lithology along, 49
    maps of, 45
    structural analysis of, 52
    Tacaná graben and, 52
Coatán River fault, 53–55
Coatepeque caldera, 8, 9
Coatepeque, Lake, 23
Cocos Plate
    Caribbean Plate and. See Caribbean/Cocos/NA plate junction
    depth of, 28, 30, 35, 36, 42
    dip angle of, 17, 28, 31, 33, 42, 260
    earthquakes in, 28, 34
    evolution of, 32, 33
    geochronology of, 28–36, 42
    magma from, 28, 31–33
    magnetic studies of, 34, 35
    maps of
        magnetic anomalies, 37
        topographic, 142
        triple junction, 29
        volcanic arc evolution, 32
        volcanoes, 3
    Moho Discontinuity and, 30, 34
    Nazca Plate and, 260–61
    North American Plate and. See Caribbean/Cocos/NA plate junction
    phase diagrams for, 31
    P-T profiles of, 31–33
    slab depth, 28, 42
    structure of, 34, 36
    subduction direction, 42
    Tacaná volcanic complex and, 53
    Tehuantepec Ridge and, 28, 31, 33–36, 42
    thermal model of, 30, 31–36
Cocos Ridge, 261
coesite, MORB phase diagram for, 31
Cofre de Perote, 21
Coliblanco, 16
Colima volcano, 22, 158
Colorado River, 13
composite cones vs. cinder cones, 122
Concepción volcano
    basement of, 20
    classification of, 143
    debris avalanches from, 23
    edifice deformation in, 20
    gas emissions from, 158
    lahars from, 154
    lava flows from, 160
    maps of, 142, 159, 160
    photograph of, 144
    quiescent periods of, 156
    seismic studies of, 156–57

    structure of, 20
    volcanic explosivity index of, 156
continental arc basalt, oxygen isotope levels in, 129
Cordillera Central
    geochronology of volcanoes, 271
    lahars from, 22
    mantle beneath, 261
    Tivives debris avalanche and, 14
Cordillera de Talamanca, 238–39
Corinto River, 15, 263
Cosigüina volcano
    bombs from, 175, 178
    caldera statistics, 168
    classification of, 143
    crater lake, 168, 184
    cross-sectional model of deposits, 178
    edifice-collapse events in, 168
    eruption of (1835AD), 168–84
    eruption of (prehistoric), 182–85
    geochemistry of, 173–75, 183
    geochronology of, 177, 181–83
    hazard assessment, 185–86
    height of, 168
    lithology of, 168–81
    maps of
        digital elevation, 169, 171, 176, 185
        hazard assessment, 185
        topographic, 142
    photographs of, 174, 179–81, 183
    pumice rafts from, 178–80
    pyroclastic activity at, 148, 168–84
    quiescent periods of, 156
    stratigraphy of, 170
    volcanic explosivity index of, 156
    wind direction above, 172, 186
Cotitos area, 17
Cotopaxi volcano, 153
Coyol debris avalanche, 19
Cráter del Derrumbo, 15
crisis management plans, 159
crust
    edifice collapse and, 17
    ice loading of, 252
    magma geochemistry and, 122
    thickness of, 17, 42, 122
Cuache River, 65, 71

# D

debris avalanches. See also specific avalanches
    basement rocks and, 21, 112
    definition of, 2
    deposit area, 20
    edifice rocks and, 112
    frequency of, 21–22
    glaciers and, 20
    habitation on, 22
    H/L ratio and, 17
    vs. lahars, 2, 20, 22
    length of, 22
    "marginal facies" in, 2
    mobility of, 17, 17–20
    orientation of, 21
    rheology of, 20

    rose diagram for, 21
    studies of, 2
    topography of, 106
    tsunamis and, 23, 150
    velocity of, 22, 154
    volume of, 16, 20, 22
debris-flow facies, 2
debris flows. See lahars
Derrumbo, 262
Desenlace wedge, 42
destructiveness index, 203
diamond, MORB phase diagram for, 31
Diego de la Haya Unit, 263–65, 268–73
dikes, edifice-collapse events and, 21
disaster mitigation, definition of, 163
disasters, definition of, 162
Domo El Azufral, 61. See also Sulfur Mountain
Dóndoli Unit, 264, 267, 271
DSDP 495 site, 135
dunite, phase diagram for, 31

# E

earthquakes
    in Cocos Plate, 28, 34
    edifice collapse and, 12
    hydrothermal activity and, 156–57
    seismic wave types, 156
    swarms of, 156
    volcanic, 156–57
East Megablock Slump, 12, 18
East Pacific Rise, 36
eclogite, 31, 33
edifice-collapse events
    crustal thickness and, 17
    debris avalanches from. See debris avalanches
    earthquakes and, 12
    factors affecting, 17, 20–21, 23, 154
    frequency of, 21–22
    hydrothermal activity and, 20, 22, 60
    lahars and, 22
    lateral blasts and, 22
    lava-dome complexes and, 2
    magmatic eruptions and, 22
    material in, 2
    orientation of, 21
    phreatic eruptions and, 22
    slab descent angle and, 23
    trench location and, 21
edifice, volcano
    basement and, 20, 21
    collapse of. See edifice-collapse events
    deformation of, 20–21, 157
    geochemistry and, 125–28, 130
    height of, 17
    isotope levels and, 129
    magma system and, 122, 136
    mineral sealing in, 60
    types of, 143
    water within, 20
El Aguila area, 46
El Aguila wedge, 42, 49
El Armado, 236
El Chichón volcano
    geochronology of, 33

magma from, 28, 31, 34
maps of, 29, 37, 41
phase diagrams for, 31
$P\text{-}T$ profile for, 31
structure of, 53
Tehuantepec Ridge and, 36
thermal model of, 31, 34
El Chonco volcano, 12, 153
El Cráter debris avalanche, 10–12, 18
elevation, volcano edifice collapse and, 17
Elia River
    maps of, 237, 243, 245, 246
    Turrialba volcano and, 239, 242, 251
El Pacayal volcano, 10
El Palmar village, 89, 90
El Palomar River, 10
El Patrocino village, 7
El Pital River, 12
El Retiro Tuff, 190. *See also* Masaya Tuff
El Tambor River, 61, 87–93
El Tigre Island, 180, 184
El Tumbador area, 3, 8
El Valle volcano, 260
epidote, MORB phase diagram for, 31
eruptions
    column height, 210, 218–21
    magma discharge rate and, 210
    paroxysmal phase of, 156
    precursors of, 156
    VEI, 156
Escuintla City, 5
Escuintla debris avalanche
    area of, 6, 18, 107
    classification of, 114
    geochemistry of, 112–19
    geochronology of, 6, 18, 107
    geological setting of, 106
    H/L ratio, 18
    length of, 18, 107
    Los Chocoyos ash and, 6, 107
    maps of, 106, 107
    orientation of, 6, 18
    origin of, 6, 110–19
    petrographic analysis of, 110–12, 118
    photographs of, 6, 108
    photomicrographs of, 111
    satellite images of, 5
    structure of, 6
    topography of, 107
    volume of, 6, 18, 107
    width of, 107
Estero de Mateo, 96
Etna, Mount
    column height of, 221
    discharge rate of, 221
    eruption of (122BC), 210
    thickness decay in, 217, 218

## F

Farallon Plate, 53
Filete Cresta Montosa
    in Cosigüina eruption of (1835AD), 174
    creation of, 183
    geological setting of, 168

    hazard assessment, 185
    maps of, 169, 171, 176, 185
Finca El Canada, 65, 71
Finca El Faro, 65, 71
Finca Filadelfia area, 90
Finca La Florida area, 61, 64, 70
Finca Liebres, 262
Finca Parador Los Trece, 65, 71
Finca La Quina, 65, 71
Finca San Juan Patzulin Stream, 65, 71
Finca Zajul area, 46
Fontana Tephra, 149, 150, 210–22
forearc trough basalt, oxygen isotope levels in, 129
fractionation
    in closed vs. open system, 133–34
    evaporative, 75
    hydrothermal activity and, 75
    magma system and, 133–34, 136
    petrographic features of, 269
    Raleigh crystallization equation, 133
fracture zones, serpentinization process and, 36
Fuego volcano
    ballistic blocks from, 203
    block-and-ash flows from, 110
    cross-sectional model of, 53
    debris avalanches from, 4–6, 18, 109, 112–14, 118–19
    geochemistry of, 112–15, 119
    geochronology of, 109
    geological setting of, 6, 108, 109
    maps of, 3, 106, 110
    petrographic analysis of, 110–12
    photographs of, 109
    satellite images of, 5
Fuentes Georginas, 64, 70
Fukutoku-Okanoba volcano, 144

## G

Galapagos spreading center, 133
garnet, phase diagrams for, 31
Georginas stream, 64, 70
glaciers
    lahars and, 20, 22
    volcanic activity and, 252
González Unit, 264, 267
Granada cinder cone, 260
Granada Islands, 10
granulite, MORB phase diagram for, 31
greenschist, MORB phase diagram for, 31
Grenada Basin, 21
Guacalate River, 88, 107, 112
Guácimo River
    maps of, 237, 243, 245, 246
    Turrialba volcano and, 239, 247, 251
Guatemala graben, 52, 53, 123
Guayabito River
    maps of, 237, 243, 245, 246
    Turrialba volcano and, 251
Guayabo caldera, 13
Guayabo River, 251
Gurdián lava flow, 268

## H

harzburgite, phase diagram for, 31, 33
Hato de Volcán debris avalanche, 16, 19
Hawaiian Islands, Pu'u 'O'o eruptions in, 157
hazard maps, 159
heat flux
    equation for, 220
    as eruption precursor, 156
    satellite images of, 158
high field strength elements, 273
H/L ratio, definition of, 17
Hurricane Mitch, 143, 154, 162
hydrothermal activity
    earthquakes and, 156–57
    edifice collapse and, 20, 22, 60
    fractionation and, 75
    isotopic values and, 72, 75
    lateral blasts and, 22
    lava flows and, 134
    oxygen isotopes and, 134
    sulfur dioxide/halogen ratio and, 158

## I

ignimbrites
    definition of, 143
    formation of, 147–148
Ilopango caldera, 9, 181
Inde Dam Spring, 65, 71
Ipala graben
    cinder cone formation and, 122, 136
    cross-sectional model of, 53
    maps of, 123
    structural geometry model of, 52
Ipala volcano, 123
Irazú volcano
    area of, 225, 260
    bombs from, 227, 267
    Cervantes lava flow. *See* Cervantes lava flow
    cross-sectional model of, 266
    debris avalanches from, 15, 19, 21, 260–61, 272
    earthquakes at, 260, 271
    eruptions of, 225–34, 260, 267
    geochemistry of, 226, 261, 264, 268–73
    geochronology of, 227–28, 233, 264, 265, 273–74
    geological setting of, 225, 260–61
    geothermobarometry of, 271–72
    glaciation of, 20
    Harker variation diagrams for, 269
    hazard assessment of, 233–34, 273
    height of, 225, 260
    hydrothermal activity at, 272
    lahars from, 15, 250, 260
    lava flows from, 225
    lithology of, 226–28, 262–68
    magma chambers in, 233, 261, 271, 272
    magma from, 233, 260, 268–73
    maps of
        ash falls, 227
        debris avalanches, 13, 15
        geographic, 3, 226
        geologic, 263
        isopach, 227

pyroclast distribution, 228
regional sketch, 237, 262
tectonic, 261
tephra, 229, 241
pyroclastic activity at, 234, 250–51
quiescent periods of, 233, 273
stratigraphy of, 230–32, 264, 266
structure of, 261, 267
Turrialba volcano and, 227, 237, 251
volcanic explosivity index of, 226, 229, 233, 273
wind direction above, 227
iron, sulfate-rich water and, 66
Ixhuatán volcano
Cerro Los Achiotes and, 8, 18
debris avalanches from, 7, 18
lateral blast from, 8
map of, 3
Miraflores volcano and, 7
Mount St. Helens and, 8
satellite images of, 5
Ixpatz River, 87–88, 91–95, 102
Ixtapa Graben, 53
Izu-Bonin subduction zone, 28

## J

jadeite, MORB phase diagram for, 31
Jalpatagua fault, 123
Jesús María River, 14
Jiménez River, 237
Jinotepe pyroclastic deposit, 148
Jocotan fault, 123
Jumaytepeque volcano, 123
Jumay volcano, 123

## K

Karymskoye, Lake, 151
Kiglapait Mountains, 133
Kliuchevskoy volcano, 144
Koloula igneous complex, 133
Krakatau volcano, 150
Kuril subduction zone, 28

## L

La Concepción Tephra
area of, 202
characteristics of, 205
composition of, 191
destructiveness index for, 202
distribution of, 206
Los Cedros Tephra and, 149
magnitude of eruptions, 202
maps of, 199, 201, 202
Masaya Triple Layer and, 149
Masaya Tuff and, 191, 195
photographs of, 194
San Antonio Tephra and, 191
stratigraphy of, 193
volcanic explosivity index of, 202
volume of, 202
La Democracia debris avalanche
area of, 18, 107
classification of, 114

geochemistry of, 112–15, 117–19
geochronology of, 6, 18, 108
H/L ratio, 18
length of, 18, 107
maps of, 106, 109, 110
orientation of, 18
origin of, 6, 110–19
petrographic analysis of, 112
photomicrographs of, 111
satellite images of, 5
volume of, 18, 107
width of, 107
La Fortuna debris avalanche, 13, 14, 19
Lago de Guija, 123
Lago de Managua
debris avalanches and, 23
depth of, 151
maps of
digital elevation, 147, 190
isopach, 147
pyroclastic activity, 148
topographic, 142
tsunamis in, 23, 151
Lago de Nicaragua
debris avalanches and, 10–12, 23
depth of, 151
maps of
Concepción volcano hazard assessment, 160
digital elevation, 147
isopach, 147
pyroclastic activity, 148
topographic, 142
tsunamis in, 23, 150
Laguna de Apoyo, 144, 153
Laguna Asososca, 144, 153
Laguna Ixpatz, 94
Laguna Madre Vieja, 96
Laguna de Mesa, 94
Laguna de Oc, 87, 94
Laguna de La Pepesca, 94
Laguna Xiloá
magma from, 145
maps of, 147, 148
tsunamis in, 151
Laguneta El Castano, 96
Laguneta Guiscoyol, 96
lahar facies, 2, 20
lahars
vs. debris avalanches, 2, 20, 22
definition of, 153
edifice-collapse events and, 22
flow behavior of, 153
formation of
dams and, 22
fluvial system aggradation and, 88
glaciers and, 20, 22
ingredients for, 86
lake-derived, 22, 110
lava-dome complexes and, 86
liquefaction and, 22
persistent vs. transient, 86
precipitation and, 153
mobility of, 20
vs. pyroclastic density currents, 86, 204
rheology of, 20

LAHARZ model, 159
La Laguna cones, 262
La Madera volcano, 142
LANDSAT coverage cycle, 158
Las Isletas debris avalanches
age of, 10, 18
area of, 18
composition of, 10
El Cráter debris avalanche and, 10
H/L ratio of, 18
length of, 18
orientation of, 18
synthetic aperture radar image of, 11
tsunamis from, 155
volume of, 18
Las Lagunas area, 16
Las Majadas River, 2, 5
Las Mesas debris avalanche, 14, 19
Las Pilas volcano, 151
Lassen, 75
Las Sierras Formation
composition of, 190
Fontana Tephra and, 210
geochronology of, 190, 210
Las Isletas debris avalanche and, 10
source of, 215
Las Viboras volcano, 122, 123
lateral blasts, edifice collapse and, 22
lava-dome collapse
block-and-ash flows from, 147
lahars from, 86
pyroclastic density currents from, 147
lava-dome complexes
collapse of. See lava-dome collapse
edifice collapse and, 2
edifice type and, 143
photograph of, 16
lava flows
edifice type and, 143
fluvial system aggradation and, 99–103
hydrothermal activity and, 134
La Vega del Volcán village, 49
lawsonite, MORB phase diagram for, 31
lead, isotope levels in cinder cones, 122, 137
Lempa River, 9, 10
Limón Basin, 236
Llano del Pinal valley, 4
Lolotique area, 10, 18
Loma El Ojochito
in Cosigüina eruption of (1835AD), 178
geological setting of, 168
hazard assessment, 186
maps of, 169, 171, 176, 185
Loma San Juan
geological setting of, 168
hazard assessment, 186
maps of, 169, 171, 176, 185
Los Achiotes debris avalanche, 8, 18
Los Cedros Tephra, 149
Los Chocoyos ash deposits, 6, 107, 108
Los Esclavos River, 5, 7–8
Los Pozos area, 17

## M

maars
  formation of, 145
  synthetic aperture radar image of, 11
MacKenney Cone, 6, 7
Maderas volcano, 20, 143, 154
magma
  barium/niobium ratio and, 31
  basaltic density, 203
  binary mixing equation, 135
  boron and, 123
  in closed systems, 122, 130–34
  column height and, 221
  crust and, 122
  discharge rate of, 210
  earthquakes and, 156–57
  edifice and, 122, 136
  fractionation and, 133–4, 136
  geochemistry of, 122
  high field strength elements in, 273
  neodymium and, 122
  in open systems, 122, 130–34, 137–38, 210
  rare earth elements in, 273
  rubidium/neodymium ratio and, 123
  sodium in, 61, 77
  strain rate and, 157
  strontium and, 122
  temperature of, 271
  tilt measurements and, 157
magnesium/calcium ratio, 63, 71, 77
magnitude of eruption equation, 203
Malacate village, 49
Managua
  area of, 205
  earthquakes in, 143
  geological setting of, 150
  La Concepción Tephra in, 206
  Masaya Triple Layer in, 206
  Masaya Tuff in, 206
  population, 205
  stratigraphy of, 149
  Ticuantepe Lapilli in, 206
Managua Series, 190
Margaritas River, 8
"marginal facies." See lahar facies
Maribios Range
  Las Pilas volcano, 151
  lava flows from, 151
  San Cristóbal volcanic complex. See San Cristóbal volcanic complex
  Telica volcano. See Telica volcano
Marinalá River, 7
Masaya Lapilli Bed. See Fontana Tephra
Masaya Triple Layer
  area of, 202
  characteristics of, 205
  Chiltepe Tephra and, 149, 195
  composition of, 191–93
  destructiveness index for, 202
  distribution of, 206, 215
  Fontana Tephra and, 215
  geochemistry of, 193
  geochronology of, 149, 191, 196
  La Concepción Tephra and, 149

magnitude of eruptions, 202
maps of, 201, 202
photographs of, 196
San Antonio Tephra and, 191
stratigraphy of, 195
volcanic explosivity index of, 202
volume of, 202
Masaya Tuff
  area of, 202–3
  Chiltepe Tephra and, 149, 195
  composition of, 195–98
  creation of, 190, 198, 205
  description of, 193–95
  destructiveness index for, 202
  distribution of, 204, 206
  La Concepción Tephra and, 191, 195
  magnitude of eruptions, 202
  maps of, 201, 202
  Masaya Triple Layer and, 195
  photographs of, 196, 197, 198, 200
  stratigraphy of, 197
  volcanic explosivity index of, 202
  volume of, 202
Masaya volcano
  Apoyo Tephra, 215
  destructiveness index for, 203
  Fontana Tephra, 149, 150, 210–22
  formation of, 189
  gas emissions from, 152–53, 158
  geochronology of, 149
  geological setting of, 10, 190, 210
  hazard assessment of, 203–6
  La Concepción Tephra. See La Concepción Tephra
  lake in, 198
  lapilli beds, 190, 199
  lava flows from, 151
  magma from, 143, 145, 260
  magnitude of eruptions, 146, 203
  Managua Series and, 190
  maps of
    digital elevation, 147, 190
    Fontana Tephra, 211, 215, 216
    hazard assessment, 206
    isopach, 147, 201
    phreatomagmatic eruptions, 148
    political, 204
    pyroclastic activity, 148
    topographic, 142
  Masaya Triple Layer. See Masaya Triple Layer
  Masaya Tuff. See Masaya Tuff
  microgravity monitoring of, 157
  pyroclastic activity at, 146–50
  quiescent periods of, 156
  risk scenario for, 159
  San Antonio Tephra. See San Antonio Tephra
  satellite images of, 153
  seismic studies of, 156–57
  sulfur dioxide/halogen ratio at, 158
  Ticuantepe Lapilli. See Ticuantepe Lapilli
  volcanic explosivity index of, 156, 203
Mateare Tephra, 149, 151, 152
Maya Block, 3
megablocks, definition of, 2
Merapi area, 86
Mercedes River, 243, 245, 246, 251

Mesa River, 94
Meseta volcano, 4–6, 108–15, 118–19
metamorphic facies, MORB phase diagram for, 31
Metapa River, 7
Méxican Volcanic Belt, Central, 29, 32–33
Méxican Volcanic Belt, Eastern, 21
Michatoya River, 107
microgravity monitoring, 157
Middle American Trench
  bend-faulting at, 36
  Central American volcanic arc and, 33
  Central Méxican Volcanic Belt and, 33
  Chiapanecan volcanic arc and, 28, 33–36
  Cocos Plate geotherm, 29
  edifice failure direction and, 21
  geochronology of, 28, 33
  magnetic studies of, 34, 35
  maps of
    digital elevation, 41
    magnetic anomalies, 37
    tectonic, 261
    triple junction, 29
    volcanoes, 3
  Moho Discontinuity and, 30, 35, 36
  Tehuantepec Ridge and, 28, 36, 42
  thermal model of, 30, 31–36
mid-oceanic-ridge basalt (MORB)
  gas triangular diagram for, 73
  oxygen isotope levels in, 129
  phase diagram for, 31
  vs. subduction zone basalt, 129
Miraflores debris avalanche, 7–8, 18
Miraflores volcano, 7
Miravalles volcano
  debris avalanches from, 13–14, 19
  geochronology of, 14
  lahars from, 13
  lava flows from, 13, 14
  maps of, 3, 13, 261
  pyroclastic activity at, 13, 14
Mixcun block-and-ash flow, 42, 44, 52
mixed-facies material, definition of, 2
Moho Discontinuity
  Caribbean/Cocos/NA plate junction and, 30
  Central American volcanic arc and, 30
  Chiapanecan volcanic arc and, 30, 34, 35
  Middle American Trench and, 30, 35
Mombacho volcano
  basement of, 20
  debris avalanches from, 10–12, 18, 23, 155
  edifice collapse on, 12, 150, 155
  gas emissions from, 158
  lahars from, 155
  maps of, 3, 11, 142
  photograph of, 144
  structure of, 11, 20
  synthetic aperture radar image of, 11
Momotombo volcano
  debris avalanches from, 23
  gas emissions from, 158
  lava flows from, 151
  maps of, 142, 169
  pyroclastic activity at, 148, 182
  seismic studies of, 156–57
monitoring of volcanoes, 156

Monte Perla area, 46
Motagua-Cayman Fracture Zone, 3
Motagua-Polochic fault zone, 40, 41, 61
Motozintla fault, 40
Moyuta volcano, 5, 8, 123
Muxbal debris avalanche
    composition of, 2, 50
    geochronology of, 2
    geological setting of, 51
    hydrothermal activity and, 54
    orientation of, 18, 54
    in Tacaná stratigraphy, 44

## N

Nahualate River, 61
Naranja River, 8
NASA satellite coverage cycle, 158
natroalunite, 77
Nazca Plate
    Cocos Plate and, 260–61
    maps of, 3, 261
Nejapa maar
    aerial photograph of, 150
    magma from, 145, 260
    maps of, 147, 148
Nejapa-Miraflores lineament
    magma from vents along, 150
    maps of, 147, 148, 190
neodymium
    in Central American volcanic arc, 123, 124, 137
    magma composition and, 122
Neuvo California, 16, 17
Nicaragua depression, 190
Nimá River, 61, 87–93
nitrogen/argon ratio, 72
nitrogen/helium ratio, 72
normalized difference vegetation index (NDVI), 90–92
North American (NA) Plate
    Caribbean Plate and. See Caribbean/Cocos/NA plate junction
    Cascadia subduction zone, 34, 75
    Cocos Plate and. See Caribbean/Cocos/NA plate junction
    El Chichón magma and, 28
    maps of, 3, 29, 34, 37
Nyos, Lake, 153

## O

Oaxaca area, 28, 31
oceanic arc basalt, oxygen isotope levels in, 129
Oc River, 87, 94
11 de Abril village, 46
Ometepe Island
    Concepción volcano. See Concepción volcano
    maps of, 142, 160
open magmatic systems, 122, 130–34, 137–38, 210
Orosito River, 62
Orosí volcanic complex, 3, 12–13, 19, 261

## P

Pacaya volcano
    Amatitlán caldera and, 6
    debris avalanches from, 6–7, 18
    geochronology of, 7
    magmatic eruptions at, 22
    maps of, 3, 123
    pyroclastic surges from, 7
    satellite images of, 5
Pacific Ocean
    Cocos Plate. See Cocos Plate
    East Pacific Rise, 36
    Izu-Bonin subduction zone, 28
    Middle American Trench. See Middle American Trench
    Middle Valley isotopic studies, 135
    Nazca Plate. See Nazca Plate
    Tonga subduction zone, 28
Pacific Plate, map of, 261
Panama Fracture Zone, 261
Panteleón River, 107, 109
Paraíso Unit, 263–65
paroxysmal phase, 156
Patria River, 263
Pecul volcano. See Santo Tomás volcano
peridotite
    phase diagrams for, 31
    phase transformation of, 31, 33
    serpentinization process and, 34
phreatomagmatic eruptions
    ash from, 250
    definition of, 145
    edifice-collapse events and, 22
    thickness decay in, 218
Pico de Orizaba volcano, 20, 41
Pico de Piedra Unit, 15, 263–65
Pinabet spring, 63, 64, 70
Pinatubo, Mount
    destructiveness index for, 146, 203
    lahars from, 86, 178
    magnitude of eruption, 146
Plan de Las Ardillas, 42–45, 47–52, 56
Platanar volcano
    Chocosuela caldera and, 14
    debris avalanches from, 14, 19
    maps of, 3, 13, 261
Playa Hermosa lava flow, 264
Plinian eruptions
    of basaltic magma, 210
    edifice type and, 143
    thickness decay in, 217–18
plume-sedimentation model, 220, 221
Poás volcano
    magma from, 260
    maps of, 261
    vent gas studies, 72, 73
Poligono, 65, 71
Portales Mud Pot
    gas emissions from, 62, 74
    isotopic studies of, 70
    water, studies of, 64
Porvenir volcano, 14
prehnit, MORB phase diagram for, 31
Prusia debris avalanche, 19
Puerto San José, 5
pumice eruptions, edifice type and, 143
pumpellyite, MORB phase diagram for, 31
Punta San José, 169, 171, 176, 177
Pu'u 'O'o eruptions, 157
Pyle model, 217–18, 221
pyroclastic density currents
    block-and-ash flows. See block-and-ash flows
    definition of, 147
    dynamic pressure of, 148
    edifice type and, 143
    explosive vs. non-explosive, 145
    formation of, 147
    hazards from, 204
    vs. lahars, 86, 204
    from lava-dome collapse, 147
    tephra. See tephra
    tsunamis and, 150
    types of, 147–48
    velocity of, 148

## Q

Quebrada Grande debris avalanche, 12, 19
Quebrada Grande River, 12
Quebrada Playa Trinidad, 110, 118
Quemados, 262
Quetzaltenango basin, 62
Quetzaltenango (city), 4, 5, 62

## R

radar interferometry, 158
Rainier, Mount, 20
rare earth elements, 273
Retana caldera, 122, 123
Reventado River
    Irazú volcano and, 226, 234
    maps of, 227, 229, 262
Reventado Unit, 15, 263–65
Reventazón River, 16, 237, 262
Rincón de la Vieja volcano, 3, 13, 19, 261
Río Birrís debris avalanche, 15, 19
Río Chiriquí Viejo debris avalanche, 19
Río Costa Rica debris avalanche, 15, 19
Río Grande de Tárcoles, 14
Río Las Majadas debris avalanches, 2, 18
Río Metapa debris avalanches, 5–7, 18
risk assessment process, 159
Roca River, 247, 251
rubidium/neodymium ratio, 123, 124

## S

St. Elena-Hess Fracture Zone, 3
St. Helens, Mount
    Cerro Quemado and, 4
    debris avalanches from, 150
    destructiveness index for, 146, 203
    gas isotope ratios in, 75
    Ixhuatán volcano and, 8
    lahars from, 22, 86
    lateral blast from, 4, 8, 22
    magnitude of eruption, 146
    Spirit Lake tsunami, 150
    tilt measurements at, 157
Salitral area, 12
Samalá River
    aggradation of, 89, 91–94
    discharge rate of, 88

flow path of, 62, 88, 94
isotopic studies of, 70
Ixpatz River and, 88
lahars along, 94
maps of
    aggradation of, 90, 94, 96
    geographic, 61
    satellite images of, 87, 93
NDVI studies of, 91–92, 95, 97
Nimá River and, 89
satellite images of, 87, 95, 97
springs, studies of, 64, 76
San Antonio Tephra
  area of, 202
  characteristics of, 205
  composition of, 191
  destructiveness index for, 202
  La Concepción Tephra and, 191
  magnitude of eruptions, 202
  maps of, 199, 201
  Masaya Triple Layer and, 191
  Mateare Tephra and, 149
  photographs of, 192
  stratigraphy of, 192
  volcanic explosivity index of, 202
  volume of, 202
  Xiloá Tephra and, 149, 191
San Antonio volcano
  block-and-ash flows from, 42, 44, 52
  construction of, 56
  cross-sectional model of, 43
  debris avalanches from, 2, 42, 44
  edifice collapse on, 2
  geochemistry of, 52
  geochronology of, 2, 52
  height of, 42
  lahars from, 2, 42
  lava dome in, 42, 44
  lava flows from, 52
  map of, 3
  Mixcun block-and-ash flow, 42, 44, 52
  Peléan eruption of, 42, 56
  photograph of, 43
  pyroclastic flow/surges from, 2, 42
  trend of, 52
San Cristóbal volcanic complex
  Casita volcano. *See* Casita volcano
  classification of, 143
  debris avalanches from, 12
  El Chonco volcano, 12, 153
  gas emissions from, 152–53, 158
  lahars from, 12, 154
  lava flows from, 151
  maps of, 3, 142, 169
  photograph of, 153
  satellite studies of, 158
  seismic studies of, 156–57
San Felipe area, 86
San Gerardo lava flow, 264
San Jerónimo Unit, 262–65, 271
San Judas Formation, 190. *See also* Masaya Triple Layer
San Martin volcano, 29, 30, 31, 41
San Miguel volcano, 3, 10, 18
San Pedro volcano, 5, 23

San Rafael caldera, 42–54, 56
San Rafael River
  Agua Caliente debris avalanche and, 2, 51
  geochronology of, 46
  lahar deposits along, 2
  lithology along, 43–46
  maps of, 45
  photograph of, 46
San Salvador Urbina, 52
Santa Ana volcano
  block-and-ash flows from, 9
  debris avalanches from, 8–9, 18, 23
  geochronology of, 8
  lava flows from, 8–9
  magmatic eruptions at, 9, 22
  maps of, 3, 9, 123
  satellite images of, 8
Santa María volcano
  Caliente vent, 61
  climate studies of, 61–62, 88–89, 99–102
  eruption of (1902AD), 44, 60–61, 78, 86, 94, 181–82
  eruption of (1922AD), 61, 86
  fluvial system aggradation, 86–103
  gas emissions from, 61, 62, 72–77
  geochronology of, 60
  geological setting of, 2, 60
  height of, 60
  hydrothermal activity in, 62, 66, 75–79
  isotopic studies of, 67, 70–77
  lahars from, 61, 86, 88, 94
  lava dome in. *See* Santiaguito lava-dome complex
  lava flows from, 91, 94–96, 98
  magma geochemistry, 61
  maps of
    digital elevation, 4
    geographic, 61
    satellite images of, 87
    Walker sample sites, 63
    water types, 67
  NDVI studies of, 91–92
  pyroclastic activity at, 60–61
  Santiaguito lava-dome complex. *See* Santiaguito lava-dome complex
  Sapper fumarole, 61
  satellite images of, 5, 87, 92
  water, studies of, 62–79
Santa Rosa area, 14, 16, 19, 46
Santiaguito lava-dome complex
  climate studies of, 88–89, 99–102
  collapse of, 12, 78
  construction of, 61
  drainage basin, 88
  extrusion rate at, 86, 96–103
  fluvial system aggradation, 86–103
  isotopic studies of, 67, 70, 77
  lahars from, 86, 88
  lava flows from, 91, 94–96, 98
  magma chamber in, 78
  maps of
    digital elevation, 4
    geographic, 61
    satellite images of, 87
    Walker sample sites, 63

    water types, 67
  NDVI studies of, 91–92
  satellite images of, 86–88, 92
  water, studies of, 62–65, 77
Santo Domingo, 43, 45, 51
Santo Tomás volcano, 5, 61, 62
San Vicente volcano, 3, 9–10, 18
San Vito debris avalanche, 17, 19
Sapper fumarole, 61
Sapper Unit, 264–65, 268–73
scoria cones
  composition of, 143
  synthetic aperture radar image of, 11
Segregado dam, 23
Semeru area, 86
serpentine, phase diagram for, 31
Shiveluch volcano, 22
Sibinal caldera
  Coatán River fault and, 54
  geological setting of, 49–50
  lithology of, 50
  maps of, 45
  Tacaná graben and, 52
Sibinal Pumice fall, 50
Sierra de Apaneca, 8, 9
Sierra Madre volcanic arc, 28–31, 33–36
Siete Orejas volcano
  Cerro Quemado and, 4
  debris avalanches from, 18
  hydrothermal activity in, 79
  maps of
    digital elevation, 4
    geographic, 3, 61
    Walker sample sites, 63
    water types, 67
  morphology of, 8
  Orosito River and, 62
  Plinian eruption of, 62
  satellite images of, 5
  water, studies of, 77
slab descent angle, edifice collapse and, 23
sodium
  bicarbonate-rich water and, 63–66
  in magma, 61, 77
Soufrière Hills volcano, 12
South American Plate, 3
Southeast debris avalanche, 12, 18
Spearman rank correlation coefficient, 136
spinel, phase diagram for, 31
strain rate, earthquakes and, 157
stratovolcano, definition of, 143
Strombolian eruptions
  of basaltic magma, 210
  from scoria cones, 143
  thickness decay in, 217, 218
strontium, magma composition and, 122
subduction zones
  anisotropy in, 28
  basalt, oxygen isotope levels in, 129
  magnetic anomalies above, 34
  serpentinization process in, 34
  shear-wave splitting in, 28
Suchiate River, 45, 52
Suchiate River fault, 53–55
Suchitan volcano, 123

Sucio River, 234, 263
sulfate-rich water
  formation of, 77
  iron and, 66
sulfur/chloride ratio, 63, 77
sulfur dioxide
  daily release in Nicaragua, 152
  eruptions and, 157–58
  satellite remote sensing of, 158
sulfur dioxide/halogen ratio, 158
Sulfur Mountain, 61–63, 67, 74–76
Surtseyan eruptions, 210

## T

Tacaná area
  geochronology of, 44, 46–47
  lithology of, 43–46
  maps of, 3, 29, 41, 45
  stratigraphy of, 44
Tacaná graben, 52, 53
Tacaná volcanic complex
  age of, 43
  Agua Caliente wedge, 42, 49
  ash fall in, 44
  basement of, 42, 44, 46
  block-and-ash flows from, 2, 44, 49, 51, 54
  Chamalecón couleé, 50
  Chanjale caldera. See Chanjale caldera
  Chiapanecan volcanic arc and, 28
  Chichuj volcano. See Chichuj volcano
  cross-sectional model of, 43, 53
  debris avalanches from
    Agua Caliente. See Agua Caliente debris avalanche
    Muxbal. See Muxbal debris avalanche
    Río Las Majadas, 2, 18
    in San Rafael caldera, 49
    in Sibinal caldera, 50
    stratigraphy of, 44
  Desenlace wedge, 42
  edifice collapse in, 2, 42
  El Aguila wedge, 42, 49
  evolution of, 53–56
  geochemistry of, 48–49, 52
  geochronology of, 40, 44, 46–51
  geomorphology of, 42
  gravimetric anomaly in, 52
  height of, 2, 42
  lahars from, 2, 49
  lava domes in, 44
  lava flows from, 44, 46–51
  magmatic eruptions at, 22
  maps of
    digital elevation, 41, 45
    geographic, 3
    hazard zonation, 40–42
    triple junction, 29
  phase diagram for, 31
  photographs of, 43, 46, 50
  Plan de las Ardillas, 42–45, 47–52, 56
  P-T profile for, 31
  pyroclastic flows/surges from, 2, 43
  REE diagram for, 49, 52
  rose diagrams for, 55

San Antonio volcano. See San Antonio volcano
San Rafael caldera, 42–54, 56
Sibinal caldera. See Sibinal caldera
slab depth beneath, 42
structure of, 42, 52
Tacaná volcano, 40, 42–44, 49–52, 54–55
thermal model of, 30
trench, distance from, 42
Tacaná volcano, 40, 42–44, 49–52, 54–55
Tahual volcano, 3, 8
talc, phase diagrams for, 31
Talquian village, 51, 52
Tambora volcano, 146
Tarawera volcano
  column height of, 221
  discharge rate of, 221
  eruption of (1886AD), 210
  thickness decay in, 217, 218
  warning time, 222
Taupo volcano, 146
Tecoluca debris avalanche, 9, 10, 18
Tecuamburro volcano
  debris avalanches from, 7, 18
  maps of, 3, 123
  Miraflores volcano and, 7
  satellite images of, 5
Tegucigalpa volcano, 122
Tehuantepec Ridge
  Cocos Plate and, 28, 31, 33–36, 42
  crustal thickness at, 42
  earthquake magnitude, 42
  East Pacific Rise and, 36
  El Chichón volcano and, 36
  height of, 42
  magnetic studies of, 36
  maps of, 29, 37, 41
  Middle American Trench and, 28, 36, 42
  MORB phase diagram for, 31
  subduction of, 36, 42
Telica volcano
  gas emissions from, 152–53, 158
  lava flows from, 151
  maps of, 142, 169
  pyroclastic activity at, 182
  satellite studies of, 158
  seismic studies of, 156–57
Tenorio volcano
  debris avalanches from, 14, 19
  maps of, 3, 13, 261
tephra
  banding in, 272
  destructive potential of, 146, 203
  hazards from, 203
  lahars and, 86
  magma discharge rate and, 210
  modeling of dispersal, 159
tephrite, 49
thorium/uranium ratio, 261
Ticuantepe Lapilli
  area of, 202, 203
  description of, 199
  destructiveness index for, 202
  distribution of, 206
  geochemistry of, 199
  magnitude of eruptions, 202

maps of, 201
photographs of, 197, 200
stratigraphy of, 199
volcanic explosivity index of, 202
volume of, 202, 203
Tiendilla, 236
Tierra Blanca debris avalanche, 15, 19
Tierra Blanca lava flow, 268, 270, 271
Tierras Morenas debris avalanche, 14, 19
Tiribi River, 234
Tiscapa maar, 145, 148, 150
TITAN-2D model, 159
Tivives debris avalanche, 13, 14–15, 19, 23
Tochab River, 51
Tolimán volcano, 5, 23
Tonga subduction zone, 28
toreva blocks, definition of, 2
Toro Amarillo River
  Irazú volcano and, 234
  maps of
    geologic, 263
    pyroclastic deposits, 243, 245, 246
    regional sketch, 237
    tephra, 241
  Turrialba volcano and, 250, 251, 265
tremors, definition of, 156
Tristán Unit, 264, 267–73
tsunamis
  debris avalanches and, 23, 150
  formation of, 150
  pyroclastic density currents and, 150
  velocity of, 151
Tuís River, 15–16
Turrialba River, 16
Turrialba volcano
  bombs from, 250
  debris avalanches from, 15–16, 19, 21, 238, 251–52
  eruptions of, 227, 231, 233, 234, 236–52
  gas emissions from, 247–51
  geochemistry of, 236, 253–56
  geochronology of, 236–47
  geological setting of, 12, 236
  glaciation of, 20, 238–39, 252
  hazard assessment of, 251
  Irazú volcano and, 227, 237, 251
  lahars from, 236
  lava flows from, 234, 239–41
  lithology of, 238–47
  maps of
    debris avalanches, 13
    geographic, 3, 226, 236
    pyroclastic deposits, 243, 245, 246
    regional sketch, 237
    tectonic, 261
    tephra, 241
  photographs of, 238–40, 248–49
  pyroclastic activity at, 236, 239, 242, 250–51
  quiescent periods of, 252
  seismic studies of, 251
  stratigraphy of, 244
  structure of, 236
  volcanic explosivity index of, 233
Tuxtla Volcanic Field
  evolution of, 33

magma from, 28, 31
maps of, 29, 32
MORB phase diagram for, 31
*P-T* profile for, 31
thermal model of, 36

## U

Unicit Tephra, 149
Union Roja, 52
Unzen volcano, 204
Upala debris avalanche, 14, 19

## V

Virilla-Durazno River, 234
volcanic explosivity index, 156
volcanic hazard, definition of, 162
volcanic risk, definition of, 162
volume, volcano, edifice collapse and, 17
vulnerability, definition of, 162

## W

Western Central Valley debris avalanche, 14
Woods model, 221

## X

Xela caldera, 62, 79
Xiloá Tephra, 149, 191
Xiloá volcano, 190

## Y

Yepocapa vent, 108–10, 119

## Z

Zapatera Island, 11
Zapatera volcano, 12, 142
zeolite, MORB phase diagram for, 31
zoisite, MORB phase diagram for, 31
Zunil fault zone, 62, 78–79
Zunil volcano area
    debris avalanches from, 5
    gas geochemistry, 74–79
    geological setting of, 62
    hydrothermal activity in, 78–79
    isotopic studies of, 67, 70
    maps of, 3, 61, 63, 67
    morphology of, 8
    satellite images of, 5
    water, studies of, 62–67, 70, 72–77
    Xela caldera and, 79